COMPUTATIONAL FLUID DYNAMICS
The Basics with Applications

McGraw-Hill Series in Mechanical Engineering

Consulting Editors

Jack P. Holman, *Southern Methodist University*

John R. Lloyd, *Michigan State University*

Anderson: *Computational Fluid Dynamics: The Basics with Applications*
Anderson: *Modern Compressible Flow: With Historical Perspective*
Arora: *Introduction to Optimum Design*
Bray and Stanley: *Nondestructive Evaluation: A Tool for Design, Manufacturing, and Service*
Burton: *Introduction to Dynamic Systems Analysis*
Culp: *Principles of Energy Conversion*
Dally: *Packaging of Electronic Systems: A Mechanical Engineering Approach*
Dieter: *Engineering Design: A Materials and Processing Approach*
Driels: *Linear Control Systems Engineering*
Eckert and Drake: *Analysis of Heat and Mass Transfer*
Edwards and McKee: *Fundamentals of Mechanical Component Design*
Gebhart: *Heat Conduction and Mass Diffusion*
Gibson: *Principles of Composite Material Mechanics*
Hamrock: *Fundamentals of Fluid Film Lubrication*
Heywood: *Internal Combustion Engine Fundamentals*
Hinze: *Turbulence*
Holman: *Experimental Methods for Engineers*
Howell and Buckius: *Fundamentals of Engineering Thermodynamics*
Hutton: *Applied Mechanical Vibrations*
Juvinall: *Engineering Considerations of Stress, Strain, and Strength*
Kane and Levinson: *Dynamics: Theory and Applications*
Kays and Crawford: *Convective Heat and Mass Transfer*
Kelly: *Fundamentals of Mechanical Vibrations*
Kimbrell: *Kinematics Analysis and Synthesis*
Kreider and Rabl: *Heating and Cooling of Buildings*
Martin: *Kinematics and Dynamics of Machines*
Modest: *Radiative Heat Transfer*
Norton: *Design of Machinery*
Phelan: *Fundamentals of Mechanical Design*
Raven: *Automatic Control Engineering*
Reddy: *An Introduction to the Finite Element Method*
Rosenberg and Karnopp: *Introduction to Physical Systems Dynamics*
Schlichting: *Boundary-Layer Theory*
Shames: *Mechanics of Fluids*
Sherman: *Viscous Flow*
Shigley: *Kinematic Analysis of Mechanisms*

Shigley and Mischke: *Mechanical Engineering Design*
Shigley and Uicker: *Theory of Machines and Mechanisms*
Stiffler: *Design with Microprocessors for Mechanical Engineers*
Stoecker and Jones: *Refrigeration and Air Conditioning*
Ullman: *The Mechanical Design Process*
Vanderplaats: *Numerical Optimization: Techniques for Engineering Design, with Applications*
Wark: *Advanced Thermodynamics for Engineers*
White: *Viscous Fluid Flow*
Zeid: *CAD/CAM Theory and Practice*

McGraw-Hill Series in Aeronautical and Aerospace Engineering

Consulting Editor

John D. Anderson, Jr., *University of Maryland*

Anderson: *Computational Fluid Dynamics: The Basics with Applications*
Anderson: *Fundamentals of Aerodynamics*
Anderson: *Hypersonic and High Temperature Gas Dynamics*
Anderson: *Introduction to Flight*
Anderson: *Modern Compressible Flow: With Historical Perspective*
Burton: *Introduction to Dynamic Systems Analysis*
D'Azzo and Houpis: *Linear Control System Analysis and Design*
Donaldson: *Analysis of Aircraft Structures: An Introduction*
Gibson: *Principles of Composite Material Mechanics*
Kane, Likins, and Levinson: *Spacecraft Dynamics*
Katz and Plotkin: *Low-Speed Aerodynamics: From Wing Theory to Panel Methods*
Nelson: *Flight Stability and Automatic Control*
Peery and Azar: *Aircraft Structures*
Rivello: *Theory and Analysis of Flight Structures*
Schlichting: *Boundary Layer Theory*
White: *Viscous Fluid Flow*
Wiesel: *Spaceflight Dynamics*

Also Available from McGraw-Hill

Schaum's Outline Series in Mechanical Engineering

Most outlines include basic theory, definitions and hundreds of example problems solved in step-by-step detail, and supplementary problems with answers.

Related titles on the current list include:

Acoustics
Continuum Mechanics
Engineering Economics
Engineering Mechanics
Fluid Dynamics
Fluid Mechanics & Hydraulics
Heat Transfer
Lagrangian Dynamics
Machine Design
Mathematical Handbook of Formulas & Tables
Mechanical Vibrations
Operations Research
Statics & Mechanics of Materials
Strength of Materials
Theoretical Mechanics
Thermodynamics for Engineers
Thermodynamics with Chemical Applications

Schaum's Solved Problems Books

Each title in this series is a complete and expert source of solved problems with solutions worked out in step-by-step detail.

Related titles on the current list include:

3000 Solved Problems in Calculus
2500 Solved Problems in Differential Equations
2500 Solved Problems in Fluid Mechanics & Hydraulics
1000 Solved Problems in Heat Transfer
3000 Solved Problems in Linear Algebra
2000 Solved Problems in Mechanical Engineering Thermodynamics
2000 Solved Problems in Numerical Analysis
700 Solved Problems in Vector Mechanics for Engineers: Dynamics
800 Solved Problems in Vector Mechanics for Engineers: Statics

Available at most college bookstores, or for a complete list of titles and prices, write to:
Schaum Division
McGraw-Hill, Inc.
1221 Avenue of the Americas
New York, NY 10020

COMPUTATIONAL FLUID DYNAMICS
The Basics with Applications

John D. Anderson, Jr.
Department of Aerospace Engineering
University of Maryland

McGraw-Hill, Inc.

New York   St. Louis   San Francisco   Auckland   Bogotá   Caracas
Lisbon   London   Madrid   Mexico City   Milan   Montreal
New Delhi   San Juan   Singapore   Sydney   Tokyo   Toronto

This book was set in Times Roman.
The editors were John J. Corrigan and Eleanor Castellano;
the production supervisor was Denise L. Puryear.
The cover was designed by Rafael Hernandez.
R. R. Donnelley & Sons Company was printer and binder.

COMPUTATIONAL FLUID DYNAMICS
The Basics with Applications

Copyright © 1995 by McGraw-Hill, Inc. All rights reserved. Printed in the United States of America. Except as permitted under the United States Copyright Act of 1976, no part of this publication may be reproduced or distributed in any form or by any means, or stored in a data base or retrieval system, without the prior written permission of the publisher.

This book is printed on acid-free paper.

5 6 7 8 9 0 DOC DOC 9 0 9

ISBN 0-07-001685-2

Library of Congress Cataloging-in-Publication Data
Anderson, John David.
   Computational fluid dynamics: the basics with applications / John
D. Anderson, Jr.
      p.   cm. — (McGraw-Hill series in mechanical engineering—McGraw-Hill series in   aeronautical and aerospace engineering)
   Includes bibliographical references and index.
   ISBN 0-07-001685-2
   1. Fluid dynamics—Data processing.   I. Title.   II. Series.
QA911.A58   1995
532′.05′015118—dc20   94-21237

ABOUT THE AUTHOR

John D. Anderson, Jr., was born in Lancaster, Pennsylvania, on October 1, 1937. He attended the University of Florida, graduating in 1959 with high honors and a Bachelor of Aeronautical Engineering Degree. From 1959 to 1962, he was a lieutenant and task scientist at the Aerospace Research Laboratory at Wright-Patterson Air Force Base. From 1962 to 1966, he attended the Ohio State University under the National Science Foundation and NASA Fellowships, graduating with a Ph.D. in aeronautical and astronautical engineering. In 1966 he joined the U.S. Naval Ordnance Laboratory as Chief of the Hypersonic Group. In 1973, he became Chairman of the Department of Aerospace Engineering at the University of Maryland, and since 1980 has been professor of Aerospace Engineering at Maryland. In 1982, he was designated a Distinguished Scholar/Teacher by the University. During 1986–1987, while on sabbatical from the university, Dr. Anderson occupied the Charles Lindbergh chair at the National Air and Space Museum of the Smithsonian Institution. He continues with the Museum in a part-time appointment as special assistant for aerodynamics. In addition to his appointment in aerospace engineering, in 1993 he was elected to the faculty of the Committee on the History and Philosophy of Science at Maryland.

Dr. Anderson has published five books: *Gasdynamic Lasers: An Introduction,* Academic Press (1976), and with McGraw-Hill, *Introduction to Flight,* 3d edition (1989), *Modern Compressible Flow,* 2d Edition (1990), *Fundamentals of Aerodynamics,* 2d edition (1991), and *Hypersonic and High Temperature Gas Dynamics* (1989). He is the author of over 100 papers on radiative gasdynamics, reentry aerothermodynamics, gas dynamic and chemical lasers, computational fluid dynamics, applied aerodynamics, hypersonic flow, and the history of aerodynamics. Dr. Anderson is in *Who's Who in America*, and is a Fellow of the American Institute of Aeronautics and Astronautics (AIAA). He is also a Fellow of the Washington Academy of Sciences, and a member of Tau Beta Pi, Sigma Tau, Phi Kappa Phi, Phi Eta Sigma, The American Society for Engineering Education (ASEE), The Society for the History of Technology, and the History of Science Society. He has received the Lee Atwood Award for excellence in Aerospace Engineering Education from the AIAA and the ASEE.

To Sarah-Allen, Katherine, and Elizabeth
for all their love and understanding

CONTENTS

Preface   xix

Part I Basic Thoughts and Equations

1 **Philosophy of Computational Fluid Dynamics**   3
 1.1 Computational Fluid Dynamics: Why?   4
 1.2 Computational Fluid Dynamics as a Research Tool   6
 1.3 Computational Fluid Dynamics as a Design Tool   9
 1.4 The Impact of Computational Fluid Dynamics—Some Other Examples   13
  1.4.1 Automobile and Engine Applications   14
  1.4.2 Industrial Manufacturing Applications   17
  1.4.3 Civil Engineering Applications   19
  1.4.4 Environmental Engineering Applications   20
  1.4.5 Naval Architecture Applications (Submarine Example)   22
 1.5 Computational Fluid Dynamics: What Is It?   23
 1.6 The Purpose of This Book   32

2 **The Governing Equations of Fluid Dynamics: Their Derivation, a Discussion of Their Physical Meaning, and a Presentation of Forms Particularly Suitable to CFD**   37
 2.1 Introduction   38
 2.2 Models of the Flow   40
  2.2.1 Finite Control Volume   41
  2.2.2 Infinitesimal Fluid Element   42
  2.2.3 Some Comments   42
 2.3 The Substantial Derivative (Time Rate of Change Following a Moving Fluid Element   43
 2.4 The Divergence of the Velocity: Its Physical Meaning   47
  2.4.1 A Comment   48

xi

	2.5	The Continuity Equation	49
		2.5.1 Model of the Finite Control Volume Fixed in Space	49
		2.5.2 Model of the Finite Control Volume Moving with the Fluid	51
		2.5.3 Model of an Infinitesimally Small Element Fixed in Space	53
		2.5.4 Model of an Infinitesimally Small Fluid Element Moving with the Flow	55
		2.5.5 All the Equations Are One: Some Manipulations	56
		2.5.6 Integral versus Differential Form of the Equations: An Important Comment	60
	2.6	The Momentum Equation	60
	2.7	The Energy Equation	66
	2.8	Summary of the Governing Equations for Fluid Dynamics: With Comments	75
		2.8.1 Equations for Viscous Flow (the Navier–Stokes Equations)	75
		2.8.2 Equations for Inviscid Flow (the Euler Equations)	77
		2.8.3 Comments on the Governing Equations	78
	2.9	Physical Boundary Conditions	80
	2.10	Forms of the Governing Equations Particularly Suited for CFD: Comments on the Conservation Form, Shock Fitting, and Shock Capturing	82
	2.11	Summary	92
		Problems	93
3	**Mathematical Behavior of Partial Differential Equations: The Impact on CFD**		**95**
	3.1	Introduction	95
	3.2	Classification of Quasi-Linear Partial Differential Equations	97
	3.3	A General Method of Determining the Classification of Partial Differential Equations: The Eigenvalue Method	102
	3.4	General Behavior of the Different Classes of Partial Differential Equations: Impact on Physical and Computational Fluid Dynamics	105
		3.4.1 Hyperbolic Equations	106
		3.4.2 Parabolic Equations	111
		3.4.3 Elliptic Equations	117
		3.4.4 Some Comments: The Supersonic Blunt Body Problem Revisited	119
	3.5	Well-Posed Problems	120
	3.6	Summary	121
		Problems	121

Part II Basics of the Numerics

4	**Basic Aspects of Discretization**		**125**
	4.1	Introduction	125
	4.2	Introduction to Finite Differences	128

4.3	Difference Equations	142
4.4	Explicit and Implicit Approaches: Definitions and Contrasts	145
4.5	Errors and an Analysis of Stability	153
	4.5.1 Stability Analysis: A Broader Perspective	165
4.6	Summary	165
	GUIDEPOST	166
	Problems	167

5 Grids with Appropriate Transformations 168

5.1	Introduction	168
5.2	General Transformation of the Equations	171
5.2	Metrics and Jacobians	178
5.4	Form of the Governing Equations Particularly Suited for CFD Revisited: The Transformed Version	183
5.5	A Comment	186
5.6	Stretched (Compressed) Grids	186
5.7	Boundary-Fitted Coordinate Systems; Elliptic Grid Generation	192
	GUIDEPOST	193
5.8	Adaptive Grids	200
5.9	Some Modern Developments in Grid Generation	208
5.10	Some Modern Developments in Finite-Volume Mesh Generation: Unstructured Meshes and a Return to Cartesian Meshes	210
5.11	Summary	212
	Problems	215

6 Some Simple CFD Techniques: A Beginning 216

6.1	Introduction	216
6.2	The Lax-Wendroff Technique	217
6.3	MacCormack's Technique	222
	GUIDEPOST	223
6.4	Some Comments: Viscous Flows, Conservation Form, and Space Marching	225
	6.4.1 Viscous Flows	225
	6.4.2 Conservation Form	225
	6.4.3 Space Marching	226
6.5	The Relaxation Technique and Its Use with Low-Speed Inviscid Flow	229
6.6	Aspects of Numerical Dissipation and Dispersion; Artificial Viscosity	232
6.7	The Alternating-Direction-Implicit (ADI) Technique	243
6.8	The Pressure Correction Technique: Application to Incompressible Viscous Flow	247
	6.8.1 Some Comments on the Incompressible Navier–Stokes Equations	248

6.8.2	Some Comments on Central Differencing of the Incompressible Navier–Stokes Equations; The Need for a Staggered Grid	250
6.8.3	The Philosophy of the Pressure Correction Method	253
6.8.4	The Pressure Correction Formula	254
6.8.5	The Numerical Procedure: The SIMPLE Algorithm	261
6.8.6	Boundary Conditions for the Pressure Correction Method	262

GUIDEPOST   264

6.9 Some Computer Graphic Techniques Used in CFD   264
     6.9.1   *xy* Plots   264
     6.9.2   Contour Plots   265
     6.9.3   Vector and Streamline Plots   270
     6.9.4   Scatter Plots   273
     6.9.5   Mesh Plots   273
     6.9.6   Composite Plots   274
     6.9.7   Summary on Computer Graphics   274
6.10 Summary   277
Problems   278

Part III   Some Applications

7   Numerical Solutions of Quasi-One-Dimensional Nozzle Flows   283

7.1 Introduction: The Format for Chapters in Part III   283
7.2 Introduction to the Physical Problem: Subsonic-Supersonic Insentropic Flow   285
7.3 CFD Solution of Subsonic-Supersonic Isentropic Nozzle Flow: MacCormack's Technique   288
     7.3.1   The Setup   288
     7.3.2   Intermediate Results: The First Few Steps   308
     7.3.3   Final Numerical Results: The Steady-State Solution   313
7.4 CFD Solution of Purely Subsonic Isentropic Nozzle Flow   325
     7.4.1   The Setup: Boundary and Initial Conditions   327
     7.4.2   Final Numerical Results: MacCormack's Technique   330
     7.4.3   The Anatomy of a Failed Solution   325
7.5 The Subsonic-Supersonic Isentropic Nozzle Solution Revisited: The Use of the Governing Equations in Conservation Form   336
     7.5.1   The Basic Equations in Conservation Form   337
     7.5.2   The Setup   340
     7.5.3   Intermediate Calculations: The First Time Step   345
     7.5.4   Final Numerical Results: The Steady State Solution   351

		7.6	A Case with Shock Capturing	356
		7.6.1	The Setup	358
		7.6.2	The Intermediate Time-Marching Procedure: The Need for Artificial Viscosity	363
		7.6.3	Numerical Results	364
	7.7	Summary		372

8 Numerical Solution of a Two-Dimensional Supersonic Flow: Prandtl-Meyer Expansion Wave — 374

- 8.1 Introduction — 374
- 8.2 Introduction to the Physical Problem: Prandtl-Meyer Expansion Wave—Exact Analytical Solution — 376
- 8.3 The Numerical Solution of a Prandtl-Meyer Expansion Wave Flow Field — 377
  - 8.3.1 The Governing Equations — 377
  - 8.3.2 The Setup — 386
  - 8.3.3 Intermediate Results — 397
  - 8.3.4 Final Results — 407
- 8.4 Summary — 414

9 Incompressible Couette Flow: Numerical Solutions by Means of an Implicit Method and the Pressure Correction Method — 416

- 9.1 Introduction — 416
- 9.2 The Physical Problem and Its Exact Analytical Solution — 417
- 9.3 The Numerical Approach: Implicit Crank-Nicholson Technique — 420
  - 9.3.1 The Numerical Formulation — 421
  - 9.3.2 The Setup — 425
  - 9.3.3 Intermediate Results — 426
  - 9.3.4 Final Results — 430
- 9.4 Another Numerical Approach: The Pressure Correction Method — 435
  - 9.4.1 The Setup — 436
  - 9.4.2 Results — 442
- 9.5 Summary — 445
- Problem — 446

10 Supersonic Flow over a Flat Plate: Numerical Solution by Solving the Complete Navier–Stokes Equations — 447

- 10.1 Introduction — 447
- 10.2 The Physical Problem — 449
- 10.3 The Numerical Approach: Explicit Finite-Difference Solution of the Two-Dimensional Complete Navier–Stokes Equations — 450
  - 10.3.1 The Governing Flow Equations — 450
  - 10.3.2 The Setup — 452

		10.3.3	The Finite-Difference Equations	453
		10.3.4	Calculation of Step Sizes in Space and Time	455
		10.3.5	Initial and Boundary Conditions	457
	10.4	Organization of Your Navier–Stokes Code		459
		10.4.1	Overview	459
		10.4.2	The Main Program	461
		10.4.3	The MacCormack Subroutine	463
		10.4.4	Final Remarks	466
	10.5	Final Numerical Results: The Steady State-Solution		466
	10.6	Summary		474

Part IV Other Topics

11 Some Advanced Topics in Modern CFD: A Discussion   479

	11.1	Introduction		479
	11.2	The Conservation Form of the Governing Flow Equations Revisited: The Jacobians of the System		480
		11.2.1	Specialization to One-Dimensional Flow	482
		11.2.2	Interim Summary	489
	11.3	Additional Considerations for Implicit Methods		489
		11.3.1	Linearization of the Equations: The Beam and Warming Method	490
		11.3.2	The Multidimensional Problem: Approximate Factorization	492
		11.3.3	Block Tridiagonal Matrices	496
		11.3.4	Interim Summary	497
	11.4	Upwind Schemes		497
		11.4.1	Flux-Vector Splitting	500
		11.4.2	The Godunov Approach	502
		11.4.3	General Comment	507
	11.5	Second-Order Upwind Schemes		507
	11.6	High-Resolution Schemes: TVD and Flux Limiters		509
	11.7	Some Results		510
	11.8	Multigrid Method		513
	11.9	Summary		514
		Problems		514

12 The Future of CFD   515

	12.1	The Importance of CFD Revisited	515
	12.2	Computer Graphics in CFD	516
	12.3	The Future of CFD: Enhancing the Design Process	517
	12.4	The Future of CFD: Enhancing Understanding	526
	12.5	Conclusion	533

**Appendix A Thomas' Algorithm for the
Solution of a Tridiagonal System of Equations** 534

References 539

Index 543

PREFACE

This computational fluid dynamics (CFD) book is *truly for beginners*. If you have never studied CFD before, if you have never worked in the area, and if you have no real idea as to what the discipline is all about, then this book is for you. Absolutely no prior knowledge of CFD is assumed on your part—only your desire to learn something about the subject is taken for granted.

The author's single-minded purpose in writing this book is to provide a simple, satisfying, and motivational approach toward presenting the subject to the reader who is learning about CFD for the first time. In the workplace, CFD is today a mathematically sophisticated discipline. In turn, in the universities it is generally considered to be a graduate-level subject; the existing textbooks and most of the professional development short courses are pitched at the graduate level. The present book is a *precursor* to these activities. It is intended to "break the ice" for the reader. This book is *unique* in that it is intended to be read and mastered *before* you go on to any of the other existing textbooks in the field, *before* you take any regular short courses in the discipline, and *before* you endeavor to read the existing literature. The hallmarks of the present book are *simplicity* and *motivation*. It is intended to prepare you for the more sophisticated presentations elsewhere—to give you an overall appreciation for the basic philosophy and ideas which will then make the more sophisticated presentations more meaningful to you later on. The mathematical level and the prior background in fluid dynamics assumed in this book are equivalent to those of a college senior in engineering or physical science. Indeed, this book is targeted primarily for use as a one-semester, senior-level course in CFD; it may also be useful in a preliminary, first-level graduate course.

There are no role models for a book on CFD at the undergraduate level; when you ask ten different people about what form such a book should take, you get ten different answers. This book is the author's answer, as imperfect as it may be, formulated after many years of thought and teaching experience. Of course, to achieve the goals stated above, the author has made some hard choices in picking and arranging the material in this book. It is *not* a state-of-the-art treatment of the modern, sophisticated CFD of today. Such a treatment would blow the uninitiated reader completely out of the water. This author knows; he has seen it happen over and over again, where a student who wants to learn about CFD is totally turned off by the advanced treatments and becomes unmotivated toward continuing further. Indeed, the purpose of this book is to prepare the reader to benefit from such advanced treatments *at a later date*. The present book provides a general perspective on CFD; its purpose is to turn you, the reader, on to the subject, not to intimidate you. Therefore, the material in this book is predominately an intuitive, physically oriented approach to CFD. A CFD expert, when examining this book, may at first think that some of it is "old-fashioned," because some of the material covered here was the state of the art in 1980. But this is the point: the older, tried-and-proven ideas form a wonderfully intuitive and meaningful learning experience for the uninitiated reader. With the background provided by this book, the reader can then progress to the more sophisticated aspects of CFD in graduate school and in the workplace. However, to increase the slope of the reader's learning curve, state-of-the-art CFD techniques *are* discussed in Chap. 11, and some very recent and powerful examples of CFD calculations are reviewed in Chap. 12. In this fashion, when you finish the last page of this book, you are already well on your way to the next level of sophistication in the discipline.

This book is in part the product of the author's experience in teaching a one-week short course titled "Introduction to Computational Fluid Dynamics," for the past ten years at the von Karman Institute for Fluid Dynamics (VKI) in Belgium, and in recent years also for Rolls-Royce in England. With this experience, this author has discovered much of what it takes to present the elementary concepts of CFD in a manner which is acceptable, productive, and motivational to the first-time student. The present book directly reflects the author's experience in this regard. The author gives special thanks to Dr. John Wendt, Director of the VKI, who first realized the need for such an introductory treatment of CFD, and who a decade ago galvanized the present author into preparing such a course at VKI. Over the ensuing years, the demand for this "Introduction to Computational Fluid Dynamics" course has been way beyond our wildest dreams. Recently, a book containing the VKI course notes has been published; it is *Computational Fluid Dynamics: An Introduction,* edited by John F. Wendt, Springer-Verlag, 1992. The present book is a greatly expanded sequel to this VKI book, aimed at a much more extensive presentation of CFD pertinent to a one-semester classroom course, but keeping within the basic spirit of simplicity and motivation.

This book is organized into four major parts. Part I introduces the basic thoughts and philosophy associated with CFD, along with an extensive discussion of the governing equations of fluid dynamics. It is vitally important for a student of CFD to fully understand, and feel comfortable with, the basic physical equations; they are the lifeblood of CFD. The author feels so strongly about this need to fully understand and appreciate the governing equations that every effort has been made to thoroughly derive and discuss these equations in Chap. 2. In a sense, Chap. 2 stands independently as a "mini course" in the governing equations. Experience has shown that students of CFD come from quite varied backgrounds; in turn, their understanding of the governing equations of fluid dynamics ranges across the spectrum from virtually none to adequate. Students from the whole range of this spectrum have continually thanked the author for presenting the material in Chap. 2; those from the "virtually none" extreme are very appreciative of the opportunity to become comfortable with these equations, and those from the "adequate" extreme are very happy to have an integrated presentation and comprehensive review that strips away any mystery about the myriad of different forms of the governing equations. Chapter 2 emphasizes the philosophy that, to be a good computational fluid dynamicist, you must first be a good fluid dynamicist.

In Part II, the fundamental aspects of numerical discretization of the governing equations are developed; the discretization of the partial differential equations (finite-difference approach) is covered in detail. Here is where the basic numerics are introduced and where several popular numerical techniques for solving flow problems are presented. The finite-volume discretization of the integral form of the equations is covered via several homework problems.

Part III contains applications of CFD to four classic fluid dynamic problems with well-known, exact analytical solutions, which are used as a basis for comparison with the numerical CFD results. Clearly, the *real-world applications of CFD* are to problems that do *not* have known analytical solutions; indeed, CFD is our mechanism for solving flow problems that cannot be solved in any other way. However, in the present book, which is intended to introduce the reader to the *basic aspects* of CFD, nothing is gained by choosing applications where it is difficult to check the validity of the results; rather everything is gained by choosing simple flows with analytical solutions so that the reader can fundamentally see the strengths and weaknesses of a given computational technique against the background of a known, exact analytical solution. Each application is worked in great detail so that the reader can see the direct use of much of the CFD fundamentals which are presented in Parts I and II. The reader is also encouraged to write his or her own computer programs to solve these same problems, and to check the results given in Chaps. 7 to 10. In a real sense, although the subject of this book is *computational* fluid dynamics, it is also a vehicle for the reader to become more thoroughly acquainted with fluid dynamics per se. This author has intentionally emphasized the physical aspects of various flow problems in order to enhance the reader's overall understanding. In some respect, this is an example of the adage that a student really learns the material of course N when he or she takes course $N + 1$. In terms of some aspects of basic fluid dynamics, the present book represents course $N + 1$.

Part IV deals with some topics which are more advanced than those discussed earlier in the book but which constitute the essence of modern state-of-the-art algorithms and applications in CFD. It is well beyond the scope of this book to present the *details* of such advanced topics—they await your attention in your future studies. Instead, such aspects are simply *discussed* in Chap. 11 just to give you a preview of coming attractions in your future studies. The purpose of Chap. 11 is just to acquaint you with some of the ideas and vocabulary of the most modern CFD techniques being developed today. Also, Chap. 12 examines the future of

CFD, giving some very recent examples of pioneering applications; Chap. 12 somewhat closes the loop of this book by extending some of the motivational ideas first discussed in Chap. 1.

The matter of *computer programing per se* was another hard choice faced by the author. Should detailed computer listings be included in this book as an aid to the reader's computer programing and as a recognition of the importance of efficient and modular programing for CFD? The decision was *no*, with the exception of a computer listing for Thomas' algorithm contained in the solution for Couette flow and listed in App. A. There are good and bad programming techniques, and it behooves the reader to become familiar and adept with efficient programming. However, this is *not the role of the present book*. Rather, you are encouraged to tackle the applications in Part III by writing your own programs as you see fit, and not following any prescribed listing provided by the author. This is assumed to be part of your learning process. The author wants you to get your own hands "dirty" with CFD by writing your own programs; it is a vital part of the learning process at this stage of your CFD education. On the other hand, detailed computer listings for all the applications discussed in Part III *are listed in the Solutions Manual for this book*. This is done as a service to classroom instructors. In turn, the instructors are free to release to their students any or all of these listings as deemed appropriate.

Something needs to be said about *computer graphics*. It was suggested by one reviewer that some aspects of computer graphics be mentioned in the present book. It is a good suggestion. Therefore, in Chap. 6 an entire section is devoted to explaining and illustrating the different computer graphic techniques commonly used in CFD. Also, examples of results presented in standard computer graphic format are sprinkled throughout the book.

Something also needs to be said about the role of *homework problems* in an introductory, senior-level CFD course, and therefore about homework problems in the present book. This is a serious consideration, and one over which the author has mulled for a considerable time. The actual applications of CFD—even the simplest techniques as addressed in this book—require a substantial learning period before the reader can actually do a reasonable calculation. Therefore, in the early chapters of this book, there is not much opportunity for the reader to practice making calculations via homework exercises. This is a departure from the more typical undergraduate engineering course, where the student is usually immersed in the "learning by doing" process through the immediate assignment of homework problems. Insead, the reader of this book is immersed in first learning the basic vocabulary, philosophy, ideas, and concepts of CFD before he or she finally encounters applications—the subject of Part III. Indeed, in these applications the reader is finally encouraged to set up calculations and to get the experience of doing some CFD work himself or herself. Even here, these applications are more on the scale of small *computer projects* rather than homework problems per se. Even the reviewers of this book are divided as to whether or not homework problems should be included; exactly half the reviewers said yes, but the others implied that such problems are not necessary. This author has taken some middle ground. There are homework problems in this book, but not very many. They are included in several chapters to help the reader think about the details of some of the concepts being discussed in the text. Because there are no established role models for a book in CFD at the undergraduate level for which the present book is aimed, the author prefers to leave the generation of large numbers of appropriate homework problems to the ingenuity of the readers and instructors—you will want to exercise your own creativity in this regard.

This book is in keeping with the author's earlier books in that every effort has been made to discuss the material in an easy-to-understand writing style. This book will *talk to you* in a conversational style in order to expedite your understanding of material that sometimes is not all that easy to understand.

As stated earlier, a unique aspect of this book is its intended use in *undergraduate* programs in engineering and physical science. Since the seventeenth century, science and engineering have developed along two parallel tracks: one dealing with pure experiment and the other dealing with pure theory. Indeed, today's undergraduate engineering and science curricula reflect this tradition; they give the student a solid background in both experimental and theoretical techniques. However, in the technical world of today, computational mechanics has emerged as a new third approach, along with those of experiment and theory. Every graduate will in some form or another be touched by computational mechanics in the future. Therefore, in terms of fluid dynamics, it is *essential* that CFD be added to the curriculum at the *undergraduate* level in order to round out the three-approach world of today. This book is intended to expedite the teaching of CFD at the undergraduate level and, it is hoped, to make it as pleasant and painless as possible to both student and teacher.

A word about the *flavor* of this book. The author is an aerodynamicist, and there is some natural tendency to discuss aeronautically related problems. However, CFD is *interdisciplinary*, cutting across the fields of aerospace, mechanical, civil, chemical, and even electrical engineering, as well as physics and chemistry. While writing this book, the author had readers from all these areas in mind. Indeed, in the CFD short courses taught by this author, students from all the above disciplines have attended and enjoyed the experience. Therefore, this book contains material related to other disciplines well beyond that of aerospace engineering. In particular, mechanical and civil engineers will find numerous familiar applications discussed in Chap. 1 and will find the ADI and pressure correction techniques discussed in Chap. 6 to be of particular interest. Indeed, the application of the pressure correction technique for the solution of a viscous incompressible flow in Chap. 9 is aimed squarely at mechanical and civil engineers. However, no matter what the application may be, please keep in mind that the material in this book is *generic* and that readers from many fields are welcome.

What about the sequence of material presented in this book? Can the reader hop around and cut out some material he or she may not have time to cover, say in a given one-semester course? The answer is essentially yes. Although the author has composed this book such that consecutive reading of all the material in sequence will result in the broadest understanding of CFD at the introductory level, he recognizes that many times the reader and/or instructor does not have that luxury.

Therefore, at strategic locations throughout the book, specifically highlighted GUIDEPOSTS appear which instruct the reader where to go in the book and what to do in order to specifically tailor the material as he or she so desires. The location of these GUIDEPOSTS is also shown in the table of contents, for ready reference.

The author wishes to give special thanks to Col. Wayne Halgren, professor of aeronautics at the U.S. Air Force Academy. Colonel Halgren took the time to study the manuscript of this book, to organize it for a one-semester senior course at the Academy, and to field-test it in the classroom during the spring of 1993. Then he graciously donated his time to visit with the author at College Park in order to share his experiences during this field test. Such information coming from an independent source was invaluable, and a number of features contained in this book came out of this interaction. The fact that Wayne was one of this author's doctoral students several years ago served to strengthen this interaction. This author is proud to have been blessed with such quality students.

The author also wishes to thank all his colleagues in the CFD community for many invigorating discussions on what constitutes an elementary presentation of CFD, and especially the following reviewers of this manuscript: Ahmed Busnaina, Clarkson University; Chien-Pin Chen, University of Alabama–Huntsville; George S. Dulikravich, Pennsylania State University; Ira Jacobson, University of Virginia; Osama A. Kandil, Old Dominion University; James McDonough, University of Kentucky; Thomas J. Mueller, University of Notre Dame; Richard Pletcher, Iowa State University; Paavo Repri, Florida Institute of Technology; P. L. Roe, University of Michigan–Ann Arbor; Christopher Rutland, University of Wisconsin; Joe F. Thompson, Mississippi State University; and Susan Ying, Florida State University. This book is, in part, a product of those discussions. Also, special thanks go to Ms. Susan Cunningham, who was the author's personal word processor for the detailed preparation of this manuscript. Sue loves to type equations—she should have had a lot of fun with this book. Of course, special appreciation goes to two important institutions in the author's life—the University of Maryland for providing the necessary intellectual atmosphere for producing such a book, and my wife, Sarah-Allen, for providing the necessary atmosphere of understanding and support during the untold amount of hours at home required for writing this book. To all of you, I say a most heartfelt thank you.

So, let's get on with it! I wish you a productive trail of happy reading and happy computing. Have fun (and I really mean that).

John D. Anderson, Jr.

COMPUTATIONAL FLUID DYNAMICS
The Basics with Applications

PART I

BASIC THOUGHTS AND EQUATIONS

In Part I, we introduce some of the basic philosophy and ideas of computational fluid dynamics to serve as a springboard for the rest of the book. We also derive and discuss the basic governing equations of fluid dynamics under the premise that these equations are the physical foundation stones upon which all computational fluid dynamics is based. Before we can understand and apply any aspect of computational fluid dynamics, we must fully appreciate the governing equations—the mathematical form and what physics they are describing. All this is the essence of Part I.

CHAPTER 1

PHILOSOPHY OF COMPUTATIONAL FLUID DYNAMICS

All the mathematical sciences are founded on relations between physical laws and laws of numbers, so that the aim of exact science is to reduce the problems of nature to the determination of quantities by operations with numbers.

James Clerk Maxwell, 1856

In the late 1970's, this approach (the use of supercomputers to solve aerodynamic problems) began to pay off. One early success was the experimental NASA aircraft called HiMAT (Highly Maneuverable Aircraft Technology), designed to test concepts of high maneuverability for the next generation of fighter planes. Wind tunnel tests of a preliminary design for HiMAT showed that it would have unacceptable drag at speeds near the speed of sound; if built that way the plane would be unable to provide any useful data. The cost of redesigning it in further wind tunnel tests would have been around $150,000 and would have unacceptably delayed the project. Instead, the wing was redesigned by a computer at a cost of $6,000.

Paul E. Ceruzzi, Curator, National Air and Space Museum, in *Beyond the Limits*, The MIT Press, 1989

1.1 COMPUTATIONAL FLUID DYNAMICS: WHY?

The time: early in the twenty-first century. *The place*: a major airport anywhere in the world. *The event*: a sleek and beautiful aircraft roles down the runway, takes off, and rapidly climbs out of sight. Within minutes, this same aircraft has accelerated to hypersonic speed; still within the atmosphere, its powerful supersonic combustion ramjet engines* continue to propel the aircraft to a velocity near 26,000 ft/s—orbital velocity—and the vehicle simply coasts into low earth orbit. Is this the stuff of dreams? Not really; indeed, this is the concept of a transatmospheric vehicle, which has been the subject of study in several countries during the 1980s and 1990s. In particular, one design for such a vehicle is shown in Fig. 1.1, which is an artist's concept for the National Aerospace Plane (NASP), the subject of an intensive study project in the United States since the mid-1980s. Anyone steeped in the history of aeronautics, where the major thrust has always been to fly faster and higher, knows that such vehicles will someday be a reality. But they will be made a reality only when computational fluid dynamics has developed to the point where the complete three-dimensional flowfield over the vehicle and through the engines can be computed expeditiously with accuracy and reliability. Unfortunately, ground test facilities—wind tunnels—do not exist in all the flight regimes covered by such hypersonic flight. We have no wind tunnels that can simultaneously simulate the higher Mach numbers and high flowfield temperatures to be encountered by transatmospheric vehicles, and the prospects for such wind tunnels in the twenty-first century are not encouraging. Hence, the major player in the design of such vehicles is computational fluid dynamics. It is for this reason, as well as many others, why computational fluid dynamics—*the subject of this book*—is so important in the modern practice of fluid dynamics.†

Computational fluid dynamics constitutes a new "third approach" in the philosophical study and development of the whole discipline of fluid dynamics. In the seventeenth century, the foundations for *experimental* fluid dynamics were laid in France and England. The eighteenth and nineteenth centuries saw the gradual development of *theoretical* fluid dynamics, again primarily in Europe. (See Refs. 3–5 for presentations of the historical evolution of fluid dynamics and aerodynamics.) As a result, throughout most of the twentieth century the study and practice of fluid dynamics (indeed, all of physical science and engineering) involved the use of pure theory on the one hand and pure experiment on the other hand. If you were learning

---

* A supersonic combustion ramjet engine, SCRAMJET for short, is an air-breathing ramjet engine where the flow through the engine remains above Mach 1 in all sections of the engine, including the combustor. Fuel is injected into the supersonic airstream in the combustor, and combustion takes place in the supersonic flow. This is in contrast to a conventional ramjet or gas turbine engine, where the flow in the combustor is at a low subsonic Mach number.

† For a basic introduction to the principles of hypersonic flight, see chap. 10 of Ref. 1. For an in-depth presentation of these principles, see Ref. 2.

FIG. 1.1
Artist's conception of an aerospace plane. (*NASA*)

fluid dynamics as recently as, say, 1960, you would have been operating in the "two-approach world" of theory and experiment. However, the advent of the high-speed digital computer combined with the development of accurate numerical algorithms for solving physical problems on these computers has revolutionized the way we study and practice fluid dynamics today. It has introduced a fundamentally important new third approach in fluid dynamics—the approach of *computational fluid dynamics*. As sketched in Fig. 1.2, computational fluid dynamics is today an *equal* partner with pure theory and pure experiment in the analysis and solution of fluid dynamic problems. And this is no flash in the pan—computational fluid dynamics will continue to play this role indefinitely, for as long as our advanced human civilization exists. Therefore, by studying computational fluid dynamics today, you are participating in an awesome and historic revolution, truly a measure of the importance of the subject matter of this book.

However, to keep things in perspective, computational fluid dynamics provides a new third approach—but nothing more than that. It nicely and synergistically complements the other two approaches of pure theory and pure experiment, but it will never *replace* either of these approaches (as sometimes suggested). There will always be a need for theory and experiment. The future advancement of fluid dynamics will rest upon a proper balance of all three

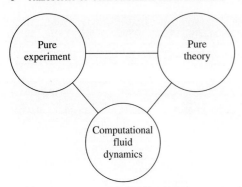

FIG. 1.2
The "three dimensions" of fluid dynamics.

approaches, with computational fluid dynamics helping to interpret and understand the results of theory and experiment, and vice versa.

Finally, we note that computational fluid dynamics is commonplace enough today that the acronym CFD is universally accepted for the phrase "computational fluid dynamics." We will use this acronym throughout the remainder of this book.

1.2 COMPUTATIONAL FLUID DYNAMICS AS A RESEARCH TOOL

Computational fluid dynamic results are directly analogous to wind tunnel results obtained in a laboratory—they both represent sets of data for given flow configurations at different Mach numbers, Reynolds numbers, etc. However, unlike a wind tunnel, which is generally a heavy, unwieldy device, a computer program (say in the form of floppy disks) is something you can carry around in your hand. Or better yet, a source program in the memory of a given computer can be accessed remotely by people on terminals that can be thousands of miles away from the computer itself. A computer program is, therefore, a readily transportable tool, a "transportable wind tunnel."

Carrying this analogy further, a computer program is a tool with which you can carry out *numerical experiments*. For example, assume that you have a program which calculates the viscous, subsonic, compressible flow over an airfoil, such as that shown in Fig. 1.3. Such a computer program was developed by Kothari and Anderson (Ref. 6); this program solves the complete two-dimensional Navier-Stokes equations for viscous flow by means of a finite-difference numerical technique. The Navier-Stokes equations, as well as other governing equations for the physical aspects of fluid flow, are developed in Chap. 2. The computational techniques employed in the solution by Kothari and Anderson in Ref. 6 are standard approaches—all of which are covered in subsequent chapters of this book. Therefore, by the time you finish this book, you will have all the knowledge necessary to construct, among many other examples, solutions of the Navier-Stokes equations for compressible flows over airfoils, just as described in Ref. 6. Now, assuming that you have such a program, you can carry out some interesting *experiments* with it—experiments which in every sense of the word are analogous to those you could carry out (in principle) in a wind tunnel, except the experiments you

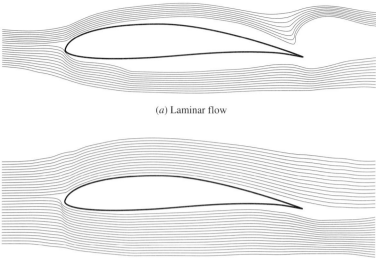

FIG. 1.3
Example of a CFD numerical experiment. (*a*) Instantaneous streamlines over a Wortmann airfoil (FX63-137) for laminar flow. Re = 100,000; M_∞ = 0.5; zero angle of attack. The laminar flow is unsteady; this picture corresponds to only one instant in time. (*b*) Streamlines over the same airfoil for the same conditions except that the flow is turbulent.

perform with the computer program are *numerical experiments*. To provide a more concrete understanding of this philosophy, let us examine one of these numerical experiments, gleaned from the work of Ref. 6.

This is an example of a numerical experiment that can elucidate physical aspects of a flow field in a manner not achievable in a real laboratory experiment. For example, consider the subsonic compressible flow over the Wortmann airfoil shown in Fig. 1.3. *Question*: What are the differences between laminar and turbulent flow over this airfoil for Re = 100,000? For the computer program, this is a straightforward question—it is just a matter of making one run with the turbulence model switched off (laminar flow), another run with the turbulence model switched on (turbulent flow), and then comparing the two sets of results. In this fashion you can dabble with Mother Nature simply by turning a switch in the computer program—something you cannot do quite as readily (if at all) in the wind tunnel. For example, in Fig. 1.3*a* the flow is completely laminar. Note that the calculated flow is separated over both the top and bottom surfaces of the airfoil, even though the angle of attack is zero. Such separated flow is characteristic of the low Reynolds number regime considered here (Re = 100,000), as discussed in Refs. 6 and 7. Moreover, the CFD calculations show that this laminar, separated flow is *unsteady*. The numerical technique used to calculate these flows is a time-marching method, using a time-accurate finite-difference solution of the unsteady Navier-Stokes equations. (The philosophy and numerical details associated with time-marching solutions will be discussed in subsequent chapters.) The streamlines shown in Fig.

1.3a are simply a "snapshot" of this unsteady flow at a given instant in time. In contrast, Fig. 1.3b illustrates the calculated streamlines when a turbulence model is "turned on" within the computer program. Note that the calculated turbulent flow is *attached flow*; moreover, the resulting flow is *steady*. Comparing Fig. 1.3a and b, we see that the laminar and turbulent flows are quite different; moreover, this CFD numerical experiment allows us to study in detail the physical differences between the laminar and turbulent flows, all other parameters being equal, in a fashion impossible to obtain in an actual laboratory experiment.

Numerical experiments, carried out in parallel with physical experiments in the laboratory, can sometimes be used to help *interpret* such physical experiments, and even to ascertain a basic phenomenological aspect of the experiments which is not apparent from the laboratory data. The laminar/turbulent comparison reflected in part in Fig. 1.3a and b is such a case. This comparison has even more implications, as follows. Consider Fig. 1.4, which is a plot of experimental wind tunnel data (the open symbols) for lift coefficient c_l as a function of angle of attack for the same Wortman airfoil. The experimental data were obtained by Dr. Thomas Mueller and his colleages at Notre Dame University. (See Ref. 7.) The solid symbols in Fig. 1.4 pertain to the CFD results at zero angle of attack, as described in Ref. 6. Note that there are two distinct sets of CFD results shown here. The solid circle represents a mean of the laminar flow results, with the brackets representing the amplitude of the unsteady fluctuations in c_l due to the unsteady separated flow, as previously illustrated in Fig. 1.3a. Note that the laminar flow value of c_l is not even close to the experimental measurements at $\alpha = 0$. On the other hand, the solid square gives the turbulent flow result, which corresponds to steady flow as previously illustrated in Fig. 1.3b. The turbulent value of c_l is in close agreement

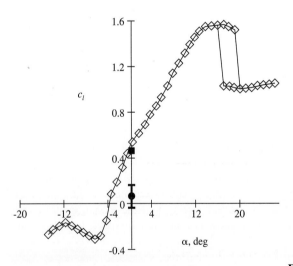

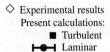

◇ Experimental results
Present calculations:
■ Turbulent
├●┤ Laminar

FIG. 1.4
Example of a CFD numerical experiment. Lift coefficient versus angle of attack for a Wortmann airfoil. Re = 100,000; $M_\infty = 0.5$.

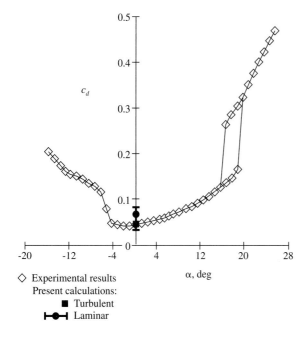

FIG. 1.5
Example of a CFD numerical experiment. Drag coefficient versus angle of attack for a Wortmann airfoil. Re = 100,000; $M_\infty = 0.5$.

with the experimental data. This comparison is reinforced by the results shown in Fig. 1.5, which is a plot of the airfoil drag coefficient versus angle of attack. The open symbols are Mueller's experimental data, and the solid symbols at $\alpha = 0$ are the CFD results. The fluctuating laminar values of the computed c_d are given by the solid circle and the amplitude bars; the agreement with experiment is poor. On the other hand, the solid square represents the steady turbulent result; the agreement with experiment is excellent for this case. The importance of this result goes beyond just a simple comparison between experiment and computation. During the course of the wind tunnel experiments, there was some uncertainty, based on the experimental observations themselves, as to whether or not the flow was laminar or turbulent. However, examing the comparisons with the CFD data shown in Figs. 1.4 and 1.5, we have to conclude that the flow over the airfoil in the wind tunnel was indeed turbulent, because the turbulent CFD results agreed with experiment whereas the laminar CFD results were far off. Here is a beautiful example of how CFD can work harmoniously with experiment—not just providing a quantitative comparison, but also in this case providing a means to *interpret* a basic phenomenological aspect of the experimental conditions. Here is a graphic example of the value of *numerical experiments* carried out within the framework of CFD.

1.3 COMPUTATIONAL FLUID DYNAMICS AS A DESIGN TOOL

In 1950, there was no CFD in the way that we think of it today. In 1970, there was CFD, but the type of computers and algorithms that existed at that time limited all practical solutions essentially to two-dimensional flows. The real world of fluid dynamic machines—compressors, turbines, flow ducts, airplanes, etc.—is mainly a *three*-dimensional world. In 1970, the storage and speed capacity of digital computers were not sufficient to allow CFD to operate in any practical fashion in this three-dimensional world. By 1990, however, this story had changed substantially. In today's CFD, three-dimensional flow field solutions are abundant; they may not be routine in the sense that a great deal of human and computer resources are still frequently needed to successfully carry out such three-dimensional solutions for applications like the flow over a complete airplane configuration, but such solutions are becoming more and more prevalent within industry and government facilities. Indeed, some computer programs for the calculation of three-dimensional flows have become industry standards, resulting in their use as a tool in the design process. In this section, we will examine one such example, just to emphasize the point.

Modern high-speed aircraft, such as the Northrop F-20 shown in Fig. 1.6, with their complicated transonic aerodynamic flow patterns, are fertile ground for the use of CFD as a design tool. Figure 1.6 illustrates the detailed pressure coefficient variation over the surface of the F-20 at a nearly sonic freestream Mach number M_∞ of 0.95 and an angle of attack α of 8°. These are CFD results obtained by Bush, Jager, and Bergman (Ref. 9), using a finite-volume explicit numerical scheme developed by Jameson et al. (Ref. 10). In Fig. 1.6a, the contours of pressure coefficient are shown over the planform of the F-20; a contour line represents a locus of constant pressure, and hence regions where the contour lines cluster together are regions of large pressure gradients. In particular, the heavily clustered band that appears at the wing trailing edge and wraps around the fuselage just downstream of the trailing edge connotes a transonic shock wave at that location. Other regions involving local shock waves and expansions are clearly shown in Fig. 1.6. In addition, the local chordwise variations of the pressure coefficient over the top and bottom of the wing section are shown for five different spanwise stations in Fig. 1.6b to f. Here, the CFD calculations, which involve the solution of the Euler equations (see Chap. 2), are given by the solid curves and are compared with experimental data denoted by the solid squares and circles. Note that there is reasonable agreement between the calculations and experiment. However, the major point made by the results in Fig. 1.6 relative to our discussion is this: CFD provides a means to calculate the detailed flow field around a complete airplane configuration, including the pressure distribution over the three-dimensional surface. This knowledge of the pressure distribution is necessary for structural engineers, who need to know the detailed distribution of aerodynamic loads over the aircraft in order to properly design the structure of the airframe. This knowledge is also necessary for aerodynamicists, who obtain the lift and pressure drag by integrating the pressure distribution over the surface (see Ref. 8 for details of such an integration). Moreover, the CFD results provide information about the vortices which are formed at the juncture of the fuselage strakes and the wing leading edge, such as shown in Fig. 1.7, also obtained from Ref. 9. Here, the values of M_∞ and α are 0.26 and 25°, respectively. Knowing where these vortices go and how they interact with other parts of the airplane is essential to the overall aerodynamic design of the airplane.

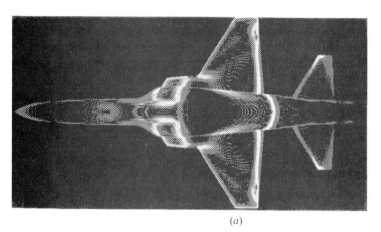

(a)

FIG. 1.6
An example of the calculation of the aerodynamic flow field for a complete airplane configuration, the Northrop F-20 fighter. (a) Contours of pressure coefficient over the entire upper surface of the airplane are shown. (b)–(f) The graphs give the pressure coefficient variation over the top and bottom of the wing at various spanwise locations, denoted by η, which is the spanwise location referenced to the semispan of the wing. (*From Ref. 9.*)

In short, CFD is playing a strong role as a design tool. Along with its role as a research tool as described in Sec. 1.2, CFD has become a powerful influence on the way fluid dynamicists and aerodynamicists do business. Of course, one of the purposes of this book is to introduce you to means of developing and using this power.

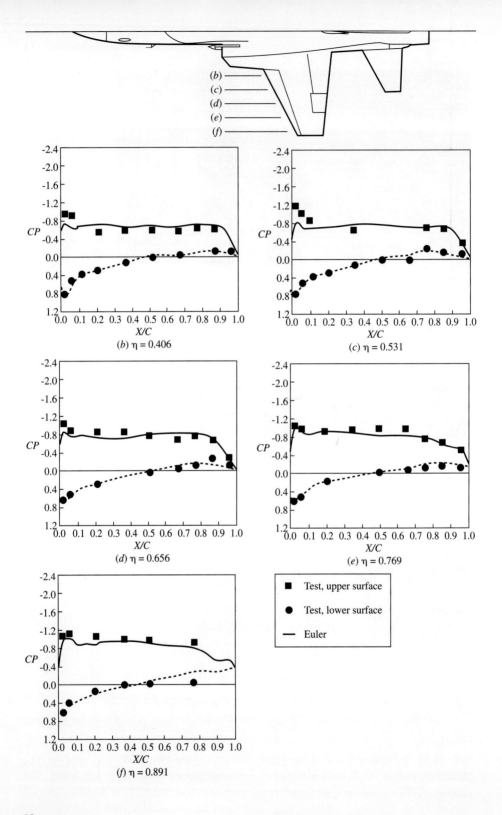

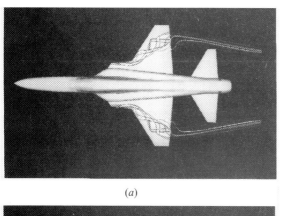

(a)

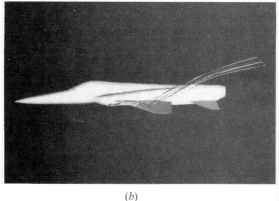

(b)

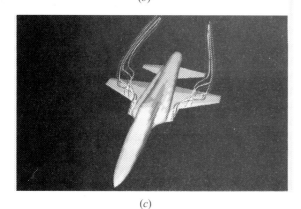

(c)

FIG. 1.7
The wing vortex for the F-20, obtained from CFD calculations. (*a*) Top view; (*b*) side view; (*c*) isometric view. (*From Ref. 9.*)

1.4 THE IMPACT OF COMPUTATIONAL FLUID DYNAMICS—SOME OTHER EXAMPLES

Historically, the early development of CFD in the 1960s and 1970s was driven by the needs of the aerospace community. Indeed, the examples of CFD applications described in Sec. 1.1 to 1.3 are from this community. However, modern CFD cuts across *all* disciplines where the flow of a fluid is important. The purpose of this section is to highlight some of these other, *nonaerospace*, applications of CFD.

1.4.1 Automobile and Engine Applications

To improve the performance of modern cars and trucks (environmental quality, fuel economy, etc.), the automobile industry has accelerated its use of high-technology research and design tools. One of these tools is CFD. Whether it is the study of the external flow over the body of a vehicle, or the internal flow through the engine, CFD is helping automotive engineers to better understand the physical flow processes, and in turn to design improved vehicles. Let us examine several such examples.

The calculation of the external airflow over a car is exemplified by the paths of air particles shown in Fig. 1.8. The outline of the left half of the car is shown by the mesh distributed over its surface, and the white streaks are the calculated paths of various air particles moving over the car from left to right. These particle paths were calculated by means of a finite-volume CFD algorithm. The calculations were made over a discrete three-dimensional mesh distributed in the space around the car; that portion of the mesh on the center plane of symmetry of the car is illustrated in Fig. 1.9. Note that one of the coordinate lines of the mesh is fitted to the body surface, a so-called boundary-fitted coordinate system. (Such coordinate systems are discussed in Sec. 5.7.) Figures 1.8 and 1.9 are taken from a study by C. T. Shaw of Jaguar Cars Limited (Ref. 58). Another example of the calculation of the external flow over a car is the work of Matsunaga et al. (Ref. 59). Figure 1.10 shows contours of vorticity in the flow field over a car, obtained from the finite-difference calculations described in Ref. 59.* (Aspects of finite-difference methods are discussed throughout this book, beginning with Chap. 4.) Here, the calculations are made on a three-dimensional rectangular grid, a portion of which is shown in Fig. 1.11. The fundamentals of grid generation—an important aspect of CFD—are discussed in Chap. 5, and special mention of cartesian, or rectangular, grids wrapped around complex three-dimensional bodies is made in Sec. 5.10.

The calculation of the internal flow inside an internal combustion engine such as that used in automobiles is exemplified by the work of Griffin et al. (Ref. 60). Here, the unsteady flow field inside the cylinder of a four-stroke Otto-cycle engine was calculated by means of a time-marching finite-difference method. (Time-marching methods are discussed in various chapters of this book.) The finite-difference grid for the cylinder is shown in Fig. 1.12. The piston crosshatched at the bottom of Fig. 1.12 moves up and down inside the cylinder during the intake, compression, power, and exhaust strokes; the intake valves open and close appropriately; and an unsteady, recirculating flow field is established inside the cylinder. A calculated velocity pattern in the valve plane when the piston is near the bottom of its stroke (bottom dead center) during the intake stroke is shown in Fig. 1.13. These early calculations were the first application of CFD to the study of flow

---

* Recall that vorticity is defined in fluid dynamics as the vector quantity $\nabla \times \mathbf{V}$, which is equal to twice the instantaneous angular velocity of a fluid element. Contours of the x component of vorticity (in the flow direction) are shown in Fig. 1.10.

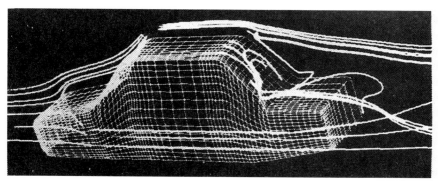

FIG. 1.8
Computed particle paths in the airflow over an automobile. Flow is moving from left to right. (*From C. T. Shaw, Ref. 58. Reprinted with permission from SAE SP-747, 1988,* © *1988, Society of Automotive Engineers, Inc.*)

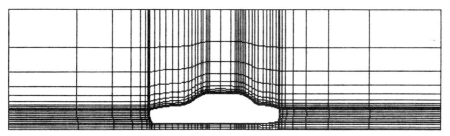

FIG. 1.9
That portion of the computational mesh on the center plane of symmetry used for the calculation shown in Fig. 1.8. (*From Ref. 58. Reprinted with permission from SAE SP-747, 1988,* © *1988 Society of Automotive Engineers, Inc.*)

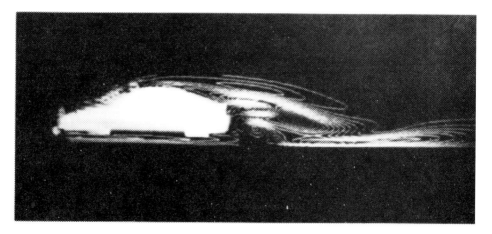

FIG. 1.10
Computed contours of the x component of vorticity in the airflow over an automobile. Flow is moving from left to right. Results are shown in a vertical plane displaced 40% of the width of the car from the center plane. (*From Matsunaga et al. Ref. 59. Reprinted with permission from SAE SP-908, 1992,* © *1992 Society of Automotive Engineers, Inc.*)

16 PHILOSOPHY OF COMPUTATIONAL FLUID DYNAMICS

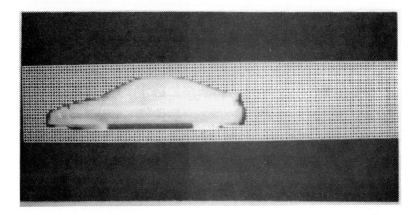

FIG. 1.11
A portion of a rectangular (cartesian) grid wrapped around a car, used for the caculations shown in Fig. 1.10. (*From Ref. 59. Reprinted with permission from SAE SP-908, 1992, © 1992 Society of Automotive Engineers, Inc.*)

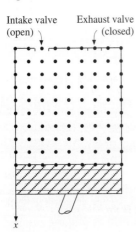

FIG. 1.12
A portion of the grid in the valve plane in cylindrical coordinates for a piston-cylinder arrangement studied in Ref. 60. Only about half the grid points in the valve plane are shown (for clarity).

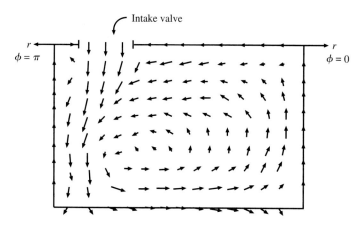

FIG. 1.13
Velocity pattern in the valve plane near bottom dead center of the intake stroke for a piston-cylinder arrangement in an internal combustion engine. (*From Ref. 60.*)

fields inside internal combustion engines. Today, the massive power of modern CFD is being applied by automotive engineers to study all aspects of the details of internal combustion engine flow fields, including combustion, turbulence, and coupling with the manifold and exhaust pipes.

As an example of the sophistication of modern CFD applications to a gas turbine engine, Fig. 1.14 illustrates a finite-volume mesh which is wrapped around both the *external* region outside the engine *and* the *internal* passages through the compressor, the combustor, the turbine, etc. (Grids and meshes are discussed in Sec. 5.10.) This complex mesh is generated by researchers at the Center for Computational Field Simulation at Mississippi State University and is a precursor to a coupled external-internal CFD calculation of the complete flow process associated with a gas turbine. In the author's opinion, this is one of the most complex and interesting CFD grids generated to date, and it clearly underscores the importance of CFD to the automotive and the gas turbine industry.

1.4.2 Industrial Manufacturing Applications

Here we will give just two examples of the myriad CFD applications in manufacturing.

Figure 1.15 shows a mold being filled with liquid modular cast iron. The liquid iron flow field is calculated as a function of time. The liquid iron is introduced into the cavity through two side gates at the right, one at the center and the other at the bottom of the mold. Shown in Fig. 1.15 are CFD results for the velocity field calculated from a finite-volume algorithm; results are illustrated for three values of time during the filling process: an early time just after the two gates are opened (top figure), a slightly later time as the two streams surge into the cavity (center figure). and yet a later time when the two streams are impinging on each

FIG. 1.14
A zonal mesh which simultaneously covers the external region around a jet engine and the internal passages through the engine. (*Courtesy of the Center for Computational Fluid Simulation, Mississippi State University.*)

other (bottom figure). These calculations were made by Mampaey and Xu at the WTCM Foundry Research Center in Belgium (Ref. 61). Such CFD calculations give a more detailed understanding of the real flow behavior of the liquid metal during mold filling and contribute to the design of improved casting techniques.

A second example of CFD in manufacturing processes is that pertaining to the manufacture of ceramic composite materials. One method of production involves the chemical vapor infiltration technique wherein a gaseous material flows through a porous substrate, depositing material on the substrate fibers and eventually forming a continuous matrix for the composite. Of particular interest is the rate and manner in which the compound silicon carbide, SiC, is deposited within the space around the fibers. Recently, Steijsiger et al. (Ref. 62) have use CFD to model SiC deposition in a chemical vapor deposition reactor. The computational mesh distribution within the reactor is shown in Fig. 1.16. The computed streamline pattern inside the reactor is shown in Fig. 1.17. Here, a gaseous mixture of CH_3SiCl_3 and H_2 flows into the reactor from a pipe at the bottom. The ensuing chemical reaction produces SiC, which then deposits on the walls of the reactor. The calculations shown in Fig. 1.17

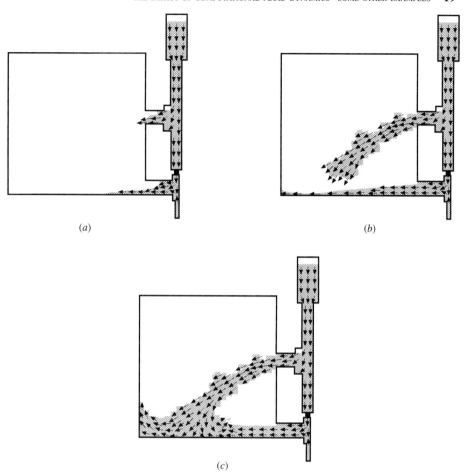

FIG. 1.15
Computed results at three different times for the velocity field set up by liquid iron flowing into a mold from two gates on the right side of the mold. (*After Ref. 61.*)

are from a finite-volume solution of the governing flow equations, and they represent an application of CFD as a research tool, contributing information of direct application to manufacturing.

1.4.3 Civil Engineering Applications

Problems involving the rheology of rivers, lakes, estuaries, etc., are also the subject of investigations using CFD. One such example is the pumping of mud from an underwater mud capture reservoir, as sketched in Fig. 1.18. Here, a layer of water sits on top of a layer of mud, and a portion of the mud is trapped and is being sucked away at the bottom left. This is only half the figure, the other half being a mirror image, forming in total a symmetrical mud reservoir. The vertical line of symmetry

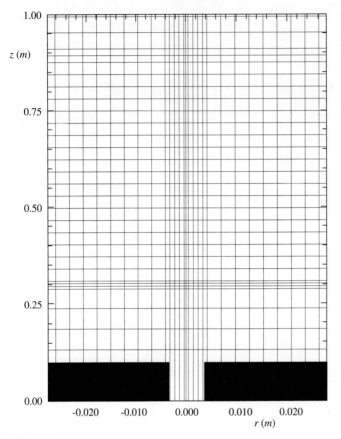

FIG. 1.16
Finite-volume mesh for the calculation of the flow in a chemical vapor deposition reactor. (*After Ref. 62.*)

is the vertical line at the left of Fig. 1.18. As the mud is sucked away at the bottom left, a crater is formed in the mud layer which fills with water. The only motion of the water is caused by the filling of this crater. The computed velocity field in both the water and mud at a certain instant in time is shown in Fig. 1.19, where the magnitude of the velocity vectors are scaled against the arrow designated as 1 cm/s. These results are from the calculations of Toorman and Berlamont as given in Ref. 63. These results contribute to the design of underwater dredging operations, such as the major offshore dredging and beach reclamation project carried out at Ocean City, Maryland, in the early 1990s.

1.4.4 Environmental Engineering Applications

The discipline of heating, air conditioning, and general air circulation through buildings have all come under the spell of CFD. For example, consider the propane-

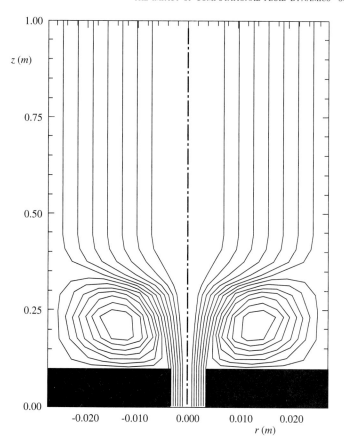

FIG. 1.17
Computed streamline pattern for the flow of CH_3SiCl_3 and H_2 into a chemical vapor deposition reactor. (*After Ref. 62.*)

burning furnace sketched in Fig. 1.20, taken from Ref. 64. The calculated velocity field through this furnace is shown in Fig. 1.21; the velocity vectors emanating from grid points in a perpendicular vertical plane through the furnace are shown. These results are from the finite-difference calculations made by Bai and Fuchs (Ref. 64). Such CFD applications provide information for the design of furnaces with increased thermal efficiency and reduced emissions of pollutants.

A calculation of the flow from an air conditioner is illustrated in Fig. 1.22 and 1.23. A schematic of a room module with the air supply forced through a supply slot in the middle of the ceiling and return exhaust ducts at both corners of the ceiling is given in Fig. 1.22. A finite-volume CFD calculation of the velocity field showing the air circulation pattern in the room is given in Fig. 1.23. These calculations were made by McGuirk and Whittle (Ref. 65).

An interesting application of CFD for the calculation of air currents throughout a building was made by Alamdari et al. (Ref. 66). Figure 1.24 shows the cross section of an office building with two symmetrical halves connected by a

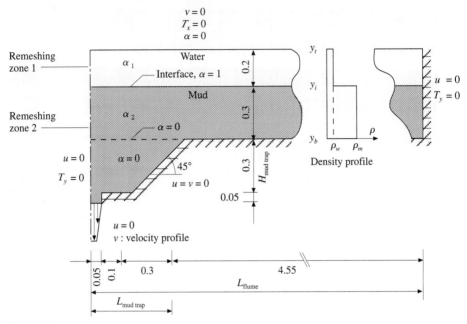

FIG. 1.18
Mud and water layers in a model of a mud trap, with mud being suctioned away at the bottom left. (*After Ref. 63.*)

passageway. Each half has a large, glazed atrium, in keeping with a popular trend in architectural design. These atria, in connection with suitable locations for air inlet and exhausts, provide a natural ventilation system which is cost- and energy-efficient. A typical wintertime simulation of the velocity field in a cross section through the entrance hall, as calculated from a finite-volume CFD algorithm, is shown in Fig. 1.25.

1.4.5 Naval Architecture Applications (Submarine Example)

Computational fluid dynamics is a major tool in solving hydrodynamic problems associated with ships, submarines, torpedos, etc. An example of a CFD application to submarines is illustrated in Fig. 1.26 and 1.27. These calculations were made by the Science Applications International Corporation and were provided to the author by Dr. Nils Salveson of SAIC. Figure 1.26 shows the multizonal grid used for the flow calculations over a generic submarine hull. (Such zonal grids are discussed in Sec. 5.9.) The three-dimensional Navier-Stokes equations for an incompressible flow are solved, including a turbulence model, for the flow over this submarine. Some results for the local streamline pattern at the stern of the submarine are given in Fig. 1.27. Flow is moving from left to right. Here we see an example of a numerical experiment, following the philosophy set forth in Sec. 1.2. The upper half of the figure shows the streamline *with* a propeller, and the lower half shows the

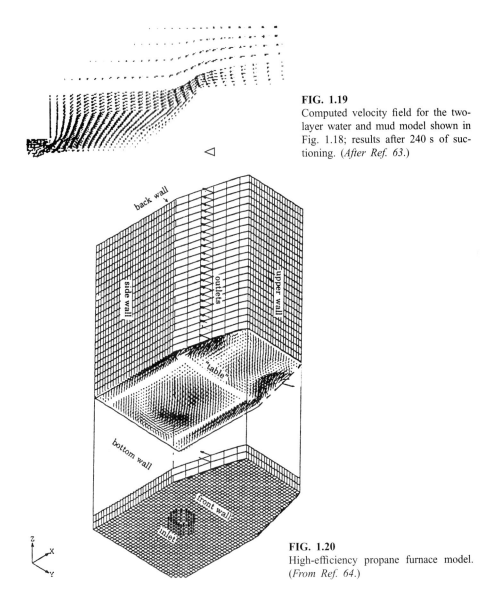

FIG. 1.19
Computed velocity field for the two-layer water and mud model shown in Fig. 1.18; results after 240 s of suctioning. (*After Ref. 63.*)

FIG. 1.20
High-efficiency propane furnace model. (*From Ref. 64.*)

streamlines *without* a propeller. In the latter case, flow separation is observed at the first corner, whereas with the propeller no flow separation takes place.

1.5 COMPUTATIONAL FLUID DYNAMICS: WHAT IS IT?

Question: What is CFD? To answer this question, we note that the physical aspects of any fluid flow are governed by three fundamental principles: (1) mass is conserved; (2) Newton's second law (force = mass × acceleration); and (3) energy

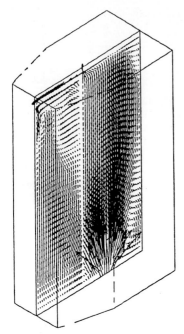

FIG. 1.21
Computed velocity field for the flow through the furnace in Fig. 1.20. Velocity vectors are shown emanating from points in a vertical perpendicular plane through the furnace. (*From Ref. 64.*)

is conserved. These fundamental physical principles can be expressed in terms of basic mathematical equations, which in their most general form are either integral equations or partial differential equations. These equations and their derivation are the subject of Chap. 2. Computational fluid dynamics is the art of replacing the integrals or the partial derivatives (as the case may be) in these equations with discretized algebraic forms, which in turn are solved to obtain *numbers* for the flow

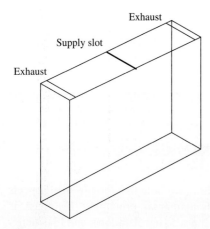

FIG. 1.22
Schematic of a room module with air supply and exhaust ducts in the ceiling. (*After Ref. 65.*)

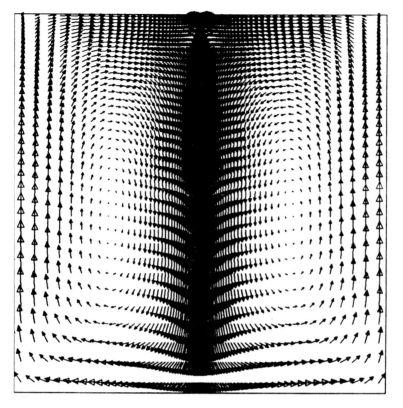

FIG. 1.23
Velocity vector pattern for the room sketched in Fig. 1.22. (*Reprinted by permission of the Council of the Institution of Mechanical Engineers from Ref. 65. On behalf of the Institution of Mechanical Engineers, United Kingdom.*)

field values at discrete points in time and/or space. The end product of CFD is indeed a collection of numbers, in contrast to a closed-form analytical solution. However, in the long run, the objective of most engineering analyses, closed form or otherwise, is a quantitative description of the problem, i.e., *numbers*. (It would be appropriate at this state to review the quote by Maxwell given at the start of this chapter.)

FIG. 1.24
Sketch of an office building. (*Reprinted by permission of the Council of the Institution of Mechanical Engineers from Ref. 66. On behalf of the Institution of Mechanical Engineers, United Kingdom.*)

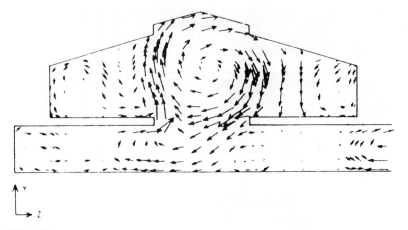

FIG. 1.25
A simulation of the wintertime airflow velocity vector field in the office building sketched in Fig. 1.24. (*Reprinted by permission of the Council of the Institution of Mechanical Engineers from Ref. 66. On behalf of the Institution of Mechanical Engineers, United Kingdom.*)

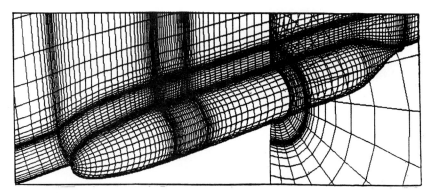

FIG. 1.26
Multizone grid for the calculation of water flow over a generic submarine hull. (*Courtesy of the Science Applications International Corporation (SAIC) and Dr. Nils Salvesen.*)

Of course, the instrument which has allowed the practical growth of CFD is the high-speed digital computer. CFD solutions generally require the repetitive manipulation of many thousands, even millions, of numbers, a task that is humanly impossible without the aid of a computer. Therefore, advances in CFD, and its applications to problems of more and more detail and sophistication, are intimately related to advances in computer hardware, particularly in regard to storage and execution speed. This is why the strongest force driving the development of new supercomputers is coming from the CFD community. Indeed, the advancement in large mainframe computers has been phenomenal over the past three decades. This is nicely illustrated by the variation of relative computation cost (for a given calculation) with years, as plotted in Fig. 1.28, taken from the definitive survey by

COMPUTATIONAL FLUID DYNAMICS: WHAT IS IT?   27

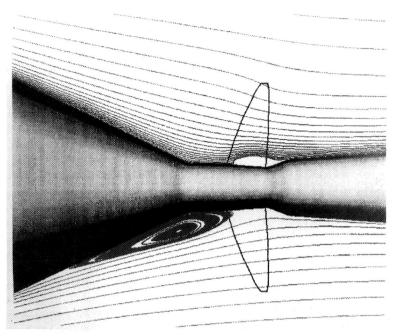

FIG. 1.27
Streamline patterns in the stern region of a submarine. The upper half of the figure illustrates the streamline pattern with a propeller; the lower half illustrates the streamline pattern without a propeller. (*Courtesy of SAIC and Dr. Nils Salvesen.*)

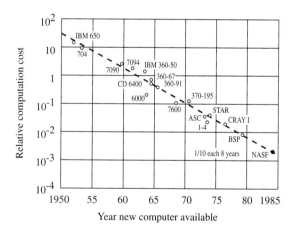

FIG. 1.28
Variation of relative cost of a given computation with time, reflecting the improvements in computer hardware over the years. (*After Ref. 11.*)

Chapman (Ref. 11). The data points on this graph correspond to specific computers, starting with the venerable IBM 650 in 1953, continuing through the development of the pioneering supercomputer, the CRAY I, in 1976, and extrapolating to the National Aerodynamic Simulator, a facility which was installed at the NASA Ames Research Laboratory in the late 1980s. Today, even more spectacular advances are

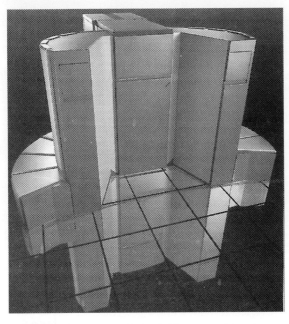

FIG. 1.29
An example of a modern supercomputer, the CRAY Y-MP. (*Cray Research, Inc.*)

being made in supercomputer architecture. An example of a supercomputer is the CRAY Y-MP, shown in Fig. 1.29; this machine has 32 million words of directly addressable central memory with an additional 512 million words available in a companion SSD (solid state device). The execution speed is close to one gigaflop (10^9 floating-point operations per second); this should be compared to the one-megaflop computers of the 1970s. Moreover, new concepts in computer architecture are emerging. The earlier high-speed digital computers were serial machines, capable of one computational operation at a time; hence, all computations had to "get in line" to be processed. The finite speed of electrons, close to the speed of light, poses an inherent limitation on the ultimate execution speed of such serial computers. To detour around this limitation, two computer architectures are now being used:

1. *Vector processors*, a configuration that allows a string of *identical* operations on an array of numbers *simultaneously*, thus saving both time and memory
2. *Parallel processors*, a configuration that is really two or more fully functioning central processing units (CPUs), each of which can handle different instruction and data streams and which can execute separate parts of a program simultaneously, working independently or in concert with other CPUs which belong to the same machine

Vector processors are in widespread use today, and parallel processors are rapidly coming on the scene. For example, the new Connection Machines, which are massively parallel processors, are now in use by many agencies. Should you choose to solve any problems in your professional future using CFD, and these problems are of any sophistication and complexity, the probability is high that you will be using either a vector computer or a parallel processor.

Why is CFD so important in the modern study and solution of problems in fluid mechanics, and why should you be motivated to learn something about CFD? In essence, Sec. 1.1 to 1.4 were devoted to some answers to this question, but we explicitly ask the question here in order to give another example of the revolution that CFD has wrought in modern fluid dynamics—an example that will serve as a focal point for some of our discussion in subsequent chapters.

Specifically, consider the flow field over a blunt-nosed body moving at supersonic or hypersonic speeds, as sketched in Fig. 1.30. The interest in such bodies is driven by the fact that aerodynamic heating to the nose is considerably reduced for blunt bodies compared to sharp-nosed bodies; this is why the Mercury and Apollo space capsules were so blunt and one of the reasons why the space shuttle has a blunt nose and wings with blunt leading edges. As shown in Fig. 1.30, there is a strong, curved bow shock wave which sits in front of the blunt nose, detached from the nose by the distance δ, called the *shock detachment distance*. The calculation of this flow field, including the shape and location of the shock wave, was one of the most perplexing aerodynamic problems of the 1950s and 1960s. Millions of research dollars were spent to solve this supersonic blunt body problem—to no avail.

What was causing the difficulty? Why was the flow field over a blunt body moving at supersonic and hypersonic speeds so hard to calculate? The answer rests basically in the sketch shown in Fig. 1.30. The region of flow behind the nearly normal portion of the shock wave, in the vicinity of the centerline, is locally

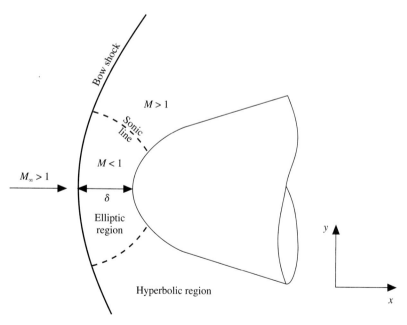

FIG. 1.30
Schematic of the flowfield over a supersonic blunt-nosed body.

subsonic, whereas further downstream, behind the weaker, more oblique part of the bow shock, the flow is locally supersonic. The dividing line between the subsonic and supersonic regions is called the *sonic line*, as sketched in Fig. 1.30. If the flow is assumed to be inviscid, i.e., neglecting the dissipative transport processes of viscosity and thermal conduction, the governing flow equations are the Euler equations (to be derived in Chap. 2). Although these equations are the same no matter whether the flow is locally subsonic or supersonic, their *mathematical behavior* is different in the two regions. In the steady subsonic region, the Euler equations exhibit a behavior that is associated with *elliptic* partial differential equations, whereas in the steady supersonic region, the mathematical behavior of the Euler equations is totally different, namely, that of *hyperbolic* partial differential equations. Such mathematical behavior, the definition of elliptic and hyperbolic equations, and the associated consequences to flow field analysis are discussed in Chap. 3. The change in the mathematical behavior of the governing equations from elliptic in the subsonic region to hyperbolic in the supersonic region made a consistent mathematical analysis which included both regions virtually impossible. Numerical techniques that worked for the subsonic region fell apart in the supersonic region, and techniques for the supersonic region broke down in the subsonic flow. Techniques were developed for just the subsonic portion, and other techniques (such as the standard method of characteristics) were developed for the supersonic region. Unfortunately, the proper patching of these different techniques through the transonic region around the sonic line was extremely difficult. Hence, as late as the mid-1960s, no uniformly valid aerodynamic technique existed to treat the entire flow field over a supersonic blunt body.

However, in 1966, a breakthrough occurred in the blunt body problem. Using the developing power of CFD at that time, and employing the concept of a *time-dependent* approach to the steady state, Moretti and Abbett (Ref. 12) at the Polytechnic Institute of Brooklyn (now the Polytechnic University) obtained a numerical, finite-difference solution to the supersonic blunt body problem which constituted the first practical, straightforward, engineering solution for this flow. After 1966, the blunt body problem was no longer a real "problem." Industry and government laboratories quickly adopted this computational technique for their blunt body analyses. Perhaps the most striking aspect of this comparison is that the supersonic blunt body problem, which was one of the most serious, most difficult, and most researched theoretical aerodynamic problems of the 1950s and 1960s, is today assigned as a *homework problem* in a CFD graduate course at the University of Maryland.

Therein lies an example of the power of CFD combined with an algorithm which properly takes into account the mathematical behavior of the governing flow equations. Here is an answer to the questions asked earlier, namely, why is CFD so important in the modern study of fluid dynamics, and why should you be motivated to learn something about CFD? We have just seen an example where CFD and proper algorithm development revolutionized the treatment of a given flow problem, turning it from a virtual unsolvable problem into a standard, everyday analysis in the nature of an extended homework problem. It is this power of CFD which is a compelling reason for you to study the subject.

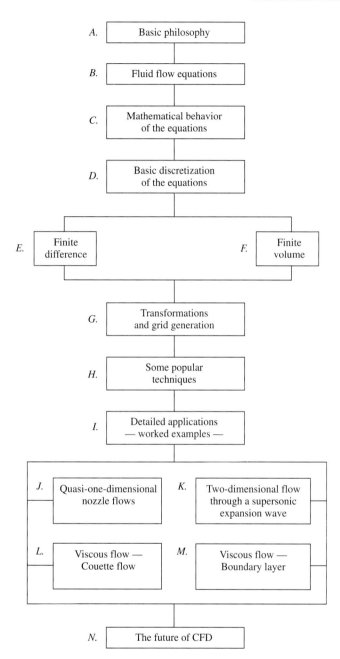

FIG. 1.31
Road map for the book.

1.6 THE PURPOSE OF THIS BOOK

The previous discussions have been intended to put you in a proper frame of mind regarding the overall nature of CFD and to provide a certain incentive to forge ahead to the remaining chapters. As you read on, you will find sandwiched between the covers of this book a very basic, elementary, and tutorial presentation of CFD, emphasizing the fundamentals, surveying a number of solution techniques, and treating various applications ranging from low-speed incompressible flow to high-speed compressible flow. This book is truly an *introduction* to CFD; it is aimed at the completely uninitiated reader, a reader who has little or no experience in CFD. There are presently several very good texts on CFD at the graduate level, such as the standard text by Anderson, Tannehill, and Pletcher (Ref. 13) and the more recent books by Fletcher (Refs. 14 and 15) and by Hirsch (Refs. 16 and 17). A concise and readable presentation is given by Hoffman (Ref. 18). The present book is aimed at a level one notch below that of Refs. 13 to 17. Here we assume on the part of the reader a physical understanding of general fluid dynamics equivalent to most junior-level courses in mechanical and aerospace engineering and a mathematical understanding equivalent to basic calculus and elementary differential equations. This is intended to be a "first book" in the development of your thought processes in CFD. Its purpose is fourfold; it is to provide you with:

1. Some insight into the power and philosophy of CFD
2. An understanding of the governing equations of fluid dynamics in forms particularly suitable to CFD
3. A familiarity with some solution techniques
4. A working vocabulary in the discipline.

By the time you finish this book, this author hopes that you will be well-prepared to launch into more advanced treatises (such as Refs. 13–17), to begin reading the literature in CFD, to follow more sophisticated state-of-the-art presentations, and to begin the direct application of CFD to your special areas of concern. If one or more of the above is what you want, then you and the author share a common purpose—simply move on to Chap. 2 and keep reading.

A road map for the material covered in this book is given in Fig. 1.31. The purpose of this road map is to help chart the course for our thinking and to see how the material flows in some logical fashion. It is the author's experience that when a student is learning a new subject, there is a tendency to get lost in the details and to lose sight of the big picture. Figure 1.31 is the big picture for our discussion on CFD; we will frequently be referring to this road map in subsequent chapters simply to touch base and to remind ourselves where the details fit into the overall scheme of CFD. If at any stage you feel somewhat lost in regard to what we are doing, please remember to refer to this central road map in Fig. 1.31. In addition, localized road maps will be included in most chapters to provide guidance for the flow of ideas in each chapter, in the same spirit as Fig. 1.31 provides guidance for the complete book. In particular, referring to Fig. 1.31, note that blocks *A* through *C* represent some basic thoughts and equations which are common to all of CFD; indeed, the material of the present chapter is represented by block *A*. After these basic aspects are understood and mastered, we will discuss the standard ways of discretizing the fundamental equations to make them amenable to numerical solution (blocks *D–F*) as well as the important aspects of grid transformation (block *G*). After describing some popular techniques for carrying out numerical solutions of the equations (block *H*), we will cover a number of specific applications in some detail in order to clearly illustrate the techniques (blocks *I–M*). Finally, we will discuss the current state of the art as well as the future of CFD (block *N*). Let us now proceed to work our way through this road map, moving on to block *B*, which is the subject of the next chapter.

Finally, Fig. 1.32*a* to *f* contains diagrams that illustrate the flow of various concepts from parts I and II into the applications discussed in Part III. At this stage, simply note that these figures exist; we will refer to them at appropriate times in our discussions. They are located here simply for convenience and to indicate to you that there *is* a logical flow of the basic ideas from Parts I and II into the applications in Part III.

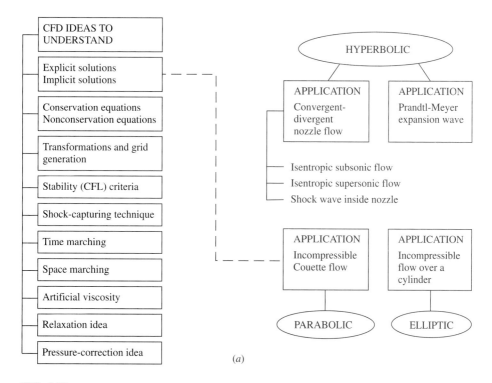

FIG. 1.32
(*a*) Flow of ideas to the incompressible Couette flow application (implicit solution). (*b*) Flow of ideas to the nozzle flow application (isentropic flow). (*c*) Flow of ideas to the nozzle flow application (with shock wave). (*d*) Flow of ideas to the Couette flow application (pressure correction solution). (*e*) Flow of ideas to the Prandtl-Meyer expansion wave application. (*f*) Flow of ideas to the incompressible inviscid cylinder flow application (relaxation solution).

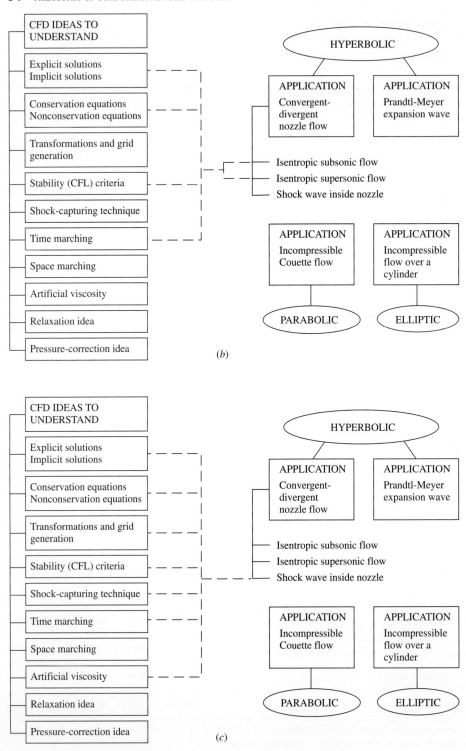

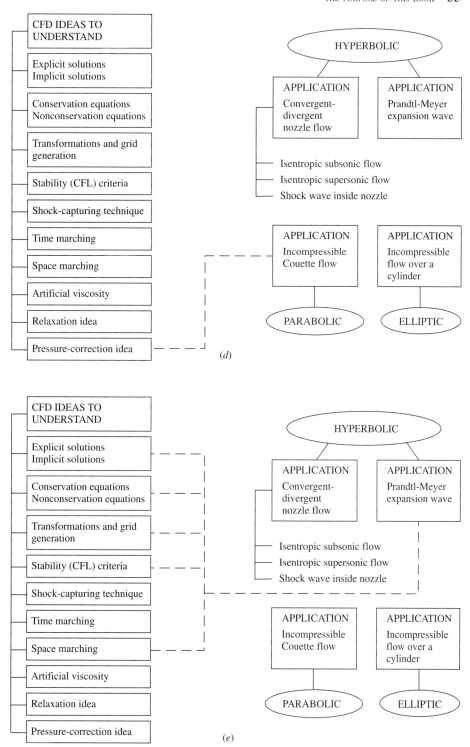

36 PHILOSOPHY OF COMPUTATIONAL FLUID DYNAMICS

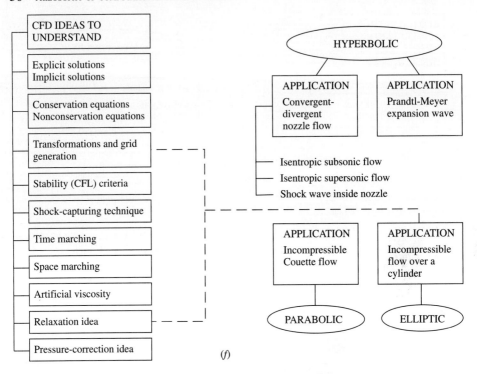

(f)

CHAPTER 2

THE GOVERNING EQUATIONS OF FLUID DYNAMICS: THEIR DERIVATION, A DISCUSSION OF THEIR PHYSICAL MEANING, AND A PRESENTATION OF FORMS PARTICULARLY SUITABLE TO CFD

A fluid is any body whose parts yield to any force impressed on it, and by yielding, are easily moved among themselves.

Isaac Newton, from Section V, Book II of the *Principia*, 1687

We are to admit no more causes of natural

> *things than such as are both true and sufficient to explain their appearances . . . . To this purpose the philosophers say that Nature does nothing in vain, and more is in vain when less will serve; for Nature is pleased with simplicity, and affects not the pomp of superfluous causes."*
>
> Isaac Newton, from Rule I, Book II of the *Principia*, 1687

2.1 INTRODUCTION

All of CFD, in one form or another, is based on the fundamental governing equations of fluid dynamics—the continuity, momentum, and energy equations. *These equations speak physics*. They are the mathematical statements of three fundamental physical principles upon which all of fluid dynamics is based:

1. Mass is conserved.
2. Newton's second law, **F** = **ma**.
3. Energy is conserved.

The purpose of this chapter is to derive and discuss these equations.

The reason for taking the time and space to derive the governing equations of fluid dynamics in this book is threefold:

1. Because all of CFD is based on these equations, it is important for each student to feel very comfortable with these equations before continuing further with his or her studies, and certainly before embarking on any application of CFD to a particular problem.
2. This author assumes that the readers of this book come from varied background and experience. Some of you may not be totally familiar with these equations, whereas others may use them every day. It is hoped that this chapter will be some enlightenment for the former and be an interesting review for the latter.
3. The governing equations can be obtained in various different forms. For most aerodynamic theory, the particular form of the equations makes little difference. However, for a given algorithm in CFD, the use of the equations in one form may lead to success, whereas the use of an alternate form may result in oscillations (wiggles) in the numerical results, incorrect results, or even instability. Therefore, in the world of CFD, the various forms of the equations are of vital interest. In turn, it is important to *derive* these equations in order to point out their differences and similarities, and to reflect on possible implications in their application to CFD.

The reader is warned in advance that this chapter may appear to be "wall-to-wall" equations. However, do not be misled. This chapter is one of the most important in the book. It is driven by the question: If you do not physically understand the meaning and significance of each of these equations—indeed, of each *term* in these equations—then how can you even hope to properly interpret the CFD results obtained by numerically solving these equations? The purpose of this chapter is to squarely address this question. Here, we hope to present the development of these equations and to discuss their significance in such detail that you will begin to feel very comfortable with all forms of all the governing equations of fluid flow. Experience has shown that beginning students find these equations sometimes complex and mystifying. This chapter is designed to take the mystery out of these equations for the reader and to replace it with solid understanding.

The road map for this chapter is given in Fig. 2.1. Notice the flow of ideas as portrayed in this map. All of fluid dynamics is based on the three fundamental

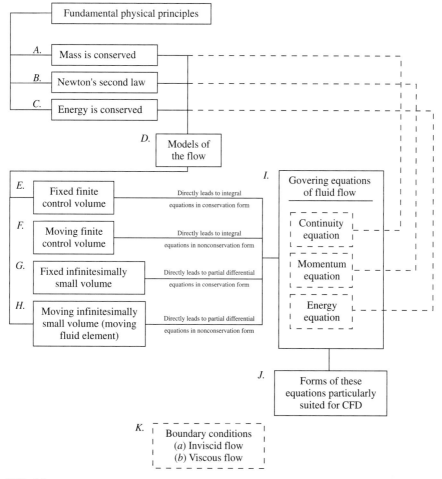

FIG. 2.1
Road map for Chapter 2.

physical principles itemized at the top left of Fig. 2.1. These physical principles are applied to a model of the flow; in turn, this application results in equations which are mathematical statements of the particular physical principles involved, namely, the continuity, momentum, and energy equations. Each different model of the flow (bottom left in Fig. 2.1) directly produces a different mathematical statement of the governing equations, some in conservation form and others in nonconservation form. (The distinction between these two different forms of the governing equations will be made clear by the end of this chapter.) After the continuity, momentum and energy equations are obtained (the large box at the lower right side of Fig. 2.1), forms particularly suited for use in formulating CFD solutions will be delineated (small box at lower right side of Fig. 2.1). Finally, the physical boundary conditions and their appropriate mathematical statements will be developed. The governing equations must be solved subject to these boundary conditions. The *physical* aspects of the boundary conditions are fundamentally independent of the forms of the governing equations, and hence the box representing the boundary conditions stands by itself at the bottom of Fig. 2.1, unconnected to any of the other boxes in the road map. (However, the appropriate *numerical* form of the physical boundary conditions is dependent on the particular mathematical form of the governing equations as well as the particular numerical algorithm used to solve these equations.) Such matters will be discussed as they naturally arise throughout this book. The road map given in Fig. 2.1 will be helpful in guiding our flow of ideas in this chapter. Also, when you finish this chapter, it would be useful to return to Fig. 2.1 to help consolidate your thoughts before proceeding to the next chapter.

2.2 MODELS OF THE FLOW

In obtaining the basic equations of fluid motion, the following philosophy is always followed:

1. Choose the appropriate fundamental physical principles from the law of physics, such as:
   a. Mass is conserved.
   b. $\mathbf{F} = \mathbf{ma}$ (Newton's second law).
   c. Energy is conserved.
2. Apply these physical principles to a suitable model of the flow.
3. From this application, extract the mathematical equations which embody such physical principles.

This section deals with item 2 above, namely, the definition of a suitable model of the flow. This is not a trivial consideration. A solid body is rather easy to see and define; on the other hand, a fluid is a "squishy" substance that is hard to grab hold of. If a solid body is in translational motion, the velocity of each part of the body is the same; on the other hand, if a fluid is in motion, the velocity may be different at each location in the fluid. How then do we visualize a moving fluid so as to apply to it the fundamental physical principles?

For a continuum fluid, the answer is to construct one of the four models described below.

2.2.1 Finite Control Volume

Consider a general flow field as represented by the streamlines in Fig. 2.2a. Let us imagine a closed volume drawn within a *finite* region of the flow. This volume defines a *control volume* $\mathcal{V}$; a *control surface* S is defined as the closed surface which bounds the volume. The control volume may be *fixed* in space with the fluid moving through it, as shown at the left of Fig. 2.2a. Alternatively, the control volume may be moving with the fluid such that the same fluid particles are always inside it, as shown at the right of Fig. 2.2a. In either case, the control volume is a reasonably large, finite region of the flow. The fundamental physical principles are applied to the fluid inside the control volume and to the fluid crossing the control surface (if the control volume is fixed in space). Therefore, instead of looking at the whole flow field at once, with the control volume model we limit our attention to

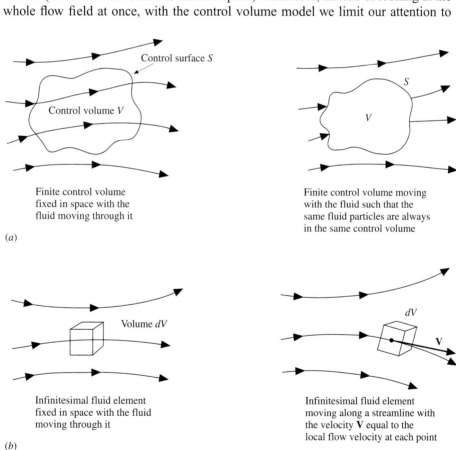

FIG. 2.2
Models of a flow. (*a*) Finite control volume approach; (*b*) infinitesimal fluid element approach.

just the fluid in the finite region of the volume itself. The fluid-flow equations that we *directly* obtain by applying the fundamental physical principles to a finite control volume are in *integral form*. These integral forms of the governing equations can be manipulated to *indirectly* obtain partial differential equations. The equations so obtained from the finite control volume fixed in space (left side of Fig. 2.2a), in either integral or partial differential form, are called the *conservation* form of the governing equations. The equations obtained from the finite control volume moving with the fluid (right side of Fig. 2.2a), in either integral or partial differential form, are called the *nonconservation* form of the governing equations.

2.2.2 Infinitesimal Fluid Element

Consider a general flow field as represented by the streamlines in Fig. 2.2b. Let us imagine an infinitesimally small fluid element in the flow with a differential volume $d\mathcal{V}$. The fluid element is infinitesimal in the same sense as differential calculus; however, it is large enough to contain a huge number of molecules so that it can be viewed as a continuous medium. The fluid element may be fixed in space with the fluid moving through it, as shown at the left of Fig. 2.2b. Alternatively, it may be moving along a streamline with a velocity vector **V** equal to the flow velocity at each point. Again, instead of looking at the whole flow field at once, the fundamental physical principles are applied to just the infinitesimally small fluid element itself. This application leads *directly* to the fundamental equations in *partial differential equation form*. Moreover, the particular partial differential equations obtained directly from the fluid element fixed in space (left side of Fig. 2.2b) are again the *conservation form* of the equations. The partial differential equations obtained *directly* from the moving fluid element (right side of Figure 2.2b) are again called the *nonconservation form* of the equations.

2.2.3 Some Comments

In the above discussion, we have introduced the idea that the governing equations can be expressed in two general forms—conservation form and nonconservation form—without even defining what this really means. Do not be flustered. At this stage in our discussion, we do not have enough insight to understand what these two different terms mean. The definition and understanding will come only while we are actually deriving the different equation forms. So just hang on; at this stage it is sufficient just to be aware of the existence of these two different forms.

In general aerodynamic theory, whether we deal with the conservation or nonconservation forms of the equations is irrelevant. Indeed, through simple manipulation, one form can be obtained from the other. However, there are cases in CFD where it is important which form we use. In fact, the nomenclature which is used to distinguish these two forms (conservation versus nonconservation) has arisen primarily in the CFD literature.

The comments made in this section become clearer after we have actually derived the governing equations. Therefore, when you finish this chapter, it would be worthwhile to reread this section.

As a final comment, in actuality the motion of a fluid is a ramification of the mean motion of its atoms and molecules. Therefore, a third model of the flow can be a microscopic approach wherein the fundamental laws of nature are applied directly to the atoms and molecules, using suitable statistical averaging to define the resulting fluid properties. This approach is in the purview of *kinetic theory*, which is a very elegant method with many advantages in the long run. However, it is beyond the scope of the present book.

2.3 THE SUBSTANTIAL DERIVATIVE (TIME RATE OF CHANGE FOLLOWING A MOVING FLUID ELEMENT)

Before deriving the governing equations, we need to establish a notation which is common in aerodynamics—that of the substantial derivative. In addition, the substantial derivative has an important physical meaning which is sometimes not fully appreciated by students of aerodynamics. A major purpose of this section is to emphasize this physical meaning. The discussion in this section follows that in Ref. 8, which should be consulted for more details.

As the model of the flow, we will adopt the picture shown at the right of Fig. 2.2b, namely, that of an infinitesimally small fluid element moving with the flow. The motion of this fluid element is shown in more detail in Fig. 2.3. Here, the fluid element is moving through cartesian space. The unit vectors along the x, y, and z axes are $\mathbf{i}$, $\mathbf{j}$, and $\mathbf{k}$, respectively. The vector velocity field in this cartesian space is given by

$$\mathbf{V} = u\mathbf{i} + v\mathbf{j} + w\mathbf{k}$$

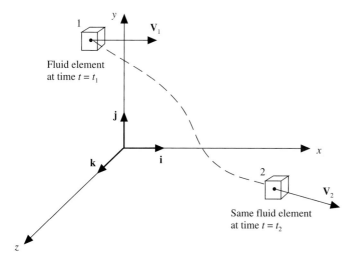

FIG. 2.3
Fluid element moving in the fluid flow—illustration for the substantial derivative

where the x, y, and z components of velocity are given, respectively, by

$$u = u(x, y, z, t)$$
$$v = v(x, y, z, t)$$
$$w = w(x, y, z, t)$$

Note that we are considering in general an *unsteady flow*, where u, v, and w are functions of both space and time t. In addition, the scalar density field is given by

$$\rho = \rho(x, y, z, t)$$

At time t_1, the fluid element is located at point 1 in Fig. 2.3. At this point and time, the density of the fluid element is

$$\rho_1 = \rho(x_1, y_1, z_1, t_1)$$

At a later time t_2, the same fluid element has moved to point 2 in Fig. 2.3. Hence, at time t_2, the density of this *same* fluid element is

$$\rho_2 = \rho(x_2, y_2, z_2, t_2)$$

Since $\rho = \rho(x, y, z, t)$, we can expand this function in a Taylor series about point 1 as follows:

$$\rho_2 = \rho_1 + \left(\frac{\partial \rho}{\partial x}\right)_1 (x_2 - x_1) + \left(\frac{\partial \rho}{\partial y}\right)_1 (y_2 - y_1) + \left(\frac{\partial \rho}{\partial z}\right)_1 (z_2 - z_1)$$
$$+ \left(\frac{\partial \rho}{\partial t}\right)_1 (t_2 - t_1) + \text{(higher-order terms)}$$

Dividing by $t_2 - t_1$ and ignoring higher-order terms, we obtain

$$\frac{\rho_2 - \rho_1}{t_2 - t_1} = \left(\frac{\partial \rho}{\partial x}\right)_1 \frac{x_2 - x_1}{t_2 - t_1} + \left(\frac{\partial \rho}{\partial y}\right)_1 \frac{y_2 - y_1}{t_2 - t_1} + \left(\frac{\partial \rho}{\partial z}\right)_1 \frac{z_2 - z_1}{t_2 - t_1} + \left(\frac{\partial \rho}{\partial t}\right)_1 \quad (2.1)$$

Examine the left side of the Eq. (2.1). This is physically the *average* time rate of change in density of the fluid element as it moves from point 1 to point 2. In the limit, as t_2 approaches t_1, this term becomes

$$\lim_{t_2 \to t_1} \frac{\rho_2 - \rho_1}{t_2 - t_1} \equiv \frac{D\rho}{Dt}$$

Here, $D\rho/Dt$ is a symbol for the *instantaneous* time rate of change of density of the fluid element as it moves through point 1. By definition, this symbol is called the *substantial derivative* D/Dt. Note that $D\rho/Dt$ is the time rate of change of density of the *given fluid element* as it moves through space. Here, our eyes are locked on the fluid element as it is moving, and we are watching the density of the element change as it moves through point 1. This is different from $(\partial \rho/\partial t)_1$, which is physically the time rate of change of density at the fixed point 1. For $(\partial \rho/\partial t)_1$, we fix our eyes on the stationary point 1 and watch the density change due to transient fluctuations in the flow field. Thus, $D\rho/Dt$ and $\partial \rho/\partial t$ are physically and numerically different quantities.

Returning to Eq. (2.1), note that

$$\lim_{t_2 \to t_1} \frac{x_2 - x_1}{t_2 - t_1} \equiv u$$

$$\lim_{t_2 \to t_1} \frac{y_2 - y_1}{t_2 - t_1} \equiv v$$

$$\lim_{t_2 \to t_1} \frac{z_2 - z_1}{t_2 - t_1} \equiv w$$

Thus, taking the limit of Eq. (2.1) as $t_2 \to t_1$, we obtain

$$\frac{D\rho}{Dt} = u\frac{\partial \rho}{\partial x} + v\frac{\partial \rho}{\partial y} + w\frac{\partial \rho}{\partial z} + \frac{\partial \rho}{\partial t} \tag{2.2}$$

Examine Eq. (2.2) closely. From it, we can obtain an expression for the substantial derivative in cartesian coordinates:

$$\frac{D}{Dt} \equiv \frac{\partial}{\partial t} + u\frac{\partial}{\partial x} + v\frac{\partial}{\partial y} + w\frac{\partial}{\partial z} \tag{2.3}$$

Furthermore, in cartesian coordinates, the vector operator ∇ is defined as

$$\nabla \equiv \mathbf{i}\frac{\partial}{\partial x} + \mathbf{j}\frac{\partial}{\partial y} + \mathbf{k}\frac{\partial}{\partial z} \tag{2.4}$$

Hence, Eq. (2.3) can be written as

$$\frac{D}{Dt} \equiv \frac{\partial}{\partial t} + (\mathbf{V} \cdot \nabla) \tag{2.5}$$

Equation (2.5) represents a definition of the substantial derivative operator in vector notation; thus, it is valid for any coordinate system.

Focusing on Eq. (2.5), we once again emphasize that D/Dt is the substantial derivative, which is physically that time rate of change following a moving fluid element; $\partial/\partial t$ is called the *local derivative*, which is physically the time rate of change at a fixed point; $\mathbf{V} \cdot \nabla$ is called the *convective derivative*, which is physically the time rate of change due to the movement of the fluid element from one location to another in the flow field where the flow properties are spatially different. The substantial derivative applies to any flow-field variable, for example, Dp/Dt, DT/Dt, Du/Dt, etc., where p and T are the static pressure and temperature, respectively. For example

$$\frac{DT}{Dt} \equiv \underbrace{\frac{\partial T}{\partial t}}_{\text{Local derivative}} + \underbrace{(\mathbf{V} \cdot \nabla)T}_{\text{Convective derivative}} \equiv \frac{\partial T}{\partial t} + u\frac{\partial T}{\partial x} + v\frac{\partial T}{\partial y} + w\frac{\partial T}{\partial z} \tag{2.6}$$

Again, Eq. (2.6) states physically that the temperature of the fluid element is changing as the element sweeps past a point in the flow because at that point the flow-field temperature itself may be fluctuating with time (the local derivative) and because the fluid element is simply on its way to another point in the flow field where the temperature is different (the convective derivative).

Consider an example which will help to reinforce the physical meaning of the substantial derivative. Imagine that you are hiking in the mountains, and you are about to enter a cave. The temperature inside the cave is cooler than outside. Thus, as you walk through the mouth of the cave, you feel a temperature decrease—this is analogous to the convective derivative in Eq. (2.6). However, imagine that, at the same time, a friend throws a snowball at you such that the snowball hits you just at the same instant you pass through the mouth of the cave. You will feel an additional, but momentary, temperature drop when the snowball hits you—this is analogous to the local derivative in Eq. (2.6). The net temperature drop you feel as you walk through the mouth of the cave is therefore a combination of both the act of moving into the cave, where it is cooler, and being struck by the snowball at the same instant—this net temperature drop is analogous to the substantial derivative in Eq. (2.6).

The purpose of the above derivation is to give you a physical feel for the substantial derivative. We could have circumvented most of the above discussion by recognizing that the substantial derivative is essentially the same as the total differential from calculus. That is, if

$$\rho = \rho(x, y, z, t)$$

then the chain rule from differential calculus gives

$$d\rho = \frac{\partial \rho}{\partial x} dx + \frac{\partial \rho}{\partial y} dy + \frac{\partial \rho}{\partial z} dz + \frac{\partial \rho}{\partial t} dt \tag{2.7}$$

From Eq. (2.7), we have

$$\frac{d\rho}{dt} = \frac{\partial \rho}{\partial t} + \frac{\partial \rho}{\partial x}\frac{dx}{dt} + \frac{\partial \rho}{\partial y}\frac{dy}{dt} + \frac{\partial \rho}{\partial z}\frac{dz}{dt} \tag{2.8}$$

Since $dx/dt = u$, $dy/dt = v$, and $dz/dt = w$, Eq. (2.8) becomes

$$\frac{d\rho}{dt} = \frac{\partial \rho}{\partial t} + u\frac{\partial \rho}{\partial x} + v\frac{\partial \rho}{\partial y} + w\frac{\partial \rho}{\partial z} \tag{2.9}$$

Comparing Eqs. (2.2) and (2.9), we see that $d\rho/dt$ and $D\rho/Dt$ are one and the same. Therefore, the substantial derivative is nothing more than a total derivative with respect to time. However, the derivation of Eq. (2.2) highlights more of the physical significance of the substantial derivative, whereas the derivation of Eq. (2.9) is more formal mathematically.*

---

* Dr. Joe Thompson of Mississippi State University points out, with some justification, that the terminology "substantial derivative" and "total derivative" are unnecessarily confusing, although this terminology is very prevalent in fluid dynamics; indeed, we have followed the standard terminology here. Based on the physical discussion in this section, Thompson suggests that the symbol $(\partial/\partial t)_{\text{fluid element}}$ be used for the appropriate derivative in lieu of D/Dt. This clearly emphasizes the meaning of time rate of change of "something" *following a moving fluid element*.

2.4 THE DIVERGENCE OF THE VELOCITY: ITS PHYSICAL MEANING

In Sec. 2.3 we examined the definition and physical meaning of the substantial derivative; this is because the governing flow equations are frequently expressed in terms of the substantial derivative, and it is important to have a physical understanding of this term. In the same vein, and as one last item before deriving the governing equations, let us consider the divergence of the velocity, $\nabla \cdot \mathbf{V}$. This term appears frequently in the equations of fluid dynamics, and it is well to consider its physical meaning.

Consider a control volume moving with the fluid as sketched on the right of Fig. 2.2a. This control volume is always made up of the same fluid particles as it moves with the flow; hence, its mass is fixed, invariant with time. However, its volume $\mathscr{V}$ and control surface S are changing with time as it moves to different regions of the flow where different values of ρ exist. That is, this moving control volume of fixed mass is constantly increasing or decreasing its volume and is changing its shape, depending on the characteristics of the flow. This control volume is shown in Fig. 2.4 at some instant in time. Consider an infinitesimal element of the surface dS moving at the local velocity $\mathbf{V}$, as shown in Fig. 2.4. The change in the volume of the control volume, $\Delta \mathscr{V}$, due to just the movement of dS over a time increment Δt is, from Fig. 2.4, equal to the volume of the long, thin cylinder with base area dS and altitude $(\mathbf{V} \Delta t) \cdot \mathbf{n}$, where $\mathbf{n}$ is a unit vector perpendicular to the surface at dS. That is,

$$\Delta \mathscr{V} = [(\mathbf{V} \Delta t) \cdot \mathbf{n}]\, dS = (\mathbf{V} \Delta t) \cdot \mathbf{dS} \qquad (2.10)$$

where the vector $\mathbf{dS}$ is defined simply as $\mathbf{dS} \equiv \mathbf{n}\, dS$. Over the time increment Δt, the total change in volume of the whole control volume is equal to the summation of Eq. (2.10) over the total control surface. In the limit as $dS \to 0$, the sum becomes the surface integral

$$\iint_S (\mathbf{V} \Delta t) \cdot \mathbf{dS}$$

If this integral is divided by Δt, the result is physically the time rate of change of the control volume, denoted by $D\mathscr{V}/Dt$; that is,

$$\frac{D\mathscr{V}}{Dt} = \frac{1}{\Delta t} \iint_S (\mathbf{V} \cdot \Delta t) \cdot \mathbf{dS} = \iint_S \mathbf{V} \cdot \mathbf{dS} \qquad (2.11)$$

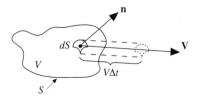

FIG. 2.4
Moving control volume used for the physical interpretation of the divergence of velocity.

48 THE GOVERNING EQUATIONS OF FLUID DYNAMICS

Note that we have written the left side of Eq. (2.11) as the substantial derivative of $\mathscr{V}$, because we are dealing with the time rate of change of the control volume *as the volume moves with the flow* (we are using the picture shown at the right of Fig. 2.2a), and this is physically what is meant by the substantial derivative. Applying the divergence theorem from vector calculus to the right side of Eq. (2.11), we obtain

$$\frac{D\mathscr{V}}{Dt} = \iiint_{\mathscr{V}} (\nabla \cdot \mathbf{V}) \, d\mathscr{V} \tag{2.12}$$

Now, let us imagine that the moving control volume in Fig. 2.4 is shrunk to a very small volume $\delta\mathscr{V}$, essentially becoming an infinitesimal moving fluid element as sketched on the right of Fig. 2.2a. Then Eq. (2.12) can be written as

$$\frac{D(\delta\mathscr{V})}{Dt} = \iiint_{\delta\mathscr{V}} (\nabla \cdot \mathbf{V}) \, d\mathscr{V} \tag{2.13}$$

Assume that $\delta\mathscr{V}$ is small enough such that $\nabla \cdot \mathbf{V}$ is essentially the same value throughout $\delta\mathscr{V}$. Then the integral in Eq. (2.13), in the limit as $\delta\mathscr{V}$ shrinks to zero, is given by $(\nabla \cdot \mathbf{V}) \delta\mathscr{V}$. From Eq. (2.13), we have

$$\frac{D(\delta\mathscr{V})}{Dt} = (\nabla \cdot \mathbf{V}) \delta\mathscr{V}$$

or

$$\boxed{\nabla \cdot \mathbf{V} = \frac{1}{\delta\mathscr{V}} \frac{D(\delta\mathscr{V})}{Dt}} \tag{2.14}$$

Examine Eq. (2.14) closely. On the left side we have the divergence of the velocity; on the right side we have its physical meaning. That is,

> $\nabla \cdot \mathbf{V}$ is physically the time rate of change of the volume of a moving fluid element, per unit volume.

2.4.1 A Comment

It is useful to keep the physical meaning of the divergence of the velocity in mind when you are dealing with the governing flow equations. Indeed, this is an example of an overall philosophy which this author urges you to embrace, as follows. Imagine that we are dealing with a vector velocity $\mathbf{V}$ in cartesian (x, y, z) space. When a pure mathematician sees the symbol $\nabla \cdot \mathbf{V}$, his or her mind will most likely register the fact that $\nabla \cdot \mathbf{V} = \partial u/\partial x + \partial v/\partial y + \partial w/\partial z$. On the other hand, when a fluid dynamicist sees the symbol $\nabla \cdot \mathbf{V}$, his or her mind should register *first* the physical meaning—he or she should first see in the symbol $\nabla \cdot \mathbf{V}$ the words "time rate of change of the volume of a moving fluid element, per unit volume." Indeed, this philosophy is extrapolated to all mathematical equations and operations having to do with physical problems. Always keep in mind the physical meaning of the terms in the equations you are dealing with. In this vein, note that in the phrase "computational fluid dynamics" the word "computational" is simply an adjective to "fluid dynamics"; when you are dealing with the discipline of CFD, it is vitally important to keep the physical understanding of *fluid dynamics* uppermost in your mind. This, in part, is the purpose of the present chapter.

2.5 THE CONTINUITY EQUATION

Let us now apply the philosophy discussed in Sec. 2.2; that is, let us (1) write down a fundamental physical principle, (2) apply it to a suitable model of the flow, and (3) obtain an equation which represents the fundamental physical principle. In this section, we will treat the following case:

Physical principle: Mass is conserved.

The governing flow equation which results from the application of this physical principle to any one of the four models of the flow shown in Fig. 2.2a and b is called the *continuity equation*. Moreover, in this section we will carry out in detail the application of this physical principle using *all four* of the flow models illustrated in Fig. 2.2a and b; in this way we hope to dispel any mystery surrounding the derivation of the governing flow equation. That is, we will derive the continuity equation four different ways, obtaining in a direct fashion four different forms of the equation. Then, by indirect manipulation of these four different forms, we will show that they are all really the *same* equation. In addition, we will invoke the idea of conservation versus nonconservation forms, helping to elucidate the meaning of those words. Let us proceed.

2.5.1 Model of the Finite Control Volume Fixed in Space

Consider the flow model shown at the left of Fig. 2.2a, namely, a control volume of arbitrary shape and of finite size. The volume is fixed in space. The surface that bounds this control volume is called the control surface, as labeled in Fig. 2.2a. The fluid moves through the fixed control volume, flowing across the control surface. This flow model is shown in more detail in Fig. 2.5. At a point on the control surface in Fig. 2.5, the flow velocity is **V** and the vector elemental surface area (as defined in Sec. 2.4) is **dS**. Also let $d\mathcal{V}$ be an elemental volume inside the finite control volume. Applied to this control volume, our fundamental physical principle that mass is conserved means

$$\begin{matrix} \text{Net mass flow } out \\ \text{of control volume} \\ \text{through surface } S \end{matrix} = \begin{matrix} \text{time rate of} \\ \text{decrease of mass} \\ \text{inside control volume} \end{matrix} \quad (2.15a)$$

or

$$B = C \quad (2.15b)$$

where B and C are just convenient symbols for the left and right sides, respectively,

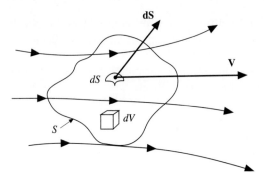

FIG. 2.5
Finite control volume fixed in space.

of Eq. (2.15a). First, let us obtain an expression for B in terms of the quantities shown in Fig. 2.5. The mass flow of a moving fluid across any fixed surface (say, in kilograms per second or slugs per second) is equal to the product of (density) × (area of surface) × (component of velocity perpendicular to the surface). Hence the elemental mass flow across the area dS is

$$\rho V_n \, dS = \rho \mathbf{V} \cdot \mathbf{dS} \qquad (2.16)$$

Examining Fig. 2.5, note that by convention, $\mathbf{dS}$ always points in a direction *out* of the control volume. Hence, when $\mathbf{V}$ also points out of the control volume (as shown in Fig. 2.5), the product $\rho \mathbf{V} \cdot \mathbf{dS}$ is *positive*. Moreover, when $\mathbf{V}$ points out of the control volume, the mass flow is physically leaving the control volume; i.e., it is an *outflow*. Hence, a positive $\rho \mathbf{V} \cdot \mathbf{dS}$ denotes an outflow. In turn, when $\mathbf{V}$ points into the control volume, $\rho \mathbf{V} \cdot \mathbf{dS}$ is *negative*. Moreover, when $\mathbf{V}$ points inward, the mass flow is physically entering the control volume; i.e., it is an *inflow*. Hence, a negative $\rho \mathbf{V} \cdot \mathbf{dS}$ denotes an inflow. The net mass flow *out* of the entire control volume through the control surface S is the summation over S of the elemental mass flow expressed in Eq. (2.16). In the limit, this becomes a surface integral, which is physically the left sides of Eqs. (2.15a) and (2.15b); that is,

$$B = \iint_S \rho \mathbf{V} \cdot \mathbf{dS} \qquad (2.17)$$

Now consider the right sides of Eqs. (2.15a) and (2.15b). The mass contained within the elemental volume $d\mathcal{V}$ is $\rho \, d\mathcal{V}$. The total mass inside the control volume is therefore

$$\iiint_\mathcal{V} \rho \, d\mathcal{V}$$

The time rate of *increase* of mass inside $\mathcal{V}$ is then

$$\frac{\partial}{\partial t} \iiint_\mathcal{V} \rho \, d\mathcal{V}$$

In turn, the time rate of decrease of mass inside $\mathscr{V}$ is the negative of the above; i.e.,

$$-\frac{\partial}{\partial t}\iiint_{\mathscr{V}} \rho \, d\mathscr{V} = C \tag{2.18}$$

Thus, substituting Eqs. (2.17) and (2.18) into (2.15b), we have

$$\iint_{S} \rho \mathbf{V} \cdot \mathbf{dS} = -\frac{\partial}{\partial t}\iiint_{\mathscr{V}} \rho \, d\mathscr{V}$$

or

$$\boxed{\frac{\partial}{\partial t}\iiint_{\mathscr{V}} \rho \, d\mathscr{V} + \iint_{S} \rho \mathbf{V} \cdot \mathbf{dS} = 0} \tag{2.19}$$

Equation (2.19) is an *integral form of the continuity equation*. It was derived on the basis of a *finite* control volume *fixed* in space. The *finite* aspect of the control volume is why the equation is obtained directly in *integral* form. The fact that the control volume was *fixed in space* leads to the specific integral form given by Eq. (2.19), which is called the *conservation form*. The forms of the governing flow equations that are *directly* obtained from a flow model which is fixed in space are, by definition, called the conservation form.

Now consider Fig. 2.6, which shows the same four flow models given in Fig. 2.2a and b. However, in Fig. 2.6 the specific form of the continuity equation obtained *directly* from each model is displayed underneath the sketch of the particular model. In this subsection, we have just finished the derivation of Eq. (2.19) using the model of a finite control volume fixed in space. Therefore, Eq. (2.19) is displayed in box I just below the sketch of this model in Fig. 2.6. In the following subsections, we will derive the remaining three equations which appear in boxes II to IV in Fig. 2.6. Then, we will show, by manipulation, that the equations in all four boxes are simply different forms of the same equation; i.e., we will connect all four equations by the paths *A* through *D* illustrated in Fig. 2.6. As stated earlier, we hope that these derivations, along with the flow of logic diagramed in Fig. 2.6, will take the mystery out of the different forms of the governing equations.

2.5.2 Model of the Finite Control Volume Moving with the Fluid

Consider the flow model shown at the right of Fig. 2.2a, namely, a control volume of finite size moving with the fluid. This control volume, as it moves with the fluid, is always composed of the same identifiable elements of mass; i.e., the moving control volume has a *fixed mass*. On the other hand, as this fixed mass moves downstream, the shape and volume of the finite control volume can, in general, change. Consider an infinitesimally small element of volume $d\mathscr{V}$ inside this finite

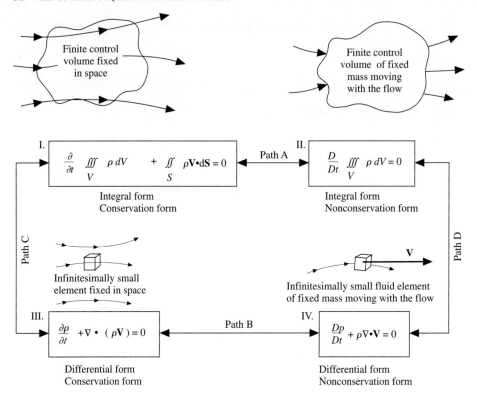

FIG. 2.6
The different forms of the continuity equation, their relationship to the different models of the flow, and the schematic emphasis that all four equations are essentially the same—they can each be obtained from the other.

control volume; the mass of this small element is $\rho \, d\mathcal{V}$, where ρ is the local density. Then, the total mass of the finite control volume is given by

$$\text{Mass} = \iiint_{\mathcal{V}} \rho \, d\mathcal{V} \qquad (2.20)$$

In Eq. (2.20), the volume integral is taken over the whole moving control volume $\mathcal{V}$. However, keep in mind that, here, $\mathcal{V}$ is changing as the control volume moves downstream. On the other hand, the physical principle that mass is conserved, when applied to this model of the flow, simply states that the mass in Eq. (2.20) is a constant as the control volume moves with the flow. Now recall the physical meaning of the substantial derivative as discussed in Sec. 2.3; it expresses the time rate of change of any property of a fluid element as it moves with the flow. Since our finite control volume is made up of an infinite number of infinitesimally small fluid elements, all with a fixed, unchanging mass, and hence all with substantial derivatives of these unchanging masses equal to zero, we can write for the finite control volume, from Eq. (2.20),

$$\boxed{\frac{D}{Dt}\iiint_{\mathscr{V}} \rho \, d\mathscr{V} = 0} \qquad (2.21)$$

Equation (2.21) is an *integral form of the continuity equation*, different from that expressed in Eq. (2.19). It was derived on the basis of a *finite* control volume *moving* with the fluid. The *finite* aspect of the control volume is why the equation is obtained *directly* in *integral* form. The fact that the control volume is *moving* with the fluid leads to the specific integral form given by Eq. (2.21), which is called the *nonconservation form*. The forms of the governing flow equations that are *directly* obtained from a flow model which is moving with the flow are, by definition, called the nonconservation form.

Equation (2.21) is displayed in box II in Fig. 2.6. Although the integral forms of the equations in boxes I and II are different, they can be shown by indirect manipulation (path *A*) to be the same equation. This will be discussed in Sec. 2.5.5.

2.5.3 Model of an Infinitesimally Small Element Fixed in Space

Consider the flow model shown at the left of Fig. 2.2*b*, namely, an infinitesimally small element fixed in space, with the fluid moving through it. This flow model is shown in more detail in Fig. 2.7. Here, for convenience we adopt a cartesian coordinate system, where the velocity and density are functions of (x, y, z) space and time *t*. Fixed in this (x, y, z) space is an infinitesimally small element of sides *dx*, *dy*, and *dz* (Fig. 2.7*a*). There is mass flow through this fixed element, as shown in Fig. 2.7*b*. Consider the left and right faces of the element which are perpendicular to the *x* axis. The area of these faces is *dy dz*. The mass flow through the left face is $(\rho u) \, dy \, dz$. Since the velocity and density are functions of spatial location, the values of the mass flux across the right face will be different from that across the left face; indeed, the difference in mass flux between the two faces is simply $[\partial(\rho u)/\partial x] \, dx$. Thus, the mass flow across the right face can be expressed as $\{\rho u + [\partial(\rho u)/\partial x] \, dx\} \, dy \, dz$. The mass flow across both the left and right faces is shown in Fig. 2.7*b*. In a similar vein, the mass flow through both the bottom and top faces, which are perpendicular to the *y* axis, is $(\rho v) \, dx \, dz$ and $\{\rho v + [\partial(\rho v)/\partial y] \, dy\} \, dx \, dz$, respectively. The mass flow through both the front and back faces, which are perpendicular to the *z* axis, is $(\rho w) \, dx \, dy$ and $\{\rho w + [\partial(\rho w)/\partial z] \, dz\} \, dx \, dy$, respectively. Note that *u*, *v*, and *w* are positive, by convention, in the positive *x*, *y*, and *z* directions, respectively. Hence, the arrows in Fig. 2.7 show the contributions to the inflow and outflow of mass through the sides of the fixed element. If we denote a net outflow of mass as a positive quantity, then from Fig. 2.7, we have

Net outflow in x direction:

$$\left[\rho u + \frac{\partial(\rho u)}{\partial x} \, dx\right] dy \, dz - (\rho u) \, dy \, dz = \frac{\partial(\rho u)}{\partial x} \, dx \, dy \, dz$$

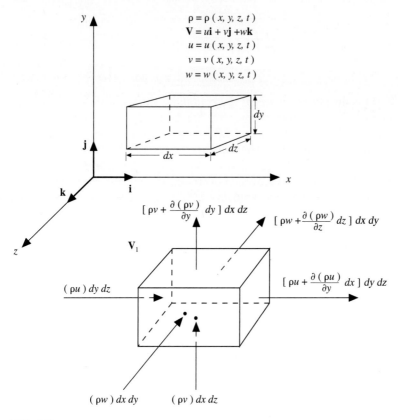

FIG. 2.7
Model of the infinitesimally small element fixed in space and a diagram of the mass fluxes through the various faces of the element—for a derivation of the continuity equation.

Net outflow in y direction:

$$\left[\rho v + \frac{\partial(\rho v)}{\partial y} dy\right] dx\, dz - (\rho v)\, dx\, dz = \frac{\partial(\rho v)}{\partial y} dx\, dy\, dz$$

Net outflow in z direction:

$$\left[\rho w + \frac{\partial(\rho w)}{\partial z} dz\right] dx\, dy - (\rho w)\, dx\, dy = \frac{\partial(\rho w)}{\partial z} dx\, dy\, dz$$

Hence, the net mass flow out of the element is given by

$$\text{Net mass flow} = \left[\frac{\partial(\rho u)}{\partial x} + \frac{\partial(\rho v)}{\partial y} + \frac{\partial(\rho w)}{\partial z}\right] dx\, dy\, dz \qquad (2.22)$$

The total mass of fluid in the infinitesimally small element is $\rho\,(dx\, dy\, dz)$; hence the time rate of increase of mass inside the element is given by

$$\text{Time rate of mass increase} = \frac{\partial \rho}{\partial t}(dx\,dy\,dz) \tag{2.23}$$

The physical principle that mass is conserved, when applied to the fixed element in Fig. 2.7, can be expressed in words as follows: the net mass flow out of the element must equal the time rate of *decrease* of mass inside the element. Denoting the mass decrease by a negative quantity, this statement can be expressed in terms of Eqs. (2.22) and (2.23) as

$$\left[\frac{\partial(\rho u)}{\partial x} + \frac{\partial(\rho v)}{\partial y} + \frac{\partial(\rho w)}{\partial z}\right] dx\,dy\,dz = -\frac{\partial \rho}{\partial t}(dx\,dy\,dz)$$

or
$$\frac{\partial \rho}{\partial t} + \left[\frac{\partial(\rho u)}{\partial x} + \frac{\partial(\rho v)}{\partial y} + \frac{\partial(\rho w)}{\partial z}\right] = 0 \tag{2.24}$$

In Eq. (2.24), the term in brackets is simply $\nabla \cdot (\rho \mathbf{V})$. Thus, Eq. (2.24) becomes

$$\boxed{\frac{\partial \rho}{\partial t} + \nabla \cdot (\rho \mathbf{V}) = 0} \tag{2.25}$$

Equation (2.25) is a *partial differential equation form of the continuity equation*. It was derived on the basis of an *infinitesimally small* element *fixed* in space. The *infinitesimally small* aspect of the element is why the equation is obtained directly in *partial differential equation* form. The fact that the element was *fixed in space* leads to the specific differential form given by Eq. (2.25), which is called the *conservation form*. As stated earlier, the forms of the governing flow equations that are *directly* obtained from a flow model which is fixed in space are, by definition, called the conservation form.

Equation (2.25) is displayed in box III in Fig. 2.6. It is the form that most directly stems from the model of an infinitesimally small element fixed in space. On the other hand, it can also be obtained by indirect manipulation from either of the integral equations displayed in boxes I and II, as will be shown in Sec. 2.2.5.

2.5.4 Model of an Infinitesimally Small Fluid Element Moving with the Flow

Consider the flow model shown at the right of Fig. 2.2*b*, namely, an infinitesimally small fluid element moving with the flow. This fluid element has a fixed mass, but in general its shape and volume will change as it moves downstream. Denote the fixed mass and variable volume of this moving fluid element by δm and $\delta \mathscr{V}$, respectively. Then

$$\delta m = \rho\, \delta \mathscr{V} \tag{2.26}$$

Since mass is conserved, we can state that the time rate of change of the mass of the fluid element is zero as the element moves along with the flow. Invoking the physical meaning of the substantial derivative discussed in Sec. 2.3, we have

$$\frac{D(\delta m)}{Dt} = 0 \qquad (2.27)$$

Combining Eqs. (2.26) and (2.27), we have

$$\frac{D(\rho\,\delta\mathcal{V})}{Dt} = \delta\mathcal{V}\frac{D\rho}{Dt} + \rho\frac{D(\delta\mathcal{V})}{Dt} = 0$$

or

$$\frac{D\rho}{Dt} + \rho\left[\frac{1}{\delta\mathcal{V}}\frac{D(\delta\mathcal{V})}{Dt}\right] = 0 \qquad (2.28)$$

We recognize the term in brackets in Eq. (2.28) as the physical meaning of $\nabla \cdot \mathbf{V}$, discussed in Sec. 2.4 and given in Eq. (2.14). Hence, combining Eqs. (2.14) and (2.28), we obtain

$$\boxed{\frac{D\rho}{Dt} + \rho\nabla \cdot \mathbf{V} = 0} \qquad (2.29)$$

Equation (2.29) is a partial differential equation form of the continuity equation, different from that expressed by Eq. (2.25). It was derived on the basis of an *infinitesimally small* fluid element *moving* with the flow. Once again, the infinitesimally small aspect of the fluid element is why the equation is obtained directly in *partial differential equation* form. The fact that the element is *moving* with the flow leads to the specific differential form given by Eq. (2.29), which is called the *nonconservation form*. As stated earlier, the forms of the governing flow equations that are *directly* obtained from a flow model which is moving with the flow are, by definition, called the nonconservation form.

Equation (2.29) is displayed in box IV in Fig. 2.6. It is the form that most directly stems from the model of an infinitesimally small fluid element moving with the flow. On the other hand, it can also be obtained by indirect manipulation from any of the equations in the other boxes in Fig. 2.6. It is now appropriate to examine this indirect manipulation.

2.5.5 All the Equations Are One: Some Manipulations

Examining Fig. 2.6, we see four different forms of the continuity equation, each one a direct product of the flow model used in its derivation. Two of the forms are integral equations; the other two are partial differential equations. Two of the equations are in conservation form; the other two are in nonconservation form. However, these four equations are not fundamentally different equations; rather, they are four different forms of the *same* equation, namely, the continuity equation. Any of these four forms can be derived by manipulation from any of the others. This is symbolized by paths *A* through *D* sketched in Fig. 2.6. For a better understanding of the meaning and significance of the governing flow equations, we need to examine the details of these different paths. This is the purpose of the present subsection.

First, let us examine how the partial differential equation form can be obtained from the integral equation form; i.e., let us examine path C in Fig. 2.6. Repeating Eq. (2.19),

$$\frac{\partial}{\partial t}\iiint_{\mathcal{V}} \rho\, d\mathcal{V} + \iint_{S} \rho \mathbf{V} \cdot d\mathbf{S} = 0 \qquad (2.19)$$

Since the control volume used for the derivation of Eq. (2.19) is fixed in space, the limits of integration for the integrals in Eq. (2.19) are constant, and hence the time derivative $\partial/\partial t$ can be placed inside the integral.

$$\iiint_{\mathcal{V}} \frac{\partial \rho}{\partial t}\, d\mathcal{V} + \iint_{S} \rho \mathbf{V} \cdot d\mathbf{S} = 0 \qquad (2.30)$$

Applying the divergence theorem from vector calculus, the surface integral in Eq. (2.30) can be expressed as a volume integral:

$$\iint_{S} (\rho \mathbf{V}) \cdot d\mathbf{S} = \iiint_{\mathcal{V}} \nabla \cdot (\rho \mathbf{V})\, d\mathcal{V} \qquad (2.31)$$

Substituting Eq. (2.31) into (2.30), we have

$$\iiint_{\mathcal{V}} \frac{\partial \rho}{\partial t}\, d\mathcal{V} + \iiint_{\mathcal{V}} \nabla \cdot (\rho \mathbf{V})\, d\mathcal{V} = 0$$

or

$$\iiint_{\mathcal{V}} \left[\frac{\partial \rho}{\partial t} + \nabla \cdot (\rho \mathbf{V})\right] d\mathcal{V} = 0 \qquad (2.32)$$

Since the finite control volume is *arbitrarily* drawn in space, the only way for the integral in Eq. (2.32) to equal zero is for the integrand to be zero at every point within the control volume. Hence, from Eq. (2.32)

$$\boxed{\frac{\partial \rho}{\partial t} + \nabla \cdot (\rho \mathbf{V}) = 0} \qquad (2.33)$$

Equation (2.33) is precisely the continuity equation in partial differential equation form that is displayed in box III in Fig. 2.6. Hence, we have shown how the integral form in box I can, after some manipulation, yield the differential form in box III. Again, note that both the equations in boxes I and III are in conservation form; the above manipulation does not change that situation.

Next, let us examine a manipulation that does change the conservation form to the nonconservation form. Specifically let us take the differential equation in box III and convert it to the differential equation in box IV. Consider the vector identity involving the divergence of the product of a scalar times a vector, such as

$$\nabla \cdot (\rho \mathbf{V}) \equiv (\rho \nabla \cdot \mathbf{V}) + (\mathbf{V} \cdot \nabla \rho) \tag{2.34}$$

In words, the divergence of a scalar times a vector is equal to the scalar times the divergence of the vector plus the vector dotted into the gradient of the scalar. (See any good text on vector analysis for a presentation of this identity, such as Ref. 19.) Substituting Eq. (2.34) into Eq. (2.33), we obtain

$$\frac{\partial \rho}{\partial t} + (\mathbf{V} \cdot \nabla \rho) + (\rho \nabla \cdot \mathbf{V}) = 0 \tag{2.35}$$

The first two terms on the left side of Eq. (2.35) are simply the substantial derivative of density. Hence, Eq. (2.35) becomes

$$\frac{D\rho}{Dt} + \rho \nabla \cdot \mathbf{V} = 0 \tag{2.36}$$

Equation (2.36) is precisely the equation displayed in box IV in Fig. 2.6. Hence, by a slight manipulation of the partial differential equation in box III, which is in conservation form, we obtained the partial differential equation in box IV, which is in nonconservation form.

Can the same type of change be made to the integral forms of the equations; e.g., can the equation in box II be manipulated to obtain the equation in box I? This is represented by path A in Fig. 2.6. The answer is yes; let us see how. The equation in box II is Eq. (2.21), repeated below:

$$\frac{D}{Dt} \iiint_{\mathcal{V}} \rho \, d\mathcal{V} = 0 \tag{2.21}$$

Recall in Eq. (2.21) that the volume integral is taken over the whole *moving* control volume $\mathcal{V}$ and that this volume is changing as it flows downstream. Indeed, the moving finite control volume consists of an infinite number of infinitesimally small volumes of fixed infinitesimally small mass, each of volume $d\mathcal{V}$, where the magnitude of $d\mathcal{V}$ also changes as the control volume moves downstream. Since the substantial derivative itself represents a time rate of change associated with a moving element, and the limits of integration on the volume integral in Eq. (2.21) are determined by these same moving elements, then the substantial derivative can be taken inside the integral. Hence, Eq. (2.21) can be written as

$$\frac{D}{Dt} \iiint_{\mathcal{V}} \rho \, d\mathcal{V} = \iiint_{\mathcal{V}} \frac{D(\rho \, d\mathcal{V})}{Dt} = 0 \tag{2.37}$$

Noting again that $d\mathcal{V}$ physically represents an infinitesimally small volume which itself is variable, the substantial derivative inside the integral in Eq. (2.37) is the derivative of a product of two variables, namely, ρ and $d\mathcal{V}$. The derivative must be expanded accordingly; Eq. (2.37) becomes

$$\iiint_{\mathcal{V}} \frac{D\rho}{Dt} \, d\mathcal{V} + \iiint_{\mathcal{V}} \rho \frac{D(d\mathcal{V})}{Dt} = 0$$

Dividing and multiplying the second term by $d\mathcal{V}$, we have

$$\iiint_\mathcal{V} \frac{D\rho}{Dt} d\mathcal{V} + \iiint_\mathcal{V} \rho \left[\frac{1}{d\mathcal{V}} \frac{D(d\mathcal{V})}{Dt} \right] d\mathcal{V} = 0 \qquad (2.38)$$

The physical meaning of the term inside the brackets is simply the "time rate of change of volume of an infinitesimally small fluid element per unit volume." We recall from Sec. 2.4 and Eq. (2.14) that this term is the divergence of velocity. Hence, Eq. (2.38) becomes

$$\iiint_\mathcal{V} \frac{D\rho}{Dt} d\mathcal{V} + \iiint_\mathcal{V} \rho \nabla \cdot \mathbf{V} \, d\mathcal{V} = 0 \qquad (2.39)$$

From the definition of the substantial derivative given by Eq. (2.5), the first term in Eq. (2.39) can be expanded as

$$\iiint_\mathcal{V} \frac{D\rho}{Dt} d\mathcal{V} = \iiint_\mathcal{V} \left[\frac{\partial \rho}{\partial t} + \mathbf{V} \cdot \nabla \rho \right] d\mathcal{V} \qquad (2.40)$$

Substituting Eq. (2.40) into (2.39), and writing all terms under a single volume integral, we have

$$\iiint_\mathcal{V} \left[\frac{\partial \rho}{\partial t} + \mathbf{V} \cdot \nabla \rho + \rho \nabla \cdot \mathbf{V} \right] d\mathcal{V} = 0 \qquad (2.41)$$

From the vector identity given in Eq. (2.34), the last two terms in Eq. (2.41) can be written as

$$\mathbf{V} \cdot \nabla \rho + \rho \nabla \cdot \mathbf{V} = \nabla \cdot (\rho \mathbf{V})$$

With this, Eq. (2.41) becomes

$$\iiint_\mathcal{V} \frac{\partial \rho}{\partial t} d\mathcal{V} + \iiint_\mathcal{V} \nabla \cdot (\rho \mathbf{V}) \, d\mathcal{V} = 0 \qquad (2.42)$$

Finally, employing the divergence theorem from vector analysis, which relates a surface integral to a volume integral as

$$\iiint_\mathcal{V} \nabla \cdot (\rho \mathbf{V}) \, d\mathcal{V} \equiv \iint_S \rho \mathbf{V} \cdot d\mathbf{S}$$

(again, see any good vector analysis text, such as Ref. 19), Eq. (2.42) finally results in

$$\iiint_\mathcal{V} \frac{\partial \rho}{\partial t} d\mathcal{V} + \iint_S \rho \mathbf{V} \cdot d\mathbf{S} = 0 \qquad (2.43)$$

Equation (2.43) is essentially the form of the equation in box I in Fig. 2.6.

We could go on, but we won't—for the sake of not boring you with essentially repetitive manipulations. The major point of this subsection has been made. We see that the four different equations displayed in the boxes in Fig. 2.6 in reality are not different equations at all but rather four different forms of the same equation—the continuity equation. However, each different form displayed in Fig. 2.6 comes most directly from the particular model of the flow adjacent to each equation, and hence the terms in each equation have slightly different physical implications. Also, the philosophy associated with these different forms, and how they were derived, is not limited to just the continuity equation—the same approach is used for the development of the momentum and energy equations, to follow.

2.5.6 Integral versus Differential Form of the Equations: An Important Comment

There is a subtle difference between the integral and differential forms of the governing flow equations which is best noted at this stage. The integral form of the equations allows for the presence of discontinuities inside the fixed control volume (fixed in space); there is no inherent mathematical reason to assume otherwise. However, the differential form of the governing equations assumes the flow properties are differentiable, hence continuous. This is particularly evident when we use the divergence theorem to derive the differential form from the integral form—the divergence theorem assumes mathematical continuity. This is a strong argument for the integral form of the equations to be considered *more fundamental* than the differential form. This consideration becomes of particular importance when calculating a flow with real discontinuities, such as shock waves.

2.6 THE MOMENTUM EQUATION

In this section, we apply another fundamental physical principle to a model of the flow, namely:

Physical principle: $\mathbf{F} = m\mathbf{a}$ (Newton's second law)

The resulting equation is called the *momentum equation*. Unlike the derivation of the continuity equation in Sec. 2.5, where great pains were taken to illustrate the use of all four models of the fluid and to highlight the different forms of the equations obtained therein, in the present section we will restrain ourselves and choose only one model of the flow. Specifically, we will utilize the moving fluid element model shown at the right of Fig. 2.2b because this model is particularly convenient for the derivation of the momentum equation as well as the energy equation (to be considered in Sec. 2.7). The moving fluid element model is sketched in more detail in Fig. 2.8. However, please keep in mind that the momentum and energy equations can be derived using any of the other three models of the fluid in Fig. 2.2a and b; as in the case of the continuity equation developed in Sec. 2.5, each different model of the flow leads directly to a different form of the momentum and energy equations, analogous to those for the continuity equation displayed in Fig. 2.6.

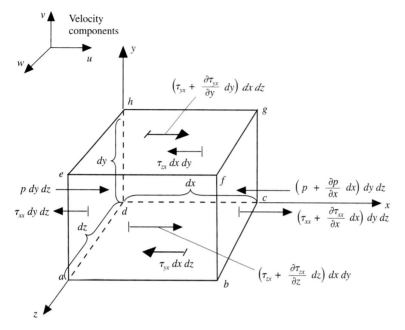

FIG. 2.8
Infinitesimally small, moving fluid element. Only the forces in the x direction are shown. Model used for the derivation of the x component of the momentum equation.

Newton's second law, expressed above, when applied to the moving fluid element in Fig. 2.8, says that the net force on the fluid element equals its mass times the acceleration of the element. This is a vector relation, and hence can be split into three scalar relations along the x, y, and z axes. Let us consider only the x component of Newton's second law,

$$F_x = ma_x \qquad (2.44)$$

where F_x and a_x are the scalar x components of the force and acceleration, respectively.

First, consider the left side of Eq. (2.44). We say that the moving fluid element experiences a force in the x direction. What is the source of this force? There are two sources:

1. *Body forces*, which act directly on the volumetric mass of the fluid element. These forces "act at a distance"; examples are gravitational, electric, and magnetic forces.

2. *Surface forces*, which act directly on the surface of the fluid element. They are due to only two sources: (*a*) the pressure distribution acting on the surface, imposed by the outside fluid surrounding the fluid element, and (*b*) the shear and normal stress distributions acting on the surface, also imposed by the outside fluid "tugging" or "pushing" on the surface by means of friction.

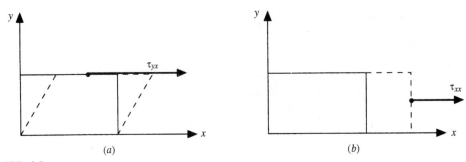

FIG. 2.9
Illustration of (*a*) shear stress (related to the time rate of charge of the shearing deformation and (*b*) normal stress (related to the time rate of charge of volume).

Let us denote the body force per unit mass acting on the fluid element by **f**, with f_x as its x component. The volume of the fluid element is ($dx\ dy\ dz$); hence,

$$\text{Body force on fluid element acting in } x \text{ direction} = \rho f_x (dx\ dy\ dz) \quad (2.45)$$

The shear and normal stresses in a fluid are related to the time rate of change of the deformation of the fluid element, as sketched in Fig. 2.9 for just the xy plane. The shear stress, denoted by τ_{xy} in Fig. 2.9*a*, is related to the time rate of change of the shearing deformation of the fluid element, whereas the normal stress, denoted by τ_{xx} in Fig. 2.9*b*, is related to the time rate of change of volume of the fluid element. As a result, both shear and normal stresses depend on velocity gradients in the flow, to be designated later. In most viscous flows, normal stresses (such as τ_{xx}) are much smaller than shear stresses and many times are neglected. Normal stresses (say τ_{xx} in the x direction) become important when the normal velocity gradients (say $\partial u/\partial x$) are very large, such as *inside* a shock wave.

The surface forces in the x-direction exerted on the fluid element are sketched in Fig. 2.8. The convention will be used here that τ_{ij} denotes a stress in the j direction exerted on a plane perpendicular to the i axis. On face *abcd*, the only force in the x direction is that due to shear stress, $\tau_{yx}\ dx\ dz$. Face *efgh* is a distance dy above face *abcd*; hence the shear force in the x direction on face *efgh* is $[\tau_{yx} + (\partial \tau_{yx}/\partial y)\ dy]\ dx\ dz$. Note the directions of the shear force on faces *abcd* and *efgh*; on the bottom face, τ_{yx} is to the left (the negative x direction), whereas on the top face, $\tau_{yz} + (\partial \tau_{yx}/\partial y)\ dy$ is to the right (the positive x direction). These directions are consistent with the convention that positive increases in all three components of velocity, u, v, and w, occur in the positive directions of the axes. For example, in Fig. 2.8, u increases in the positive y direction. Therefore, concentrating on face *efgh*, u is higher just above the face than on the face; this causes a "tugging" action which tries to pull the fluid element in the positive x direction (to the right) as shown in Fig. 2.8. In turn, concentrating on face *abcd*, u is lower just beneath the face than on the face; this causes a retarding or dragging action on the fluid element, which acts in the negative x direction (to the left) as shown in Fig. 2.8. The directions of all the other viscous stresses shown in Fig. 2.8, including τ_{xx}, can be justified in a like fashion. Specifically on face *dcgh*, τ_{zx} acts in the negative x-direction, whereas on face *abfe*, $\tau_{zx} + (\partial\tau_{zx}/\partial z)\,dz$ acts in the positive x direction. On face *adhe*, which is perpendicular to the x axis, the only forces in the x direction are the pressure force $p\,dy\,dz$, which always acts in the direction *into* the fluid element, and $\tau_{xx}\,dy\,dz$, which is in the negative x direction. In Fig. 2.8, the reason why τ_{xx} on face *adhe* is to the left hinges on the convention mentioned earlier for the direction of increasing velocity. Here, by convention, a positive increase in u takes place in the positive x direction. Hence, the value of u just to the left face of *adhe* is smaller than the value of u on the face itself. As a result, the viscous action of the normal stress acts as a "suction" on face *adhe*; i.e., there is a dragging action toward the left that wants to retard the motion of the fluid element. In contrast, on face *bcgf*, the pressure force $[p + \partial p/\partial x)\,dx]\,dy\,dz$ presses inward on the fluid element (in the negative x direction), and because the value of u just to the right of the face *bcgf* is larger than the value of u on the face, there is a "suction" due to the viscous normal stress which tries to pull the element to the right (in the positive x direction) with a force equal to $[\tau_{xx} + (\partial\tau_{xx}/\partial x)\,dx]\,dy\,dz$.

With the above in mind, for the moving fluid element we can write

$$\begin{aligned}\text{Net surface force} \\ \text{in } x \text{ direction}\end{aligned} = \left[p - \left(p + \frac{\partial p}{\partial x}\,dx\right)\right] dy\,dz$$

$$+ \left[\left(\tau_{xx} + \frac{\partial \tau_{xx}}{\partial x}\,dx\right) - \tau_{xx}\right] dy\,dz + \left[\left(\tau_{yx} + \frac{\partial \tau_{yx}}{\partial y}\,dy\right) - \tau_{yx}\right] dx\,dz$$

$$+ \left[\left(\tau_{zx} + \frac{\partial \tau_{zx}}{\partial z}\,dz\right) - \tau_{zx}\right] dx\,dy \qquad (2.46)$$

The total force in the x direction F_x, is given by the sum of Eqs. (2.45) and (2.46). Adding, and cancelling terms, we obtain

$$F_x = \left[-\frac{\partial p}{\partial x} + \frac{\partial \tau_{xx}}{\partial x} + \frac{\partial \tau_{yx}}{\partial y} + \frac{\partial \tau_{zx}}{\partial z}\right] dx\,dy\,dz + \rho f_x\,dx\,dy\,dz \qquad (2.47)$$

Equation (2.47) represents the left-hand side of Eq. (2.44).

To summarize and reinforce the physical significance of the force on a moving fluid element, let us display Newton's second law in diagramatic form as follows:

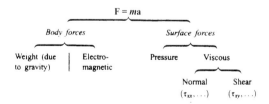

Considering the right-hand side of Eq. (2.44), recall that the mass of the fluid element is fixed and is equal to

$$m = \rho\,dx\,dy\,dz \qquad (2.48)$$

Also, recall that the acceleration of the fluid element is the time rate of change of its velocity. Hence, the component of acceleration in the x direction, denoted by a_x, is simply the time rate of change of u; since we are following a moving fluid element, this time rate of change is given by the substantial derivative. Thus,

$$a_x = \frac{Du}{Dt} \tag{2.49}$$

Combining Eqs. (2.44) and (2.47) to (2.49), we obtain

$$\boxed{\rho \frac{Du}{Dt} = -\frac{\partial p}{\partial x} + \frac{\partial \tau_{xx}}{\partial x} + \frac{\partial \tau_{yx}}{\partial y} + \frac{\partial \tau_{zx}}{\partial z} + \rho f_x} \tag{2.50a}$$

which is the x component of the momentum equation for a viscous flow. In a similar fashion, the y and z components can be obtained as

$$\boxed{\rho \frac{Dv}{Dt} = -\frac{\partial p}{\partial y} + \frac{\partial \tau_{xy}}{\partial x} + \frac{\partial \tau_{yy}}{\partial y} + \frac{\partial \tau_{zy}}{\partial z} + \rho f_y} \tag{2.50b}$$

and

$$\boxed{\rho \frac{Dw}{Dt} = -\frac{\partial p}{\partial z} + \frac{\partial \tau_{xz}}{\partial x} + \frac{\partial \tau_{yz}}{\partial y} + \frac{\partial \tau_{zz}}{\partial z} + \rho f_z} \tag{2.50c}$$

Equations (2.50a) to (2.50c) are the x, y, and z components, respectively, of the momentum equation. Note that they are partial differential equations obtained *directly* from an application of the fundamental physical principle to an infinitesimal fluid element. Moreover, since this fluid element is moving with the flow, Eqs. (2.50a) to (2.50c) are in *nonconservation* form. They are scalar equations and are called the *Navier-Stokes equations* in honor of two men—the Frenchman M. Navier and the Englishman G. Stokes—who independently obtained the equations in the first half of the nineteenth century.

The Navier-Stokes equations can be obtained in conservation form as follows. Writing the left-hand side of Eq. (2.50a) in terms of the definition of the substantial derivative, we have

$$\rho \frac{Du}{Dt} = \rho \frac{\partial u}{\partial t} + \rho \mathbf{V} \cdot \nabla u \tag{2.51}$$

Also, expanding the following derivative,

$$\frac{\partial (\rho u)}{\partial t} = \rho \frac{\partial u}{\partial t} + u \frac{\partial \rho}{\partial t}$$

and rearranging, we have

$$\rho \frac{\partial u}{\partial t} = \frac{\partial (\rho u)}{\partial t} - u \frac{\partial \rho}{\partial t} \tag{2.52}$$

Recalling the vector identity for the divergence of the product of a scalar times a vector, we have

$$\nabla \cdot (\rho u \mathbf{V}) = u \nabla \cdot (\rho \mathbf{V}) + (\rho \mathbf{V}) \cdot \nabla u$$

or
$$\rho \mathbf{V} \cdot \nabla u = \nabla \cdot (\rho u \mathbf{V}) - u \nabla \cdot (\rho \mathbf{V}) \tag{2.53}$$

Substitute Eqs. (2.52) and (2.53) into (2.51).

$$\begin{aligned}\rho \frac{Du}{Dt} &= \frac{\partial(\rho u)}{\partial t} - u\frac{\partial \rho}{\partial t} - u\nabla \cdot (\rho \mathbf{V}) + \nabla \cdot (\rho u \mathbf{V}) \\ &= \frac{\partial(\rho u)}{\partial t} - u\left[\frac{\partial \rho}{\partial t} + \nabla \cdot (\rho \mathbf{V})\right] + \nabla \cdot (\rho u \mathbf{V})\end{aligned} \tag{2.54}$$

The term in brackets in Eq. (2.54) is simply the left-hand side of the continuity equation as Eq. (2.25); hence the term in brackets is zero. Thus Eq. (2.54) reduces to

$$\rho \frac{Du}{Dt} = \frac{\partial(\rho u)}{\partial t} + \nabla \cdot (\rho u \mathbf{V}) \tag{2.55}$$

Substitute Eq. (2.55) into (2.50a).

$$\boxed{\frac{\partial(\rho u)}{\partial t} + \nabla \cdot (\rho u \mathbf{V}) = -\frac{\partial p}{\partial x} + \frac{\partial \tau_{xx}}{\partial x} + \frac{\partial \tau_{yx}}{\partial y} + \frac{\partial \tau_{zx}}{\partial z} + \rho f_x} \tag{2.56a}$$

Similarly, Eqs. (2.50b) and (2.50c) can be expressed as

$$\boxed{\frac{\partial(\rho v)}{\partial t} + \nabla \cdot (\rho v \mathbf{V}) = -\frac{\partial p}{\partial y} + \frac{\partial \tau_{xy}}{\partial x} + \frac{\partial \tau_{yy}}{\partial y} + \frac{\partial \tau_{zy}}{\partial z} + \rho f_y} \tag{2.56b}$$

and

$$\boxed{\frac{\partial(\rho w)}{\partial t} + \nabla \cdot (\rho w \mathbf{V}) = -\frac{\partial p}{\partial z} + \frac{\partial \tau_{xz}}{\partial x} + \frac{\partial \tau_{yz}}{\partial y} + \frac{\partial \tau_{zz}}{\partial z} + \rho f_z} \tag{2.56c}$$

Equations (2.56a) to (2.56c) are the Navier-Stokes equations in *conservation form*.

In the late seventeenth century, Isaac Newton stated that shear stress in a fluid is proportional to the time rate of strain, i.e., velocity gradients. Such fluids are called *newtonian* fluids. (Fluids in which τ is *not* proportional to the velocity gradients are nonnewtonian fluids; blood flow is one example.) In virtually all practical aerodynamic problems, the fluid can be assumed to be newtonian. For such fluids, Stokes in 1845 obtained

$$\tau_{xx} = \lambda(\nabla \cdot \mathbf{V}) + 2\mu \frac{\partial u}{\partial x} \tag{2.57a}$$

$$\tau_{yy} = \lambda(\nabla \cdot \mathbf{V}) + 2\mu \frac{\partial v}{\partial y} \tag{2.57b}$$

$$\tau_{zz} = \lambda(\nabla \cdot \mathbf{V}) + 2\mu \frac{\partial w}{\partial z} \tag{2.57c}$$

$$\tau_{xy} = \tau_{yx} = \mu\left[\frac{\partial v}{\partial x} + \frac{\partial u}{\partial y}\right] \tag{2.57d}$$

$$\tau_{xz} = \tau_{zx} = \mu\left(\frac{\partial u}{\partial z} + \frac{\partial w}{\partial x}\right) \qquad (2.57e)$$

$$\tau_{yz} = \tau_{zy} = \mu\left(\frac{\partial w}{\partial y} + \frac{\partial v}{\partial z}\right) \qquad (2.57f)$$

where μ is the molecular viscosity coefficient and λ is the second viscosity coefficient. Stokes made the hypothesis that

$$\lambda = -\tfrac{2}{3}\mu$$

which is frequently used but which has still not been definitely confirmed to the present day.

Substituting Eqs. (2.57) into (2.56), we obtain the complete Navier-Stokes equations in conservation form;

$$\frac{\partial(\rho u)}{\partial t} + \frac{\partial(\rho u^2)}{\partial x} + \frac{\partial(\rho uv)}{\partial y} + \frac{\partial(\rho uw)}{\partial z} = -\frac{\partial p}{\partial x}$$
$$+ \frac{\partial}{\partial x}\left(\lambda \nabla \cdot \mathbf{V} + 2\mu\frac{\partial u}{\partial x}\right) + \frac{\partial}{\partial y}\left[\mu\left(\frac{\partial v}{\partial x} + \frac{\partial u}{\partial y}\right)\right]$$
$$+ \frac{\partial}{\partial z}\left[\mu\left(\frac{\partial u}{\partial z} + \frac{\partial w}{\partial x}\right)\right] + \rho f_x \qquad (2.58a)$$

$$\frac{\partial(\rho v)}{\partial t} + \frac{\partial(\rho uv)}{\partial x} + \frac{\partial(\rho v^2)}{\partial y} + \frac{\partial(\rho vw)}{\partial z} = -\frac{\partial p}{\partial y}$$
$$+ \frac{\partial}{\partial x}\left[\mu\left(\frac{\partial v}{\partial x} + \frac{\partial u}{\partial y}\right)\right] + \frac{\partial}{\partial y}\left(\lambda \nabla \cdot \mathbf{V} + 2\mu\frac{\partial v}{\partial y}\right)$$
$$+ \frac{\partial}{\partial z}\left[\mu\left(\frac{\partial w}{\partial y} + \frac{\partial v}{\partial z}\right)\right] + \rho f_y \qquad (2.58b)$$

$$\frac{\partial(\rho w)}{\partial t} + \frac{\partial(\rho uw)}{\partial x} + \frac{\partial(\rho vw)}{\partial y} + \frac{\partial(\rho w^2)}{\partial z} = -\frac{\partial p}{\partial z}$$
$$+ \frac{\partial}{\partial x}\left[\mu\left(\frac{\partial u}{\partial z} + \frac{\partial w}{\partial x}\right)\right] + \frac{\partial}{\partial y}\left[\mu\left(\frac{\partial w}{\partial y} + \frac{\partial v}{\partial z}\right)\right]$$
$$+ \frac{\partial}{\partial z}\left(\lambda \nabla \cdot \mathbf{V} + 2\mu\frac{\partial w}{\partial z}\right) + \rho f_z \qquad (2.58c)$$

2.7 THE ENERGY EQUATION

In the present section, we apply the third physical principle as itemized at the beginning of Sec. 2.1, namely,

Physical principle: Energy is conserved.

In keeping with our derivation of the Navier-Stokes equations (i.e., the momentum equation) in Sec. 2.6, we will use again the flow model of an infinitesimally small fluid element moving with the flow (as shown at the right of Fig. 2.2b). The physical principle stated above is nothing more than the first law of thermodynamics. When applied to the flow model of a fluid element moving with the flow, the first law states that

$$\begin{array}{c}\text{Rate of change}\\\text{of energy inside}\\\text{fluid element}\end{array} = \begin{array}{c}\text{Net flux of}\\\text{heat into}\\\text{element}\end{array} + \begin{array}{c}\text{Rate of work done on}\\\text{element due to}\\\text{body and surface forces}\end{array}$$

or (2.59)

$$A = B + C$$

where A, B, and C denote the respective terms above.

Let us first evaluate C; that is, let us obtain an expression for the rate of work done on the moving fluid element due to body and surface forces. It can be shown that the rate of doing work by a force exerted on a moving body is equal to the product of the force and the component of velocity in the direction of the force (see Refs. 1 and 8 for such a derivation). Hence the rate of work done by the body force acting on the fluid element moving at a velocity $\mathbf{V}$ is

$$\rho \mathbf{f} \cdot \mathbf{V}(dx\ dy\ dz)$$

With regard to the surface forces (pressure plus shear and normal stresses), consider just the forces in the x direction, shown in Fig. 2.8. The rate of work done on the moving fluid element by the pressure and shear forces in the x direction shown in Fig. 2.8 is simply the x component of velocity, u, multiplied by the forces; e.g., on face $abcd$ the rate of work done by $\tau_{yx}\ dx\ dz$ is $u\tau_{yx}\ dx\ dz$, with similar expressions for the other faces. To emphasize these energy considerations, the moving fluid element is redrawn in Fig. 2.10, where the rate of work done on each face by surface forces in the x direction is shown explicitly. To obtain the *net* rate of work done on the fluid element by the surface forces, note that forces in the positive x direction do positive work and that forces in the negative x direction do negative work. Hence, comparing the pressure forces on face $adhe$ and $bcgf$ in Fig. 2.10, the net rate of work done by pressure in the x direction is

$$\left[up - \left(up + \frac{\partial(up)}{\partial x}dx\right)\right]dy\ dz = -\frac{\partial(up)}{\partial x}dx\ dy\ dz$$

Similarly, the net rate of work done by the shear stresses in the x direction on faces $abcd$ and $efgh$ is

$$\left[\left(u\tau_{yx} + \frac{\partial(u\tau_{yx})}{\partial y}dy\right) - u\tau_{yx}\right]dx\ dz = \frac{\partial(u\tau_{yx})}{\partial y}dx\ dy\ dz$$

Considering all the surface forces shown in Fig. 2.10, the net rate of work on the moving fluid element due to these forces is simply

$$\left[-\frac{\partial(up)}{\partial x} + \frac{\partial(u\tau_{xx})}{\partial x} + \frac{\partial(u\tau_{yx})}{\partial y} + \frac{\partial(u\tau_{zx})}{\partial z}\right]dx\ dy\ dz$$

The above expression considers only surface forces in the x direction. When the surface forces in the y and z directions are also included, similar expressions are

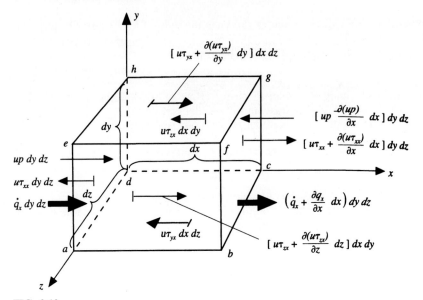

FIG. 2.10
Energy fluxes associated with an infinitesimally small, moving fluid element. For simplicity, only the fluxes in the x direction are shown. Model used for the derivation of the energy equation.

obtained. In total, the net rate of work done on the moving fluid element is the sum of the surface force contributions in the x, y, and z directions, as well as the body force contribution. This is denoted by C in Eq. (2.59) and is given by

$$C = -\left[\left(\frac{\partial(up)}{\partial x} + \frac{\partial(vp)}{\partial y} + \frac{\partial(wp)}{\partial z}\right) + \frac{\partial(u\tau_{xx})}{\partial x} + \frac{\partial(u\tau_{yx})}{\partial y}\right.$$
$$+ \frac{\partial(u\tau_{zx})}{\partial z} + \frac{\partial(v\tau_{xy})}{\partial x} + \frac{\partial(v\tau_{yy})}{\partial y} + \frac{\partial(v\tau_{zy})}{\partial z} + \frac{\partial(w\tau_{xz})}{\partial x}$$
$$\left.+ \frac{\partial(w\tau_{yz})}{\partial y} + \frac{\partial(w\tau_{zz})}{\partial z}\right] dx\,dy\,dz + \rho \mathbf{f} \cdot \mathbf{V}\, dx\,dy\,dz \quad (2.60)$$

Note in Eq. (2.60) that the first three terms on the right-hand side are simply $\nabla \cdot (p\mathbf{V})$.

Let us turn our attention to B in Eq. (2.59), i.e., the net flux of heat into the element. This heat flux is due to (1) volumetric heating such as absorption or emission of radiation and (2) heat transfer across the surface due to temperature gradients, i.e., thermal conduction. Define $\dot{q}$ as the rate of volumetric heat addition per unit mass. Noting that the mass of the moving fluid element in Fig. 2.10 is $\rho\, dx\, dy\, dz$, we obtain

$$\text{Volumetric heating of element} = \rho\dot{q}\, dx\, dy\, dz \quad (2.61)$$

In Fig. 2.10, the heat transferred by thermal conduction into the moving fluid element across face $adhe$ is $\dot{q}_x\, dy\, dz$, where $\dot{q}_x$ is the heat transferred in the x direction per unit time per unit area by thermal conduction. (The heat transfer in a given direction, when expressed in dimensions of energy per unit time per unit area perpendicular to the direction, is called the *heat flux* in that direction. Here, $\dot{q}_x$ is the heat flux in the x direction.) The heat transferred out of the element across face $bcgf$ is $[\dot{q}_x + (\partial \dot{q}_x/\partial x)\, dx]\, dy\, dz$. Thus, the net heat transferred in the x direction into the fluid element by thermal conduction is

$$\left[\dot{q}_x - \left(\dot{q}_x + \frac{\partial \dot{q}_x}{\partial x} dx \right) \right] dy\, dz = -\frac{\partial \dot{q}_x}{\partial x} dx\, dy\, dz$$

Taking into account heat transfer in the y and z directions across the other faces in Fig. 2.10, we obtain

$$\text{Heating of fluid element by thermal conduction} = -\left(\frac{\partial \dot{q}_x}{\partial x} + \frac{\partial \dot{q}_y}{\partial y} + \frac{\partial \dot{q}_z}{\partial z} \right) dx\, dy\, dz \quad (2.62)$$

The term B in Eq. (2.59) is the sum of Eqs. (2.61) and (2.62).

$$B = \left[\rho \dot{q} - \left(\frac{\partial \dot{q}_x}{\partial x} + \frac{\partial \dot{q}_y}{\partial y} + \frac{\partial \dot{q}_z}{\partial z} \right) \right] dx\, dy\, dz \quad (2.63)$$

The heat flux due to thermal conduction, from Fourier's law of heat conduction, is proportional to the local temperature gradient:

$$\dot{q}_x = -k \frac{\partial T}{\partial x} \qquad \dot{q}_y = -k \frac{\partial T}{\partial y} \qquad \dot{q}_z = -k \frac{\partial T}{\partial z}$$

where k is the thermal conductivity. Hence, Eq. (2.63) can be written

$$B = \left[\rho \dot{q} + \frac{\partial}{\partial x}\left(k \frac{\partial T}{\partial x} \right) + \frac{\partial}{\partial y}\left(k \frac{\partial T}{\partial y} \right) + \frac{\partial}{\partial z}\left(k \frac{\partial T}{\partial z} \right) \right] dx\, dy\, dz \quad (2.64)$$

Finally, the term A in Eq. (2.59) denotes the time rate of change of energy of the fluid element. Pause for a moment and ask yourself the question: the time rate of change of *what* energy? In classical thermodynamics, we generally deal with a system that is stationary; in this case, the energy that appears in the first law of thermodynamics is the *internal* energy. Let us examine more closely the source of this internal energy. If the system is a gas, the atoms and molecules are moving within the system, translating in a purely random fashion. That is, each atom or molecule has translational kinetic energy, and this energy is associated with the purely random motion of the particle. In addition, as they translate through space, molecules (not atoms) can also rotate and vibrate, adding rotational and vibrational energy to the molecule. Finally, the motion of electrons around the nuclei of the atoms or molecules adds electronic energy to the particle. The total energy of a given molecule is the sum of its translational, rotational, vibrational, and electronic energies; the total energy of each atom is the sum of its translational and electronic energy (see Ref. 2 for an extensive discussion of molecular and atomic energies). The *internal energy* of the gas system is simply the energy of each molecule or atom, summed over all the molecules and atoms in the system. This is the physical significance of the internal energy that appears in the first law of thermodynamics.

Now, return to Eq. (2.59) and examine the term labeled A. We are now considering a gaseous medium that is *in motion*; i.e., the energy term labeled A concerns the energy of a *moving* fluid element. Hence, the fluid element has two contributions to its energy:

1. The internal energy due to random molecular motion, e (per unit mass). This is the energy we have described above.
2. The kinetic energy due to translational motion of the fluid element. The kinetic energy per unit mass is simply $V^2/2$.

Hence, the moving fluid element has both *internal* and *kinetic* energy; the *sum* of these two is the "total" energy. In Eq. (2.59), the energy in the term A is the total energy, i.e., the *sum* of the internal and kinetic energies. The total energy is $e + V^2/2$. Since we are following a moving fluid element, the time rate of change of total energy per unit mass is given by the substantial derivative. Since the mass of the fluid element is $\rho\, dx\, dy\, dz$, we have

$$A = \rho \frac{D}{Dt}\left(e + \frac{V^2}{2}\right) dx\, dy\, dz \qquad (2.65)$$

The final form of the energy equation is obtained by substituting Eqs. (2.60), (2.64), and (2.65) into (2.59), obtaining

$$\begin{aligned}
\rho \frac{D}{Dt}\left(e + \frac{V^2}{2}\right) &= \rho \dot{q} + \frac{\partial}{\partial x}\left(k \frac{\partial T}{\partial x}\right) + \frac{\partial}{\partial y}\left(k \frac{\partial T}{\partial y}\right) + \frac{\partial}{\partial z}\left(k \frac{\partial T}{\partial z}\right) \\
&- \frac{\partial(up)}{\partial x} - \frac{\partial(vp)}{\partial y} - \frac{\partial(wp)}{\partial z} + \frac{\partial(u\tau_{xx})}{\partial x} + \frac{\partial(u\tau_{yx})}{\partial y} + \frac{\partial(u\tau_{zx})}{\partial z} + \frac{\partial(v\tau_{xy})}{\partial x} + \frac{\partial(v\tau_{yy})}{\partial y} \\
&+ \frac{\partial(v\tau_{zy})}{\partial z} + \frac{\partial(w\tau_{xz})}{\partial x} + \frac{\partial(w\tau_{yz})}{\partial y} + \frac{\partial(w\tau_{zz})}{\partial z} + \rho \mathbf{f} \cdot \mathbf{V}
\end{aligned}$$

(2.66)

This is the *nonconservation* form of the energy equation; also note that it is in terms of the *total* energy $e + V^2/2$. Once again, the nonconservation form results from the application of the fundamental physical principle to a *moving* fluid element.

The left-hand side of Eq. (2.66) involves the substantial derivative of the total energy, $D(e + V^2/2)/Dt$. This is just one of the many different forms of the energy equation; it is the form that comes directly from the principle of conservation of energy applied to a moving fluid element. This equation can be readily modified in two respects, as follows:

1. The left-hand side can be expressed in terms of the internal energy e alone, or the static enthalpy h alone, or the total enthalpy $h_0 = h + V^2/2$ alone. In each case,

THE ENERGY EQUATION **71**

the right-hand side of the pertinent equation also changes. [For example, in the next paragraph we will examine the necessary manipulations to change Eq. (2.66) into one involving De/Dt.]

2. For each of the different forms of the energy equation mentioned above, there are both nonconservation as well as conservation forms. The manipulation to convert one form into the other is also discussed below.

Let us start with Eq. (2.66) and first cast it in the form dealing with e only. To accomplish this, multiply Eqs. (2.50a), (2.50b), and (2.50c) by u, v, and w, respectively.

$$\rho \frac{D(u^2/2)}{Dt} = -u\frac{\partial p}{\partial x} + u\frac{\partial \tau_{xx}}{\partial x} + u\frac{\partial \tau_{yx}}{\partial y} + u\frac{\partial \tau_{zx}}{\partial z} + \rho u f_x \qquad (2.67)$$

$$\rho \frac{D(v^2/2)}{Dt} = -v\frac{\partial p}{\partial y} + v\frac{\partial \tau_{xy}}{\partial x} + v\frac{\partial \tau_{yy}}{\partial y} + v\frac{\partial \tau_{zy}}{\partial z} + \rho v f_y \qquad (2.68)$$

$$\rho \frac{D(w^2/2)}{Dt} = -w\frac{\partial p}{\partial z} + w\frac{\partial \tau_{xz}}{\partial x} + w\frac{\partial \tau_{yz}}{\partial y} + w\frac{\partial \tau_{zz}}{\partial z} + \rho w f_z \qquad (2.69)$$

Adding Eqs. (2.67) to (2.69), and noting that $u^2 + v^2 + w^2 = V^2$, we obtain

$$\rho \frac{DV^2}{Dt} = -u\frac{\partial p}{\partial x} - v\frac{\partial p}{\partial y} - w\frac{\partial p}{\partial z} + u\left(\frac{\partial \tau_{xx}}{\partial x} + \frac{\partial \tau_{yx}}{\partial y} + \frac{\partial \tau_{zx}}{\partial z}\right)$$

$$+ v\left(\frac{\partial \tau_{xy}}{\partial x} + \frac{\partial \tau_{yy}}{\partial y} + \frac{\partial \tau_{zy}}{\partial z}\right) + w\left(\frac{\partial \tau_{xz}}{\partial x} + \frac{\partial \tau_{yz}}{\partial y} + \frac{\partial \tau_{zz}}{\partial z}\right)$$

$$+ \rho(uf_x + vf_y + wf_z) \qquad (2.70)$$

Subtracting Eq. (2.70) from (2.66), and noting that $\rho \mathbf{f} \cdot \mathbf{V} = \rho(uf_x + vf_y + wf_z)$, we have

$$\rho \frac{De}{Dt} = \rho \dot{q} + \frac{\partial}{\partial x}\left(k\frac{\partial T}{\partial x}\right) + \frac{\partial}{\partial y}\left(k\frac{\partial T}{\partial y}\right) + \frac{\partial}{\partial z}\left(k\frac{\partial T}{\partial z}\right)$$

$$- p\left(\frac{\partial u}{\partial x} + \frac{\partial v}{\partial y} + \frac{\partial w}{\partial z}\right) + \tau_{xx}\frac{\partial u}{\partial x} + \tau_{yx}\frac{\partial u}{\partial y} + \tau_{zx}\frac{\partial u}{\partial z}$$

$$+ \tau_{xy}\frac{\partial v}{\partial x} + \tau_{yy}\frac{\partial v}{\partial y} + \tau_{zy}\frac{\partial v}{\partial z} + \tau_{xz}\frac{\partial w}{\partial x}$$

$$+ \tau_{yz}\frac{\partial w}{\partial y} + \tau_{zz}\frac{\partial w}{\partial z} \qquad (2.71)$$

Equation (2.71) is a form of the energy equation where the substantial derivative on the left-hand side is strictly in terms of the internal energy only. The kinetic energy and the body force terms have dropped out; indeed, it is important to emphasize that the energy equation when written in terms of e alone does not explicitly contain the body force. Also note that in comparison with Eq. (2.66), where the shear stresses and normal stresses multiplied by velocities appear *inside* the x, y, and z derivatives, in Eq. (2.71) the viscous stresses appear by themselves, multiplied by velocity gradients. Finally, we note that Eq. (2.71) is in *nonconservation* form; the manipulations which resulted in the derivation of Eq. (2.71) from Eq. (2.66) do not change that situation. By similar approaches, the energy equation can be couched also in terms of h and $h + V^2/2$; the derivations are left to you (for your leisure time).

Let us take Eq. (2.71) a few steps further. Recall from Eqs. (2.57d) to (2.57f) that $\tau_{xy} = \tau_{yx}$, $\tau_{xz} = \tau_{zx}$, and $\tau_{yz} = \tau_{zy}$. [This symmetry between the shear stresses is necessary to keep the angular velocity of a fluid element from going to infinity as the volume of the element shrinks to a point—it is associated with the moments exerted on the fluid element. See Schlichting (Ref. 20) for the details.] Hence, some of the terms in Eq. (2.71) can be factored, yielding

$$\rho \frac{De}{Dt} = \rho \dot{q} + \frac{\partial}{\partial x}\left(k \frac{\partial T}{\partial x}\right) + \frac{\partial}{\partial y}\left(k \frac{\partial T}{\partial y}\right) + \frac{\partial}{\partial z}\left(k \frac{\partial T}{\partial z}\right)$$

$$- p\left(\frac{\partial u}{\partial x} + \frac{\partial v}{\partial y} + \frac{\partial w}{\partial z}\right) + \tau_{xx}\frac{\partial u}{\partial x} + \tau_{yy}\frac{\partial v}{\partial y} + \tau_{zz}\frac{\partial w}{\partial z}$$

$$+ \tau_{yx}\left(\frac{\partial u}{\partial y} + \frac{\partial v}{\partial x}\right) + \tau_{zx}\left(\frac{\partial u}{\partial z} + \frac{\partial w}{\partial x}\right) + \tau_{zy}\left(\frac{\partial v}{\partial z} + \frac{\partial w}{\partial y}\right) \quad (2.72)$$

Appealing again to Eqs. (2.57a) to (2.57f) in order to express the viscous stresses in terms of velocity gradients, Eq. (2.72) can be written as

$$\rho \frac{De}{Dt} = \rho \dot{q} + \frac{\partial}{\partial x}\left(k \frac{\partial T}{\partial x}\right) + \frac{\partial}{\partial y}\left(k \frac{\partial T}{\partial y}\right) + \frac{\partial}{\partial z}\left(k \frac{\partial T}{\partial z}\right)$$

$$- p\left(\frac{\partial u}{\partial x} + \frac{\partial v}{\partial y} + \frac{\partial w}{\partial z}\right) + \lambda\left(\frac{\partial u}{\partial x} + \frac{\partial v}{\partial y} + \frac{\partial w}{\partial z}\right)^2$$

$$+ \mu\left[2\left(\frac{\partial u}{\partial x}\right)^2 + 2\left(\frac{\partial v}{\partial y}\right)^2 + 2\left(\frac{\partial w}{\partial z}\right)^2 + \left(\frac{\partial u}{\partial y} + \frac{\partial v}{\partial x}\right)^2\right.$$

$$\left. + \left(\frac{\partial u}{\partial z} + \frac{\partial w}{\partial x}\right)^2 + \left(\frac{\partial v}{\partial z} + \frac{\partial w}{\partial y}\right)^2\right] \quad (2.73)$$

Equation (2.73) is a form of the energy equation completely in terms of the flow-field variables. A similar substitution of Eqs. (2.57a) to (2.57f) can be made into Eq. (2.66); the resulting form of the energy equation in terms of the flow-field variables is lengthy, and to save time and space it will not be given here.

We emphasize again that only the internal energy appears in the left-hand side of Eq. (2.73). Our derivation leading to Eq. (2.73) is an example of how the left side of the energy equation can be couched in terms of different energy forms—for example, in terms of total energy in Eq. (2.66) and in terms of internal energy in Eq. (2.73). As stated earlier, other forms in terms of static enthalpy h and total enthalpy $h + V^2/2$ can be obtained by similar manipulations. (For example, see Ref. 2 for these other forms.) This is one of the aspects of the energy equation mentioned earlier, namely, that the left-hand side can be expressed in terms of different energy forms; for each of these different forms, there is also a different form of the right-hand side of the energy equation. Now, let us address another aspect of the energy equation—an aspect common to the continuity and momentum equations as well—namely, that the energy equation can be expressed in conservation form. The forms of the energy equation given by Eqs. (2.66), (2.71), and (2.73) are expressed in terms of a substantial derivative on the left-hand side; hence these are nonconservation forms. They stem directly from our model of a *moving* fluid element. However, with some manipulation, all these equations can be expressed in conservation form. Let us examine this for the case of Eq. (2.73). Consider the left-hand side of Eq. (2.73). From the definition of the substantial derivative,

$$\rho \frac{De}{Dt} = \rho \frac{\partial e}{\partial t} + \rho \mathbf{V} \cdot \nabla e \qquad (2.74)$$

However,

$$\frac{\partial(\rho e)}{\partial t} = \rho \frac{\partial e}{\partial t} + e \frac{\partial \rho}{\partial t}$$

or

$$\rho \frac{\partial e}{\partial t} = \frac{\partial(\rho e)}{\partial t} - e \frac{\partial \rho}{\partial t} \qquad (2.75)$$

From the vector identity concerning the divergence of the product of a scalar times a vector,

$$\nabla \cdot (\rho e \mathbf{V}) = e \nabla \cdot (\rho \mathbf{V}) + \rho \mathbf{V} \cdot \nabla e$$

or

$$\rho \mathbf{V} \cdot \nabla e = \nabla \cdot (\rho e \mathbf{V}) - e \nabla \cdot (\rho \mathbf{V}) \qquad (2.76)$$

Substitute Eqs. (2.75) and (2.76) into (2.74):

$$\rho \frac{De}{Dt} = \frac{\partial(\rho e)}{\partial t} - e \left[\frac{\partial \rho}{\partial t} + \nabla \cdot (\rho \mathbf{V}) \right] + \nabla \cdot (\rho e \mathbf{V}) \qquad (2.77)$$

The term in brackets in Eq. (2.77) is zero, from the continuity equation, Eq. (2.33). Thus, Eq. (2.77) becomes

$$\rho \frac{De}{Dt} = \frac{\partial(\rho e)}{\partial t} + \nabla \cdot (\rho e \mathbf{V}) \qquad (2.78)$$

Substituting Eq. (2.78) into Eq. (2.73), we have

74 THE GOVERNING EQUATIONS OF FLUID DYNAMICS

$$\frac{\partial(\rho e)}{\partial t} + \nabla \cdot (\rho e \mathbf{V}) = \rho \dot{q} + \frac{\partial}{\partial x}\left(k\frac{\partial T}{\partial x}\right) + \frac{\partial}{\partial y}\left(k\frac{\partial T}{\partial y}\right) + \frac{\partial}{\partial z}\left(k\frac{\partial T}{\partial z}\right)$$

$$-p\left(\frac{\partial u}{\partial x} + \frac{\partial v}{\partial y} + \frac{\partial w}{\partial z}\right) + \lambda\left(\frac{\partial u}{\partial x} + \frac{\partial v}{\partial y} + \frac{\partial w}{\partial z}\right)^2$$

$$+\mu\left[2\left(\frac{\partial u}{\partial x}\right)^2 + 2\left(\frac{\partial v}{\partial y}\right)^2 + 2\left(\frac{\partial w}{\partial z}\right)^2 + \left(\frac{\partial u}{\partial y} + \frac{\partial v}{\partial x}\right)^2 \right.$$

$$\left. + \left(\frac{\partial u}{\partial z} + \frac{\partial w}{\partial x}\right)^2 + \left(\frac{\partial v}{\partial z} + \frac{\partial w}{\partial y}\right)^2 \right] \quad (2.79)$$

Equation (2.79) is the *conservation* form of the energy equation, written in terms of the internal energy.

Repeating the steps from Eq. (2.74) to Eq. (2.78), except operating on the *total* energy $e + V^2/2$ instead of just the internal energy e, we obtain

$$\rho \frac{D(e + V^2/2)}{Dt} = \frac{\partial}{\partial t}\left[\rho\left(e + \frac{V^2}{2}\right)\right] + \nabla \cdot \left[\rho\left(e + \frac{V^2}{2}\right)\mathbf{V}\right] \quad (2.80)$$

Substituting Eq. (2.80) into the left-hand side of Eq. (2.66), we obtain

$$\frac{\partial}{\partial t}\left[\rho\left(e + \frac{V^2}{2}\right)\right] + \nabla \cdot \left[\rho\left(e + \frac{V^2}{2}\right)\mathbf{V}\right] = \rho \dot{q} + \frac{\partial}{\partial x}\left(k\frac{\partial T}{\partial x}\right)$$

$$+ \frac{\partial}{\partial y}\left(k\frac{\partial T}{\partial y}\right) + \frac{\partial}{\partial z}\left(k\frac{\partial T}{\partial z}\right) - \frac{\partial(up)}{\partial x} - \frac{\partial(vp)}{\partial y} - \frac{\partial(wp)}{\partial z}$$

$$+ \frac{\partial(u\tau_{xx})}{\partial x} + \frac{\partial(u\tau_{yx})}{\partial y} + \frac{\partial(u\tau_{zx})}{\partial z} + \frac{\partial(v\tau_{xy})}{\partial x} + \frac{\partial(v\tau_{yy})}{\partial y}$$

$$+ \frac{\partial(v\tau_{zy})}{\partial z} + \frac{\partial(w\tau_{xz})}{\partial x} + \frac{\partial(w\tau_{yz})}{\partial y} + \frac{\partial(w\tau_{zz})}{\partial z} + \rho \mathbf{f} \cdot \mathbf{V} \quad (2.81)$$

Equation (2.81) is the *conservation* form of the energy equation, written in terms of total energy $e + V^2/2$.

Note that the manipulations required to change the nonconservation form to the conservation form change only the left-hand side of the equations; the right-hand sides remain the same. For example, compare Eqs. (2.73) and (2.79). Both are in terms of internal energy. Equation (2.73) is in nonconservation form, and Eq. (2.79) is in conservation form. The left-hand sides are different forms, but the right-hand sides are the same. The same comparison can also be made between Eqs. (2.66) and (2.81).

2.8 SUMMARY OF THE GOVERNING EQUATIONS FOR FLUID DYNAMICS: WITH COMMENTS

By this point in our discussions, you have seen a large number of equations, and they may seem to you to "all look alike." Equations by themselves can be tiring, and this chapter would seem to be "wall-to-wall" equations. However, *all* of theoretical and computational fluid dynamics is based on these equations, and therefore it is absolutely *essential* that you are familiar with them and that you understand their physical significance. That is why we have spent so much time and effort in deriving the governing equations.

Considering this time and effort, it is important to now summarize the important forms of these equations and to sit back and digest them. First of all, now is a good time to reflect back to the chapter road map in Fig. 2.1. We have already traveled our way through about 80 percent of this map. Starting at the top of Fig. 2.1, we have taken the three fundamental principles on which all of fluid dynamics is based (boxes A–C) and applied these to various models of the flow (boxes D–H). We have seen how each model of the flow leads *directly* to a particular form of the governing equation (the routes from left to right at the bottom center of Fig. 2.1, from boxes E–H to box I). We have also seen how these particular forms can be reexpressed by suitable manipulation into other forms of the equations (as illustrated for the continuity equation in Fig. 2.6). All routes lead to box I at the right of Fig. 2.1, which represents the basic continuity, momentum, and energy equations in all their glorious forms. In our present discussion, this is where we are now. For emphasis and clarity, in this section, we summarize those equations represented by box I.

2.8.1 Equations for Viscous Flow (the Navier-Stokes Equations)

A viscous flow is one where the transport phenomena of friction, thermal conduction, and/or mass diffusion are included. These transport phenomena are dissipative—they always increase the entropy of the flow. The equations that have been derived and discussed up to this point in the present chapter apply to such a viscous flow, with the exception that mass diffusion is not included. Mass diffusion occurs when there are concentration gradients of different chemical species in the flow. An example is a nonhomogeneous mixture of nonreacting gases, such as the flow field associated with the injection of helium through a hole or slot into a primary stream of air. Another example is a chemically reacting gas, such as the dissociation of air that occurs in the high-temperature flow over hypersonic vehicles; in such flows, concentration gradients are induced by different rates of reaction and/or by the prominence of different types of reactions in different parts of the flow at different pressures and temperatures. Chemically reacting flows as well as nonhomogenous flows are discussed at length in Ref. 2. These types of flows are not treated in the present book, simply for clarity. Our purpose here is to discuss the basic aspects of CFD—we choose not to obscure the computational aspects by carrying along the extra complications and physics associated with chemically reacting flows. For this reason, diffusion is not included in the equations in this book. See Ref. 2 for an in-depth discussion of chemically reacting flows and especially for a discussion of the physical and numerical effects of mass diffusion.

With the above restrictions in mind, the governing equations for an unsteady, three-dimensional, compressible, viscous flow are:

Continuity equation

Nonconservation form

$$\frac{D\rho}{Dt} + \rho \nabla \cdot \mathbf{V} = 0 \qquad (2.29)$$

Conservation form

$$\frac{\partial \rho}{\partial t} + \nabla \cdot (\rho \mathbf{V}) = 0 \qquad (2.30)$$

Momentum equations

Nonconservation form

x component : $\quad \rho \dfrac{Du}{Dt} = -\dfrac{\partial p}{\partial x} + \dfrac{\partial \tau_{xx}}{\partial x} + \dfrac{\partial \tau_{yx}}{\partial y} + \dfrac{\partial \tau_{zx}}{\partial z} + \rho f_x \qquad (2.50a)$

y component : $\quad \rho \dfrac{Dv}{Dt} = -\dfrac{\partial p}{\partial y} + \dfrac{\partial \tau_{xy}}{\partial x} + \dfrac{\partial \tau_{yy}}{\partial y} + \dfrac{\partial \tau_{zy}}{\partial z} + \rho f_y \qquad (2.50b)$

z component : $\quad \rho \dfrac{Dw}{Dt} = -\dfrac{\partial p}{\partial z} + \dfrac{\partial \tau_{xz}}{\partial x} + \dfrac{\partial \tau_{yz}}{\partial y} + \dfrac{\partial \tau_{zz}}{\partial z} + \rho f_z \qquad (2.50c)$

Conservation form
x component:

$$\frac{\partial(\rho u)}{\partial t} + \nabla \cdot (\rho u \mathbf{V}) = -\frac{\partial p}{\partial x} + \frac{\partial \tau_{xx}}{\partial x} + \frac{\partial \tau_{yx}}{\partial y} + \frac{\partial \tau_{zx}}{\partial z} + \rho f_x \qquad (2.56a)$$

y component:

$$\frac{\partial(\rho v)}{\partial t} + \nabla \cdot (\rho v \mathbf{V}) = -\frac{\partial p}{\partial y} + \frac{\partial \tau_{xy}}{\partial x} + \frac{\partial \tau_{yy}}{\partial y} + \frac{\partial \tau_{zy}}{\partial z} + \rho f_y \qquad (2.56b)$$

z component:

$$\frac{\partial(\rho w)}{\partial t} + \nabla \cdot (\rho w \mathbf{V}) = -\frac{\partial p}{\partial z} + \frac{\partial \tau_{xz}}{\partial x} + \frac{\partial \tau_{yz}}{\partial y} + \frac{\partial \tau_{zz}}{\partial z} + \rho f_z \qquad (2.56c)$$

Energy equation

Nonconservation form

$$\rho \frac{D}{Dt}\left(e + \frac{V^2}{2}\right) = \rho \dot{q} + \frac{\partial}{\partial x}\left(k\frac{\partial T}{\partial x}\right) + \frac{\partial}{\partial y}\left(k\frac{\partial T}{\partial y}\right) + \frac{\partial}{\partial z}\left(k\frac{\partial T}{\partial z}\right)$$
$$- \frac{\partial(up)}{\partial x} - \frac{\partial(vp)}{\partial y} - \frac{\partial(wp)}{\partial z} + \frac{\partial(u\tau_{xx})}{\partial x}$$
$$+ \frac{\partial(u\tau_{yx})}{\partial y} + \frac{\partial(u\tau_{zx})}{\partial z} + \frac{\partial(v\tau_{xy})}{\partial x} + \frac{\partial(v\tau_{yy})}{\partial y}$$
$$+ \frac{\partial(v\tau_{zy})}{\partial z} + \frac{\partial(w\tau_{xz})}{\partial x} + \frac{\partial(w\tau_{yz})}{\partial y} + \frac{\partial(w\tau_{zz})}{\partial z} + \rho \mathbf{f} \cdot \mathbf{V} \quad (2.66)$$

Conservation form

$$\frac{\partial}{\partial t}\left[\rho\left(e + \frac{V^2}{2}\right)\right] + \nabla \cdot \left[\rho\left(e + \frac{V^2}{2}\right)\mathbf{V}\right] = \rho \dot{q} + \frac{\partial}{\partial x}\left(k\frac{\partial T}{\partial x}\right) + \frac{\partial}{\partial y}\left(k\frac{\partial T}{\partial y}\right)$$
$$+ \frac{\partial}{\partial z}\left(k\frac{\partial T}{\partial z}\right) - \frac{\partial(up)}{\partial x} - \frac{\partial(vp)}{\partial y} - \frac{\partial(wp)}{\partial z} + \frac{\partial(u\tau_{xx})}{\partial x}$$
$$+ \frac{\partial(u\tau_{yx})}{\partial y} + \frac{\partial(u\tau_{zx})}{\partial z} + \frac{\partial(v\tau_{xy})}{\partial x} + \frac{\partial(v\tau_{yy})}{\partial y}$$
$$+ \frac{\partial(v\tau_{zy})}{\partial z} + \frac{\partial(w\tau_{xz})}{\partial x} + \frac{\partial(w\tau_{yz})}{\partial y} + \frac{\partial(w\tau_{zz})}{\partial z} + \rho \mathbf{f} \cdot \mathbf{V} \quad (2.81)$$

2.8.2 Equations for Inviscid Flow (the Euler Equations)

Inviscid flow is, by definition, a flow where the dissipative, transport phenomena of viscosity, mass diffusion, and thermal conductivity are *neglected*. If we take the equations listed in Sec. 2.8.1 and simply drop all the terms involving friction and thermal conduction, we then have the equations for an inviscid flow. The resulting equations for an unsteady, three-dimensional, compressible inviscid flow are displayed below.

Continuity equation

Nonconservation form

$$\frac{D\rho}{Dt} + \rho \nabla \cdot \mathbf{V} = 0 \quad (2.82a)$$

Conservation form

$$\frac{\partial \rho}{\partial t} + \nabla \cdot (\rho \mathbf{V}) = 0 \quad (2.82b)$$

Momentum equations

Nonconservation form

x component : $$\rho \frac{Du}{Dt} = -\frac{\partial p}{\partial x} + \rho f_x \quad (2.83a)$$

y component : $$\rho \frac{Dv}{Dt} = -\frac{\partial p}{\partial y} + \rho f_y \quad (2.83b)$$

z component : $$\rho \frac{Dw}{Dt} = -\frac{\partial p}{\partial z} + \rho f_z \quad (2.83c)$$

Conservation form

x component : $$\frac{\partial(\rho u)}{\partial t} + \nabla \cdot (\rho u \mathbf{V}) = -\frac{\partial p}{\partial x} + \rho f_x \quad (2.84a)$$

y component : $$\frac{\partial(\rho v)}{\partial t} + \nabla \cdot (\rho v \mathbf{V}) = -\frac{\partial p}{\partial y} + \rho f_y \quad (2.84b)$$

z component : $$\frac{\partial(\rho w)}{\partial t} + \nabla \cdot (\rho w \mathbf{V}) = -\frac{\partial p}{\partial z} + \rho f_z \quad (2.84c)$$

Energy equation

Nonconservation form

$$\rho \frac{D}{Dt}\left(e + \frac{V^2}{2}\right) = \rho \dot{q} - \frac{\partial(up)}{\partial x} - \frac{\partial(vp)}{\partial y} - \frac{\partial(wp)}{\partial z} + \rho \mathbf{f} \cdot \mathbf{V} \quad (2.85)$$

Conservation form

$$\frac{\partial}{\partial t}\left[\rho\left(e + \frac{V^2}{2}\right)\right] + \nabla \cdot \left[\rho\left(e + \frac{V^2}{2}\right)\mathbf{V}\right] = \rho \dot{q} - \frac{\partial(up)}{\partial x} - \frac{\partial(vp)}{\partial y} - \frac{\partial(wp)}{\partial z} + \rho \mathbf{f} \cdot \mathbf{V}$$
$$(2.86)$$

2.8.3 Comments on the Governing Equations

Surveying all the equations summarized in Sec. 2.8.1 and 2.8.2, several comments and observations can be made, as follows:

1. They are a coupled system of nonlinear partial differential equations, and hence are very difficult to solve analytically. To date, there is no general closed-form solution to these equations. (This does not mean that no general solution exists—we just have not been able to find one.)
2. For the momentum and energy equations, the difference between the nonconservation and conservation forms of the equations is just the left-hand side.

The right-hand side of the equations in the two different forms is the same.
3. Note that the conservation forms of the equations contain terms on the left-hand side which include the divergence of some quantity, such as $\nabla \cdot (\rho \mathbf{V})$ or $\nabla \cdot (\rho u \mathbf{V})$. For this reason, the conservation form of the governing equations is sometimes called the *divergence form*.
4. The normal and shear stress terms in these equations are functions of the velocity gradients, as given by Eqs. (2.57a–b).
5. Examine the equations in Sec. 2.8.1 and 2.8.2 closely. Count the number of unknown, dependent variables in each section. In both cases, we have five equations in terms of six unknown flow-field variables ρ, p, u, v, w, e. In aerodynamics, it is generally reasonable to assume the gas is a perfect gas (which assumes that intermolecular forces are negligible—see Refs. 1, 8, and 21). For a perfect gas, the equation of state is

$$p = \rho R T$$

where R is the specific gas constant. This equation is sometimes labeled the *thermal equation of state*. This provides a sixth equation, but it also introduces a seventh unknown, namely, temperature T. A seventh equation to close the entire system must be a thermodynamic relation between state variables. For example,

$$e = e(T, p)$$

For a calorically perfect gas (constant specific heats), this relation would be

$$e = c_v T$$

where c_v is the specific heat at constant volume. This equation is sometimes labeled the *caloric equation of state*.

6. In Sec. 2.6, the momentum equations for a viscous flow were identified as the *Navier-Stokes equations*, which is historically accurate. However, in the modern CFD literature, this terminology has been expanded to include the *entire system* of flow equations for the solution of a viscous flow—continuity and energy as well as momentum. Therefore, when the CFD literature discusses a numerical solution to the "complete Navier-Stokes equations," it usually is referring to a numerical solution of the *complete system of equations*, say, for example Eqs. (2.33), (2.56a) to (2.56c), and (2.81). In this sense, in the CFD literature, a "Navier-Stokes solution" simply means a solution of a *viscous flow* problem using the *full governing equations*. This is why the entire block of equations summarized in Sec. 2.8.1 is labeled as the Navier-Stokes equations. This author suspects that the CFD usage of this nomenclature will soon seep through all of fluid dynamics. For this reason, and because the subject of this book is CFD, we will follow this nomenclature. That is, when we refer to the Navier-Stokes equations, we will mean the whole system of equations, such as summarized in Sec. 2.8.1.

7. In a similar vein, the equations for inviscid flow in Sec. 2.8.2 are labeled as the *Euler equations*. Historically, Euler derived the continuity and momentum equations in 1753; he did not deal with the energy equation—indeed, he had

very little to work with because the science of thermodynamics is a nineteenth century product. Therefore, on a strictly historical basis, only the continuity and momentum equations can be labeled as the Euler equations. Indeed, in much of the fluid dynamics literature, just the momentum equations for an inviscid flow, e.g., Eqs. (2.83a) to (2.83c), are labeled as the Euler equations. However, in the modern CFD literature, solutions to the complete system of equations for an inviscid flow, e.g., the equations summarized in Sec. 2.8.2, are called *Euler solutions*, and the whole system of equations—continuity, momentum, and energy—are called the Euler equations. We will follow this nomenclature in the present book.

2.9 PHYSICAL BOUNDARY CONDITIONS

The equations given above govern the flow of a fluid. They are the same equations whether the flow is, for example, over a Boeing 747, through a subsonic wind tunnel, or past a windmill. However, the flow fields are quite *different* for these cases, although the governing equations are the *same*. Why? Where does the difference enter? The answer is through the *boundary conditions*, which are quite different for each of the above examples. The boundary conditions, and sometimes the initial conditions, dictate the particular solutions to be obtained from the governing equations. When the geometric shape of a Boeing 747 is treated, when certain physical boundary conditions are applied on that particular geometric surface, and when the appropriate boundary conditions associated with the freestream far ahead of the airplane are invoked, then the resulting solution of the governing partial differential equations will yield the flow field over the Boeing 747. This is in contrast to the flow-field solutions that would be obtained for a windmill if the geometric shape and freestream conditions pertinent to the windmill were treated. Hence, once we have the governing flow equations as described in the previous sections, then the real *driver* for any particular solution is the *boundary conditions*. This has particular significance in CFD; any numerical solution of the governing flow equations must be made to see a strong and compelling numerical representation of the proper boundary conditions.

First, let us review the proper physical boundary conditions for a viscous flow. Here, the boundary condition on a surface assumes zero relative velocity between the surface and the gas immediately at the surface. This is called the *no-slip* condition. If the surface is stationary, with the flow moving past it, then

$$u = v = w = 0 \quad \text{at the surface (for a viscous flow)} \quad (2.87)$$

In addition, there is an analogous "no-slip" condition associated with the temperature at the surface. If the material temperature of the surface is denoted by T_w (the wall temperature), then the temperature of the fluid layer immediately in contact with the surface is also T_w. If in a given problem the wall temperature is known, then the proper boundary condition on the gas temperature T is

$$T = T_w \quad \text{(at the wall)} \quad (2.88)$$

On the other hand, if the wall temperature is not known, e.g., if it is changing as a function of time due to aerodynamic heat transfer to or from the surface, then the Fourier law of heat conduction provides the boundary condition at the surface. If we let $\dot{q}_w$ denote the instantaneous heat flux to the wall, then from the Fourier law

$$\dot{q}_w = -\left(k\frac{\partial T}{\partial n}\right)_w \quad \text{(at the wall)} \tag{2.89}$$

where n denotes the direction normal to the wall. Here, the surface material is responding to the heat transfer to the wall, $\dot{q}_w$, hence changing T_w, which in turn affects $\dot{q}_w$. This general, unsteady heat transfer problem must be solved by treating the viscous flow and the thermal response of the wall material simultaneously. This type of boundary condition, as far as the flow is concerned, is a boundary condition on the temperature *gradient* at the wall, in contrast to stipulating the wall temperature itself as the boundary condition. That is, from Eq. (2.89),

$$\left(\frac{\partial T}{\partial n}\right)_w = -\frac{\dot{q}_w}{k} \quad \text{(at the wall)} \tag{2.90}$$

Finally, when the wall temperature becomes such that there is *no heat transfer to the surface*, this wall temperature, by definition, is called the *adiabatic wall temperature* T_{aw}. The proper boundary condition for the adiabatic wall case comes from Eq. (2.90) with $\dot{q}_w = 0$, by definition. Hence, for an adiabatic wall, the boundary condition is

$$\left(\frac{\partial T}{\partial n}\right)_w = 0 \quad \text{(at the wall)} \tag{2.91}$$

Once again, we see that the wall boundary condition is the stipulation of the temperature *gradient* at the wall; the actual adiabatic wall temperature T_{aw} then falls out as part of the flow-field solution.

Of all the temperature boundary conditions stated above, that of a fixed wall temperature [Eq. (2.88)] is the easiest to apply, with that of an adiabatic wall [Eq. (2.91)] being the next easiest. These two different cases represent two extreme ends of the general problem, which is that associated with the boundary condition given by Eq. (2.90). However, the general problem, which involves the coupled solution of the flow field with the thermal response of the surface material, is by far the most difficult to set up. For these reasons, the vast majority of viscous flow solutions assume either a constant wall temperature or an adiabatic wall. In summary, if Eq. (2.88) is used as the boundary condition, then the temperature gradient at the wall, $(\partial T/\partial w)_n$, and hence $\dot{q}_w$ fall out as part of the solution. If Eq. (2.91) is used as the boundary condition, then T_{aw} falls out as part of the solution. If Eq. (2.90) is used as the boundary condition, along with a coupled solution with the thermal response of the material, then T_w and $(\partial T/\partial n)_w$ fall out as part of the solution.

Finally, we note that the only physical boundary conditions along a wall for a continuum viscous flow are the no-slip conditions discussed above; these boundary conditions are associated with velocity and temperature at the wall. Other flow properties, such as pressure and density at the wall, fall out as part of the solution.

For an *inviscid flow*, there is no friction to promote its "sticking" to the surface. Hence, the flow velocity at the wall is a finite, nonzero value. Moreover, for a nonporous wall, there can be no mass flow into or out of the wall; this means that the flow velocity vector immediately adjacent to the wall must be *tangent* to the wall. If **n** is a unit normal vector at a point on the surface, the wall boundary condition can be given as

$$\mathbf{V} \cdot \mathbf{n} = 0 \quad \text{(at the surface)} \quad (2.92)$$

Equation (2.92) is simply a statement that the component of velocity perpendicular to the wall is zero; i.e., the flow at the surface is tangent to the wall. This is the only surface boundary condition for an inviscid flow. The magnitude of the velocity, as well as values of the fluid temperature, pressure, and density at the wall, falls out as part of the solution.

Depending on the problem at hand, whether it be viscous or inviscid, there are various types of boundary conditions elsewhere in the flow, away from the surface boundary. For example, for flow through a duct of fixed shape, there are boundary conditions which pertain to the inflow and outflow boundaries, such as at the inlet and exit of the duct. If the problem involves an aerodynamic body immersed in a known freestream, then the boundary conditions applied at a distance infinitely far upstream, above, below, and downstream of the body are simply that of the given freestream conditions.

The boundary conditions discussed above are *physical boundary conditions* imposed by nature. In CFD we have an additional concern, namely, the *proper numerical implementation of these physical boundary conditions*. In the same sense as the real flow field is dictated by the physical boundary conditions, the computed flow field is driven by the numerical formulation designed to simulate these boundary conditions. The subject of proper and accurate boundary conditions in CFD is very important and is the subject of much current CFD research. We will return to this matter at appropriate stages in this book.

2.10 FORMS OF THE GOVERNING EQUATIONS PARTICULARLY SUITED FOR CFD: COMMENTS ON THE CONSERVATION FORM, SHOCK FITTING, AND SHOCK CAPTURING

In this section, we finally address the significance of the conservation versus the nonconservation forms of the governing flow equations vis-à-vis applications of CFD. In the historical development of these equations, there was no reason for a preference of one form over the other; indeed, theoretical fluid dynamics evolved quite well over the last few centuries without paying any attention to this matter. This is reflected in *all* the general fluid dynamics and aerodynamics textbooks up to the early 1980s, where this author defies you to find any mention of, or reference to, conservation versus nonconservation forms—the equations are there, but they are simply not identified in these terms. The labeling of the governing equations as either conservation or nonconservation form grew out of modern CFD, as well as concern for when one form or the other should be used for a given CFD application. Let us address this matter from two perspectives.

The first perspective is simply that the conservation form of the governing equations provides a numerical and computer programing *convenience* in that the continuity, momentum, and energy equations in conservation form can all be expressed by the same generic equation. This can help to simplify and organize the logic in a given computer program. To prepare us for this generic form, note that all the previous equations in *conservation form* have a divergence term on the left-hand side. These terms involve the divergence of the *flux* of some physical quantity, such as

From Eq. (2.33) : $\rho \mathbf{V}$ mass flux

From Eq. (2.56a) : $\rho u \mathbf{V}$ flux of x component of momentum

From Eq. (2.56b) : $\rho v \mathbf{V}$ flux of y component of momentum

From Eq. (2.56c) : $\rho w \mathbf{V}$ flux of z component of momentum

From Eq. (2.79) : $\rho e \mathbf{V}$ flux of internal energy

From Eq. (2.81) : $\rho \left(e + \dfrac{V^2}{2} \right) \mathbf{V}$ flux of total energy

Recall that the conservation form of the equations was obtained *directly* from a control volume which was *fixed in space* rather than moving with the fluid. When the volume is fixed in space, we are concerned with the *flux* of mass, momentum, and energy into and out of the volume. In this case, the *fluxes* themselves become important dependent variables in the equations, rather than just the primitive variables such as p, ρ, $\mathbf{V}$.

Let us pursue this idea further. Examine the *conservation* form of *all* the governing equations—continuity, momentum, and energy. This is perhaps most conveniently done by returning to Secs. 2.8.1 and 2.8.2, where the governing equations for viscous and inviscid flows, respectively, are compactly summarized. Looking at the conservation forms, we note that they all have the same generic form, given by

$$\boxed{\frac{\partial U}{\partial t} + \frac{\partial F}{\partial x} + \frac{\partial G}{\partial y} + \frac{\partial H}{\partial z} = J} \qquad (2.93)$$

Equation (2.93) can represent the *entire system* of governing equations in conservation form if U, F, G, H, and J are interpreted as column vectors, given by

$$U = \begin{Bmatrix} \rho \\ \rho u \\ \rho v \\ \rho w \\ \rho\left(e + \dfrac{V^2}{2}\right) \end{Bmatrix} \qquad (2.94)$$

$$F = \begin{Bmatrix} \rho u \\ \rho u^2 + p - \tau_{xx} \\ \rho v u - \tau_{xy} \\ \rho w u - \tau_{xz} \\ \rho\left(e + \dfrac{V^2}{2}\right)u + pu - k\dfrac{\partial T}{\partial x} - u\tau_{xx} - v\tau_{xy} - w\tau_{xz} \end{Bmatrix} \qquad (2.95)$$

$$G = \begin{Bmatrix} \rho v \\ \rho u v - \tau_{yx} \\ \rho v^2 + p - \tau_{yy} \\ \rho w v - \tau_{yz} \\ \rho\left(e + \dfrac{V^2}{2}\right)v + pv - k\dfrac{\partial T}{\partial y} - u\tau_{yx} - v\tau_{yy} - w\tau_{yz} \end{Bmatrix} \qquad (2.96)$$

$$H = \begin{Bmatrix} \rho w \\ \rho u w - \tau_{zx} \\ \rho v w - \tau_{zy} \\ \rho w^2 + p - \tau_{zz} \\ \rho\left(e + \dfrac{V^2}{2}\right)w + pw - k\dfrac{\partial T}{\partial z} - u\tau_{zx} - v\tau_{zy} - w\tau_{zz} \end{Bmatrix} \qquad (2.97)$$

$$J = \begin{Bmatrix} 0 \\ \rho f_x \\ \rho f_y \\ \rho f_z \\ \rho(u f_x + v f_y + w f_z) + \rho \dot{q} \end{Bmatrix} \qquad (2.98)$$

In Eq. (2.93), the column vectors F, G, and H are called the *flux terms* (or *flux vectors*), and J represents a *source term* (which is zero if body forces and volumetric heating are negligible.) The column vector U is called the *solution vector*, for reasons to be stated shortly. To help yourself get used to this generic equation written in terms of column vectors, note that the first elements of the U, F, G, H, and J vectors, when added together via Eq. (2.93), reproduce the continuity equation. The second elements of the U, F, G, H, and J vectors, when added together via Eq. (2.93), reproduce the x-momentum equation, and so forth. Indeed, Eq. (2.93) is simply one large column vector equation which represents the whole system of governing equations.

Let us explore the ramifications of Eq. (2.93) further. It is written with a time derivative $\partial U/\partial t$; hence it applies to an *unsteady flow*. In a given problem, the actual transients in an unsteady flow may be of primary interest. In other problems, a steady-state solution may be desired but wherein the best manner to solve for this steady state is to solve the unsteady equations and let the steady state be approached asymptotically at large times. (This approach is sometimes called the *time-dependent* solution of steady flows; the solution of the supersonic blunt body problem as discussed at the end of Sec. 1.5 is one such example.) We will be exploring such matters in depth in Part III of this book dealing with applications of CFD; we mention them here only in passing. For either an inherent transient solution, or a time-dependent solution leading to a steady state, the solution of Eq. (2.93) takes the form of a *time-marching* solution, i.e., where the dependent flow-field variables are solved progressively in steps of time. For such a time-marching solution, we isolate $\partial U/\partial t$ by rearranging Eq. (2.93) as

$$\frac{\partial U}{\partial t} = J - \frac{\partial F}{\partial x} - \frac{\partial G}{\partial y} - \frac{\partial H}{\partial z} \tag{2.99}$$

In Eq. (2.99), U is called the solution vector because the elements in U (ρ, ρu, ρv, etc.) are the dependent variables which are usually obtained numerically in steps of time; the spatial derivatives on the right side of Eq. (2.99) are considered in some fashion as known, say from the previous time step. Please note that, in this formalism, it is the elements of U which are obtained computationally; i.e., numbers are directly obtained for the density ρ and the *products* ρu, ρv, ρw, and $\rho(e + V^2/2)$. These are called the *flux variables*. This is in contrast to u, v, w, and e by themselves, which are examples of *primitive variables*. Hence, in a computational solution of an unsteady flow problem using Eq. (2.99), the dependent variables are the elements of the U vector as displayed in Eq. (2.94), that is, ρ, ρu, ρv, ρw, and $\rho(e + V^2/2)$. Of course, once numbers are known for these dependent variables (which includes ρ by itself), obtaining the primitive variables is simple:

$$\rho = \rho \tag{2.100}$$

$$u = \frac{\rho u}{\rho} \tag{2.101}$$

$$v = \frac{\rho v}{\rho} \tag{2.102}$$

$$w = \frac{\rho w}{\rho} \tag{2.103}$$

$$e = \frac{\rho(e + V^2/2)}{\rho} - \frac{u^2 + v^2 + w^2}{2} \tag{2.104}$$

For example, the first element of the U vector is ρ itself; a number for ρ is obtained by solving Eq. (2.99). The second element of U is ρu; a number for the *product* ρu is obtained by solving Eq. (2.99). In turn, a number for the primitive variable u is easily obtained from Eq. (2.101) by taking the number obtained for ρu and dividing it by the number obtained for ρ. The same approach can be used to obtain the primitive variables v, w, and e from the numbers for the flux variables, as shown by Eqs. (2.102) to (2.104).

For an *inviscid* flow, Eqs. (2.93) and (2.99) remain the same, except that the elements of the column vectors are simplified. Examining the conservation form of the inviscid equations summarized in Sec. 2.8.2, we find that

$$U = \begin{Bmatrix} \rho \\ \rho u \\ \rho v \\ \rho w \\ \rho\left(e + \dfrac{V^2}{2}\right) \end{Bmatrix} \qquad (2.105)$$

$$F = \begin{Bmatrix} \rho u \\ \rho u^2 + p \\ \rho u v \\ \rho u w \\ \rho u\left(e + \dfrac{V^2}{2}\right) + pu \end{Bmatrix} \qquad (2.106)$$

$$G = \begin{Bmatrix} \rho v \\ \rho u v \\ \rho v^2 + p \\ \rho w v \\ \rho v\left(e + \dfrac{V^2}{2}\right) + pv \end{Bmatrix} \qquad (2.107)$$

$$H = \begin{Bmatrix} \rho w \\ \rho u w \\ \rho v w \\ \rho w^2 + p \\ \rho w\left(e + \dfrac{V^2}{2}\right) + pw \end{Bmatrix} \qquad (2.108)$$

$$J = \begin{Bmatrix} 0 \\ \rho f_x \\ \rho f_y \\ \rho f_z \\ \rho(u f_x + v f_y + w f_z) + \rho \dot{q} \end{Bmatrix} \qquad (2.109)$$

For the numerical solution of an unsteady inviscid flow, once again the solution vector is U, and the dependent variables for which numbers are directly obtained are ρ, ρu, ρv, ρw, and $\rho(e + V^2/2)$.

In CFD, marching solutions are not limited to marching just in time. Under certain circumstances, steady-state flows can also be solved by marching in a given spatial direction. The circumstances that allow the use of a spatially marching solution depend on the mathematical properties of the governing equation, and will be developed in later chapters, beginning with Chap. 3. For our purposes at present, simply imagine that we are dealing with a steady flow, for which $\partial U/\partial t = 0$ in Eq. (2.93). If a marching solution in the x direction is allowed, then Eq. (2.93) is rearranged as

$$\frac{\partial F}{\partial x} = J - \frac{\partial G}{\partial y} - \frac{\partial H}{\partial z} \qquad (2.110)$$

Here, F becomes the "solution" vector; we can imagine that the terms on the right side of Eq. (2.110) are known, say by evaluation at the previous step, i.e., at the previous upstream x location. This leaves the elements of the F vector at the next step, i.e., at the next downstream x location, as the unknowns. For simplicity, let us assume that we are dealing with an inviscid flow. In such a case, the dependent variables are the elements of F as displayed in Eq. (2.106), namely, ρu, $\rho u^2 + p$, $\rho u v$, $\rho u w$, and $\rho u(e + V^2/2) + pu$. The numerical solution of Eq. (2.110) yields numbers for these dependent variables, called the flux variables. From these dependent variables, it is possible to obtain the primitive variables, although the algebra is more complex than in our previously discussed case for unsteady flow. To see this more clearly, let us denote the flux variables which appear as elements of F as displayed in Eq. (2.106) by

$$\rho u = c_1 \qquad (2.111a)$$

$$\rho u^2 + p = c_2 \qquad (2.111b)$$

$$\rho u v = c_3 \qquad (2.111c)$$

$$\rho u w = c_4 \qquad (2.111d)$$

$$\rho u\left(e + \frac{u^2 + v^2 + w^2}{2}\right) + pu = c_5 \qquad (2.111e)$$

A numerical solution of Eq. (2.110) for an inviscid flow yields numbers for c_1, c_2, c_3, c_4, and c_5 at specific points throughout the flow. Consider just one of those points. The numerical solution yields numbers for the right-hand sides of Eqs. (2.111a) to (2.111e) at that point. In turn, Eqs. (2.111a) to (2.111e) can be solved simultaneously for the primitive variables ρ, u, v, w, and p, and e at that point. Note that we have six unknowns. To Eqs. (2.111a) to (2.111e) must be added a thermodynamic state relation; for a system in thermodynamic equilibrium, this relation can be of the generic form

$$e = e(p, \rho) \tag{2.112a}$$

Indeed, if we are dealing with a calorifically perfect gas, i.e., a gas with constant specific heats (see, for example, Ref. 21), this state relation is $e = c_v T$, with $c_v = R/(\gamma - 1)$, where R is the specific gas constant. Also involving the perfect gas equation of state, $p = \rho R T$, we have

$$e = c_v T = \frac{RT}{\gamma - 1} = \frac{R}{\gamma - 1} \frac{p}{\rho R} \tag{2.112b}$$

or

$$e = \frac{1}{\gamma - 1} \frac{p}{\rho}$$

Equations (2.111a) to (2.111e) and Eq. (2.112b) constitute all six equations from which the six unknown primitive variables can be obtained. The algebra necessary to solve these six equations for explicit relations for ρ, u, v, w, p, and e individually in terms of the known c_1, c_2, c_3, c_4, and c_5 is left to you as Prob. 2.1. Finally, we note that the algebra is even more complex when we consider a viscous flow, where the solution yields numbers for the elements of the F vector as displayed in Eq. (2.95). Here, we also have to contend with the viscous stresses, and the decoding for the primitive variables becomes yet more involved.

We have emphasized the distinction between nonconservation and conservation forms of the governing equations. Let us now expand the definition of the conservation form into two categories: *strong* and *weak*. Notice that the governing equations, when written in the form of Eq. (2.93), have no flow variables outside the single x, y, z, and t derivatives. Indeed, the terms in Eq. (2.93) have everything buried inside these derivatives. The flow equations in the form of Eq. (2.93) are said to be in *strong* conservation form. In contrast, examine the form of Eqs. (2.56a) to (2.56c) and (2.81). These equations have a number of x, y, and z derivatives explicitly appearing on the right-hand side. These are the *weak* conservation form of the equations.

At the beginning of this section, we stated that the matter of conservation versus nonconservation forms of the governing equations within the framework of CFD would be discussed from two perspectives. The ensuing material discussed the first perspective—that the conservation form provides a numerical and computer convenience due to the generic form of Eq. (2.93). Now, let us consider the second perspective—one that is much more compelling than the first. Also, this second perspective is intertwined with two distinct and different philosophical approaches for the calculation of flows with shock waves, namely, the shock-fitting approach and the shock-capturing approach. Let us first define these different approaches to handling shock waves. In flow fields involving shock waves, there are sharp, discontinuous changes in the primitive flow-field variables p, ρ, u, T, etc., across the shocks. Many computations of flows with shocks are designed to have the shock waves appear naturally within the computational space as a direct result of the overall flow-field solution, i.e., as a direct result of the general algorithm, without any special treatment to take care of the shocks themselves. Such approaches are called *shock-capturing methods*. This is in contrast to the alternate approach, where shock waves are *explicitly* introduced into the flow-field solution, the exact Rankine-Hugoniot relations for changes across a shock are used to relate the flow immediately ahead of and behind the shock, and the governing flow equations are used to calculate the remainder of the flow field between the shock and some other boundary, such as the surface of an aerodynamic body. This approach is called the *shock-fitting method*. These two different approaches are illustrated in Figs. 2.11 and 2.12. In Fig. 2.11, the computational domain for calculating the supersonic flow over the body extends both upstream and downstream of the nose. The shock wave is allowed to form within the computational domain as a consequence of the general flow-field algorithm, without any special shock relations being introduced. In this manner, the shock wave is "captured" within the domain by means of the computational solution of the governing partial differential equations. Therefore,

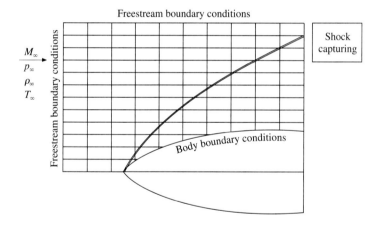

FIG. 2.11
Grid for the shock-capturing approach.

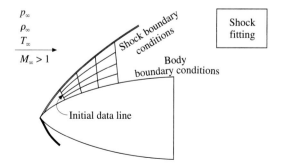

FIG. 2.12
Grid for the shock-fitting approach.

Fig. 2.11 is an example of the *shock-capturing method*. In contrast, Fig. 2.12 illustrates the same flow problem, except that now the computational domain is the flow between the shock and the body. The shock wave is introduced directly into the solution as an explicit discontinuity, and the standard oblique shock relations (the Rankine-Hugoniot relations, see Ref. 21) are used to fit the freestream supersonic flow ahead of the shock to the flow computed by the partial differential equations downstream of the shock. Therefore, Fig. 2.12 is an example of the *shock-fitting method*. There are advantages and disadvantages of both methods. For example, the shock-capturing method is ideal for complex flow problems involving shock waves for which we do not know either the location or number of shocks. Here, the shocks simply form within the computational domain as nature would have it. Moreover, this takes place without requiring any special treatment of the shock within the algorithm and hence simplifies the computer programming. However, a disadvantage of this approach is that the shocks are generally smeared over a finite number of grid points in the computational mesh, and hence the numerically obtained shock thickness bears no relation whatsoever to the actual physical shock thickness, and the *precise* location of the shock discontinuity is uncertain within a few mesh sizes. In contrast, the advantage of the shock-fitting method is that the shock is always treated as a discontinuity, and its location is well-defined numerically. However, for a given problem you have to know in advance approximately where to put the shock waves, and how many there are. For complex flows, this can be a distinct disadvantage. Therefore, there are pros and cons associated with both shock-capturing and shock-fitting methods, and both have been employed extensively in CFD. In fact, a combination of these two methods is possible, wherein a shock-capturing approach during the course of the solution is used to predict the formation and approximate location of shocks, and then these shocks are fit with explicit discontinuities midway through the solution. Another combination is to fit shocks explicitly in those parts of a flow field where you know in advance they occur and to employ a shock-capturing method for the remainder of the flow field in order to generate shocks that you cannot predict in advance.

Again, what does all of this discussion have to do with the conservation form of the governing equations as given by Eq. (2.93)? Simply this, For the *shock-capturing* method, experience has shown that the *conservation form* of the governing equations should be used. When the conservation form is used, the computed flow-field results are generally smooth and stable. However, when the nonconservation form is used for a shock-capturing solution, the computed flow-field results usually exhibit unsatisfactory spatial oscillations (wiggles) upstream and downstream of the shock wave, the shocks may appear in the wrong location, and the solution may even become unstable. In contrast, for the shock-fitting method, satisfactory results are usually obtained for either form of the equations, conservation or nonconservation.

Why is the use of the conservation form of the equations so important for the shock-capturing method? The answer can be seen by considering the flow across a normal shock wave, as illustrated in Fig. 2.13. Consider the density distribution across the shock, as sketched in Fig. 2.13*a*. Clearly, there is a discontinuous increase in ρ across the shock. If the nonconservation form of the governing equations were

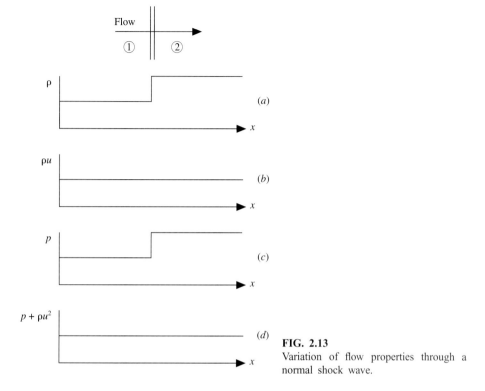

FIG. 2.13
Variation of flow properties through a normal shock wave.

used to calculate this flow, where the primary dependent variables are the primitive variables such as ρ and p, then the equations would see a large discontinuity in the dependent variable ρ. This in turn would compound the numerical errors associated with the calculation of ρ. On the other hand, recall the continuity equation for a normal shock wave (see Ref. 8 and 21):

$$\rho_1 u_1 = \rho_2 u_2 \qquad (2.113)$$

From Eq. (2.113), the *mass flux* ρu is *constant* across the shock wave, as illustrated in Fig. 2.13b. The conservation form of the governing equations uses the product ρu as a dependent variable, and hence the conservation form of the equations see *no* discontinuity in this dependent variable in the normal direction across the shock wave. In turn, the numerical accuracy and stability of the solution should be greatly enhanced. To reinforce this discussion, consider the momentum equation across a normal shock wave (Refs. 8 and 21):

$$p_1 + \rho_1 u_1^2 = p_2 + \rho_2 u_2^2 \qquad (2.114)$$

As shown in Fig. 2.13c, the pressure itself is discontinuous across the shock; however, from Eq. (2.114) the flux variable $p + \rho u^2$ is constant across the shock. This is illustrated in Fig. 2.13d. Examining the inviscid flow equations in the conservation form given by Eq. (2.93) with the flux vectors displayed as Eqs.

(2.105) to (2.109), we clearly see from the F vector in Eq. (2.106) that the quantity $p + \rho u^2$ is one of the dependent variables. Therefore, the conservation form of the equations would see no discontinuity in this dependent variable in a normal direction across the shock. Although this example of the flow across a normal shock wave is somewhat simplistic, it serves to explain why the use of the conservation form of the governing equations is so important for calculations using the shock-capturing method. Because the conservation form uses flux variables as the dependent variables and because the changes in these flux variables are either zero or small across a shock wave, the numerical quality of a shock-capturing method will be enhanced by the use of the conservation form in contrast to the nonconservation form, which uses the primitive variables as dependent variables.

In summary, the previous discussion is one of the primary reasons why CFD makes a distinction between the two forms of the governing equations, conservation and nonconservation. And this is why we have gone to great lengths in this chapter to derive these different forms, to explain what basic physical models lead to the different forms, and why we should be aware of the differences between the two forms. Again, we emphasize that the distinction CFD places between conservation and nonconservation forms of the equation is an outgrowth of the realities of *numerical* solutions—it is germane to CFD only. In the world of purely theoretical fluid dynamics, we could not care less.

Finally, recall the discussion in Sec. 2.5.6 regarding a fundamental difference between the integral form and the differential form of the equations. The integral form does not require mathematical continuity, whereas the differential form assumes mathematical continuity. This situation imposes extreme conditions on a solution with shock waves when the differential equations are used, no matter what form they take. In contrast, a formulation that deals directly with the integral form, such as the finite-volume method, is fundamentally more appropriate for such flows. For reasons such as these, the integral form of the governing equations can be considered as *more fundamental* than the differential form.

2.11 SUMMARY

This is a book on CFD; however, to this stage in our discussion we have yet to address any computational techniques. The reason is straightforward; before we can develop any computational solutions to any problem, we have to have the correct governing equations, with an in-depth physical understanding of what these equations mean. Such has been the purpose of this chapter. At this stage, it is well to return to the road map given in Fig. 2.1. Study this road map carefully, fixing in your mind the various aspects of our discussions that pertain to each box in Fig. 2.1. Focus especially on the governing equations in boxes I and J, which are summarized in Sec. 2.8 and 2.10, respectively. These equations are the "bread and butter" of CFD—learn them well.

Part I of this book has to do with the basic thoughts and equations which are the foundation of CFD (indeed, the foundation of all theoretical fluid dynamics). We are not quite finished with these basic thoughts. The partial differential equations for continuity, momentum, and energy in a fluid flow (like any system of partial differential equations) have certain mathematical behavior. This behavior may be different from one case to another, depending, for example, on the local Mach number of the flow—the same equations may have totally different mathematical behavior depending on whether the flow is locally subsonic or supersonic. (The supersonic blunt body problem described in Sec. 1.5 was, for a long time, the victim of this totally different mathematical behavior in locally subsonic and supersonic regions.) The behavior may be different depending on whether we are dealing with the Euler equations (for an inviscid flow) or the Navier-Stokes equations (for a fully viscous flow). The behavior may also be different depending on whether the flow is unsteady or steady. Of course, as you may suspect, any differences in mathematical behavior of these equations reflect different physical behavior as well. What does all this mean? The answer is contained in the next chapter—simply read on.

PROBLEMS

2.1. In conjunction with the spatially marching solutions of Eq. (2.110) for an inviscid flow, the elements of the solution vector F are given in Eqs. (2.111a) to (2.111e) as $\rho u = c_1$, $\rho u^2 + p = c_2$, $\rho u v = c_3$, $\rho u w = c_4$, and $\rho u [e + (u^2 + v^2 + w^2)/2] + pu = c_5$. Derive expressions for the primitive variables ρ, u, v, w, and p in terms of c_1, c_2, c_3, c_4, and c_5. Assume a calorifically perfect gas (with constant γ).

Answer:

$$\rho = \frac{-B \pm \sqrt{B^2 - 4AC}}{2A}$$

where

$$A = \frac{1}{2}\left(\frac{c_3^2}{c_1} + \frac{c_4^2}{c_1}\right) - c_5$$

$$B = \frac{\gamma c_1 c_2}{\gamma - 1}$$

$$C = -\frac{(\gamma + 1)c_1^3}{2(\gamma - 1)}$$

$$u = \frac{c_1}{\rho}$$

$$v = \frac{c_3}{c_1}$$

$$w = \frac{c_4}{c_1}$$

$$p = c_2 - \rho u^2$$

2.2. Derive the momentum and energy equations for a viscous flow in integral form. Show that all three conservation equations—continuity momentum, and energy—can be put in a single generic integral form.

CHAPTER 3

MATHEMATICAL BEHAVIOR OF PARTIAL DIFFERENTIAL EQUATIONS: THE IMPACT ON CFD

No knowledge can be certain, if it is not based upon mathematics or upon some other knowledge which is itself based upon the mathematical sciences.

Leonardo da Vinci (1425–1519)

Mathematics is the queen of the sciences.

Carl Friedrich Gauss, 1856

3.1 INTRODUCTION

A "rose is a rose is a rose . . ." as Gertrude Stein wrote. In turn, a partial differential equation is a partial differential equation is a partial differential equation—*or is it?* In this chapter, we will emphasize the answer—*not really.* We will find that, beyond just finding a solution to a given partial differential equation, we must be aware that such solutions have mathematical behavior which can be quite different from one circumstance to another. The same governing flow equations, when solved in one region of a flow field, can exhibit completely different solutions in another region, even though the equations themselves remain identically the same equations. It is just their *mathematical* behavior that is different. This mysterious aspect of differential equations was alluded to strongly in Sec. 2.11. The purpose of the present chapter is to remove (we hope) some of the mystery.

The governing equations of fluid dynamics derived in Chap. 2 are either integral forms [such as Eq. (2.19) obtained directly from a finite control volume] or partial differential equations [such as Eq. (2.25) obtained directly from an infinitesimal fluid element]. Before taking up a study of numerical methods for the solution of these equations, it is useful to examine some mathematical properties of partial differential equations themselves. Any valid numerical solution of the equations should exhibit the property of obeying the general mathematical properties of the governing equations.

Examine the governing partial differential equations of fluid dynamics as derived in Chap. 2. Note that in all cases the *highest-order* derivatives occur *linearly*; i.e., there are no products or exponentials of the highest-order derivatives—they appear by themselves, multiplied by coefficients which are functions of the dependent variables themselves. Such a system of equations is called a *quasi-linear system*. For example, for inviscid flows, examining the equations in Sec. 2.8.2 we find that the highest-order derivatives are first-order, and all of them appear linearly. For viscous flows, examining the equations in Sec. 2.8.1 we find the highest-order derivatives are second-order, and they always occur linearly. For this reason, in the next section, let us examine some mathematical properties of a system of quasi-linear partial differential equations. In the process, we will establish a classification of three types of partial differential equations—all three of which are encountered in fluid dynamics.

Finally, the road map for this chapter is given in Fig. 3.1. Here, we map out a fairly straightforward course. We will discuss two separate techniques for determining the classification of partial differential equations: the method using Cramer's rule, described in Sec. 3.2, and the eigenvalue method described in Sec. 3.3. Both these methods lead to the same results. We will see that many partial differential equations can be classified as either hyperbolic, parabolic, or elliptic; these definitions as well as many other details will be given in Sec. 3.2. Other equations are of a "mixed" type. We will then contrast the mathematical behavior of solutions of these different classes of equations, giving examples from actual fluid dynamic flows.

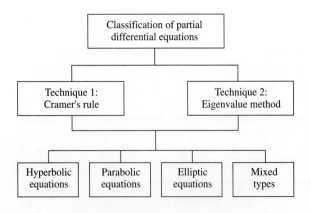

FIG. 3.1
Road map for Chap. 3.

3.2 CLASSIFICATION OF QUASI-LINEAR PARTIAL DIFFERENTIAL EQUATIONS

For simplicity, let us consider a fairly simple system of quasi-linear equations. They will *not* be the flow equations, but they are similar in some respects. Therefore, this section serves as a simplified example.

Consider the system of quasi-linear equations given below.

$$a_1 \frac{\partial u}{\partial x} + b_1 \frac{\partial u}{\partial y} + c_1 \frac{\partial v}{\partial x} + d_1 \frac{\partial v}{\partial y} = f_1 \quad (3.1a)$$

$$a_2 \frac{\partial u}{\partial x} + b_2 \frac{\partial u}{\partial y} + c_2 \frac{\partial v}{\partial x} + d_2 \frac{\partial v}{\partial y} = f_2 \quad (3.1b)$$

where u and v are the dependent variables, functions of x and y, and the coefficients $a_1, a_2, b_1, b_2, c_2, c_2, d_1, d_2, f_1$, and f_2 can be functions of x, y, u, and v. Furthermore, u and v are *continuous* functions of x and y; we can imagine that u and v represent a continuous velocity field throughout the xy space. At any given point in the xy space, there is a unique value of u and a unique value of v; moreover, the derivatives of u and v, $\partial u/\partial x, \partial u/\partial y, \partial v/\partial x, \partial v/\partial y$, are finite values at this given point. We could imagine going into this flow field if it were set up in the laboratory and measuring both the velocities and their derivatives at any given point.

However, we are now going to make a strange statement. Consider any point in the xy plane, such as point P in Fig. 3.2. Let us seek the lines (or directions) through this point (if any exist) along which the *derivatives* of u and v are *indeterminant* and across which may be discontinuous. This sounds almost contradictory to our earlier statement in the previous paragraph, but it is not. If you are confused, just hang on for the next few paragraphs. These special lines that we are seeking are called *characteristic lines*. To find such lines, we recall that u and v are continuous functions of x and y and write their total differentials as

$$du = \frac{\partial u}{\partial x} dx + \frac{\partial u}{\partial y} dy \quad (3.2a)$$

$$dv = \frac{\partial v}{\partial x} dx + \frac{\partial v}{\partial y} dy \quad (3.2b)$$

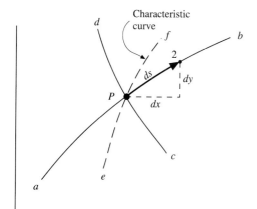

FIG. 3.2
Illustration of a characteristic curve.

Equations (3.1a) and (3.1b) and (3.2a) and (3.2b) constitute a system of four linear equations with four unknowns ($\partial u/\partial x$, $\partial u/\partial y$, $\partial v/\partial x$, and $\partial v/\partial y$). These equations can be written in matrix form as

$$\begin{bmatrix} a_1 & b_1 & c_1 & d_1 \\ a_2 & b_2 & c_2 & d_2 \\ dx & dy & 0 & 0 \\ 0 & 0 & dx & dy \end{bmatrix} \begin{bmatrix} \partial u/\partial x \\ \partial u/\partial y \\ \partial v/\partial x \\ \partial v/\partial y \end{bmatrix} = \begin{bmatrix} f_1 \\ f_2 \\ du \\ dv \end{bmatrix} \qquad (3.3)$$

Let [A] denote the coefficient matrix.

$$[A] \equiv \begin{bmatrix} a_1 & b_1 & c_1 & d_1 \\ a_2 & b_2 & c_2 & d_2 \\ dx & dy & 0 & 0 \\ 0 & 0 & dx & dy \end{bmatrix} \qquad (3.4)$$

Let us solve Eq. (3.3) for the unknown $\partial u/\partial x$, using Cramer's rule. To do this, we define the matrix [B] as the matrix [A] with its first column replaced by the column vector on the right-hand side of Eq. (3.3), i.e.,

$$[B] = \begin{bmatrix} f_1 & b_1 & c_1 & d_1 \\ f_2 & b_2 & c_2 & d_2 \\ du & dy & 0 & 0 \\ dv & 0 & dx & dy \end{bmatrix} \qquad (3.5)$$

Denoting the determinants of [A] and [B] by $|A|$ and $|B|$, respectively, Cramer's rule gives the solution for $\partial u/\partial x$ as

$$\frac{\partial u}{\partial x} = \frac{|B|}{|A|} \qquad (3.6)$$

To obtain an actual number for $\partial u/\partial x$ from Eq. (3.6), we have to establish values of du, dv, dx, and dy that appear in the matrices [A] and [B]. But what does this mean? What are dx, dy, du, and dv? To answer this question, examine Fig. 3.2. Imagine a curve ab drawn through point P in an *abritrary* direction. Let us move an infinitesimally small distance away from point P, following the curve ab, say to point 2. This small distance is denoted by ds in Fig. 3.2 and is the distance between points P and 2. The change in x associated with moving from point P to point 2 is $dx = x_2 - x_p$, and the associated change in y is $dy = y_2 - y_p$. These are the values of dx and dy that appear in matrices [A] and [B] as displayed in Eqs. (3.4) and (3.5). In addition, the values of u and v at point 2 are different than at point P; they have changed by the amounts $du = u_2 - u_p$ and $dv = v_2 - v_p$. These are the values of du and dv that appear in matrix [B] as displayed in Eq. (3.5). Inserting these numbers for dx, dy, du, and dv in Eqs. (3.4) and (3.5), we will obtain a solution for $\partial u/\partial x$ from Eq. (3.6) in the limiting case as dx and dy go to zero. Now, draw *another* arbitrary curve thorugh P in Fig. 3.2, say curve cd. We could go through the same scenario, i.e., move an infinitesimally small distance ds away from point P along curve cd,. and obtain the corresponding values of dx, dy, du, and dv. These values will, of course, be different than those used earlier because we are moving in a different direction away from point P, namely, this time along curve cd rather than curve ab. However, when these different values for dx, dy, du, and dv are inserted into Eqs. (3.4) and (3.5), in the limiting case as dx and dy go to zero, the *same* value for $\partial u/\partial x$ is obtained from Eq. (3.6) as obtained earlier. Indeed, this has to be the case, since the value of $\partial u/\partial x$ at point P is a fixed value, a *point value* if you wish, that has nothing inherently to do with "directions through point P." We have only used the idea of "directions through point P" to allow us to obtain a solution for $\partial u/\partial x$ from Cramer's rule using Eq. (3.6) The "direction" chosen is purely arbitrary, such as curves ab and cd in Fig. 3.2.

However, there is one major *exception* to this formalism. What happens if we choose to move in a direction away from point P such that $|A|$ in Eq. (3.6) is zero? In Fig. 3.2, let ef be such a direction. Then in Eq. (3.6) the denominator is zero, and the calculation of $\partial u/\partial x$ using this particular direction ef through point P is not possible. At best, we have to say that $\partial u/\partial x$ is *indeterminant* when we choose this direction. By definition, curve ef is called a *characteristic curve* (or a *characteristic line*) through point P. In this sense, we have now explained the statement made earlier that might have seemed strange. Namely, if we consider any point P in the xy plane, let us seek the lines or directions through this point (if any exist) along which the derivatives of u and v are *indeterminant* and across which may even be discontinuous. We now know that if we pick just the right direction through point P such that dx and dy are just the right values to make $|A| = 0$ in Eq. (3.6), then we have found the lines we were seeking—we have found the characteristic lines. In this case, such characteristic lines indeed do exist, and we can find them by setting

$$|A| = 0 \qquad (3.7)$$

Note that the characteristice lines are independent of whether we are solving Eq. (3.3) for $\partial u/\partial x$, or $\partial u/\partial y$, or $\partial v/\partial x$, or $\partial v/\partial y$; in all four cases, $|A|$ is the same demoninator for Cramer's rule, and Eq. (3.7) defines the same characteristic lines.

When the characteristic lines do exist for a given system of equations, note that they are identifiable curves in the xy plane, such as curve ef sketched in Fig. 3.2. Therefore, we should be able to calculate the equations of these curves, and especially the slopes of the curves at point P. This calculation is readily made from Eq. (3.7). Recalling the elements of $|A|$ from Eq. (3.4), we have

$$\begin{vmatrix} a_1 & b_1 & c_1 & d_1 \\ a_2 & b_2 & c_2 & d_2 \\ dx & dy & 0 & 0 \\ 0 & 0 & dx & dy \end{vmatrix} = 0$$

Expanding the determinant, we have

$$(a_1 c_2 - a_2 c_1)(dy)^2 - (a_1 d_2 - a_2 d_1 + b_1 c_2 - b_2 c_1)\, dx\, dy \\ + (b_1 d_2 - b_2 d_1)(dx)^2 = 0 \quad (3.8)$$

Divide Eq. (3.8) by $(dx)^2$.

$$(a_1c_2 - a_2c_1)\left(\frac{dy}{dx}\right)^2 - (a_1d_2 - a_2d_1 + b_1c_2 - b_2c_1)\frac{dy}{dx} + (b_1d_2 - b_2d_1) = 0 \tag{3.9}$$

Equation (3.9) is a quadratic equation in dy/dx. For any point in the xy plane, the solution of Eq. (3.9) will give the slopes of the lines along which the derivatives of u and v are indeterminant. Why? Because Eq. (3.9) was obtained by setting $|A| = 0$, which from the matrix Eq. (3.3) ensures that the solutions for the derivatives $\partial u/\partial x$, $\partial u/\partial y$, $\partial v/\partial x$, and $\partial v/\partial y$, are, at best, indeterminant. As stated earlier, these lines in the xy space along which the derivatives of u and v are indeterminant are called the *characteristic lines* for the system of equations given by Eqs. (3.1a) and (3.1b).

In Eq. (3.9), let

$$a = (a_1c_2 - a_2c_1)$$
$$b = -(a_1d_2 - a_2d_1 + b_1c_2 - b_2c_1)$$
$$c = (b_1d_2 - b_2d_1)$$

Then Eq. (3.9) can be written as

$$a\left(\frac{dy}{dx}\right)^2 + b\frac{dy}{dx} + c = 0 \tag{3.10}$$

Equation (3.10) can, in principle, be integrated to give $y = y(x)$, which is the equation of a characteristic curve in the xy plane. However, for our purposes, we are interested in only the *slopes* of the characteristics through point P in Fig. 3.2. Hence, from the quadratic formula:

$$\frac{dy}{dx} = \frac{-b \pm \sqrt{b^2 - 4ac}}{2a} \tag{3.11}$$

Equation (3.11) gives the direction of the characteristic lines through a given point in the xy plane, such as point P in Fig. 3.2. These lines have a different nature, depending on the value of the discriminant in Eq. (3.11). Denote the discriminant by D.

$$D = b^2 - 4ac \tag{3.12}$$

The mathematical classification of the system of equations given by Eqs. (3.1a) and (3.1b) is determined by the value of D. Specifically:

If $D > 0$ Two real and distinct characteristics exist through each point in the xy plane. The system of equations given by Eqs. (3.1a) and (3.1b) is called *hyperbolic*.

If $D = 0$ Here the system of Eqs. (3.1a) and (3.1b) is called *parabolic*.

If $D < 0$ The characteristic lines are imaginary. The system of Eqs. (3.1a) and (3.1b) is called *elliptic*.

The classification of quasi-linear partial differential equations as either *elliptic, parabolic,* or *hyperbolic* is common in the analysis of such equations.

It is this classification which has been the major focus of this section. These three classes of equations have totally different behavior, as will be discussed shortly. The origin of the words "elliptic," "parabolic," or "hyperbolic" used to label these equations is simply a direct analogy with the case for conic sections. The general equation for a conic section from analytic geometry is

$$ax^2 + bxy + cy^2 + dx + ey + f = 0$$

where, if

$$b^2 - 4ac > 0 \quad \text{the conic is a hyperbola}$$
$$b^2 - 4ac = 0 \quad \text{the conic is a parabola}$$
$$b^2 - 4ac < 0 \quad \text{the conic is an ellipse}$$

We could end this section here, as far as our purposes in this book are concerned. However, the temptation to extend the thoughts in this section one more step is too overwhelming, since it pertains to one of the classic methods in the solution of compressible flow problems—the method of characteristics. Return to Eq. (3.6). Note that, if only $|A|$ were zero, then $\partial u/\partial x$ would be infinite. However, the definition of a characteristic line states that $\partial u/\partial x$ be indeterminant along the characteristic, not infinite. Thus, for $\partial u/\partial x$ to be indeterminant, $|B|$ in Eq. (3.6) must also be zero. Then, $\partial u/\partial x$ is of the form

$$\frac{\partial u}{\partial x} = \frac{|B|}{|A|} = \frac{0}{0} \tag{3.13}$$

namely, an indeterminant form which can have a finite value. Hence, from Eq. (3.5)

$$|B| = \begin{vmatrix} f_1 & b_1 & c_1 & d_1 \\ f_2 & b_2 & c_2 & d_2 \\ du & dy & 0 & 0 \\ dv & 0 & dx & dy \end{vmatrix} = 0 \tag{3.14}$$

Expansion of the determinant in Eq. (3.14) yields an *ordinary differential equation* in terms of du and dv, where dx and dy are restricted to hold along a characteristic line (see Prob. 3.1). [Since $|B| = 0$ is a direct consequence of $|A| = 0$ from Eq. (3.13), then whatever relation is derived from setting $|B| = 0$ must be restricted to hold along a characteristic line.] The equation for the dependent variables u and v which comes from Eq. (3.14) is called the *compatibility* equation. It is an equation involving the unknown dependent variables which holds *only* along the characteristic line; the advantage of this compatibility equation is that it is in one less dimension than the original partial differential equations. Since the model equations treated in this section [Eqs. (3.1a) and (3.1b)] are partial differential equations in two dimensions, then the compatibility equation is in one dimension—hence it is an ordinary differential equation—and the "one dimension" is along the characteristic direction. Since ordinary differential equations are in general simpler to solve than partial differential equations, then the compatibility equations provide some advantage. This leads to a solution technique for the original system of equations

[Eqs. (3.1a) and (3.1b)] wherein the characteristic lines are constructed in the xy space, and the simpler compatibility equations are solved along these characteristics. This technique is called the *method of characteristics*. In general, the successful implementation of the method of characteristics requires at least two characteristic directions through any point in the xy plane, with different compatibility equations applicable to each different characteristic line; i.e., the method of characteristics is useful for the solution of hyperbolic partial differential equations only. This method is highly developed for the solution of inviscid supersonic flows, for which the system of governing flow equations is hyperbolic. The practical implementation of the method of characteristics requires the use of a high-speed digital computer and therefore may legitimately be considered a part of CFD. However, the method of characteristics is a well-known classical technique for the solution of inviscid supersonic flows, and therefore we will not consider it in any detail in this book. For more information, see Ref. 21.

3.3 A GENERAL METHOD OF DETERMINING THE CLASSIFICATION OF PARTIAL DIFFERENTIAL EQUATIONS: THE EIGENVALUE METHOD

In Sec. 3.2 we developed a method based on Cramer's rule for analyzing a system of quasi-linear equations in order to determine the classification of those equations. However, there is a more general and slightly more sophisticated method for assessing the classification of quasi-linear partial differential equations based on the *eigenvalues* of the system. This approach is developed in the present section. In the process, we will be using some basic matrix notation and manipulation, which is assumed to be familiar to most junior or senior engineering and science students. For a basic review of matrix algebra, see, for example, Ref. 22.

The eigenvalue method is based on a display of the system of partial differential equations written in column vector form. For example, let us assume that f_1 and f_2 in Eqs. (3.1a) and (3.1b) are zero for simplicity, such that the equations become

$$a_1 \frac{\partial u}{\partial x} + b_1 \frac{\partial u}{\partial y} + c_1 \frac{\partial v}{\partial x} + d_1 \frac{\partial v}{\partial y} = 0 \qquad (3.15a)$$

$$a_2 \frac{\partial u}{\partial x} + b_2 \frac{\partial u}{\partial y} + c_2 \frac{\partial v}{\partial x} + d_2 \frac{\partial v}{\partial y} = 0 \qquad (3.15b)$$

Defining W as the column vector

$$W = \begin{Bmatrix} u \\ v \end{Bmatrix}$$

the system of equations given by Eqs. (3.15a) and (3.15b) can be written as

$$\begin{bmatrix} a_1 & c_1 \\ a_2 & c_2 \end{bmatrix} \frac{\partial W}{\partial x} + \begin{bmatrix} b_1 & d_1 \\ b_2 & d_2 \end{bmatrix} \frac{\partial W}{\partial y} = 0 \qquad (3.16)$$

or
$$[K]\frac{\partial W}{\partial x} + [M]\frac{\partial W}{\partial y} = 0 \tag{3.17}$$

where $[K]$ and $[M]$ are the appropriate 2×2 matrices in Eq. (3.16). Multiplying Eq. (3.17) by the inverse of $[K]$, we have

$$\frac{\partial W}{\partial x} + [K]^{-1}[M]\frac{\partial W}{\partial y} = 0 \tag{3.18}$$

or
$$\frac{\partial W}{\partial x} + [N]\frac{\partial W}{\partial y} = 0 \tag{3.19}$$

where by definition $[N] = [K]^{-1}[M]$. With the system of equations written in the form of Eq. (3.19), the eigenvalues of $[N]$ determine the classification of the system. If the eigenvalues are all real, the equations are hyperbolic. If the eigenvalues are all complex, the equations are elliptic. This statement is made without proof; see Ref. 23 for more details.

Example 3.1. We will illustrate this procedure using an actual system of equations from fluid dynamics. Consider the irrotational, two-dimensional, inviscid, steady flow of a compressible gas. If the flow field is only slightly perturbed from its freestream conditions, such as the flow over a thin body at small angles of attack, and if the freestream Mach number is either subsonic or supersonic (but *not* transonic or hypersonic), the governing continuity, momentum, and energy equations can be reduced to the system

$$(1 - M_\infty^2)\frac{\partial u'}{\partial x} + \frac{\partial v'}{\partial y} = 0 \tag{3.20}$$

$$\frac{\partial u'}{\partial y} - \frac{\partial v'}{\partial x} = 0 \tag{3.21}$$

where u' and v' are small perturbation velocities, measured relative to the freestream velocity. For example

$$u = V_\infty + u'$$
$$v = v'$$

Also in Eq. (3.20), M_∞ is the freestream Mach number; it can be subsonic or supersonic. Equation (3.21) is a statement that the flow is irrotational. For the derivation of Eqs. (3.20) and (3.21), and a major discussion of the physical aspects of these equations, see chap. 9 of Ref. 21 or chap. 11 of Ref. 8. However, for our purposes here, we simply use these equations as an example of a system of quasi-linear equations. Indeed, Eqs. (3.20) and (3.21) are precisely linear equations; these equations have been the foundation of numerous linearized aerodynamic analyses in the past.

Question: How do we classify Eqs. (3.20) and (3.21)? First, let us utilize the method developed in Sec. 3.2. Comparing Eqs. (3.20) and (3.21) with the standard form given by Eqs. (3.1*a*) and (3.1*b*), we have [in terms of the nomenclature of Eqs. (3.1*a*) and (3.1*b*)]

$$a_1 = 1 - M_\infty^2 \qquad a_2 = 0$$
$$b_1 = 0 \qquad b_2 = 1$$
$$c_1 = 0 \qquad c_2 = -1$$
$$d_1 = 1 \qquad d_2 = 0$$

Restating these values in terms of a, b, and c as given in Eq. (3.10), we have

$$a = -(1 - M_\infty^2)$$
$$b = 0$$
$$c = -1$$

Hence, Eq. (3.11) yields

$$\frac{dy}{dx} = \frac{\pm\sqrt{-4(1 - M_\infty^2)}}{-2(1 - M_\infty^2)} = \frac{\pm\sqrt{4(M_\infty^2 - 1)}}{2(M_\infty^2 - 1)} = \pm\frac{1}{\sqrt{M_\infty^2 - 1}} \quad (3.22)$$

Examining Eq. (3.22) for the case of supersonic flow, $M_\infty > 1$, we see that there are two real characteristic directions through each point, one with slope $= (M_\infty^2 - 1)^{-1/2}$ and the other with slope $= -(M_\infty^2 - 1)^{-1/2}$. Hence, for $M_\infty > 1$, the system of equations given by Eqs. (3.20) and (3.21) is *hyperbolic*. On the other hand, if $M_\infty < 1$, then the characteristics are imaginary, and the equations are *elliptic*.

Now, let us employ the eigenvalue method. Written in the form of Eq. (3.16), Eqs. (3.20) and (3.21) are

$$\begin{bmatrix} 1 - M_\infty^2 & 0 \\ 0 & -1 \end{bmatrix} \frac{\partial W}{\partial x} + \begin{bmatrix} 0 & 1 \\ 1 & 0 \end{bmatrix} \frac{\partial W}{\partial y} = 0$$

or

$$[K]\frac{\partial W}{\partial x} + [M]\frac{\partial W}{\partial y} = 0$$

where

$$W = \begin{Bmatrix} u' \\ v' \end{Bmatrix}$$

To find $[K]^{-1}$, we first replace the elements of $[K]$ with their cofactors, yielding

$$\begin{bmatrix} -1 & 0 \\ 0 & 1 - M_\infty^2 \end{bmatrix}$$

The transpose of the above is also

$$\begin{bmatrix} -1 & 0 \\ 0 & 1 - M_\infty^2 \end{bmatrix}$$

The determinant of $[K]$ is $-(1 - M_\infty^2)$. Hence,

$$[K]^{-1} = -\frac{1}{1 - M_\infty^2}\begin{bmatrix} -1 & 0 \\ 0 & 1 - M_\infty^2 \end{bmatrix}$$

or

$$[K]^{-1} = \begin{bmatrix} \dfrac{1}{1 - M_\infty^2} & 0 \\ 0 & -1 \end{bmatrix}$$

In turn,

$$[N] = [K]^{-1}[M] = \begin{bmatrix} \frac{1}{1-M_\infty^2} & 0 \\ 0 & -1 \end{bmatrix} \begin{bmatrix} 0 & 1 \\ 1 & 0 \end{bmatrix} = \begin{bmatrix} 0 & \frac{1}{1-M_\infty^2} \\ -1 & 0 \end{bmatrix}$$

This is the matrix $[N]$ in the form given by Eq. (3.19). Hence, we wish to examine the eigenvalues of $[N]$, denoted by λ. These are found by setting

$$||[N] - \lambda[I]|| = 0 \tag{3.23}$$

where $[I]$ is the identity matrix. Hence

$$\begin{vmatrix} -\lambda & \frac{1}{1-M_\infty^2} \\ -1 & -\lambda \end{vmatrix} = 0$$

Expanding the determinant, we have

$$\lambda^2 + \frac{1}{1-M_\infty^2} = 0$$

or

$$\lambda = \pm\sqrt{\frac{1}{M_\infty^2 - 1}} \tag{3.24}$$

Equation (3.24) yields precisely the same result as obtained in Eq. (3.22). Indeed, the eigenvalues of $[N]$ are precisely the slopes of the characteristic lines. Moreover, from our rule stated above, if $M_\infty > 1$, then from Eq. (3.24) the eigenvalues are all real, and the system of equations given by Eqs. (3.20) and (3.21) is *hyperbolic*. If $M_\infty < 1$, then from Eq. (3.24) the eigenvalues are all imaginary, and the system of equations is *elliptic*. This illustrates how the eigenvalue method can be used to classify partial differential equations.

As a final note in this section, things are not always so clear-cut. For some systems of equations, the eigenvalues may be a mix of both real and complex values. In this case the system is neither hyperbolic nor elliptic. The mathematical behavior of such equations then exhibits a mixed hyperbolic-elliptic nature. Consequently, please keep in mind that systems of partial differential equations cannot always be conveniently placed in just one of the classifications of hyperbolic, parabolic, or elliptic; sometimes the equations have mixed behavior, as mentioned above.

3.4 GENERAL BEHAVIOR OF THE DIFFERENT CLASSES OF PARTIAL DIFFERENTIAL EQUATIONS: IMPACT ON PHYSICAL AND COMPUTATIONAL FLUID DYNAMICS

In the previous sections, we discussed the *classification* of partial differential equations, leading to the *definition* of hyperbolic, parabolic, and elliptic equations. Why do we care about making such a distinction? What difference does it make in terms of the analysis of a fluid dynamic problem whether the governing equations are hyperbolic, parabolic, elliptic, or of some mixed nature? The answers to these questions are the subject of the present section. The answers rest on the fact that each type of equation has a different *mathematical behavior*, and this reflects different *physical behavior* of the flow fields as well. In turn, this implies that different computational methods should be used for solving equations associated with the different classifications. This is a basic fact of life in CFD, and it is the reason why we are discussing such matters before we address any particular numerical techniques.

The mathematical behavior of partial differential equations is a lengthy subject whose details can be found in many advanced mathematics textbooks, such as Refs. 19 and 24. In the present section, we will simply discuss, without proof, some of the essential features of the behavior of hyperbolic, parabolic, and elliptic partial differential equations and will relate this behavior to the physics of the flow and to its impact on CFD.

3.4.1 Hyperbolic Equations

To begin with, let us consider a hyperbolic equation in two independent variables x and y. The xy plane is sketched in Fig. 3.3. Consider a given point P in this plane. Since we are dealing with a hyperbolic equation, there are two real characteristic curves through point P; these are labeled as left- and right-running characteristics, respectively. (The nomenclature "left- and right-running" stems from the following idea. Imagine that you place Fig. 3.3 on the floor and that you stand on point P, facing in the general x direction. You have to turn your head to the left to see one characteristic curve running out in front of you—this is the left-running characteristic. Similarly, you have to turn your head to the right to see the other

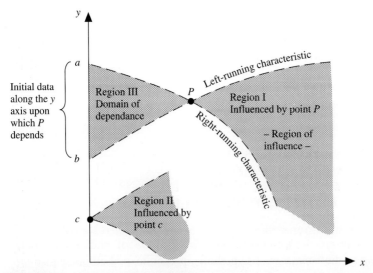

FIG. 3.3
Domain and boundaries for the solution of hyperbolic equations. Two-dimensional steady flow.

characteristic running out in front of you—this is the right-running characteristic.) A significance of these characteristics is that information at point P influences only the region between the two characteristics. For example, if in Fig. 3.3 we jabbed point P with a pin, i.e., if we set up a small disturbance at point P, then this disturbance is felt at every point within region I in Fig. 3.3, *but only in that region*. In this sense, region I is defined as the *region of influence* of point P. Now imagine the two characteristics through P extended backward to the y axis. That portion of the y axis which is intercepted by the two characteristics is labeled ab. This has a corollary effect on boundary conditions for hyperbolic equations. For example, assume that boundary conditions are specified on the y axis ($x = 0$). That is, the dependent variables u and v are known along the y axis. Then the solution can be obtained by "marching forward" in the distance x, starting from the given boundary. However, the solution for u and v at point P will depend *only* on that part of the boundary between a and b, as shown in Fig. 3.3. Information at point c, which is outside the interval ab, is propagated along characteristics through c and influences only region II in Figure 3.3. Point P is outside region II, and hence does not feel the information from point c. Point P depends on only that part of the boundary which is intercepted by and included between the two retreating characteristic lines through point P, that is, interval ab. For this reason, the region to the left of point P, region III in Fig. 3.3, is called the *domain of dependence* of point P; that is, properties at P depend only on what is happening in region III.

In terms of CFD, the computation of flow fields that are governed by hyperbolic equations is set up as "marching" solutions. The algorithm is designed to start with the given initial conditions, say the y axis in Fig. 3.3, and sequentially calculate the flow field, step by step, marching in the x direction.

In fluid dynamics, the following types of flows are governed by hyperbolic partial differential equations and hence exhibit the behavior described above.

STEADY INVISCID SUPERSONIC FLOW. If the flow is two-dimensional, the behavior is like that already discussed in Fig. 3.3. Imagine a supersonic flow over a two-dimensional circular-arc airfoil as sketched in Fig. 3.4; the airfoil can be at an angle of attack α, but α must not be so large as to cause the leading-edge shock wave to become detached, or else there will be pockets of locally subsonic flow. (In a steady flow field, any pockets of subsonic flow will be governed by elliptic equations, and the downstream marching procedure originally established for the solution of the hyperbolic equations will be mathematically ill-posed—the computer program usually "blows-up" under such conditions.) To be more specific, reflect again on the Euler equations given in Sec. 2.8.2, i.e., Eqs. (2.82) to (2.86). When written for a steady flow in either conservation or nonconservation form, these equations are hyperbolic when the local Mach number is supersonic (see Example 3.1). [A case in point is Eq. (3.20), which is derived from the Euler equations for the special case of irrotational flow with small perturbations. We proved in Sec. 3.3 that this equation is hyperbolic when $M_\infty > 1$. A more general analysis of the eigenvalues associated with the general Euler equations for a steady flow demonstrates that this system of equations is hyperbolic at every point where the *local* Mach number > 1.] Hence, in Figs. 3.4, where the flow is assumed to be

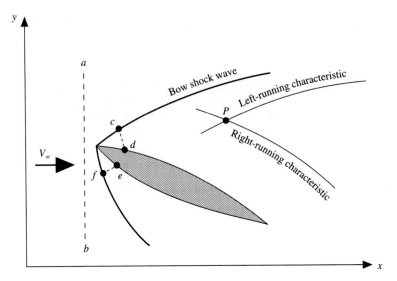

FIG. 3.4
Illustration of initial data lines for the method of characteristics.

locally supersonic everywhere, the entire flow field is governed by hyperbolic equations. The general flow direction is in the x direction. Therefore, the flow field can be computed starting with given initial data at some location in the flow and then solving the governing equations numerically, marching step by step, in the general x direction downstream of the initial data. The location of the initial data line is influenced somewhat by whether shock capturing or shock fitting is being used in the calculation. (Recall the discussion of shock capturing and shock fitting in Sec. 2.10.) If shock capturing is being used, line ab upstream of the body can be used as the initial data line, where the initial data are simply freestream conditions along ab. If shock fitting is being used, lines cd and ef just downstream of the nose, and reaching across the flow field from the body surface to the shock surface, can be used as the initial data lines. In this case, the initial data usually specified along cd or ef are that associated with a classical solution of the oblique shock flow over a wedge, with a wedge angle equal to the body angle at the nose relative to the freestream direction. See Ref. 21 for such classical wedge solutions. The results of these classical solutions yield a set of constant properties along cd and another different set of constant properties along ef. In turn, the remainder of the flow field in Fig. 3.4 is calculated by marching downstream from these initial data lines. These matters will be made clearer when we discuss actual applications in Part III of this book.

To extend the above discussion to *three-dimensional*, steady, supersonic, inviscid flows, consider the picture shown in Fig. 3.5. In this three-dimensional *xyz* space, the characteristics are *surfaces*, as sketched in Fig. 3.5. Consider point P at a given (xyz) location. Information at P influences the shaded volume contained within the advancing characteristic surface. In addition, if the yz plane is an initial data surface, then only that portion of the initial data shown as the crosshatched area

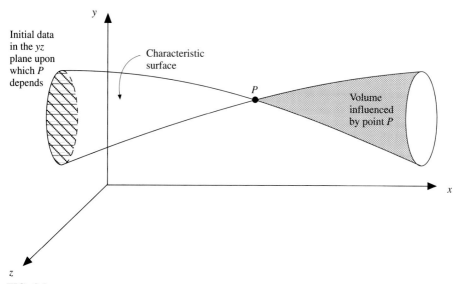

FIG. 3.5
Domain and boundaries for the solution of hyperbolic equations. Three-dimensional steady flow.

in the xy plane, intercepted by the retreating characteristic surface, has any affect on P. In Fig. 3.5, the dependent variables are solved by starting with data given in the yz plane and "marching" in the x direction. For an inviscid supersonic flow problem, the general flow direction would also be in the x direction.

UNSTEADY, INVISCID FLOW. Examine again the Euler equations summarized in Sec. 2.8.2. If the time derivatives in these equations are finite, as would be the case of an *unsteady* flow, then the governing equations are hyperbolic, no matter whether the flow is locally subsonic or supersonic. More precisely, we say that such flows are hyperbolic *with respect to time*. (The classification of the unsteady Euler equations as hyperbolic with respect to time is derived in Sec. 11.2.1.) This implies that in such unsteady flows, no matter whether we have one, two, or three spatial dimensions, the marching direction is always the *time* direction. Let us examine this more closely. For one-dimensional unsteady flow, consider point P in the xt plane shown in Fig. 3.6. Once again, the region influenced by P is the shaded area between the two advancing characteristics through P. The x axis ($t = 0$) is the initial data line. The interval ab is the only portion of the initial data along the x axis upon which the solution at P depends. Extending these thoughts for two-dimensional unsteady flow, consider point P in the xyt space as shown in Fig. 3.7. The region influenced by P and the portion of the boundary in the xy plane upon which the solution at P depends are shown in this figure. Starting with known initial data in the xy plane, the solution "marches" forward in time. Indeed, the extension to three-dimensional unsteady flow is made in the same fashion, although we cannot easily draw a sketch of this case since we are dealing with four independent variables. In this case, the full three-dimensional Euler equations summarized in Sec. 2.8.2 are utilized, and the solution is still marched forward in time.

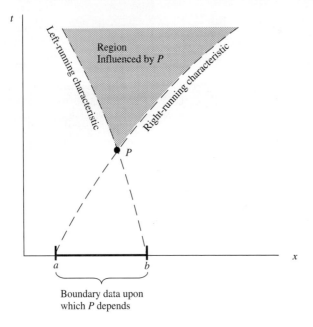

FIG. 3.6
Domain and boundaries for the solution of hyperbolic equations. One-dimensional unsteady flow.

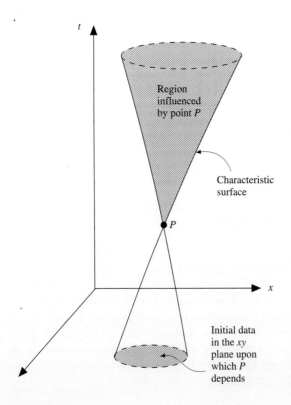

FIG. 3.7
Domain and boundaries for the solution of hyperbolic equations. Two-dimensional unsteady flow.

When do we encounter unsteady, inviscid flow? The classic case of one-dimensional wave motion in a duct is one example; here, we are truly interested in the transient variations (see, for example, chap. 7 of Ref. 21). The two-dimentional unsteady flow over a flapping or plunging airfoil is another example. However, by far the most common use of unsteady time-marching solutions in CFD is to ultimately obtain a steady flow result in the limit of large times, as long as the boundary conditions are time-invariant. Here, the time marching is simply a means to the end—the end being a steady-state flow field. At first glance, this would seem inefficient. Why calculate a steady flow by going to the trouble of introducing time as another independent variable? The answer is that sometimes this is the only way to have a well-posed problem and hence the only way to obtain the steady-state solution computationally. The solution of the supersonic blunt body problem described in Sec. 1.5 is one such case. We will see many other examples of this approach in Part III.

3.4.2 Parabolic Equations

Let us consider a parabolic equation in two independent variables x and y. The xy plane is sketched in Fig. 3.8. Consider a given point P in this plane. Since we are dealing with a parabolic equation, there is only one characteristic direction through point P. Furthermore, in Fig. 3.8, assume that initial conditions are given along the line ac and that boundary conditions are known along curves ab and cd. The characteristic direction is given by a vertical line through P. Then, information at P influences the entire region on one side of the vertical characteristic and contained within the two boundaries; i.e., if we jab P with a needle, the effect of this jab is felt throughout the shaded region shown in Fig. 3.8. Parabolic equations, like hyperbolic

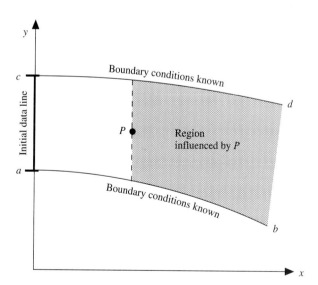

FIG. 3.8
Domain and boundaries for the solution of parabolic equations in two dimensions.

equations as discussed in Sec. 3.4.1, lend themselves to marching solutions. Starting with the initial data line *ac*, the solution between the boundaries *cd* and *ab* is obtained by marching in the general *x* direction. The extension to the case of three dimensions is straightforward, as sketched in Fig. 3.9. Here, the parabolic equation has three independent variables, *x*, *y*, and *z*. Consider point *P* located in this space. Assume that the initial conditions are given over the area *abcd* in the *yz* plane. Furthermore, assume boundary conditions given along the four surfaces *abgh*, *cdef*, *ahed*, and *bgfc*, which extend in the general *x* direction away from the perimeter of the initial data surface. Then, information at *P* influences the entire three-dimensional region to the right of *P*, contained within the boundary surfaces. This region is crosshatched in Fig. 3.9. Starting with the initial data plane *abcd*, the solution is marched in the general *x* direction. Again, please make special note that parabolic equations lend themselves to marching-type solutions, analogous to that of hyperbolic equations.

What types of fluid dynamic flow fields are governed by parabolic equations? Before answering this question, recall that the whole analysis is based upon the governing equations derived in Chap. 2—the most general form of which are the Navier-Stokes equations. Throughout the evolution of classical fluid dynamic theory, various simplified (and usually approximate) forms of the Navier-Stokes equations have been used, depending on the particular flow field to be analyzed. Although the Navier-Stokes equations themselves exhibit a mixed mathematical behavior, many of the approximate forms derived from the Navier-Stokes equations are parabolic equations. Therefore, when we ask what types of fluid dynamic flow fields are governed by parabolic equations, we are really asking what types of approximate flow-field *models* are governed by parabolic equations. (Indeed, we

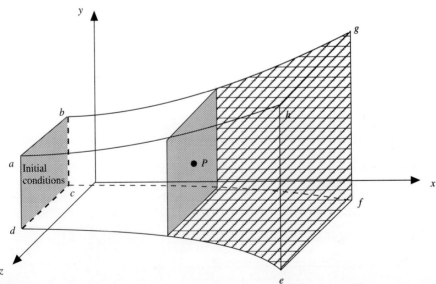

FIG. 3.9
Domain and boundaries for the solution of parabolic equations in three dimensions.

were implicitly following a similar philosophy in Sec. 3.4.1 dealing with hyperbolic equations, because the examples given in the two subsections all pertained to the Euler equations, which are a simplified version of the Navier-Stokes equations when applied to an inviscid flow.) If we delve into the various approximate forms of the Navier-Stokes equations, then the following types of flow-field *models* are governed by parabolic equations.

STEADY BOUNDARY-LAYER FLOWS. The concept of dividing a general flow field into two regions, (1) a thin layer adjacent to any solid surface wherein all the viscous effects are contained and (2) an inviscid flow outside this thin viscous layer, was one of the most profound developments in fluid dynamics. It was presented by Ludwig Prandtl at the Third Congress of Mathematicians at Heidelberg, Germany, in 1904. The thin viscous layer adjacent to a surface is called a *boundary layer*. It is assumed that you have been introduced to the idea of boundary layers; if not, you are referred to the introductory discussion in chap. 17 of Ref. 8. A schematic of the boundary layer on a generic aerodynamic body is given in Fig. 3.10. Under the combined assumptions that his boundary layer is *thin* and that the Reynolds number, Re, based on body length L is large (Re = $\rho_\infty V_\infty L/\mu_\infty$), the Navier-Stokes equations reduce to an approximate set of equations called the *boundary-layer equations*. Suffice it to say here that the boundary-layer equations are *parabolic*

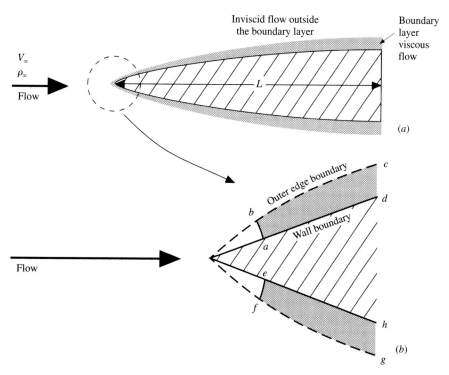

FIG. 3.10
Schematic of a boundary-layer flow.

equations. These equations describe in an approximate (but usually sufficiently accurate) fashion the flow in the thin shaded region sketched along the surface of the body in Fig. 3.10a. Here, all the viscous effects are assumed to be contained within the thin boundary layer, and the rest of the flow outside the boundary layer is inviscid. Since the boundary-layer equations are parabolic, they can be solved by a marching technique; starting from initial data at the nose of the body, the boundary-layer equations are solved by marching downstream in the s direction, where s is the distance along the surface of the body measured from the nose, as shown in Fig. 3.10a. A detail of the nose region is shown in Fig. 3.10b. Here, initial conditions are given along lines *ab* and *ef* across the boundary layer. These initial conditions are obtained from an independent, specialized solution of the boundary-layer equations, such as a self-similar solution for a flat wedge surface (if the body in Fig. 3.10 is two-dimensional) or for a sharp cone (if the body is axisymmetric). Then, starting from these initial conditions, the boundary-layer equations are solved by marching downstream from lines *ab* and *ef*. Curves *ad* and *eh* represent one boundary, namely, that along the surface at which the no-slip boundary conditions described in Sec. 2.9 are applied. Curves *bc* and *fg* represent the other boundary, namely, the outer edge of the boundary layer at which the (usually) known inviscid flow conditions are applied. It is a tenet of first-order boundary-layer theory that the inviscid flow conditions along *bc* and *fg* are the same as those obtained along the body surface in a purely inviscid solution of the flow. In summary, examining Fig. 3.10, because the boundary-layer equations are parabolic, they are solved by marching downstream in the s direction from an initial data line, while at the same time satisfying the wall and outer-edge boundary conditions at each s location.

"PARABOLIZED" VISCOUS FLOWS. What happens when the boundary layer is not thin, indeed, when the entire flow field of interest is fully viscous? An example is shown in Fig. 3.11, where a supersonic flow over a pointed-nose body is shown. If the Reynolds number is low enough, the viscous effects will reach well into the flow field far away from the surface. Indeed, the flow field between the shock wave and

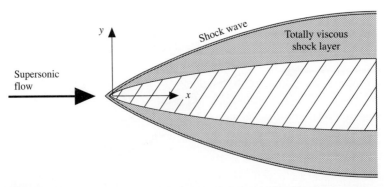

FIG. 3.11
Schematic of a totally viscous shock layer.

the body surface might be totally viscous. For this case, a boundary-layer solution is inappropriate; the boundary-layer equations are not valid for this flow. On the other hand, if the flow field does not exhibit any regions of localized, reversed, separated flow in the streamwise direction, still another simplified version of the Navier-Stokes equations may suffice. For example, if all the viscous terms in Eqs. (2.58a) to (2.58c) and Eq. (2.81) that involve derivatives in the streamwise direction [such terms as $(\partial/\partial x) (\lambda \nabla \cdot \mathbf{V} + 2\mu \, \partial u/\partial x)$, $(\partial/\partial y) (\mu \, \partial v/\partial x)$, and $(\partial/\partial x) (k \, \partial T/\partial x)$] are assumed to be small and can be neglected, and if the flow is assumed to be steady, then the resulting equations are called the *parabolized Navier-Stokes* (PNS) *equations*. This is because the resulting simplified version of the Navier-Stokes equations exhibit parabolic mathematical behavior. The PNS equations are derived in Ref. 13 and are displayed and discussed in Ref. 2, to cite just two of many sources. The advantages of the PNS equations are (1) they are simpler, i.e., contain less terms, than the full Navier-Stokes equations and (2) they can be solved by means of a downstream marching procedure. On the other hand, because the viscous terms involving derivatives in the flow direction have been neglected, and these derivatives represent the physical mechanism by which information is fed *upstream* due to viscous action in the flow, then the PNS equations are not appropriate for the calculation of viscous flows that involve regions of flow separation in the streamwise direction. This is a severe limitation for some applications. In spite of this drawback, the downstream marching aspect of the PNS equations is such a compelling advantage that this methodology is in widespread use. The type of well-behaved, fully viscous flow sketched in Fig. 3.11 is perfect for a PNS solution, and the accuracy of such solutions is usually quite acceptable.

UNSTEADY THERMAL CONDUCTION. Consider a stationary fluid (liquid or gas) wherein heat is transferred by thermal conduction. Moreover, assume that the temperature gradients in the fluid are changing as a function of time; this can be imagined as due to a time-varying wall temperature, for example. Although this example is not a flow per se, the governing conduction heat transfer equation is easily obtained from Eq. (2.73) applied to the case where $\mathbf{V} = 0$. In this case, Eq. (2.73) becomes

$$\rho \frac{\partial e}{\partial t} = \rho \dot{q} + \frac{\partial}{\partial x}\left(k \frac{\partial T}{\partial x}\right) + \frac{\partial}{\partial y}\left(k \frac{\partial T}{\partial y}\right) + \frac{\partial}{\partial z}\left(k \frac{\partial T}{\partial z}\right) \qquad (3.25)$$

Furthermore, if there is no volumetric heat addition ($\dot{q} = 0$), and assuming the state relation $e = c_v T$, Eq. (3.25) becomes

$$\frac{\partial T}{\partial t} = \frac{1}{\rho c_v}\left[\frac{\partial}{\partial x}\left(k \frac{\partial T}{\partial x}\right) + \frac{\partial}{\partial y}\left(k \frac{\partial T}{\partial y}\right) + \frac{\partial}{\partial z}\left(k \frac{\partial T}{\partial z}\right)\right] \qquad (3.26)$$

Equation (3.26) is the governing equation for the timewise and spatial variation of T throughout the fluid; it is *parabolic* with respect to time, which allows a time-

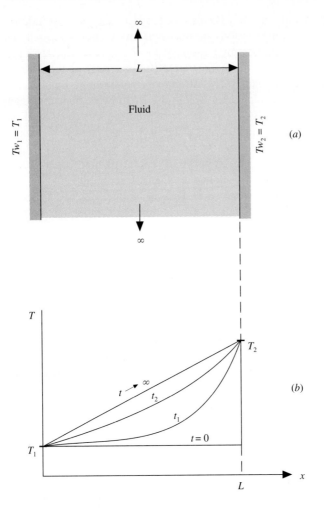

FIG. 3.12
Typical transient temperature distributions in a constant property fluid (constant ρ, c_v, and k), starting from an impulsive increase in T_{W_2} from T_1 to T_2 at time zero.

marching solution to the heat conduction problem. If we further assume that k is constant, then Eq. (3.26) is written as

$$\frac{\partial T}{\partial t} = \alpha \nabla^2 T \tag{3.27}$$

where α = *thermal diffusivity* = $k/\rho c_v$. On a physical basis, α is an index of a fluid element's ability to conduct energy due to the thermal conduction compared to its ability to retain this energy, i.e., its capacity to absorb heat. Equation (3.27) is the well-known *heat conduction equation*; again, it is a parabolic equation.

A typical solution of the heat conduction equation is sketched qualitatively in Fig. 3.12. Here we have a thermally conducting, semi-infinite fluid contained between two parallel walls separated by a distance L. We assume that the fluid is initially at constant temperature throughout, at a value $T = T_1$, and in equilibrium with both walls where initially $T_{w_1} = T_{w_2} = T_1$. Now assume that at time $t = 0$ the temperature at the right-hand wall is impulsively increased to $T_{w_2} = T_2$, while that of the left-hand wall is held fixed at $T_{w_1} = T_1$. There will be an unsteady change in the fluid temperature as a result of this impulsive increase in wall temperature; the transient temperature distributions are governed by Eq. (3.27) written in one spatial dimension, i.e.,

$$\frac{\partial T}{\partial t} = \alpha \frac{\partial^2 T}{\partial x^2} \tag{3.28}$$

Several instantaneous distributions of T versus x are sketched in Fig. 3.12, starting with the constant initial temperature at $t = 0$, and then progressing through increasing time, $t_2 > t_1 > 0$, with the final steady-state distributions given by the linear variation at infinite time.

3.4.3 Elliptic Equations

Let us consider an elliptic equation in two independent variables x and y. The xy plane is sketched in Fig. 3.13. Recall from Sec. 3.2 that the characteristic curves for an elliptic equation are imaginary—for the most part, the methodology associated with the method of characteristics is therefore useless for the solution of elliptic equations. For elliptic equations, there are no limited regions of influence or domains of dependence; rather, information is propagated everywhere in all directions. For example, consider point P in the xy plane sketched in Fig. 3.13. Assume that the domain of the problem is defined as the rectangle $abcd$ shown in Fig. 3.13 and that P is located somewhere inside this closed domain. This is already in contrast to the rather open domains considered in Figs. 3.3 and 3.8 for hyperbolic and parabolic equations, respectively. Now assume that we jab point P in Fig. 3.13 with a needle; i.e., we introduce a disturbance at point P. The major mathematical characteristic of elliptic equations is that this disturbance is felt *everywhere* throughout the domain. Furthermore, because point P influences all points in the domain, then in turn the solution at point P is influenced by the *entire* closed boundary $abcd$. Therefore, the solution at point P must be carried out *simultaneously* with the solution at all other points in the domain. This is in stark contrast to the "marching" solutions germane to parabolic and hyperbolic equations. For this reason, problems involving elliptic equations are frequently called *jury* problems, because the solution within the domain depends on the *total* boundary

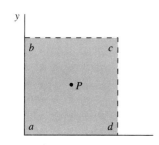

FIG. 3.13
Domain and boundaries for the solution of elliptic equations in two dimensions.

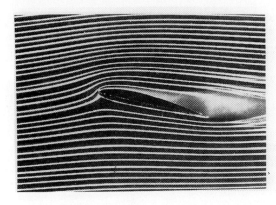

FIG. 3.14
Smoke-flow photograph of the low-speed, subsonic flow over an airfoil. (*Courtesy of Hikaru Ito, Meiji University, Japan.*)

domain, boundary conditions must be applied over the entire boundary *abcd*. These boundary conditions can take the following forms:

1. A specification of the *dependent variables* u and v along the boundary. This type of boundary condition is called the *Dirichlet* condition.
2. A specification of *derivatives* of the dependent variables, such as $\partial u/\partial x$, along the boundary. This type of boundary condition is called the *Neumann* condition.
3. A mix of both Dirichlet and Newmann conditions.

What types of flow are governed by elliptic equations? We will consider two such flows, as follows.

STEADY, SUBSONIC, INVISCID FLOW. The key, operable word here is "subsonic." In a subsonic flow, disturbances (which travel at the speed of sound, or faster) can physically work their way upstream for as far as they want—theoretically, a finite disturbance in an inviscid subsonic flow (no dissipation due to friction, thermal combustion, or mass diffusion) will propagate to infinity in all directions. For example, many of you are familiar with the streamline patterns for subsonic flows over airfoils, such as shown in the photograph in Fig. 3.14. Notice how the streamlines in front of the airfoil are deflected upward and those behind the airfoil are deflected downward. The disturbances introduced by the presence of the airfoil in the subsonic flow are felt throughout the entire flow field, including far upstream. Figure 3.14 is a physical picture consistent with the mathematical behavior of elliptic equations. Inviscid flow is governed by the Euler equations [Eqs. (2.82)–(2.86)] in turn, the methods of Sec. 3.2 and 3.3 show that the *steady* Euler equations are elliptic when the local Mach number is less than unity (see Example 3.1). Hence, the presence of an airfoil in a subsonic, inviscid flow should be felt everywhere throughout the flow, and Fig. 3.14 is an example of such behavior.

INCOMPRESSIBLE INVISCID FLOW. In reality, an incompressible flow is a limiting case of a subsonic flow wherein the Mach number goes to zero. (The Mach number is defined as $M = V/a$, where a is the speed of sound. In a theoretically precise incompressible flow, the compressibility is zero, and hence the speed of sound is infinite. If a is infinitely large, then $M = 0$, even though V is finite.)

Therefore, it is no surprise that an incompressible inviscid flow is governed by elliptic equations; indeed, such flows are the "queen" of elliptic behavior. All the behavior of steady, subsonic, inviscid flows described above carries over to the case of incompressible inviscid flow, and with a *stronger* effect at that.

3.4.4 Some Comments: The Supersonic Blunt Body Problem Revisited

One of the most important problems in modern high-speed aerodynamics is the solution of the inviscid flow over a supersonic or hypersonic blunt body. Some background on this problem was provided in Sec. 1.5, where the difficulty associated with obtaining a solution for the mixed subsonic-supersonic steady flow was underscored. It would be very pertinent to reread the last half of Sec. 1.5 before progressing further. Also, turn back to Fig. 1.30. There, along with the related text, you will find a discussion of the mixed nature of the supersonic blunt body flow field, where the locally subsonic flow is identified as an elliptic region and the locally supersonic flow is identified as a hyperbolic region. The problem in solving this steady, inviscid flow field is due entirely to the extreme difficulty in obtaining a solution technique that is valid in both regions. Now, with our vantage point after our discussions of the mathematical behavior of partial differential equations in the present chapter, we can fully understand and appreciate the source of this difficulty. Because of the totally different mathematical behavior of elliptic and hyperbolic equations, the sudden change in nature of the Euler equations across the sonic line virtually precludes any practical steady flow solution of the blunt body problem involving a uniform treatment of both the subsonic and supersonic regions. However, in Sec. 1.5, a breakthrough in this problem was mentioned, which took place in the middle 1960s. We are now in a position to understand the nature of this breakthrough. Recall from Fig. 3.7 that *unsteady* inviscid flow is governed by hyperbolic equations no matter whether the flow is locally subsonic or supersonic. This provides the following opportunity. Starting with rather arbitrary initial conditions for the flow field in the xy plane in Fig. 1.30, solve the *unsteady*, two-dimensional inviscid flow equations, marching forward in time as sketched in Fig. 3.7. At large times, the solution approaches a steady state, where the time derivatives of the flow variables approach zero. This steady state is the desired result, and what you have when you approach this steady state is a solution for the *entire* flow field *including* both the subsonic and supersonic regions. Moreover, this solution is obtained with the same uniform method throughout the entire flow. The above discussion gives the elementary philosophy of the *time-dependent technique* for the solution of flow problems. Its practical numerical implementation by Moretti and Abbett in 1966 (see Ref. 12) constituted the major scientific breakthrough for the solution of the supersonic blunt body problem as discussed in Sec. 1.5. At first glance, the use of an additional independent variable, namely, time, may seem like extra baggage, but nothing could be further from the truth. Without introducing time as an independent variable, the problem cannot be solved. By introducing time as an independent variable, the governing Euler equations become hyperbolic with respect to time, thus allowing a straightforward marching solution in time, with the proper steady flow results appearing in the limit of large times. For the blunt body problem, this steady state obtained at large times is the desired result—the time-marching procedure is simply a means to that end. *Here is a classic example of the importance of understanding the mathematical behavior of various types of partial differential equations.* In the blunt body problem, an intelligent application of this understanding finally resulted in a practical solution, whereas none existed before.

The time-marching approach described here, where the final steady state at large times is the primary goal, is widely used in modern CFD for a whole host of different applications—it is in no way unique to the blunt body problem. For example, the mathematical behavior of the full Navier-Stokes equations for unsteady flow is not easily placed in a single category. Rather, the Navier-Stokes equations have both parabolic and elliptic behavior. The parabolic behavior is through the time derivatives of velocity and internal energy, much in the same manner as the heat conduction equation, Eq. (3.27), is parabolic via the time derivative of T. The partially elliptic behavior stems from the viscous terms, which provide a mechanism for feeding information upstream in the flow. However, in spite of the mixed nature of the Navier-Stokes equations, a time-marching solution is well-posed; most of the existing numerical solutions to the full, compressible Navier-Stokes equations use the time-marching methodology.

3.5 WELL-POSED PROBLEMS

We end this chapter with a definition—but a definition that we are in a position to appreciate. In the solution of partial differential equations, it is sometimes easy to attempt a solution using incorrect or insufficient boundary and initial conditions. Whether the solution is being attempted analytically or numerically, such an "ill-posed" problem will usually lead to spurious results at best and no solution at worst. The supersonic blunt body problem discussed above is a classic example. When considering the mixed subsonic-supersonic flow from a *steady flow* point of view, any attempt to obtain a uniformly valid solution procedure for both regions was ill-posed.

Therefore, we define a well-posed problem as follows: If the solution to a partial differential equation exists and is unique, and if the solution depends continuously upon the initial and boundary conditions, then the problem is *well-posed*. In CFD, it is important that you establish that your problem is well-posed before you attempt to carry out a numerical solution. When the blunt body problem was set up using the unsteady Euler equations, and a time-marching procedure was employed to go to the steady state at large times starting with essentially arbitrary assumed initial conditions at time $t = 0$, the problem suddenly became well-posed.

3.6 SUMMARY

Examine again the road map sketched in Fig. 3.1, and think about the rather straightforward course we mapped in order to discuss the mathematical behavior of various types of partial differential equations. There are two standard methods for determining the mathematical behavior of a given equation: the Cramer's rule approach and the eigenvalue method, described in Sec. 3.2 and 3.3, respectively. Many equations can be distinctly classified as either hyperbolic, parabolic, or elliptic. Others, such as the unsteady Navier-Stokes equations, have mixed behavior. The major mathematical behavior of hyperbolic and parabolic equations is that they lend themselves quite well to marching solutions, beginning from a known initial data plane or line. In contrast, elliptic equations do not. For elliptic equations, the flow variables at a given point must be solved simultaneously with the flow variables at all other points. Of course, the different mathematical behavior of elliptic compared to parabolic and hyperbolic equations is a direct reflection of the different *physical* behavior of the flows described by these equations.

Finally, we note that the present chapter brings us to an end of Part I of this book. We have examined and derived some of the basic thoughts and equations essential to an understanding and application of CFD. We are now ready to move on to an emphasis on the numerical aspects of CFD—the subject of Part II.

PROBLEMS

3.1. By expanding the determinant in Eq. (3.14), obtain the compatibility equation which holds along the characterisitc lines.

3.2. The discussion in Unsteady Thermal Conduction (a subsection of Sec. 3.4.2) stated, without proof, that the heat conduction equation given by Eqs. (3.26) or (3.27) are parabolic equations. For simplicity, consider the one-dimensional heat conduction equation

$$\frac{\partial T}{\partial t} = \alpha \frac{\partial^2 T}{\partial x^2}$$

Prove that this equation is a parabolic equation.

3.3. Consider Laplace's equation, given by

$$\frac{\partial^2 \phi}{\partial x^2} + \frac{\partial^2 \phi}{\partial y^2} = 0$$

Show that this is an elliptic equation.

3.4. Show that the second-order wave equation

$$\frac{\partial^2 u}{\partial t^2} = c^2 \frac{\partial^2 u}{\partial x^2}$$

is a hyperbolic equation.

3.5. Show that the first-order wave equation

$$\frac{\partial u}{\partial t} + c \frac{\partial u}{\partial x} = 0$$

is a hyperbolic equation.

PART II

BASICS OF THE NUMERICS

In Part I we discussed the philosophy of CFD, derived and examined carefully the governing equations of motion for fluid dynamics, and contrasted the mathematical behavior of various types of partial differential equations. Such background is essential to CFD, even though we have not as yet mentioned the first thing about numerical techniques. However, now is the time! In Part II, we emphasize the basic aspects of the *numerics* of CFD. Here we will present some of the basic aspects of discretization, i.e., how to replace the partial derivatives (or integrals) in the governing equations of motion with discrete *numbers*. Discretization of the partial differential equations is called *finite differences*, and discretization of the integral form of the equations is called *finite volumes*. Furthermore, many applications of numerical solutions involve sophisticated coordinate systems and grid networks laid out in these systems. Sometimes, the use of such coordinate systems requires that the governing equations be suitably transformed into these systems. Thus, another aspect to be discussed—one that is due entirely to the need for dealing with sometimes fancy coordinate systems in CFD—is that dealing with *transformations* and *grid generation*. All the above matters come under the general heading of the basic numerics of CFD—the subject of Part II.

CHAPTER 4

BASIC ASPECTS OF DISCRETIZATION

Numerical precision is the very soul of science.

Sir D'Arcy Wentworth Thompson, Scottish biologist
and natural scientist, 1917

4.1 INTRODUCTION

The word "discretization" requires some explanation. Obviously, it comes from "discrete," defined in *The American Heritage Dictionary of the English Language* as "constituting a separate thing; individual; distinct; consisting of unconnected distinct parts." However, the word "discretization" cannot be found in the same dictionary; it cannot be found in *Webster's New World Dictionary* either. The fact that it does not appear in two of the most popular dictionaries of today implies, at the very least, that it is a rather new and esoteric word. Indeed, it seems to be unique to the literature of numerical analysis, first being introduced in the German literature in 1955 by W. R. Wasow, carried on by Ames in 1965 in his classic book on partial differential equations (Ref. 24), and recently embraced by the CFD community as found in Refs. 13, 14, and 16. In essence, discretization is the process by which a closed-form mathematical expression, such as a function or a differential or integral equation involving functions, all of which are viewed as having an infinite continuum of values throughout some domain, is approximated by analogous (but different) expressions which prescribe values at only a finite number of discrete points or volumes in the domain. This may sound a bit mysterious, so let us elaborate for the sake of clarity. Also, we will single out partial differential equations for purposes of discussion. Therefore, the remainder of this introductory section dwells on the meaning of "discretization."

Analytical solutions of partial differential equations involve closed-form expressions which give the variation of the dependent variables *continuously* throughout the domain. In contrast, numerical solutions can give answers at only *discrete points* in the domain, called *grid points*. For example, consider Fig. 4.1, which shows a section of a discrete grid in the *xy* plane. For convenience, let us assume that the spacing of the grid points in the *x* direction is uniform and given by Δx and that the spacing of the points in the *y* direction is also uniform and given by Δy, as shown in Fig. 4.1. In general, Δx and Δy are different. Indeed, it is not absolutely necessary that Δx or Δy be uniform; we could deal with totally unequal spacing in both directions, where Δx is a different value between each successive pairs of grid points, and similarly for Δy. However, the majority of CFD applications involve numerical solutions on a grid which contains uniform spacing in each direction, because this greatly simplifies the programming of the solution, saves storage space, and usually results in greater accuracy. This uniform spacing does not have to occur in the physical *xy* space; as is frequently done in CFD, the numerical calculations are carried out in a transformed computational space which has uniform spacing in the transformed independent variables but which corresponds to nonuniform spacing in the physical plane. These matters will be discussed in detail in Chap. 5. In any event, in this chapter we will assume uniform spacing in each coordinate direction but not necessarily equal spacing for both directions; i.e., we will assume Δx and Δy to be constants, but Δx does not have to equal Δy. (We should note that recent research in CFD has focused on *unstructured* grids, where the grid points are placed in the flow field in a very irregular fashion; this is in contrast to a *structured* grid which reflects some type of consistent geometrical regularity. Figure 4.1 is an example of a structured grid. Some aspects of unstructured grids will be discussed in Chap. 5.)

Returning to Fig. 4.1, the grid points are identified by an index *i* which runs in the *x* direction and an index *j* which runs in the *y* direction. Hence, if (i, j) is the index for point *P* in Fig. 4.1, then the point immediately to the right of *P* is labeled as $(i + 1, j)$, the point immediately to the left is $(i - 1, j)$, the point directly above is $(i, j + 1)$, and the point directly below is $(i, j - 1)$.

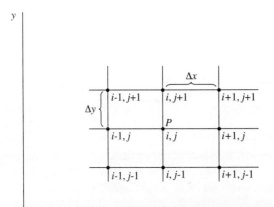

FIG. 4.1 Discrete grid points.

We are now in a position to elaborate on the word "discretization." Imagine that we have a two-dimensional flow field which is governed by the Navier-Stokes equations, or as the case may be, the Euler equations, as derived in Chap. 2. These are partial differential equations. An analytical solution of these equations would provide, in principle, closed-form expressions for u, v, p, ρ, etc., as functions of x and y, which could be used to give values of the flow-field variables at *any* point we wish to choose in the flow, i.e., at any of the infinite number of (x, y) points in the domain. On the other hand, if the partial derivatives in the governing equations are replaced by approximate algebraic difference quotients (to be derived in the next section), where the algebraic difference quotients are expressed strictly in terms of the flow-field variables at two or more of the discrete grid points shown in Fig. 4.1, then the partial differential equations are totally replaced by a system of *algebraic* equations which can be solved for the values of the flow-field variables at the discrete grid points *only*. In this sense, the original partial differential equations have been discretized. Moreover, this method of discretization is called the *method of finite differences*. Finite-difference solutions are widely employed in CFD, and hence much of this chapter will be devoted to matters concerning finite differences.

So this is what discretization means. All methods in CFD utilize some form of discretization. The purpose of this chapter is to derive and discuss the more common forms of discretization in use today for finite-difference applications. This constitutes one of the three main headings in Fig. 4.2, which is the road map for this

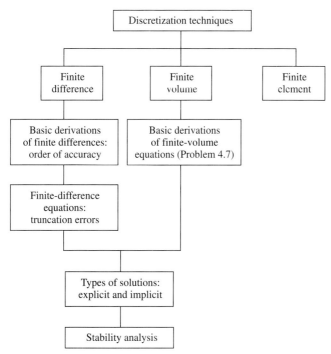

FIG. 4.2
Road map for Chap. 4.

chapter. The second and third main headings are labeled finite volume and finite element, respectively. Both finite-volume and finite-element methods have been in widespread use in computational mechanics for years. However, we will not discuss finite-volume or finite-element methods in this book, mainly because of length constraints. The essential aspects of finite volume discretization are dealt with via Problem 4.7 at the end of this chapter. It is important to note that CFD can be approached using any of the three main types of discretization: finite difference, finite volume, or finite element, as displayed in Fig. 4.2.

Examining the road map in Fig. 4.2 further, the purpose of the present chapter is to construct the basic discretization formulas for finite differences, while at the same time addressing the order of accuracy of these formulas. The road map in Fig. 4.2 gives us our marching orders—let's go to it!

4.2 INTRODUCTION TO FINITE DIFFERENCES

Here, we are interested in replacing a partial derivative with a suitable algebraic difference quotient, i.e., a *finite difference*. Most common finite-difference representations of derivatives are based on Taylor's series expansions. For example, referring to Fig. 4.1, if $u_{i,j}$ denotes the x component of velocity at point (i, j), then the velocity $u_{i+1, j}$ at point $(i + 1, j)$ can be expressed in terms of a Taylor series expanded about point (i, j) as follows:

$$u_{i+1,j} = u_{i,j} + \left(\frac{\partial u}{\partial x}\right)_{i,j} \Delta x + \left(\frac{\partial^2 u}{\partial x^2}\right)_{i,j} \frac{(\Delta x)^2}{2} + \left(\frac{\partial^3 u}{\partial x^3}\right)_{i,j} \frac{(\Delta x)^3}{6} + \cdots \quad (4.1)$$

Equation (4.1) is mathematically an exact expression for $u_{i+1, j}$ if (1) the number of terms is infinite and the series converges and/or (2) $\Delta x \to 0$.

> **Example 4.1.** Since some readers may not be totally comfortable with the concept of a Taylor series, we will review some aspects in this example.
>
> First, consider a continuous function of x, namely, $f(x)$, with all derivatives defined at x. Then, the value of f at a location $x + \Delta x$ can be estimated from a Taylor series expanded about point x, that is,
>
> $$f(x + \Delta x) = f(x) + \frac{\partial f}{\partial x} \Delta x + \frac{\partial^2 f}{\partial x^2} \frac{(\Delta x)^2}{2} + \cdots \frac{\partial^n f}{\partial x^n} \frac{(\Delta x)^n}{n!} + \cdots \quad (E.1)$$
>
> [Note in Eq. (E.1) that we continue to use the partial derivative nomenclature to be consistent with Eq. (4.1), although for a function of one variable, the derivatives in Eq. (E.1) are really ordinary derivatives.] The significance of Eq. (E.1) is diagramed in Fig. E4.1. Assume that we know the value of f at x (point 1 in Fig. E4.1); we want to calculate the value of f at $x + \Delta x$ (point 2 in Fig. E4.1) using Eq. (E.1). Examining the right-hand side of Eq. (E4.1), we see that the first term, $f(x)$, is not a good guess for $f(x + \Delta x)$, unless, of course, the function $f(x)$ is a horizontal line between points 1 and 2. An improved guess is made by approximately accounting for the slope of the curve at point 1, which is the role of the second term, $\partial f/\partial x \, \Delta x$, in Eq. (E.1). To obtain an even better estimate of f at $x + \Delta x$, the third term, $\partial^2 f/\partial x^2 \, (\Delta x)^2/2$, is added, which approximately accounts for the curvature between points 1 and 2. In general, to obtain

INTRODUCTION TO FINITE DIFFERENCES **129**

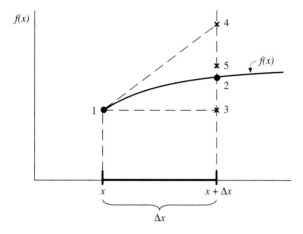

FIG. E4.1
Illustration of behavior of the first three terms in a Taylor series (for Example 4.1).

more accuracy, additional higher-order terms must be included. Indeed, Eq. (E.1) becomes an *exact* representation of $f(x + \Delta x)$ only when an infinite number of terms is carried on the right-hand side. To examine some numbers, let

$$f(x) = \sin 2\pi x$$
$$\text{At } x = 0.2: \quad f(x) = 0.9511 \quad \text{(E.2)}$$

This *exact* value of $f(0.2)$ corresponds to point 1 in Fig. E4.1. Now, let $\Delta x = 0.02$. We wish to evaluate $f(x + \Delta x) = f(0.22)$. From Eq. (E.2), we have the *exact* value:

$$\text{At } x = 0.22: \quad f(x) = 0.9823$$

This corresponds to point 2 in Fig. E4.1. Now, let us *estimate* $f(0.22)$ using Eq. (E.1). Using just the first term on the right-hand side of Eq. (E.1), we have

$$f(0.22) \approx f(0.2) = 0.9511$$

This corresponds to point 3 in Fig. E4.1. The percentage error in this estimate is $[(0.9823 - 0.9511)/0.9823] \times 100 = 3.176$ percent. Using two terms in Eq. (E.1),

$$f(x + \Delta x) \approx f(x) + \frac{\partial f}{\partial x} \Delta x$$
$$f(0.22) \approx f(0.2) + 2\pi \cos [2\pi(0.2)](0.02)$$
$$\approx 0.9511 + 0.388 = 0.9899$$

This corresponds to point 4 in Fig. E4.1. The percentage error in this estimate is $[0.9899 - 0.9823)/0.9823] \times 100 = 0.775$ percent. This is much closer than the previous estimate. Finally, to obtain yet an even better estimate, let us use three terms in Eq. (E.1).

$$f(x + \Delta x) \approx f(x) + \frac{\partial f}{\partial x} \Delta x + \frac{\partial^2 f}{\partial x^2} \frac{(\Delta x)^2}{2}$$

$$f(0.22) \approx f(0.2) + 2\pi \cos[2\pi(0.2)](0.02) - 4\pi^2 \sin[2\pi(0.2)] \frac{(0.02)^2}{2}$$

$$\approx 0.9511 + 0.0388 - 0.0075$$

$$\approx 0.9824$$

This corresponds to point 5 in Fig. E4.1. The percentage error in this estimate is $[(0.9824 - 0.9823)/0.9823] \times 100 = 0.01$ percent. *This is a very close estimate of $f(0.22)$ using just three terms in the Taylor series given by Eq. (E.1).*

Let us now return to Eq. (4.1) and pursue our discussion of finite-difference representations of derivatives. Solving Eq. (4.1) for $(\partial u/\partial x)_{i,j}$, we obtain

$$\left(\frac{\partial u}{\partial x}\right)_{i,j} = \underbrace{\frac{u_{i+1,j} - u_{i,j}}{\Delta x}}_{\text{Finite-difference representation}} \underbrace{- \left(\frac{\partial^2 u}{\partial x^2}\right)_{i,j} \frac{\Delta x}{2} - \left(\frac{\partial^3 u}{\partial x^3}\right)_{i,j} \frac{(\Delta x)^2}{6} + \cdots}_{\text{Truncation error}} \quad (4.2)$$

In Eq. (4.2), the actual partial derivative evaluated at point (i, j) is given on the left side. The first term on the right side, namely, $(u_{i+1,j} - u_{i,j})/\Delta x$, is a finite-difference representation of the partial derivative. The remaining terms on the right side constitute the *truncation error*. That is, if we wish to *approximate* the partial derivative with the above algebraic finite-difference quotient,

$$\left(\frac{\partial u}{\partial x}\right)_{i,j} \approx \frac{u_{i+1,j} - u_{i,j}}{\Delta x} \quad (4.3)$$

then the truncation error in Eq. (4.2) tells us what is being neglected in this approximation. In Eq. (4.2), the lowest-order term in the truncation error involves Δx to the first power; hence, the finite-difference expression in Eq. (4.3) is called *first-order-accurate*. We can more formally write Eq. (4.2) as

$$\left(\frac{\partial u}{\partial x}\right)_{i,j} = \frac{u_{i+1,j} - u_{i,j}}{\Delta x} + O(\Delta x) \quad (4.4)$$

In Eq. (4.4), the symbol $O(\Delta x)$ is a formal mathematical notation which represents "terms of order Δx." Equation (4.4) is a more precise notation than Eq. (4.3), which involves the "approximately equal" notation; in Eq. (4.4) the order of magnitude of the truncation error is shown explicitly by the notation. Also referring to Fig. 4.1, note that the finite-difference expression in Eq. (4.4) uses information to the *right* of grid point (i, j); that is, it uses $u_{i+1,j}$ as well as $u_{i,j}$. No information to the left of (i, j) is used. As a result, the finite difference in Eq. (4.4) is called a *forward difference*. For this reason, we now identify the first-order-accurate difference representation for the derivative $(\partial u/\partial x)_{i,j}$ expressed by Eq. (4.4) as a *first-order forward difference*, repeated below.

$$\left(\frac{\partial u}{\partial x}\right)_{i,j} = \frac{u_{i+1,j} - u_{i,j}}{\Delta x} + O(\Delta x) \tag{4.4}$$

Let us now write a Taylor series expansion for $u_{i-1,j}$, expanded about $u_{i,j}$.

$$u_{i-1,j} = u_{i,j} + \left(\frac{\partial u}{\partial x}\right)_{i,j}(-\Delta x) + \left(\frac{\partial^2 u}{\partial x^2}\right)_{i,j}\frac{(-\Delta x)^2}{2}$$
$$+ \left(\frac{\partial^3 u}{\partial x^3}\right)_{i,j}\frac{(-\Delta x)^3}{6} + \cdots$$

or

$$u_{i-1,j} = u_{i,j} - \left(\frac{\partial u}{\partial x}\right)_{i,j}\Delta x + \left(\frac{\partial^2 u}{\partial x^2}\right)_{i,j}\frac{(\Delta x)^2}{2}$$
$$- \left(\frac{\partial^3 u}{\partial x^3}\right)_{i,j}\frac{(\Delta x)^3}{6} + \cdots \tag{4.5}$$

Solving for $(\partial u/\partial x)_{i,j}$, we obtain

$$\left(\frac{\partial u}{\partial x}\right)_{i,j} = \frac{u_{i,j} - u_{i-1,j}}{\Delta x} + O(\Delta x) \tag{4.6}$$

The information used in forming the finite-difference quotient in Eq. (4.6) comes from the *left* of grid point (i, j); that is, it uses $u_{i-1,j}$ as well as $u_{i,j}$. No information to the right of (i, j) is used. As a result, the finite difference in Eq. (4.6) is called a *rearward* (or *backward*) *difference*. Moreover, the lowest-order term in the truncation error involves Δx to the first power. As a result, the finite difference in Eq. (4.6) is called a *first-order rearward difference*.

In most applications in CFD, first-order accuracy is not sufficient. To construct a finite-difference quotient of second-order accuracy, simply subtract Eq. (4.5) from Eq. (4.1):

$$u_{i+1,j} - u_{i-1,j} = 2\left(\frac{\partial u}{\partial x}\right)_{i,j}\Delta x + 2\left(\frac{\partial^3 u}{\partial x^3}\right)_{i,j}\frac{(\Delta x)^3}{6} + \cdots \tag{4.7}$$

Equation (4.7) can be written as

$$\left(\frac{\partial u}{\partial x}\right)_{i,j} = \frac{u_{i+1,j} - u_{i-1,j}}{2\Delta x} + O(\Delta x)^2 \tag{4.8}$$

The information used in forming the finite-difference quotient in Eq. (4.8) comes from *both* sides of the grid point located at (i, j); that is, it uses $u_{i+1,j}$ as well as $u_{i-1,j}$. Grid point (i, j) falls between the two adjacent grid points. Moreover, in the truncation error in Eq. (4.7), the lowest-order terms involves $(\Delta x)^2$, which is second-order accuracy. Hence, the finite-difference quotient in Eq. (4.8) is called a *second-order central difference*.

Difference expressions for the y derivatives are obtained in exactly the same fashion. (See Prob. 4.1 and 4.2.) The results are directly analogous to the previous equations for the x derivatives. They are:

$$\left(\frac{\partial u}{\partial y}\right)_{i,j} = \begin{cases} \dfrac{u_{i,j+1} - u_{i,j}}{\Delta y} + O(\Delta y) & \text{Forward difference} \quad (4.9) \\[2mm] \dfrac{u_{i,j} - u_{i,j-1}}{\Delta y} + O(\Delta y) & \text{Rearward difference} \quad (4.10) \\[2mm] \dfrac{u_{i,j+1} - u_{i,j-1}}{2\Delta y} + O(\Delta y)^2 & \text{Central difference} \quad (4.11) \end{cases}$$

Equations (4.4), (4.6), and (4.8) to (4.11) are examples of finite-difference quotients for *first* partial derivatives. Is this all that we need for CFD? Let us return to Chap. 2 for a moment and take a look at the governing equations of motion. If we are dealing with inviscid flows only, the governing equations are the Euler equations, summarized in Sec. 2.8.2 and expressed by Eqs. (2.82) to (2.86). Note that the highest-order derivatives which appear in the Euler equations are first partial derivatives. Hence, finite differences for the first derivatives, such as those expressed by Eqs. (4.4), (4.6), and (4.8), are all that we need for the numerical solution of inviscid flows. On the other hand, if we are dealing with viscous flows, the governing equations are the Navier-Stokes equations, summarized in Sec. 2.8.1 and expressed by Eqs. (2.29), (2.50), (2.56), and (2.66). Note that the highest-order derivatives which appear in the Navier-Stokes equations are *second* partial derivatives, as reflected in the viscous terms such as $\partial \tau_{xy}/\partial x = \partial/\partial x \, [\mu(\partial v/\partial x + \partial u/\partial y)]$ which appears in Eq. (2.50b), and $\partial/\partial x \, (k \, \partial T/\partial x)$ which appears in Eq. (2.66). When expanded, these terms involve such second partial derivatives as $\partial^2 u/\partial x \, \partial y$ and $\partial^2 T/\partial x^2$, just for example. Consequently, there is a need for discretizing second-order derivatives for CFD. We can obtain such finite-difference expressions by continuing with a Taylor series analysis, as follows.

Summing the Taylor series expansions given by Eqs. (4.1) and (4.5), we have

$$u_{i+1,j} + u_{i-1,j} = 2u_{i,j} + \left(\frac{\partial^2 u}{\partial x^2}\right)_{i,j} (\Delta x)^2 + \left(\frac{\partial^4 u}{\partial x^4}\right)_{i,j} \frac{(\Delta x)^4}{12} + \cdots$$

Solving for $(\partial^2 u/\partial x^2)_{i,j}$,

$$\boxed{\left(\frac{\partial^2 u}{\partial x^2}\right)_{i,j} = \frac{u_{i+1,j} - 2u_{i,j} + u_{i-1,j}}{(\Delta x)^2} + O(\Delta x)^2} \qquad (4.12)$$

In Eq. (4.12), the first term on the right-hand side is a central finite difference for the second derivative with respect to x evaluated at grid point (i, j); from the remaining order-of-magnitude term, we see that this central difference is of second-order accuracy. An analogous expression can easily be obtained for the second derivative with respect to y, with the result that

$$\left(\frac{\partial^2 u}{\partial y^2}\right)_{i,j} = \frac{u_{i,j+1} - 2u_{i,j} + u_{i,j-1}}{(\Delta y)^2} + O(\Delta y)^2 \qquad (4.13)$$

Equations (4.12) and (4.13) are examples of *second-order central second differences*.

For the case of mixed derivatives, such as $\partial^2 u/\partial x\, \partial y$, appropriate finite-difference quotients can be found as follows. Differentiating Eq. (4.1) with respect to y, we have

$$\left(\frac{\partial u}{\partial y}\right)_{i+1,j} = \left(\frac{\partial u}{\partial y}\right)_{i,j} + \left(\frac{\partial^2 u}{\partial x\, \partial y}\right)_{i,j} \Delta x + \left(\frac{\partial^3 u}{\partial x^2\, \partial y}\right)_{i,j} \frac{(\Delta x)^2}{2} + \left(\frac{\partial^4 u}{\partial x^3\, \partial y}\right)_{i,j} \frac{(\Delta x)^3}{6} + \cdots \qquad (4.14)$$

Differentiating Eq. (4.5) with respect to y, we have

$$\left(\frac{\partial u}{\partial y}\right)_{i-1,j} = \left(\frac{\partial u}{\partial y}\right)_{i,j} - \left(\frac{\partial^2 u}{\partial x\, \partial y}\right)_{i,j} \Delta x + \left(\frac{\partial^3 u}{\partial x^2\, \partial y}\right)_{i,j} \frac{(\Delta x)^2}{2}$$
$$+ \left(\frac{\partial^4 u}{\partial x^3\, \partial y}\right)_{i,j} \frac{(\Delta x)^3}{6} + \cdots \qquad (4.15)$$

Subtracting Eq. (4.15) from Eq. (4.14) yields

$$\left(\frac{\partial u}{\partial y}\right)_{i+1,j} - \left(\frac{\partial u}{\partial y}\right)_{i-1,j} = 2\left(\frac{\partial^2 u}{\partial x\, \partial y}\right)_{i,j} \Delta x + \left(\frac{\partial^4 u}{\partial x^3\, \partial y}\right)_{i,j} \frac{(\Delta x)^3}{6} + \cdots$$

Solving for $(\partial^2 u/\partial x\, \partial y)_{i,j}$, which is the mixed derivative for which we are seeking a finite-difference expression, the above equation yields

$$\left(\frac{\partial^2 u}{\partial x\, \partial y}\right)_{i,j} = \frac{(\partial u/\partial y)_{i+1,j} - (\partial u/\partial y)_{i-1,j}}{2\Delta x} - \left(\frac{\partial^4 u}{\partial x^3\, \partial y}\right)_{i,j} \frac{(\Delta x)^2}{12} + \cdots \qquad (4.16)$$

In Eq. (4.16), the first term on the right-hand-side involves $\partial u/\partial y$, first evaluated at grid point $(i+1, j)$ and then at grid point $(i-1, j)$. Returning to the grid sketched in Fig. 4.1, we can see that $\partial u/\partial y$ at each of these two grid points can be replaced with a second-order central difference patterned after that given by Eq. (4.11) but using appropriate grid points first centered on $(i+1, j)$ and then on $(i-1, j)$. To be more specific, in Eq. (4.16) first replace $(\partial u/\partial y)_{i+1,j}$ with

$$\left(\frac{\partial u}{\partial y}\right)_{i+1,j} = \frac{u_{i+1,j+1} - u_{i+1,j-1}}{2\Delta y} + O(\Delta y)^2$$

and then replace $(\partial u/\partial y)_{i-1,j}$ with the analogous difference,

$$\left(\frac{\partial u}{\partial y}\right)_{i-1,j} = \frac{u_{i-1,j+1} - u_{i-1,j-1}}{2\Delta y} + O(\Delta y)^2$$

In this fashion, Eq. (4.16) becomes

$$\left(\frac{\partial^2 u}{\partial x\, \partial y}\right)_{i,j} = \frac{u_{i+1,j+1} - u_{i+1,j-1} - u_{i-1,j+1} + u_{i-1,j-1}}{4\Delta x\, \Delta y} + O[(\Delta x)^2, (\Delta y)^2] \quad (4.17)$$

The truncation error in Eq. (4.17) comes from Eq. (4.16), where the lowest-order neglected term is of $O(\Delta x)^2$, and from the fact that the central difference in Eq. (4.11) is of $O(\Delta y)^2$. Hence, the truncation error in Eq. (4.17) must be $O[(\Delta x)^2, (\Delta y)^2]$. Equation (4.17) gives a *second-order central difference for the mixed derivative*, $(\partial^2 u/\partial x\, \partial y)_{i,j}$.

It is important to note that when the governing flow equations are used in the form of Eq. (2.93), only first derivatives are needed, even for viscous flows. The dependent variables being differentiated are U, F, G, and H in Eq. (2.93), and only as first derivatives. Hence, these derivatives can be replaced with the appropriate finite-difference expressions for first derivatives, such as Eqs. (4.4), (4.6), and (4.8) to (4.11). In turn, some elements of F, G, and H involve viscous stresses, such as τ_{xx}, τ_{xy}, and thermal conduction terms. These terms depend on velocity or temperature gradients, which are also first derivatives. Hence, the finite-difference forms for first derivatives can also be used for the viscous terms *inside* F, G, and H. In this fashion, the need to use a finite-difference expression for second derivatives, such as Eqs. (4.12), (4.13), and (4.17), is circumvented.

To this stage, we have derived a number of different forms of finite-difference expressions for various partial derivatives. To help reinforce these finite differences in your mind, the graphical concept of *finite-difference modules* is useful. All the above difference expressions can be nicely displayed in the context of the finite-difference modules shown in Fig. 4.3. This figure is a concise review of the finite-difference forms we have discussed, as well as illustrating on a grid the specific grid points that participate in the formation of each finite difference. These participating grid points are shown by large filled circles connected by bold lines; such a schematic is called a finite-difference module. The plus and minus signs adjacent to the participating grid points remind us of whether the information at each of these points is added or subtracted to form the appropriate finite differences; similarly, a (-2) beside a grid point connotes that twice the variable at that grid point is subtracted in the formation of the finite-difference quotient. Compare the $(+)$, $(-)$, and (-2) in the finite-difference modules with the corresponding formula for the finite difference which appears to the left of each module in Fig. 4.3.

The finite-difference expressions derived in this section and displayed in Fig. 4.3 represent just the "tip of the iceberg." Many other difference approximations can be obtained for the same derivatives we treated above. In particular, more

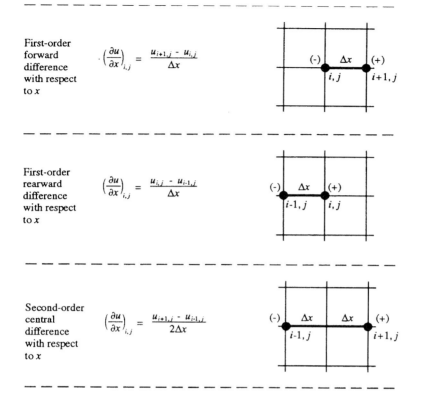

FIG. 4.3
Finite-difference expressions with their appropriate finite-difference modules.

accurate finite-difference quotients can be derived, exhibiting third-order accuracy, fourth-order accuracy, and more. Such higher-order-accurate difference quotients generally involve information at more grid points than those we have derived. For example, a fourth-order-accurate central finite-difference for $\partial^2 u/\partial x^2$ is

$$\left(\frac{\partial^2 u}{\partial x^2}\right)_{i,j} = \frac{-u_{i+2,j} + 16u_{i+1,j} - 30u_{i,j} + 16u_{i-1,j} - u_{i-2,j}}{12(\Delta x)^2} + O(\Delta x)^4 \quad (4.18)$$

Note that information at five grid points is required to form this fourth-order finite difference; compare this with Eq. (4.12), where $(\partial^2 u/\partial x^2)_{i,j}$ is represented in terms of information at only three grid points, albeit with only second-order accuracy. Equation (4.18) can be derived by repeated application of Taylor's series expanded about grid points $(i + 1, j)$, (i, j), and $(i - 1, j)$; the details are considered in Prob. 4.5. We are simply emphasizing that an almost unlimited number of finite-difference expressions can be derived with ever-increasing accuracy. In the past, second-order accuracy has been considered sufficient for most CFD applications, so the types of difference quotients we have derived in this section have been, by far, the most

136 BASIC ASPECTS OF DISCRETIZATION

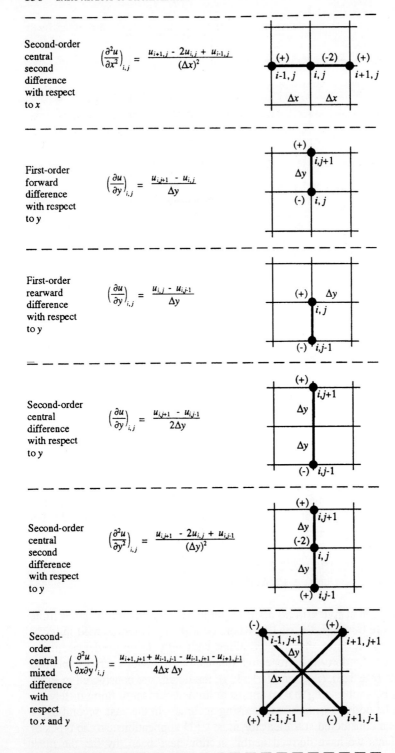

Second-order central second difference with respect to x	$\left(\dfrac{\partial^2 u}{\partial x^2}\right)_{i,j} = \dfrac{u_{i+1,j} - 2u_{i,j} + u_{i-1,j}}{(\Delta x)^2}$
First-order forward difference with respect to y	$\left(\dfrac{\partial u}{\partial y}\right)_{i,j} = \dfrac{u_{i,j+1} - u_{i,j}}{\Delta y}$
First-order rearward difference with respect to y	$\left(\dfrac{\partial u}{\partial y}\right)_{i,j} = \dfrac{u_{i,j} - u_{i,j-1}}{\Delta y}$
Second-order central difference with respect to y	$\left(\dfrac{\partial u}{\partial y}\right)_{i,j} = \dfrac{u_{i,j+1} - u_{i,j-1}}{2\Delta y}$
Second-order central second difference with respect to y	$\left(\dfrac{\partial^2 u}{\partial y^2}\right)_{i,j} = \dfrac{u_{i,j+1} - 2u_{i,j} + u_{i,j-1}}{(\Delta y)^2}$
Second-order central mixed difference with respect to x and y	$\left(\dfrac{\partial^2 u}{\partial x \partial y}\right)_{i,j} = \dfrac{u_{i+1,j+1} + u_{i-1,j-1} - u_{i-1,j+1} - u_{i+1,j-1}}{4\Delta x \, \Delta y}$

FIG. 4.3 (*continued*)

commonly used forms. The pros and cons of higher-order accuracy are as follows:

1. Higher-order-accurate difference quotients, such as displayed in Eq. (4.18), by requiring more grid points, result in more computer time required for each time wise or spatial step—a con.
2. On the other hand, a higher-order difference scheme may require a smaller number of total grid points in a flow solution to obtain comparable overall accuracy—a pro.
3. Higher-order difference schemes may result in a "higher- quality" solution, such as captured shock waves that are sharper and more distinct—also a pro. In fact, this aspect is a matter of current research in CFD.

For these reasons, the matter of what degree of accuracy is desirable for various CFD solutions is not clear-cut. Because second-order accuracy has been previously accepted in the vast majority of CFD applications, and because the purpose of this book is to present a basic introduction to the elements of CFD without undue complication, we will consider that second-order accuracy will be sufficient for our purposes in this and subsequent chapters. For a detailed tabulation of many forms of difference quotients, see pp. 44 and 45 of Ref. 13.

We have one more item of business before finishing this section on finite-difference quotients. We pose the following question: What happens at a boundary? What type of differencing is possible when we have only one direction to go, namely, the direction away from the boundary? For example, consider Fig. 4.4, which illustrates a portion of a boundary to a flow field, with the y axis perpendicular to the boundary. Let grid point 1 be on the boundary, with points 2 and 3 a distance Δy and $2\Delta y$ above the boundary, respectively. We wish to construct a finite-difference approximation for $\partial u/\partial y$ at the boundary. It is easy to construct a forward difference as

$$\left(\frac{\partial u}{\partial y}\right)_1 = \frac{u_2 - u_1}{\Delta y} + O(\Delta y) \tag{4.19}$$

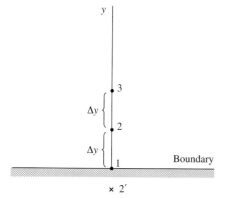

FIG. 4.4
Grid points at a boundary

which is of first-order accuracy. However, how do we obtain a result which is of second-order accuracy? Our central difference in Eq. (4.11) fails us because it requires another point beneath the boundary, such as illustrated as point 2' in Fig. 4.4. Point 2' is outside the domain of computation, and we generally have no information about u at this point. In the early days of CFD, many solutions attempted to sidestep this problem by assuming that $u_{2'} = u_2$. This is called the *reflection boundary condition*. In most cases it does not make physical sense and is just as inaccurate, if not more so, than the forward difference given by Eq. (4.19).

So we ask the question again, how do we find a second-order-accurate finite-difference at the boundary? The answer is straightforward, as we will describe here. Moreover, we will seize this occasion to illustrate an alternative approach to the construction of finite-difference quotients—alternative to the Taylor's series analyses presented earlier. We will use a *polynomial approach*, as follows. Assume at the boundary shown in Fig. 4.4 that u can be expressed by the polynomial

$$u = a + by + cy^2 \tag{4.20}$$

Applied successively to the grid points in Fig. 4.4, Eq. (4.20) yields at grid point 1 where $y = 0$,

$$u_1 = a \tag{4.21}$$

and at grid point 2 where $y = \Delta y$,

$$u_2 = a + b\,\Delta y + c(\Delta y)^2 \tag{4.22}$$

and at grid point 3 where $y = 2\Delta y$,

$$u_3 = a + b(2\Delta y) + c(2\Delta y)^2 \tag{4.23}$$

Solving Eqs. (4.21) to (4.23) for b, we obtain

$$b = \frac{-3u_1 + 4u_2 - u_3}{2\Delta y} \tag{4.24}$$

Returning to Eq. (4.20), and differentiating with respect to y,

$$\frac{\partial u}{\partial y} = b + 2cy \tag{4.25}$$

Equation (4.25), evaluated at the boundary where $y = 0$, yields

$$\left(\frac{\partial u}{\partial y}\right)_1 = b \tag{4.26}$$

Combining Eqs. (4.24) and (4.26), we obtain

$$\left(\frac{\partial u}{\partial y}\right)_1 = \frac{-3u_1 + 4u_2 - u_3}{2\Delta y} \tag{4.27}$$

Equation (4.27) is a one-sided finite-difference expression for the derivative at the boundary—called *one-sided* because it uses information only on one side of the grid point at the boundary, namely, information only *above* grid point 1 in Fig. 4.4. Also, Eq. (4.27) was derived using a polynomial expression, namely, Eq. (4.20), rather than a Taylor series representation. This illustrates an alternative approach to the formulation of finite-difference quotients; indeed, all our previous results as summarized in Fig. 4.3 could have been obtained using this polynomial approach. It remains to show the order of accuracy of Eq. (4.27). Here, we have to appeal to a Taylor series again. Consider a Taylor series expansion about the point 1.

$$u(y) = u_1 + \left(\frac{\partial u}{\partial y}\right)_1 y + \left(\frac{\partial^2 u}{\partial y^2}\right)_1 \frac{y^2}{2} + \left(\frac{\partial^3 u}{\partial y^3}\right)_1 \frac{y^3}{6} + \cdots \quad (4.28)$$

Compare Eqs. (4.28) and (4.20). Our assumed polynomial expression in Eq. (4.20) is the same as using the first three terms in the Taylor series. Hence, Eq. (4.20) is of $O(\Delta y)^3$. Now examine the numerator of Eq. (4.27); here u_1, u_2, and u_3 can all be expressed in terms of the polynomial given by Eq. (4.20). Since Eq. (4.20) is of $O(\Delta y)^3$, then the numerator of Eq. (4.27) is also of $O(\Delta y)^3$. However, in forming the derivative in Eq. (4.27), we divided by Δy, which then makes Eq. (4.27) of $O(\Delta y)^2$. Thus, we can write from Eq. (4.27)

$$\left(\frac{\partial u}{\partial y}\right)_1 = \frac{-3u_1 + 4u_2 - u_3}{2\Delta y} + O(\Delta y)^2 \quad (4.29)$$

This is our desired second-order-accurate difference quotient at the boundary.

Both Eqs. (4.19) and (4.29) are called *one-sided differences*, because they express a derivative at a point in terms of dependent variables on *only one side* of that point. Moreover, these equations are general; i.e., they are not in any way limited to application just at a boundary; they can be applied at internal grid points as well. It just so happens that we have taken advantage of our discussion of finite-difference quotients at a boundary to derive such one-sided differences. Of course, as we have seen here, one-sided differences are essentially mandatory for a representation of a derivative at a boundary, but such one-sided differences simply offer another option when applied *internally* within the domain of the overall calculations. Furthermore, Eq. (4.29) displays a one-sided finite difference of second-order accuracy; many other one-sided difference formulas for a derivative at a point can be derived with higher orders of accuracy using additional grid points to one side of that point. In some CFD applications, it is not unusual to see four- and five-point one-sided differences applied at a boundary. This is especially true for viscous flow calculations. In such calculations, the shear stress and heat transfer at the wall, due to a flow over that wall, are of particular importance. The shear stress at the wall is given by (see, for example, chap. 12 of Ref. 8)

$$\tau_w = \mu \left(\frac{\partial u}{\partial y}\right)_w \quad (4.30)$$

and the heat transfer at the wall is given by

$$q_w = k\left(\frac{\partial T}{\partial y}\right)_w \qquad (4.31)$$

In finite-difference solutions of a viscous flow (solutions of the Navier-Stokes equations, parabolized Navier-Stokes equations, the boundary-layer equations, etc.), the flow-field values of u and T are calculated at all the grid points, internal as well as boundary points. Then, after these flow-field values are obtained (by whatever algorithm is chosen, such as one of the appropriate techniques discussed in Part III of this book), the shear stress and heat transfer are calculated after the fact from Eqs. (4.30) and (4.31). Clearly, the more accurate the one-sided finite difference used to represent $(\partial u/\partial y)_w$ and $(\partial T/\partial y)_w$ in Eqs. (4.30) and (4.31), respectively, the more accurate will be the calculated results for τ_w and q_w.

Example 4.2. Consider the viscous flow of air over a flat plate. At a given station in the flow direction, the variation of the flow velocity, u, in the direction perpendicular to the plate (the y direction) is given by the expression

$$u = 1582(1 - e^{-y/L}) \qquad (E.3)$$

where L = characteristic length = 1 in. The units of u are feet per second. The viscosity coefficient $\mu = 3.7373 \times 10^{-7}$ slug/(ft · s). We use Eq. (E.3) to provide the values of u at discrete grid points equally spaced in the y direction, with $\Delta y = 0.1$ in. Specifically, we obtain from Eq. (E.3):

y, in	u, ft/s
0	0
0.10	150.54
0.20	286.77
0.30	410.03

Imagine that the values of u listed above are discrete values at the discrete grid points located at y = 0, 0.1, 0.2, and 0.3, in the same nature as would be obtained from a numerical finite-difference solution of the flow field. Indeed, assume that these discrete values of u are all that we know; we have used Eq. (E.3) just to specify these discrete values of u. Using these discrete values, calculate the shear stress at the wall τ_w three different ways, namely:

(a) Using a first-order one-sided difference

(b) Using the second-order one-sided difference given by Eq. (4.29)

(c) Using the third-order one-sided difference derived in Prob. 4.6.

Finally, compare these calculated finite-difference results with the *exact* value of τ_w as specified from Eq. (E.3).

Solution

(a) First-order difference:

$$\left(\frac{\partial u}{\partial y}\right)_{j=1} = \frac{u_{j=2} - u_{j=1}}{\Delta y}$$

$$= \frac{150.54 - 0}{0.1} = 1505.4 \text{ ft}/(\text{s} \cdot \text{in})$$

$$\tau_w = \mu \left(\frac{\partial u}{\partial y}\right)_{j=1} = (3.7373 \times 10^{-7})(1505.4)(12)$$

$$= \boxed{6.7514 \times 10^{-3} \text{ lb/ft}^2}$$

(Note that the factor 12 has been used above to convert the velocity gradient to units of ft/(s · ft), rather than per inch; in the calculation of τ_w, we must use consistent units, in this case the English engineering system of units.)

(b) Second-order difference (from Eq. 4.29):

$$\left(\frac{\partial u}{\partial y}\right)_{j=1} = \frac{-3u_{j=1} + 4u_{j=2} - u_{j=3}}{2\Delta y}$$

$$= \frac{-3(0) + 4(150.54) - 286.77}{2(0.1)}$$

$$= 1577.0 \text{ ft}/(\text{s} \cdot \text{in})$$

$$\tau_w = \mu \left(\frac{\partial u}{\partial y}\right)_{j=1} = (3.7373 \times 10^{-7})(1577.0)(12)$$

$$= \boxed{7.072 \times 10^{-3} \text{ lb/ft}^2}$$

(c) Third-order difference (from Prob. 4.6):

$$\left(\frac{\partial u}{\partial y}\right)_{j=1} = \frac{-11u_{j=1} + 18u_{j=2} - 9u_{j=3} + 2u_{j=4}}{6\Delta y}$$

$$= \frac{-11(0) + 18(150.54) - 9(286.77) + 2(410.03)}{6(0.1)}$$

$$= 1581.4 \text{ ft}/(\text{s} \cdot \text{in})$$

$$\tau_w = \mu \left(\frac{\partial u}{\partial y}\right)_{j=1} = (3.7373 \times 10^{-7})(1581.4)(12)$$

$$= \boxed{7.092 \times 10^{-3} \text{ lb/ft}^2}$$

(d) Exact value [from Eq. (E.3)]:

$$\frac{\partial u}{\partial y} = \frac{1582}{L} e^{-y/L} \tag{E.4}$$

Recalling that $L = 1$ in, then at the wall ($y = 0$), Eq. (E.4) yields

$$\left(\frac{\partial u}{\partial y}\right)_{y=0} = 1582 \text{ ft/(s} \cdot \text{in)}$$

$$\tau_w = \mu\left(\frac{\partial u}{\partial y}\right)_{y=0} = (3.7373 \times 10^{-7})(1582)(12)$$

$$= \boxed{7.095 \times 10^{-3} \text{ lb/ft}^2}$$

Important: Surveying the above results, we see that the use of progressively higher-order-accurate difference expressions gives progressively more accurate values of τ_w. Specifically, compared to the exact value of 7.095×10^{-3} lb/ft^2, we have:

Order of accuracy	τ_w lb/ft^2	% error
First order (part *a*)	6.7514×10^{-3}	4.8
Second order (part *b*)	7.072×10^{-3}	0.3
Third order (part *c*)	7.092×10^{-3}	0.04
Exact [Eq. (E.4)]	7.095×10^{-3}	0

Note from the above tabulation that the use of a second-order-accurate difference formula gives a much better result for τ_w than a simple first-order difference and that the use of a third-order difference formula further improves the accuracy, but it is less dramatic. Here is an indication that, for most finite-difference solutions, at least second-order accuracy is needed, and it turns out to be sufficient.

4.3 DIFFERENCE EQUATIONS

In Sec. 4.2, we discussed the representation of a partial derivative by means of an algebraic finite-difference quotient. Most partial differential *equations* involve a number of partial derivative terms. When all the partial derivatives in a given partial differential equation are replaced by finite-difference quotients, the resulting *algebraic* equation is called a *difference equation*, which is an algebraic representation of the partial differential equation. The essence of finite-difference solutions in CFD is to use the difference quotients derived in Sec. 4.2 (or others that are similar) to replace the partial derivatives in the governing flow equations, resulting in a system of algebraic difference equations for the dependent variables at each grid point. In the present section, we examine some of the basic aspects of a difference equation.

For simplicity, we choose to examine a partial differential equation which is less elaborate than the governing flow equations. For example, let us consider Eq. (3.28), which is the unsteady, one-dimensional heat conduction equation with constant thermal diffusivity, repeated below.

$$\frac{\partial T}{\partial t} = \alpha \frac{\partial^2 T}{\partial x^2} \tag{3.28}$$

We choose this simple equation for convenience; at this stage in our discussions there is no advantage to be obtained by dealing with the much more complex flow equations. The basic aspects of finite-difference equations to be examined in this section can just as well be developed using Eq. (3.28). As stated in the Unsteady Thermal Conduction subsection of Sec. 3.4.2, this is a parabolic partial differential equation. (See Prob. 3.2.) As such, it lends itself to a marching solution with respect to time t as discussed in Chap. 3.

Let us replace the partial derivatives in Eq. (3.28) with finite-difference quotients. Equation (3.28) has two independent variables, x and t. We consider the grid sketched in Fig. 4.5. Here, i is the running index in the x direction and n is the running index in the t direction. When one of the independent variables in a partial differential equation is a *marching* variable, such as t in Eq. (3.28), it is conventional in CFD to denote the running index for this marching variable by n and to display this index as a *superscript* in the finite-difference quotient. For example, let us replace the time derivative in Eq. (3.28) with a forward difference patterned after Eq. (4.4), i.e.,

$$\left(\frac{\partial T}{\partial t}\right)_i^n = \frac{T_i^{n+1} - T_i^n}{\Delta t} - \left(\frac{\partial^2 T}{\partial t^2}\right)_i^n \frac{\Delta t}{2} + \cdots \qquad (4.32)$$

where the truncation error is the same as that displayed in Eq. (4.2). Also, let us replace the x derivative in Eq. (3.28) with a central difference patterned after Eq. (4.12), i.e.,

$$\left(\frac{\partial^2 T}{\partial x^2}\right)_i^n = \frac{T_{i+1}^n - 2T_i^n + T_{i-1}^n}{(\Delta x)^2} - \left(\frac{\partial^4 T}{\partial x^4}\right)_i^n \frac{(\Delta x)^2}{12} + \cdots \qquad (4.33)$$

where the truncation error is the same as that displayed immediately above Eq. (4.12). Let us write Eq. (3.28) as

$$\frac{\partial T}{\partial t} - \alpha \frac{\partial^2 T}{\partial x^2} = 0 \qquad (4.34)$$

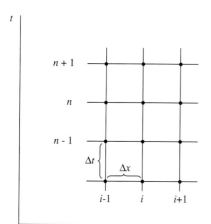

FIG. 4.5
Grid for the differencing of Eq. (3.28).

144 BASIC ASPECTS OF DISCRETIZATION

Inserting Eqs. (4.32) and (4.33) into (4.34), we have

$$\underbrace{\frac{\partial T}{\partial t} - \alpha \frac{\partial^2 T}{\partial x^2} = 0}_{\text{Partial differential equation}} = \underbrace{\frac{T_i^{n+1} - T_i^n}{\Delta t} - \frac{\alpha(T_{i+1}^n - 2T_i^n + T_{i-1}^n)}{(\Delta x)^2}}_{\text{Difference equation}}$$

$$- \underbrace{\left[-\left(\frac{\partial^2 T}{\partial t^2}\right)_i^n \frac{\Delta t}{2} + \alpha \left(\frac{\partial^4 T}{\partial x^4}\right)_i^n \frac{(\Delta x)^2}{12} + \cdots \right]}_{\text{Truncation error}} \quad (4.35)$$

Examining Eq. (4.35), the left-hand side is the original partial differential equation, the first two terms on the right-hand side are the finite-difference representation of this equation, and the terms in the vertical brackets give the truncation error for the difference equation. Writing just the difference equation from Eq. (4.35), we have

$$\frac{T_i^{n+1} - T_i^n}{\Delta t} = \frac{\alpha(T_{i+1}^n - 2T_i^n + T_{i-1}^n)}{(\Delta x)^2} \quad (4.36)$$

Equation (4.36) is a *difference equation* which represents the original partial differential equation expressed in Eq. (3.28). However, Eq. (4.36) is just an *approximation* for Eq. (3.28); since each of the finite-difference quotients used in Eq. (4.36) has a truncation error, then the final form of the difference equation has its own truncation error synthesized from the errors of each of the finite differences. The truncation error for the difference equation given by Eq. (4.36) is displayed in Eq. (4.35). Note that the truncation error for the difference equation is $O[\Delta t, (\Delta x)^2]$.

Important: The difference equation is not the same as the original partial differential equation—it is a different thing altogether. The difference equation is an *algebraic* equation, which when written at all the grid points in the domain sketched in Fig. 4.5 yields a simultaneous system of algebraic equations. In turn, by some fashion these algebraic equations are solved numerically for the dependent variable at all grid points, i.e., solved for T_i^n, T_{i+1}^n, T_i^{n+1}, T_{i+1}^{n+1}, T_i^{n+2}, etc. In principle, we can only hope that the numerical results give values for T which represent those that would be obtained from a closed-form analytical solution of the original partial differential equation, at least within the truncation error. Some confidence in this regard can be obtained if we can answer "yes" to the following question: Does the difference equation reduce to the original differential equation as the number of grid points goes to infinity, i.e., as $\Delta x \to 0$ and $\Delta t \to 0$? Examining Eq. (4.35), we note that the truncation error approaches zero, and hence the difference equation does indeed approach the original differential equation. When this is the case, the finite-difference representation of the partial differential equation is said to be *consistent*.

If the difference equation is consistent, if the numerical algorithm used to solve the difference equation is stable, and if the boundary conditions are handled in a proper numerical fashion, then the numerical solution of the difference equation should be an appropriate representation of the analytical solution of the partial differential equation, at least to within the truncation error. However, there are several big "ifs" in the above statement. These "ifs," along with the undesirable propagation of truncation errors throughout the domain, make any successful CFD solution somewhat of a challenge and sometimes as much of an "art" as it is a "science."

The purpose of this section has been to introduce the idea of difference equations. The general concept of a finite-difference solution is to represent the governing partial differential equations by means of difference equations, and to solve these difference equations for numerical values of the dependent variables at each of the discrete grid points which cover the physical domain of interest. We have not yet discussed any precise algorithms that might be used for such numerical solutions; appropriate techniques (algorithms) for solving CFD problems by the finite-difference approach will evolve as we work through Part II of this book and as we deal with specific applications in Part III.

At this stage, it is worthwhile to return to the road map in Fig. 4.2. We have discussed the material represented by the first three boxes in the left column of Fig. 4.2—we have covered the basic elements of finite differences and their use to construct difference equations. There are several other important considerations to be discussed, such as explicit versus implicit solutions, stability analyses, and numerical dissipation.

4.4 EXPLICIT AND IMPLICIT APPROACHES: DEFINITIONS AND CONTRASTS

To this point in the present chapter, we have discussed some basic elements of the finite-difference method. We have done nothing more than just create some numerical tools for future use; we have not yet described how these tools can be put to use for the solutions of CFD problems. The *way* that these tools are put together and used for a given solution can be called a CFD *technique*, and we have not yet discussed any specific techniques. Aspects of several difference techniques commonly used in CFD will be discussed in Chap. 6. However, once you choose a specific technique to solve your given problem, you will find that the technique falls into one or the other of two different general approaches, an *explicit* approach or an *implicit* approach. It is appropriate to introduce and define these two general approaches now; they represent a fundamental distinction between various numerical techniques, a distinction for which we need to have some appreciation at this stage of our discussion.

For simplicity, let us return to the one-dimensional heat conduction equation given by Eq. (3.28), repeated below.

$$\frac{\partial T}{\partial t} = \alpha \frac{\partial^2 T}{\partial x^2} \qquad (3.28)$$

146   BASIC ASPECTS OF DISCRETIZATION

We will treat Eq. (3.28) as a "model equation" for our discussion in this section; all the necessary points concerning explicit and implicit approaches can be made using this model equation without going to the extra complexity of the governing flow equations. In Sec. 4.3, we used Eq. (3.28) to illustrate what was meant by a difference equation. In particular, in that section we chose to represent $\partial T/\partial t$ with a forward difference and $\partial^2 T/\partial x^2$ with a central second difference, leading to the particular form of the difference equation given by Eq. (4.36), repeated below:

$$\frac{T_i^{n+1} - T_i^n}{\Delta t} = \frac{\alpha(T_{i+1}^n - 2T_i^n + T_{i-1}^n)}{(\Delta x)^2} \tag{4.36}$$

With some rearrangement, this equation can be written as

$$T_i^{n+1} = T_i^n + \alpha \frac{\Delta t}{(\Delta x)^2}(T_{i+1}^n - 2T_i^n + T_{i-1}^n) \tag{4.37}$$

Let us examine the implications of Eq. (3.28) and its difference equation counterpart given by Eq. (4.37). Recall from our previous discussion in Sec. 4.3 that Eq. (3.28) is a parabolic partial differential equation. Being parabolic, this equation lends itself to a *marching solution*, as described in Sec. 3.4.2. The marching variable here is time t. To be more specific, consider the finite-difference grid sketched in Fig. 4.6. Assume that T is known at all grid points at time level n. Time marching means that T at all grid points at time level $n + 1$ are calculated from the known values at time level n. When this calculation is finished, we have known values at time level $n + 1$. Then the same procedure is used to calculate T at all grid points at time level $n + 2$, using the known values at level $n + 1$. In this fashion, the solution is progressively obtained by marching in steps of time. Casting our attention to Eq. (4.37), we see a straightforward mechanism to accomplish this time marching. Notice that Eq. (4.37) is written with properties at time level n on the right-hand side

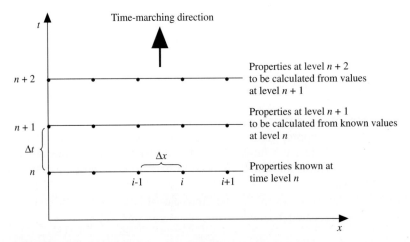

FIG. 4.6
Illustration of time marching.

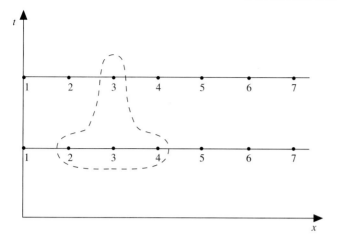

FIG. 4.7
An explicit finite-difference module.

and properties at time level $n + 1$ on the left-hand side. Recall that, within the time-marching philosophy, all properties at level n are known and those at level $n + 1$ are to be calculated. Of particular significance is that only *one unknown* appears in Eq. (4.37), namely, T_i^{n+1}. Hence, Eq. (4.37) allows for the immediate solution of T_i^{n+1} from the known properties at time level n. We have a single equation with a single unknown—nothing could be simpler. For example, consider the grid shown in Fig. 4.7, where we choose to distribute seven grid points along the x axis. Centering on grid point 2, Eq. (4.37) is written as

$$T_2^{n+1} = T_2^n + \alpha \frac{\Delta t}{(\Delta x)^2}(T_3^n - 2T_2^n + T_1^n) \qquad (4.38)$$

This allows the direct calculation of T_2^{n+1} since the quantities on the right-hand side of Eq. (4.38) are all known numbers. Then, centering on grid point 3, Eq. (4.37) is written as

$$T_3^{n+1} = T_3^n + \alpha \frac{\Delta t}{(\Delta x)^2}(T_4^n - 2T_3^n + T_2^n) \qquad (4.39)$$

This allows the direct calculation of T_3^{n+1} from the known numbers on the right-hand side of Eq. (4.39). In the same vein, by sequential application of Eq. (4.37) to grid points 4, 5, and 6, we obtain sequentially T_4^{n+1}, T_5^{n+1}, and T_6^{n+1}.

What we have just presented in the above paragraph is an example of an *explicit* approach. By definition, in an explicit approach each difference equation contains only one unknown and therefore can be solved *explicitly* for this unknown in a straightforward manner. Nothing could be simpler. This explicit approach is further illustrated by the finite-difference module contained within the dashed balloon in Fig. 4.7. Here, the module contains only one unknown at time level $n + 1$.

148 BASIC ASPECTS OF DISCRETIZATION

In regard to grid points 1 and 7 in Fig. 4.7, the marching solution of a parabolic partial differential equation presupposes the stipulation of boundary conditions. In regard to Fig. 4.7, this means that T_1 and T_7, which represent T at the left and right boundaries, respectively, are known numbers at each time level, known from the stipulated boundary conditions.

Equation (4.36) is not the only difference equation that can represent Eq. (3.28); in fact, it is only one of many different representations of the original partial differential equation. As a counterexample to the above discussion concerning the explicit approach, let us be somewhat daring and return to Eq. (3.28), this time writing the spatial difference on the right-hand side in terms of *average* properties between time levels n and $n+1$. That is, we will represent Eq. (3.28) by

$$\frac{T_i^{n+1} - T_i^n}{\Delta t} = \alpha \frac{\frac{1}{2}(T_{i+1}^{n+1} + T_{i+1}^n) + \frac{1}{2}(-2T_i^{n+1} - 2T_i^n) + \frac{1}{2}(T_{i-1}^{n+1} + T_{i-1}^n)}{(\Delta x)^2} \quad (4.40)$$

The special type of differencing employed in Eq. (4.40) is called the *Crank-Nicolson form*. (Crank-Nicolson differencing is commonly used to solve problems governed by parabolic equations. In CFD, the Crank-Nicolson form, or modified versions of it, is used frequently for finite-difference solutions of the boundary-layer equations.) Examine Eq. (4.40) closely. The unknown T_i^{n+1} is not only expressed in terms of the known quantities at time level n, namely, T_{i+1}^n, T_i^n, and T_{i-1}^n, but also in terms of other unknown quantities at time level $n+1$, namely, T_{i+1}^{n+1} and T_{i-1}^{n+1}. In other words, Eq. (4.40) represents one equation with *three* unknowns, namely, T_{i+1}^{n+1}, T_i^{n+1}, and T_{i-1}^{n+1}. Hence, Eq. (4.40) applied at a given grid point i does not stand alone; it cannot by itself result in a solution for T_i^{n+1}. Rather Eq. (4.40) must be written at all interior grid points, resulting in a *system* of algebraic equations from which the unknowns T_i^{n+1} for all i can be solved *simultaneously*. This is an example of an *implicit* approach. By definition, an implicit approach is one where the unknowns must be obtained by means of a *simultaneous solution* of the difference equations applied at *all* the grid points arrayed at a given time level. Because of this need to solve large systems of simultaneous algebraic equations, implicit methods are usually involved with the manipulations of large matrices. By now, it is easy to get the feeling that the implicit approach involves a more complex set of calculations than the explicit approach discussed earlier. In contrast to the simple explicit finite-difference module shown in Fig. 4.7, the implicit module for Eq. (4.40) is sketched in Fig. 4.8, clearly delineating the three unknowns at level $n+1$.

Let us be more specific, using the seven-point spatial grid shown in Fig. 4.8 as an example. Equation (4.40) can be rearranged to display the unknowns on the left-hand side and the known numbers on the right-hand side. The result is

$$\frac{\alpha \Delta t}{2(\Delta x)^2} T_{i-1}^{n+1} - \left[1 + \frac{\alpha \Delta t}{(\Delta x)^2}\right] T_i^{n+1} + \frac{\alpha \Delta t}{2(\Delta x)^2} T_{i+1}^{n+1}$$

$$= -T_i^n - \frac{\alpha \Delta t}{2(\Delta t)^2}(T_{i+1}^n - 2T_i^n + T_{i-1}^n) \quad (4.41)$$

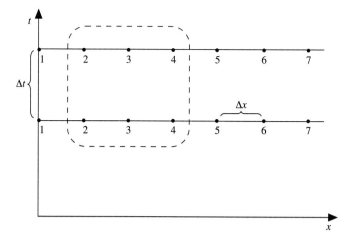

FIG. 4.8
An implicit finite-difference module.

Simplifying the nomenclature by denoting the following quantities by A, B, and K_i,

$$A = \frac{\alpha \, \Delta t}{2(\Delta x)^2}$$

$$B = 1 + \frac{\alpha \, \Delta t}{(\Delta x)^2}$$

$$K_i = -T_i^n - \frac{\alpha \, \Delta t}{2(\Delta x)^2}(T_{i+1}^n - 2T_i^n + T_{i-1}^n)$$

we can write Eq. (4.41) in the form

$$AT_{i-1}^{n+1} - BT_i^{n+1} + AT_{i+1}^{n+1} = K_i \qquad (4.42)$$

Note that K_i in Eq. (4.42) consists of properties at time level n, which are known. Hence, K_i is a known number in Eq. (4.42). Returning to Fig. 4.8, we now apply Eq. (4.42) sequentially to grid points 2 through 6.

At grid point 2: $\qquad AT_1 - BT_2 + AT_3 = K_2 \qquad (4.43)$

Here, we have dropped the superscript for convenience; it is easy to remember that T_1, T_2, and T_3 represent three values at time level $n + 1$, and K_2 is a known number as stated before. Moreover, because of the stipulated boundary conditions at grid points 1 and 7, T_1 in Eq. (4.43) is a known number. Hence, in Eq. (4.43) the term involving the known T_1 can be transferred to the right-hand side, resulting in

$$-BT_2 + AT_3 = K_2 - AT_1 \qquad (4.44)$$

Denoting $K_2 - AT_1$ by K_2', where K_2' is a known number, Eq. (4.44) is written as

$$-BT_2 + AT_3 = K_2' \qquad (4.45)$$

At grid point 3 : $\quad AT_2 - BT_3 + AT_4 = K_3 \quad$ (4.46)

At grid point 4 : $\quad AT_3 - BT_4 + AT_5 = K_4 \quad$ (4.47)

At grid point 5 : $\quad AT_4 - BT_5 + AT_6 = K_5 \quad$ (4.48)

At grid point 6 : $\quad AT_5 - BT_6 + AT_7 = K_6 \quad$ (4.49)

In Eq. (4.49), since grid point 7 is on a boundary, T_7 is known from the stipulated boundary condition. Hence, Eq. (4.49) can be rearranged as

$$AT_5 - BT_6 = K_6 - AT_7 = K'_6 \qquad (4.50)$$

where K'_6 is a known number.

Equations (4.45) to (4.48) and (4.50) are five equations for the five unknowns T_2, T_3, T_4, T_5, and T_6. This system of equations can be written in matrix form as follows.

$$\begin{bmatrix} -B & A & 0 & 0 & 0 \\ A & -B & A & 0 & 0 \\ 0 & A & -B & A & 0 \\ 0 & 0 & A & -B & A \\ 0 & 0 & 0 & A & -B \end{bmatrix} \begin{bmatrix} T_2 \\ T_3 \\ T_4 \\ T_5 \\ T_6 \end{bmatrix} = \begin{bmatrix} K'_2 \\ K_3 \\ K_4 \\ K_5 \\ K'_6 \end{bmatrix} \qquad (4.51)$$

The coefficient matrix is a *tridiagonal matrix*, defined as having nonzero elements only along the three diagonals which are marked with the three dashed lines in Eq. (4.51). The solution of the system of equations denoted by Eq. (4.51) involves the manipulation of the tridiagonal arrangement; such solutions are usually obtained using Thomas' algorithm, which has become almost standard for the treatment of tridiagonal systems of equations. A description of this algorithm is given in App. A of this book; it will be handy when we discuss the applications in Part III.

Clearly, on the basis of the above example, an implicit approach is more involved than an explicit approach. Also, this is not the whole story. The model equation we have chosen in this section, namely, Eq. (3.28), is a linear partial differential equation, and it leads to a linear difference equation, such as the forms given by Eqs. (4.37) and (4.40). On the other hand, what happens when the governing partial differential equation is nonlinear? For example, let us assume that the thermal diffusivity α in Eq. (3.28) is a function of temperature, i.e., we write from Eq. (3.28)

$$\frac{\partial T}{\partial t} = \alpha(T) \frac{\partial^2 T}{\partial x^2} \qquad (4.52)$$

Equation (4.52) is now a *nonlinear* partial differential equation. This has virtually no effect on the explicit approach, where a difference equation can be written for Eq. (4.52), analogous to Eq. (4.37), as

$$T_i^{n+1} = T_i^n + \alpha(T_i^n) \frac{\Delta t}{(\Delta x)^2} (T_{i+1}^n - 2T_i^n + T_{i-1}^n) \qquad (4.53)$$

Equation (4.53) is still linear in the single unknown T_i^{n+1}, because α is evaluated at time level n; that is, $\alpha = \alpha(T_i^n)$, where T_i^n is a known number. On the other hand, if the Crank–Nicolson method is used for Eq. (4.52), the right-hand-side is evaluated as an average between time levels n and $n + 1$, resulting in $\alpha(T)$ being represented by $\frac{1}{2}[\alpha(T_i^{n+1}) + \alpha(T_i^n)]$. The resulting difference equation is given by Eq. (4.41), with the exception that now α is replaced everywhere in that equation by $\frac{1}{2}[\alpha(T_i^{n+1}) + \alpha(T_i^n)]$. Clearly, the new difference equation involves products of the dependent variables, such as $[\alpha(T_i^{n+1})]T_i^{n+1}$, $[\alpha(T_i^{n+1})]T_{i+1}^{n+1}$, and $[\alpha(T_i^{n+1})]T_{i-1}^{n+1}$. In other words, the resulting difference equation is a *nonlinear* algebraic equation. An implicit solution would therefore demand a simultaneous solution of a large system of nonlinear equations—an exceptionally difficult task. This is a tremendous disadvantage of an implicit approach. To circumvent this problem, the difference equations are usually "linearized" in an approximate fashion. For example, if in Eq. (4.52) α is simply evaluated at time level n rather than an average between levels n and $n + 1$, then no nonlinear algebraic terms will appear in the difference equation; the resulting difference equation will be identical to Eq. (4.31), with α evaluated as $\alpha(T_i^n)$. Other linearization ploys for implicit methods appropriate to the governing flow equations are discussed in Chap. 6.

With the complexity of the implicit approach relative to the explicit approach in mind, the immediate question is: Why deal with the implicit approach at all? Why not always use an explicit approach? Unfortunately, life is not that simple. We have yet to mention the most important difference between the explicit and implicit approaches. Note that the increments Δx and Δt appear in all the above difference equations. For the explicit approach, once Δx is chosen, then Δt is *not* an independent, arbitrary choice; rather, Δt is restricted to be equal to or less than a certain value prescribed by a *stability criterion*. If Δt is taken larger than the limit imposed by the stability criterion, the time-marching procedure will quickly go unstable, and your computer program will quickly shut down due to such things as numbers going to infinity or taking the square root of a negative number. In many cases, Δt must be very small to maintain stability; this can result in long computer running times to make calculations over a given interval of time. On the other hand, there are no such stability restrictions on an implicit approach. For most implicit methods, stability can be maintained over much larger values of Δt than for a corresponding explicit method; indeed, some implicit methods are *unconditionally stable*, meaning that any value of Δt, no manner how large, will yield a stable solution. Hence, for an implicit method, considerably fewer time steps are required to cover a given interval in time compared to an explicit method. Therefore, for some applications, even though the implicit approach requires more computations per time step due to its relative complexity, the fact that considerably fewer time steps are required to cover a given interval of time actually can result in a shorter run time on the computer compared to an explicit approach.

There is a downside to the large values of Δt allowable for implicit methods. To see this, we must recall that time marching in the context of CFD is used to accomplish one or the other of the following purposes:

1. To obtain a steady-state solution by means of assuming some arbitrary initial conditions for a flow field, and then calculating the flow in steps of time, going out to a sufficiently large number of time steps until a final steady-state flow is approached at large values of time. In this situation, the final steady state is the desired result, and the time marching is simply a means to that end. The solution to the supersonic blunt body problem is a case in point, as discussed in Sec. 3.4.4.
2. To obtain an accurate timewise solution of an inherently unsteady flow, such as the time-varying flow field over a pitching airfoil or the naturally unsteady flow pattern that results for many separated flows. A case in point is the unsteady laminar separated flow over the airfoil shown in Fig. 1.3a, as discussed in Sec. 1.2. (Go back for a moment and review the short discussion in Sec. 1.2 associated with Fig. 1.3a—it will help you to form a better impression of our current discussion.)

In regard to item 1 above, the time-marching procedure does not have to be timewise-accurate; it only has to, by some means, ultimately approach the *correct* steady-state flow field. On the other hand, for item 2 above, timewise accuracy of the time-marching method is absolutely necessary—it is the time variation of the flow field that we want to solve. Here is where the downside of an implicit approach using a large value of Δt enters our considerations. Clearly, as Δt increases, so does the truncation error associated with the difference expression for the time derivative. In turn, an implicit method using large values of Δt may not accurately define the timewise variation of the flow field. In this situation, the advantage of an implicit approach may be totally negated.

So what does all this mean? It simply says that there are cases where the use of an explicit method makes the most sense and others where an implicit method is clearly the best choice. To help clarify this situation, the relative major advantages and disadvantages of these two approaches are summarized as follows.

Explicit approach

Advantage   Relatively simple to set up and program.

Disadvantage   In terms of our above example, for a given Δx, Δt must be less than some limit imposed by stability constraints. In some cases, Δt must be very small to maintain stability; this can result in long computer running times to make calculations over a given interval of t.

Implicit approach

Advantage   Stability can be maintained over much larger values of Δt, hence using considerable fewer time steps to make calculations over a given interval of t. This results in less computer time.

Disadvantage   More complicated to set up and program.

Disadvantage   Since massive matrix manipulations are usually required at each time step, the computer time per time step is much larger than in the explicit approach.

Disadvantage   Since large Δt can be taken, the truncation error is large, and the use of implicit methods to follow the exact transients (time variations of the independent variable) may not be as accurate as an explicit approach. However, for a time-dependent solution in which the steady state is the desired result, this relative timewise inaccuracy is not important.

During the period 1969 to about 1979, the vast majority of practical CFD solutions involving *marching* solutions (such as in the above example) employed explicit methods. Today, they are still the most straightforward methods for flow-field solutions. However, many of the more sophisticated CFD applications—those requiring very closely spaced grid points in some regions of the flow—would demand inordinately large computer running times due to the small marching step required. The calculation of high Reynolds number viscous flows, where extreme changes in the flow field occur close to a surface and therefore require many closely spaced points adjacent to the surface, is a case in point. This has made the advantage listed above for implicit methods very attractive, namely, the ability to use large marching steps even for a very fine grid. For this reason, implicit methods became the major focus of CFD applications in the 1980s. However, today there is an enhancement in computer architecture that may shift the emphasis back to explicit solutions, namely, the development of massively parallel processor computers such as the connection machine. (Recall the discussion on different types of modern computers in Sec. 1.5.) For such massively parallel processors, explicit calculations can be made at thousands of grid points in the flow all at the same instant on the computer. Indeed, such computers are tailor-made for explicit methods. Again, in retrospect, the choice between the explicit or the implicit approach for the solution of a given problem is not always clear; when faced with such a choice, you will have to use your best judgment. Our purpose in the present section has simply been to define the general nature of the two approaches and to constrast some of the advantages and disadvantages of both.

Finally, we note that the discussion in this section, although couched in terms of the finite-difference method, is certainly not limited to that method. Finite-volume methods also fall under the same classification; there are explicit finite-volume techniques, and there are implicit finite-volume techniques. The distinctions, advantages, and disadvantages are exactly the same as discussed throughout this section.

4.5 ERRORS AND AN ANALYSIS OF STABILITY

Some ramifications of the stability behavior of numerical solutions were raised in Sec. 4.4 in conjunction with explicit methods. There, we stated that such methods would be numerically unstable if the increment in the marching direction (Δt in our previous discussion) exceeded some prescribed value. The prescription for this maximum allowable value comes, in principle, from a formal stability analysis of the governing equations in finite-difference form. An exact stability analysis of the difference representation of the nonlinear Euler or Navier-Stokes equations does not exist. However, there are simplified approaches applied to simpler model equations which can provide some reasonable guidance. In this author's opinion, rigorous stability analyses of numerical methods is the purview of applied mathematics; it is certainly outside the scope of the present book. However, it is important for workers in CFD to have some appreciation of the nature of stability analyses and the results obtained therein. This can be achieved by discussing a simple, approximate analysis for a linear "model" equation. Such is the purpose of the present section.

Note: The stability analyses described in the following discussions are applied to specific difference equations, and hence the results pertain directly to those specific equations. In this sense, you might consider the remainder of Sec. 4.5 as a sequence of a few, rather extended, worked examples. However, these examples reflect an approach which is more general than might appear at first. Therefore, we will direct your attention to the forest as well as the trees.

For our model equation, we continue to choose the one-dimensional heat conduction equation, namely, Eq. (3.28) repeated below:

$$\frac{\partial T}{\partial t} = \alpha \frac{\partial^2 T}{\partial x^2} \tag{3.28}$$

and for the difference representation of this equation we again choose the explicit form given by Eq. (4.36), also repeated below:

$$\frac{T_i^{n+1} - T_i^n}{\Delta t} = \frac{\alpha(T_{i+1}^n - 2T_i^n + T_{i-1}^n)}{(\Delta x)^2} \tag{4.36}$$

What is this matter of stability all about? What is it that makes a given calculation go unstable? By the time you reach the end of this section, it is hoped that you will have a better idea about the answers to these questions. The answers are, for the most part, dependent upon the concept of numerical errors that are generated throughout the course of a given calculation and, more to the point, the way that these errors are *propagated* from one marching step to the next. Simply stated, if a given numerical error is *amplified* in going from one step to the next, then the calculation will become unstable; if the error does not grow, and especially if it *decreases* from one step to another, then the calculation usually has a stable behavior. Therefore, a consideration of stability must first be prefaced by a discussion on numerical errors—what they are and what they are like. Let us proceed with such a discussion.

Consider a partial differential equation, such as, for example, Eq. (3.28), given above. The numerical solution of this equation is influenced by two sources of error:

Discretization error, the difference between the exact analytical solution of the partial differential equation [for example, Eq. (3.28)] and the exact (round-off-free) solution of the corresponding difference equation [for example, Eq. (4.36)]. From our discussion in Sec. 4.3, the discretization error is simply the truncation error for the difference equation plus any errors introduced by the numerical treatment of the boundary conditions.

Round-off error, the numerical error introduced after a repetitive number of calculations in which the computer is constantly rounding the numbers to some significant figure.

If we let

A = analytical solution of partial differential equation

D = exact solution of difference equation

N = numerical solution from a real computer with finite accuracy

then

$$\text{Discretization error} = A - D$$
$$\text{Round-off error} = \epsilon = N - D \tag{4.54}$$

From Eq. (4.54), we can write

$$N = D + \epsilon \tag{4.55}$$

where, again, ϵ is the round-off error, which for the remainder of our discussion in this section we will simply call *error* for brevity. The numerical solution N must satisfy the difference equation. This is because the computer is programmed to solve the difference equation; in our example, the computer is programmed to solve Eq. (4.36), albeit the answer comes out with a round-off error cranked in. Hence from Eq. (4.36),

$$\frac{D_i^{n+1} + \epsilon_i^{n+1} - D_i^n - \epsilon_i^n}{\alpha \, \Delta t} = \frac{D_{i+1}^n + \epsilon_{i+1}^n - 2D_i^n - 2\epsilon_i^n + D_{i-1}^n + \epsilon_{i-1}^n}{(\Delta x)^2} \tag{4.56}$$

By definition, D is the *exact* solution of the difference equation; hence it exactly satisfies the difference equation. Thus, we can write

$$\frac{D_i^{n+1} - D_i^n}{\alpha \, \Delta t} = \frac{D_{i+1}^n - 2D_i^n + D_{i-1}^n}{(\Delta x)^2} \tag{4.57}$$

Subtracting Eq. (4.57) from (4.56), we have

$$\frac{\epsilon_i^{n+1} - \epsilon_i^n}{\alpha \, \Delta t} = \frac{\epsilon_{i+1}^n - 2\epsilon_i^n + \epsilon_{i-1}^n}{(\Delta x)^2} \tag{4.58}$$

From Eq. (4.58), we see that the error ϵ also satisfies the difference equation.

We now consider aspects of the *stability* of the difference equation, Eq. (4.36). If errors ϵ_i are already present at some stage of the solution of this equation (as they always are in any real computer solution), then the solution will be *stable* if the ϵ_i's shrink, or at best stay the same, as the solution progresses from step n to $n + 1$; on the other hand, if the ϵ_i's grow larger during the progression of the solution from

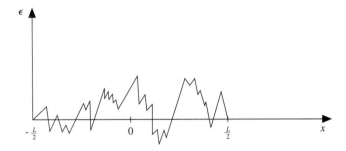

FIG. 4.9
Schematic of the variation of round-off error as a function of x.

steps n to $n + 1$, then the solution is *unstable*. That is, for a solution to be *stable*,

$$\left|\frac{\epsilon_i^{n+1}}{\epsilon_i^n}\right| \leq 1 \qquad (4.59)$$

For Eq. (4.36), let us examine under what conditions Eq. (4.59) holds.

First of all, we need to examine how the round-off error looks. For the unsteady, one-dimensional problem exemplified by Eq. (3.28), the round-off error can be plotted versus x at any given time step. For example, one such representation is sketched in Fig. 4.9. Here, we assume that the length of the domain on which the equation is being solved is denoted by L. For convenience later on we place the origin at the midpoint of the domain; hence the left boundary is located at $-L/2$ and the right boundary is at $+L/2$. The distribution of ϵ along the x axis is represented by the rather random variation sketched in Fig. 4.9. Note that $\epsilon = 0$ at $x = -L/2$ and $L/2$, because there are specified boundary values at both ends of the domain, and hence no error is introduced—the boundary values are always fed in as exact, known numbers. At any given time, the random variation of ϵ with x in Fig. 4.9 can be expressed analytically by a Fourier series as follows:

$$\epsilon(x) = \sum_m A_m e^{ik_m x} \qquad (4.60)$$

Equation (4.60) represents both a sine and a cosine series, since $e^{ik_m x} = \cos k_m x + i \sin k_m x$. Here, k_m is called the *wave number*. The real part of Eq. (4.60) represents the error. Before proceeding further, let us examine the meaning of the wave number. For simplicity, consider just the sine function plotted as a function of x as shown in Fig. 4.10. By definition, the wavelength λ is the interval over x encompassing one complete wavelength, as sketched in Fig. 4.10. Therefore, a form of this sine function which may seem more familiar to you is

$$y = \sin \frac{2\pi x}{\lambda} \qquad (4.61)$$

In the wave number notation, this would be written as

$$y = \sin k_m x \qquad (4.62)$$

ERRORS AND AN ANALYSIS OF STABILITY   157

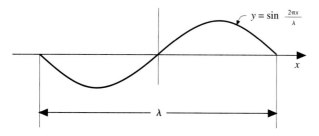

FIG. 4.10
Sine function.

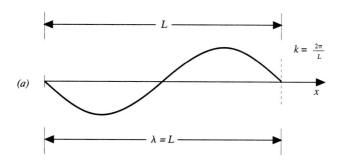

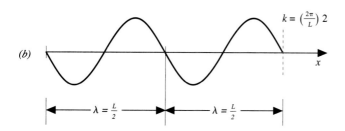

FIG. 4.11
Sine functions. (*a*) Wavelength is L. (*b*) Wavelength is $L/2$.

Comparing Eqs. (4.61) and (4.62), clearly the wave number is given by

$$k_m = \frac{2\pi}{\lambda} \tag{4.63}$$

In Eq. (4.60), the wave number k_m is written with a subscript m. It remains to explain the meaning of m. This is related to the number of waves that are fitted inside a given interval. Consider an interval along the x axis of length L, as sketched in Fig. 4.11. If one sine wave is completely fitted within this interval, its wavelength is $\lambda = L$, as shown in Fig. 4.11*a*. This sine wave is expressed by Eq. (4.62), where k_m is defined by Eq. (4.63). For this case, since $\lambda = L$, then k_m is given by $k_m = 2\pi/L$.

Now examine the case of two sine waves fitted within the interval L, as sketched in Fig. 4.11b. The wavelength of the new sine waves in Fig. 4.11b is $\lambda = L/2$. The equation of these waves is given by

$$y = \sin k_m x$$

where now $k_m = 2\pi/(L/2) = (2\pi/L)2$. Extrapolating this thinking, if three waves were fitted within the interval L, then $k_m = (2\pi/L)3$, and so forth. Therefore, we can write the wave number for these various sine waves of different wavelengths as

$$k_m = \left(\frac{2\pi}{L}\right) m \qquad m = 1, 2, 3, \qquad (4.64)$$

This illustrates the meaning of the subscript m on the wave number; it is simply equal to the number of waves fitted into the given interval L. Clearly, from Eq. (4.64), the wave number itself is proportional to the number of waves in a given interval; the higher the wave number for a given L, the more waves we have fitted inside the interval.

We are now in a better position to understand the significance of Eq. (4.60), where the summation over m denotes the sequential addition of sine and cosine functions with sequentially increasing wave numbers. That is, Eq. (4.60) is a sum of terms, each representing a higher harmonic. When taken out for an infinite number of terms, Eq. (4.60) can represent a continuous variation of ϵ as a function of x, as sketched in Fig. 4.9. However, in regard to a practical numerical solution which involves only a finite number of grid points, there is a constraint imposed on the number of terms in Eq. (4.60). To see this more clearly, consider Fig. 4.12, which shows the interval L over which the numerical calculations are being made. The largest allowable wavelength is $\lambda_{max} = L$; this is the wavelength for the first term in Eq. (4.60) and corresponds to $m = 1$. In turn, the smallest possible wavelength is that having all three zeros of the sine (or cosine) function going through three adjacent grid points, as shown in Fig. 4.12. Hence, the smallest allowable wavelength is $\lambda_{min} = 2\Delta x$. If there are $N + 1$ grid points distributed over the interval L, then there are N intervals between these grid points, and hence $\Delta x = L/N$.

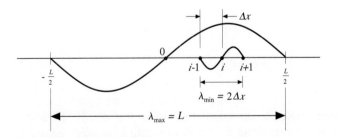

FIG. 4.12
Illustration of maximum and minimum wavelengths for the Fourier components of the round-off error.

Thus, $\lambda_{min} = 2L/N$. From Eq. (4.63),

$$k_m = \frac{2\pi}{2L/N} = \frac{2\pi}{L}\frac{N}{2} \qquad (4.65)$$

Comparing Eqs. (4.64) and (4.65), we see that m in Eq. (4.65) is equal to $N/2$. This is the highest-order harmonic allowable in Eq. (4.60). Hence, from Eq. (4.60), we have, for a grid with $N + 1$ grid points,

$$\epsilon(x) = \sum_{m=1}^{N/2} A_m e^{ik_m x} \qquad (4.66)$$

We are not finished with the representation of the round-off error. Equation (4.66) gives the *spatial* variation at a given time level n. From Eq. (4.59), for an assessment of numerical stability, we are interested in the variation of ϵ with *time*. Therefore, we extend Eq. (4.66) by assuming the amplitude A_m is a function of time.

$$\epsilon(x, t) = \sum_{m=1}^{N/2} A_m(t) e^{ik_m x} \qquad (4.67)$$

Moreover, it is reasonable to assume an exponential variation with time; errors tend to grow or diminish exponentially with time. Therefore, we write

$$\epsilon(x, t) = \sum_{m=1}^{N/2} e^{at} e^{ik_m x} \qquad (4.68)$$

where a is a constant (which may take on different values for different m's). Equation (4.68) represents a final, reasonable form for the variation of round-off error in both space and time.

After all this work to construct ϵ in terms of a truncated Fourier series with amplitudes exponentially varying with time, we now make the following observation. Since the original difference equation, Eq. (4.36), is linear and since the round-off error satisfies the same difference equation as proven by Eq. (4.58), then when Eq. (4.68) is substituted into Eq. (4.58), the behavior of each term of the series is the same as the series itself. Hence, let us deal with just one term of the series and write

$$\epsilon_m(x, t) = e^{at} e^{ik_m x} \qquad (4.69)$$

The stability characteristics can be studied using just this form for ϵ with no loss in generality. The value of our discussion leading to the more general form for ϵ given by Eq. (4.68) is to tell us what we are really dealing with in terms of the round-off error and to allow us to make the observation embodied in Eq. (4.69). Let us now proceed to find out how ϵ varies in steps of time and therefore to find out what conditions on Δt are necessary such that Eq. (4.59) is satisfied.

To begin with, substitute Eq. (4.69) into Eq. (4.58).

$$\frac{e^{a(t+\Delta t)} e^{ik_m x} - e^{at} e^{ik_m x}}{\alpha \Delta t} = \frac{e^{at} e^{ik_m(x+\Delta x)} - 2 e^{at} e^{ik_m x} + e^{at} e^{ik_m(x-\Delta x)}}{(\Delta x)^2} \qquad (4.70)$$

Divide Eq. (4.70) by $e^{at}e^{ik_m x}$.

$$\frac{e^{a\Delta t} - 1}{\alpha \Delta t} = \frac{e^{ik_m \Delta x} - 2 + e^{-ik_m \Delta x}}{(\Delta x)^2}$$

or
$$e^{a\Delta t} = 1 + \frac{\alpha \Delta t}{(\Delta x)^2}(e^{ik_m \Delta x} + e^{-ik_m \Delta x} - 2) \quad (4.71)$$

Recalling the identity

$$\cos(k_m \Delta x) = \frac{e^{ik_m \Delta x} + e^{-ik_m \Delta x}}{2}$$

Eq. (4.71) can be written as

$$e^{a\Delta t} = 1 + \frac{2\alpha \Delta t}{(\Delta x)^2}[\cos(k_m \Delta x) - 1] \quad (4.72)$$

Recalling another trigonometric identity

$$\sin^2 \frac{k_m \Delta x}{2} = \frac{1 - \cos(k_m \Delta x)}{2}$$

Eq. (4.72) becomes finally

$$e^{a\Delta t} = 1 - \frac{4\alpha \Delta t}{(\Delta x)^2} \sin^2 \frac{k_m \Delta x}{2} \quad (4.73)$$

From Eq. (4.73),

$$\frac{\epsilon_i^{n+1}}{\epsilon_i^n} = \frac{e^{a(t+\Delta t)} e^{ik_m x}}{e^{at} e^{ik_m x}} = e^{a\Delta t} \quad (4.74)$$

Combining Eqs. (4.59), (4.73), and (4.74), we have

$$\left|\frac{\epsilon_i^{n+1}}{\epsilon_i^n}\right| = |e^{a\Delta t}| = \left|1 - \frac{4\alpha \Delta t}{(\Delta x)^2} \sin^2 \frac{k_m \Delta x}{2}\right| \leq 1 \quad (4.75)$$

Equation (4.75) *must* be satisfied to have a *stable* solution, as dictated by Eq. (4.59). In Eq. (4.75) the factor

$$\left|1 - \frac{4\alpha \Delta t}{(\Delta x)^2} \sin^2 \frac{k_m \Delta x}{2}\right| \equiv G$$

is called the *amplification* factor and is denoted by G. Evaluating the inequality in Eq. (4.75), namely, $G \leq 1$, we have two possible situations which must hold simultaneously:

1. $\qquad 1 - \dfrac{4\alpha \Delta t}{\Delta x} \sin^2 \dfrac{k_m \Delta x}{2} \leq 1$

Thus

$$\frac{4\alpha \, \Delta t}{\Delta x} \sin^2 \frac{k_m \, \Delta x}{2} \geq 0 \qquad (4.76)$$

Since $4\alpha \, \Delta t/(\Delta x)^2$ is always positive, the condition expressed in Eq. (4.76) always holds.

2.
$$1 - \frac{4\alpha \, \Delta t}{\Delta x} \sin^2 \frac{k_m \, \Delta x}{2} \geq -1$$

Thus

$$\frac{4\alpha \, \Delta t}{\Delta x} \sin^2 \frac{k_m \, \Delta x}{2} - 1 \leq 1$$

For the above condition to hold,

$$\frac{\alpha \, \Delta t}{(\Delta x)^2} \leq \frac{1}{2} \qquad (4.77)$$

Equation (4.77) gives the *stability requirement* for the solution of the difference equation, Eq. (4.36), to be *stable*. Clearly, for a given Δx, the allowed value of Δt must be small enough to satisfy Eq. (4.77). Here is a stunning example of the limitation placed on the marching variable by stability considerations for explicit finite-difference models. As long as $\alpha \, \Delta t/(\Delta x)^2 \leq \frac{1}{2}$, the error will *not* grow for subsequent marching steps in t, and the numerical solution will proceed in a stable manner. On the other hand, if $\alpha \, \Delta t/(\Delta x)^2 > \frac{1}{2}$, then the error will progressively become larger and will eventually cause the numerical marching solution to "blow up" on the computer.

The above analysis is an example of a general method called the *von Neumann stability method*, which is used frequently to study the stability properties of linear difference equations.

The exact form of the stability criterion depends on the form of the difference equation. For example, let us briefly examine the stability characteristics of another simple equation, this time a hyperbolic equation. Consider the first-order wave equation (see Prob. 3.5):

$$\frac{\partial u}{\partial t} + c \frac{\partial u}{\partial x} = 0 \qquad (4.78)$$

Let us replace the spatial derivative with a central difference.

$$\frac{\partial u}{\partial x} = \frac{u_{i+1}^n - u_{i-1}^n}{2 \, \Delta x} \qquad (4.79)$$

If we replace the time derivative with a simple forward difference, then the resulting difference equation representing Eq. (4.78) would be

$$\frac{u_i^{n+1} - u_i^n}{\Delta t} = -c \frac{u_{i+1}^n - u_{i-1}^n}{2 \, \Delta x} \qquad (4.80)$$

This is about as simple a difference equation as can be obtained from Eq. (4.78); it is sometimes called the *Euler explicit form*. However, the application of the von Neumann stability analysis to Eq. (4.80) shows that Eq. (4.80) leads to an unstable solution no matter what the value of Δt is—Eq. (4.80) is therefore called *unconditionally unstable*. Instead, let us replace the time derivative with a first-order difference, where $u(t)$ is represented by an average value between grid points $i + 1$ and $i - 1$, i.e.,

$$u(t) = \tfrac{1}{2}(u_{i+1}^n + u_{i-1}^n)$$

Then

$$\frac{\partial u}{\partial t} = \frac{u_i^{n+1} - \tfrac{1}{2}(u_{i+1}^n + u_{i-1}^n)}{\Delta t} \qquad (4.81)$$

Substituting Eqs. (4.79) and (4.81) into (4.78), we have

$$u_i^{n+1} = \frac{u_{i+1}^n + u_{i-1}^n}{2} - c\frac{\Delta t}{\Delta x}\frac{u_{i+1}^n - u_{i-1}^n}{2} \qquad (4.82)$$

The differencing used in the above equation, where Eq. (4.81) is used to represent the time derivative, is called the *Lax method*, after the mathematician Peter Lax who first proposed it. If we now assume an error of the form $\epsilon_m(x, t) = e^{at}e^{ik_m t}$ as done previously and substitute this form in Eq. (4.82), the amplification factor becomes

$$e^{at} = \cos(k_m \Delta x) - iC\sin(k_m \Delta x) \qquad (4.83)$$

where $C = c\,\Delta t/\Delta x$. The stability requirement is $|e^{at}| \leq 1$, which when applied to Eq. (4.83) yields

$$C = c\frac{\Delta t}{\Delta x} \leq 1 \qquad (4.84)$$

In Eq. (4.84), C is called the *Courant number*. This equation says that $\Delta t \leq \Delta x/c$ for the numerical solution of Eq. (4.82) to be stable. Moreover, Eq. (4.84) is called the *Courant-Friedrichs-Lewy* condition, generally written as the CFL condition. It is an important stability criterion for hyperbolic equations. The CFL condition dates back to 1928; the original work can be found in Ref. 25.

The CFL condition, i.e., the Courant number must be less than or at most equal to unity, is also the stability condition which holds for the second-order wave equation (see Prob. 3.4),

$$\frac{\partial^2 u}{\partial t^2} = c^2 \frac{\partial^2 u}{\partial x^2} \qquad (4.85)$$

There is a connection between the characteristic lines associated with Eq. (4.85) and the CFL condition, a connection which helps to elucidate the physical significance of the CFL condition. Let us pursue this connection. These characteristic lines (see Sec. 3.2) for Eq. (4.85) are given by

$$x = \begin{cases} ct & \text{(right-running)} \\ -ct & \text{(left-running)} \end{cases} \qquad \begin{matrix}(4.86a)\\(4.86b)\end{matrix}$$

and are sketched in Fig. 4.13a and b. In both parts of Fig. 4.13 let point b be the intersection of the right-running characteristic through grid point $i - 1$ and the left-running characteristic through grid point $i + 1$. However, point b, determined by the intersection of characteristic lines through grid points $i - 1$ and $i + 1$, also has a slightly different significance—one associated with the CFL stability condition, which states that, at most, the Courant number $C = 1$. To see this more clearly, let $\Delta t_{C=1}$ denote the value of Δt given by Eq. (4.84) when $C = 1$. Then, from Eq. (4.84),

$$\Delta t_{C=1} = \frac{\Delta x}{c} \tag{4.87}$$

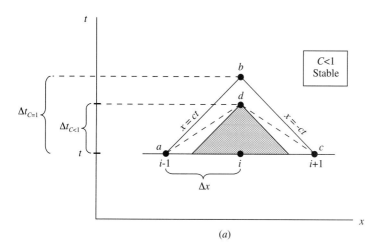

(a)

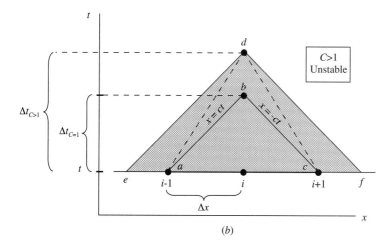

(b)

FIG. 4.13
(a) Illustration of a stable case. The numerical domain includes all the analytical domain. (b) Illustration of an unstable case. The numerical domain does *not* include all the analytical domain.

If in Fig. 4.13a and b we move a distance $\Delta t_{C=1}$ directly above grid point i, we find ourselves directly on top of point b. This is because from the characteristic lines given by Eqs. (4.86a) and (4.86b)

$$\Delta t = \pm \frac{\Delta x}{c} \qquad (4.88)$$

Obviously, the increment Δt in Eq. (4.87), having to do with the CFL condition, and the increment Δt in Eq. (4.88), having to do with the intersection of characteristic lines, are the same values, since the right-hand sides of Eqs. (4.87) and (4.88) are the same. Therefore, $\Delta t_{C=1}$ is exactly the distance between point b and grid point i in Fig. 4.13a and b. Now assume that $C < 1$, which pertains to the sketch in Fig. 4.13a. Then from Eq. (4.84), $\Delta t_{C<1} < \Delta t_{C=1}$, as shown in Fig. 4.13a. Let point d correspond to the grid point directly above point i existing at time $t + \Delta t_{C<1}$. Since properties at point d are calculated numerically from the difference equation using information at grid points $i - 1$ and $i + 1$, the *numerical domain* for point d is the triangle adc shown in Fig. 4.13a. The *analytical domain* for point d is the shaded triangle in Fig. 4.13a, defined by the characteristics through point d. These characteristics are parallel to those through point b. Note that in Fig. 4.13a, the numerical domain of point d *includes* the analytical domain. In contrast, consider the case shown in Fig. 4.13b. Here, $C > 1$. Then, from Eq. (4.84), $\Delta t_{C>1} > \Delta t_{C=1}$, as shown in Fig. 4.13b. Let point d in Fig. 4.13b correspond to the grid point directly above point i existing at time $t + \Delta t_{c>1}$. Since properties at point d are calculated numerically from the difference equation using information at grid points $i - 1$ and $i + 1$, the *numerical domain* for point d is the triangle adc shown in Fig. 4.13b. The *analytical domain* for point d is the shaded triangle in Fig. 4.13b, defined by the characteristics through point d. Note that in Fig. 4.13b, the numerical domain does not include all the analytical domain. Moreover, Fig. 4.13b is for $C > 1$, which leads to unstable behavior. Therefore, we can give the following physical interpretation of the CFL condition:

> For stability, the numerical domain must include all the analytical domain.

The above considerations dealt with *stability*. The question of *accuracy*, which is sometimes quite different, can also be examined from the point of view of Fig. 4.13. Consider a stable case, as shown in Fig. 4.13a. Note that the analytical domain of dependence for point d is the shaded triangle in Fig. 4.13a. From our discussion in Chap. 3, the properties at point d theoretically depend only on those points within the shaded triangle. However, note that the numerical grid points $i - 1$ and $i + 1$ are *outside* the domain of dependence for point d and hence *theoretically* should not influence the properties at point d. On the other hand, the *numerical* calculation of properties at point d takes information from grid points $i - 1$ and $i + 1$. This situation is exacerbated when $\Delta t_{C<1}$ is chosen to be very small, $\Delta t_{C<1} \ll \Delta t_{C=1}$. In this case, even though the calculations are stable, the results may be quite *inaccurate* due to the large mismatch between the domain of dependence of point d and the location of the actual numerical data used to calculate properties at d.

In light of the above discussion, we conclude that the Courant number must be equal to or less than unity for stability, $C \leq 1$, but at the same time it is desirable to have C as close to unity as possible for accuracy.

4.5.1. Stability Analysis: A Broader Perspective

The preceding discussion focused on the behavior of *errors* as a means to analyze the stability characteristics of a given difference equation; in particular, the behavior of the round-off error ϵ, as defined in Eq. (4.55), was studied. This might leave the incorrect impression that if we had a *perfect computer* with no round-off error, then there would be no instabilities. Such is not the case. The general concept of numerical stability is, in reality, based on the timewise behavior of the *solution* itself; it does not inherently depend on the behavior of the round-off error per se. For example, instead of considering Eq. (4.60), we could set up a more general von Neumann analysis where the *solution itself* is written as a Fourier series as follows:

$$U(x) = \sum_m V_m e^{ik_m x} \qquad (4.89)$$

where V_m is the amplitude of the mth harmonic of the solution. In turn, the amplification factor is written as

$$G = \frac{V_m^{n+1}}{V_m^n} \qquad (4.90)$$

For stability $|G| \leq 1$.

This, and other considerations of stability, are left for your future studies. Our purpose here has been simply to introduce you to the basic thought that stability considerations are important to CFD and to give you some idea, no matter how incomplete, how these considerations can be approached.

4.6 SUMMARY

"*Discretization*" has been the key word in the present chapter. We have seen how to discretize partial differential equations, including the governing flow equations. Such discretization is the foundation of finite-difference methods. In addition, via Problem 4.7, you will see how to discretize the governing flow equations in integral form. Such discretization is the foundation of finite-volume methods. Both finite-difference and finite-volume methods abound in CFD. However, keep in mind that the discretizations discussed in this chapter are simply *tools*; they do not by themselves constitute any specific *technique* for the solution of a given flow problem. A CFD *technique* is defined by what tools we choose for a solution, how and in what sequence we use these tools to pursue a solution, and how we handle the boundary conditions. Some techniques that have been popular in CFD are discussed in Chap. 6, and applications of these techniques to some classic fluid flow problems are illustrated in detail in Part III of this book. On the other hand, in the present chapter we *have* touched on a few important aspects of CFD techniques without detailing the techniques themselves. For example, we have noted that any CFD technique falls within one or the other of two general categories, explicit approaches or implicit approaches. We have discussed to some extent just what we mean by explicit and implicit approaches. Furthermore, we have touched on the stability aspects of these approaches and have examined the von Neumann stability analysis which gives us some insight to the stability restrictions and stability criterion for explicit methods. So with this chapter we have taken a giant step in the world of CFD.

Before progressing further, return to Fig. 4.2, the road map for this chapter, and make certain that you feel comfortable with the material which constitutes each block in this road map. Note that we have emphasized the finite-difference method and have chosen not to address the finite-volume or finite-element methods. Finite-element methods are still not used to any great extent in CFD; finite-difference and finite-volume methods account for about 95 percent of all practical CFD solutions. In time, this situation may change. We note that the situation is reversed in structural mechanics, when the numerical method of choice is almost always the finite-element method. The road map in Fig. 4.2 also emphasizes that the matters of implicit versus explicit methods, as well as matters of stability, are common to both the finite-difference and finite-volume methods.

Finally, we note that we are not quite ready to go directly to a discussion of CFD techniques. There remains one item of unfinished business, namely, the aspects of grid generation and the necessary transformations which it entails. This is the subject of the next chapter.

GUIDEPOST

We have reached the stage where you have enough information to actually set up a meaningful calculation for certain types of flows. This guidepost is intended for those of you who want to get on the fast track toward "getting your hands dirty" with an actual computer project. For those of you who want to add more depth to your background in CFD *before* tackling a computer project, simply continue to read on. We will delve into the matters of grid generation and transformation in the next chapter—very important material for the application discussed in Chap. 8. However, for those of you who are really tired of reading at this stage and want to work on a computer project, the following guidepost is suggested. This guidepost will take you through implicit and explicit solutions of a special incompressible viscous flow problem: Couette flow. To carry through this application, you need essentially no more information than you already have. Therefore,

Go to Secs. 9.1 to → Then go to
9.3, Couette flow solution   Prob. 9.1, explicit
(implicit).   solution of Couette flow.

The flow of information associated with the above guidepost is diagrammed in Fig. 1.32a, which you should briefly reexamine at this time. Note that only a few essential aspects are being driven home with this excursion, essentially only those associated with implicit and explicit finite-difference philosophies.

There is yet another but more lengthy option that you can take from here. Another guidepost will take you through an explicit, time-marching solution of the Euler equations for a quasi-one-dimensional nozzle flow. This excursion is more demanding of your time, but again it gives you an opportunity to tackle a computer project with essentially the information you now have. Therefore,

> Go to Sec. 6.3, MacCormacks's technique. → Then go to Sec. 7.1 to 7.4, the computer solution of isentropic subsonic and supersonic nozzle flows.

The flow of information associated with the above guidepost is diagramed in Fig. 1.32b, which you should briefly reexamine at this time. Note that the essential aspects being driven home are those of time-marching finite-difference solutions, some of which involve the concept of artificial viscosity.

Should you choose to undertake either or both of these computer projects at this stage, make certain after you are finished to return to the point we are at now and continue with your general reading in CFD. We will move on to the subject of grid generation in Chap. 5 and then cover a number of different CFD techniques in Chap. 6.

PROBLEMS

4.1. Using Taylor's series, derive first-order forward-difference and rearward-difference expressions for $\partial u/\partial y$.

4.2. Using Taylor's series, derive the second-order central difference for $\partial u/\partial y$.

4.3. Consider the function $\phi(x, y) = e^x + e^y$. Consider the point $(x, y) = (1, 1)$.
   (a) Calculate the *exact* values of $\partial \phi/\partial x$ and $\partial \phi/\partial y$ at this point.
   (b) Use first-order forward differences, with $\Delta x = \Delta y = 0.1$, to calculate approximate values of $\partial \phi/\partial x$ and $\partial \phi/\partial y$ at point (1, 1). Calculate the percentage difference when compared with the exact values from part (a).
   (c) Use first-order rearward differences, with $\Delta x = \Delta y = 0.1$, to calculate approximate values of $\partial \phi/\partial x$ and $\partial \phi/\partial y$ at point (1, 1). Calculate the percentage difference when compared with the exact values from part (a).
   (d) Use second-order central differences, with $\Delta x = \Delta y = 0.1$, to calculate approximate values for $\partial \phi/\partial x$ and $\partial \phi/\partial y$ at point (1, 1). Calculate the percentage difference when compared with the exact value from part (a).

4.4. Repeat Prob. 4.3, but with $\Delta x = \Delta y = 0.01$. Compare the accuracy of the finite-difference results obtained here with those obtained in Prob. 4.3.

4.5. Derive Eq. (4.18).

4.6. Derive the following expression, which is a third-order-accurate one-sided difference.
$$\left(\frac{\partial u}{\partial y}\right)_{i,j} = \frac{1}{6\Delta y}(-11u_{i,j} + 18u_{i,j+1} - 9u_{i,j+2} + 2u_{i,j+3})$$

4.7. Derive a discretized form of the generic integral form of the continuity, momentum, and energy equations obtained in Prob. 2.2. This discretized form is the essence of the *finite volume approach*.

CHAPTER 5

GRIDS WITH APPROPRIATE TRANSFORMATIONS

The area of numerical grid generation is relatively young in practice, although its roots in mathematics are old. This somewhat eclectic area involves the engineer's feel for physical behavior, the mathematician's understanding of functional behavior, and a lot of imagination, with perhaps a little help from Urania.

Joe F. Thompson, Z. V. A. Warsi, and C. Wayne Mastin,
from *Numerical Grid Generation*, North-Holland,
New York, 1985

5.1 INTRODUCTION

Think about the finite-difference approach discussed in Chap. 4; it requires that calculations be made over a collection of discrete grid points. The *arrangement* of these discrete points throughout the flow field is simply called a *grid*. The determination of a proper grid for the flow over or though a given geometric shape is a serious matter—one that is by no means trivial. The *way* that such a grid is determined is called *grid generation*. The matter of grid generation is a significant consideration in CFD; the type of grid you choose for a given problem can make or break the numerical solution. Because of this, grid generation has become an entity by itself in CFD; it is the subject of numerous special conferences, as well as several books (see Refs. 26 and 27).

The generation of an appropriate grid or mesh is one thing; the solution of the governing flow equations over such a grid is quite another thing. Assume that (for reasons to be discussed later) we construct a nonuniform grid in our flow field. We have seen in Sec. 4.2 that the standard finite-difference approach requires a uniform grid. We do not have a direct way of numerically solving the governing flow equations over a nonuniform grid within the context of a finite-difference method. Instead, the nonuniform grid must (somehow) be *transformed* into a uniform, rectangular grid. Moreover, along with this transformation, the governing partial differential equations must be recast so as to apply in this transformed, rectangular grid. Since the need for such grid transformations is inherent in the finite-difference method, then much of what we have to say in this chapter concerning the transformation of the governing partial differential equations pertains just to that method. Let us proceed accordingly.

If all CFD applications dealt with physical problems where a uniform, rectangular grid could be used in the physical plane, there would be no reason to alter the governing partial differential equations derived in Chap. 2. We would simply apply these equations in rectangular (x, y, z, t) space, finite difference these equations according to the difference quotients derived in Sec. 4.2 and 4.3, and calculate away, using uniform values of Δx, Δy, Δz, and Δt. However, few real problems are ever so accommodating. For example, assume we wish to calculate the flow over an airfoil, as sketched in Fig. 5.1. In Fig. 5.1, we have placed the airfoil in a rectangular grid. Note the problems with this rectangular grid:

1. Some grid points fall inside the airfoil, where they are completely out of the flow. What values of the flow properties do we ascribe to these points?
2. There are few, if any, grid points that fall on the surface of the airfoil. This is not good, because the airfoil surface is a vital boundary condition for the determination of the flow, and hence the airfoil surface must be clearly and strongly seen by the numerical solution.

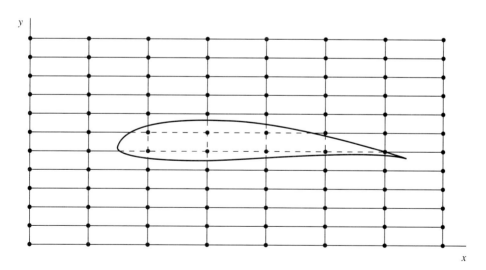

FIG. 5.1
An airfoil in a purely rectangular grid.

170 GRIDS WITH APROPRIATE TRANSFORMATIONS

As a result, we can conclude that the rectangular grid in Fig. 5.1 is not appropriate for the solution of the flow field. In contrast, a grid that *is* appropriate is sketched in Fig. 5.2a. Here we see a *nonuniform, curvilinear* grid which is literally wrapped around the airfoil. New coordinate lines ξ and η are defined such that the airfoil surface becomes a coordinate line, η = constant. This is called a boundary-fitted coordinate system, and will be discussed in detail in Sec. 5.7. The important point is that grid points *naturally* fall on the airfoil surface, as shown in Fig. 5.2a. What is equally important is that, in the physical space shown in Fig. 5.2a, the grid is *not* rectangular and is not uniformly spaced. As a consequence, the conventional difference quotients are difficult to use. What must be done is to *transform* the curvilinear grid in physical space to a *rectangular grid* in terms of ξ and η. This is

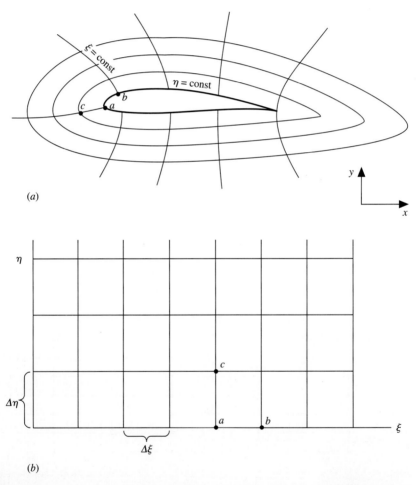

FIG. 5.2
Schematic of a boundary-fitted coordinate system. (*a*) Physical plane; (*b*) computational plane.

shown in Fig. 5.2b, which illustrates a rectangular grid in terms of ξ and η. The rectangular grid shown in Fig. 5.2b is called the *computational plane*. The transformation must be defined such that there is a one-to-one correspondence between the rectangular grid in Fig. 5.2b and the curvilinear grid in Fig. 5.2a, called the *physical plane*. For example, points a, b, and c in the physical plane (Fig. 5.2a) correspond to points a, b, and c in the computational plane, which involves uniform $\Delta \xi$ and uniform $\Delta \eta$. The governing partial differential equations are solved by a finite-difference method carried out in the computational space (Fig. 5.2b). Then the computed information is directly carried back to the physical plane via the one-to-one correspondence of grid points. Moreover, when the governing equations are solved in the computational space, they must be expressed in terms of the variables ξ and η rather than of x and y. *That is, the governing equations must be transformed from (x, y) to (ξ, η) as the new independent variables.*

The purpose of this chapter is to first describe the general transformation of the governing partial differential equations between the physical plane and the computational plane. Following this, various specific grids will be discussed. As stated earlier, this material is an example of a very active area of CFD research called *grid generation*. In this sense, we can only scratch the surface of activity in the present chapter; however, what is presented in this chapter is sufficient to give you the basic ideas and philosophy of grid generation and how it relates to the overall, larger picture of CFD in general.

The road map for this chapter is presented in Fig. 5.3. The general aspects of the transformation process are reflected in the left column of boxes, all of which then feed into the grid generation process reflected in the right column of boxes. We now proceed to examine what is meant by the derivative transformation in the next section; i.e., we proceed to the first box in the left-hand column.

5.2 GENERAL TRANSFORMATION OF THE EQUATIONS

For simplicity, we will consider a two-dimensional unsteady flow, with independent variables x, y, and t; the results for a three-dimensional unsteady flow, with independent variables x, y, z, and t, are analogous and simply involve more terms.

We will transform the independent variables in physical space (x, y, t) to a new set of independent variables in transformed space (ξ, η, τ), where

$$\xi = \xi(x, y, t) \tag{5.1a}$$

$$\eta = \eta(x, y, t) \tag{5.1b}$$

$$\tau = \tau(t) \tag{5.1c}$$

Equations (5.1a) to (5.1c) represent *the transformation*. For the time being, the transformation is written in generic form; for an actual application, the transformation represented by Eqs. (5.1a) to (5.1c) must be given as some type of *specific* analytical relation, or sometimes a specific numerical relation. In the above transformation, τ is considered a function of t only and is frequently given by $\tau = t$. This seems rather trivial; however, Eq. (5.1c) must be carried through the

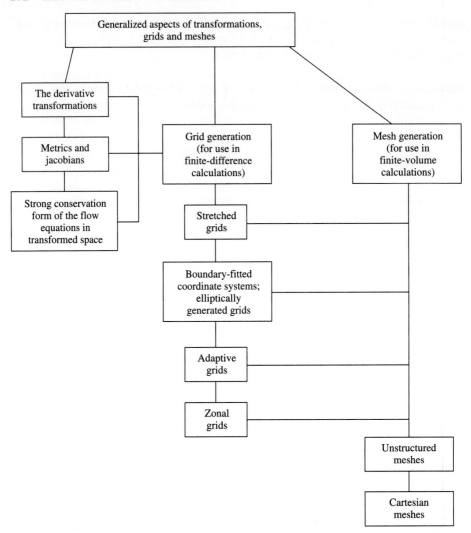

FIG. 5.3
Road map for Chapter 5.

transformation in a formal manner or else certain necessary terms will not be generated.

Consider one or more partial differential equations written in terms of x, y, and t as the independent variables—say, for example, the continuity, momentum, and energy equations derived in Chap. 2. In these equations, the independent variables appear in the form of *derivatives*, such as $\partial \rho/\partial x$, $\partial u/\partial y$, $\partial e/\partial t$. Therefore, to transform these equations from (x, y, t) space to (ξ, η, τ) space, we need a transformation for the *derivatives*; i.e., we need to replace the x, y, and t derivatives in the original partial differential equations with corresponding derivatives with respect to ξ, η, and τ. In other words, we need to replace $\partial u/\partial y$ with some combination of $\partial u/\partial \xi$, $\partial u/\partial \eta$, etc. This derivative transformation can be obtained from the original transformation given by Eqs. (5.1a) to (5.1c) as follows. From the chain rule of differential calculus, we have

$$\left(\frac{\partial}{\partial x}\right)_{y,t} = \left(\frac{\partial}{\partial \xi}\right)_{\eta,\tau}\left(\frac{\partial \xi}{\partial x}\right)_{y,t} + \left(\frac{\partial}{\partial \eta}\right)_{\xi,\tau}\left(\frac{\partial \eta}{\partial x}\right)_{y,t} + \left(\frac{\partial}{\partial \tau}\right)_{\xi,\eta}\left(\frac{\partial \tau}{\partial x}\right)_{y,t}$$

The subscripts in the above expression are added to emphasize what variables are being held constant in the partial differentiation. In our subsequent expressions, subscripts will be dropped; however, it is always useful to keep them in your mind. Thus, we will write the above expression as

$$\boxed{\frac{\partial}{\partial x} = \left(\frac{\partial}{\partial \xi}\right)\left(\frac{\partial \xi}{\partial x}\right) + \left(\frac{\partial}{\partial \eta}\right)\left(\frac{\partial \eta}{\partial x}\right)} \qquad (5.2)$$

Similarly,

$$\boxed{\frac{\partial}{\partial y} = \left(\frac{\partial}{\partial \xi}\right)\left(\frac{\partial \xi}{\partial y}\right) + \left(\frac{\partial}{\partial \eta}\right)\left(\frac{\partial \eta}{\partial y}\right)} \qquad (5.3)$$

Also,

$$\left(\frac{\partial}{\partial t}\right)_{x,y} = \left(\frac{\partial}{\partial \xi}\right)_{\eta,\tau}\left(\frac{\partial \xi}{\partial t}\right)_{x,y} + \left(\frac{\partial}{\partial \eta}\right)_{\xi,\tau}\left(\frac{\partial \eta}{\partial t}\right)_{x,y} + \left(\frac{\partial}{\partial \tau}\right)_{\xi,\eta}\left(\frac{\partial \tau}{\partial t}\right)_{x,y} \qquad (5.4)$$

or

$$\boxed{\frac{\partial}{\partial t} = \left(\frac{\partial}{\partial \xi}\right)\left(\frac{\partial \xi}{\partial t}\right) + \left(\frac{\partial}{\partial \eta}\right)\left(\frac{\partial \eta}{\partial t}\right) + \left(\frac{\partial}{\partial \tau}\right)\left(\frac{d\tau}{dt}\right)} \qquad (5.5)$$

Equations (5.2), (5.3), and (5.5) allow the *derivatives* with respect to x, y, and t to be transformed into derivatives with respect to ξ, η, and τ. For example, in the governing flow equations, such as Eqs. (2.29), (2.33), (2.50a) to (2.50c), (2.56a) to (2.56c), (2.66), and (2.81), wherever you see a derivative with respect to x, replace it with Eq. (5.2); wherever you see a derivative with respect to y, replace it with Eq. (5.3); and wherever you see a derivative with respect to t, replace it with Eq. (5.5). The *coefficients* of the derivatives with respect to ξ, η, and τ are called *metrics*; for example, $\partial\xi/\partial x$, $\partial\xi/\partial y$, $\partial\eta/\partial x$, and $\partial\eta/\partial y$ are metric terms which can be obtained from the general transformation given by Eqs. (5.1a) to (5.1c). If Eqs. (5.1a) to (5.1c) are given as closed-form analytic expressions, then the metrics can also be obtained in closed form. However, the transformation given by Eqs. (5.1a) to (5.1c) is frequently a purely numerical relationship, in which case the metrics can be evaluated by finite-difference quotients, typically central differences.

Examining the equations which govern a *viscous* flow, as derived in Chap. 2, we see that they involve second derivatives which appear in the viscous terms. Examples are Eqs. (2.58a) to (2.58c), which involve terms such as $\partial/\partial x\, (\mu\, \partial v/\partial x)$. Therefore, we need a transformation for these derivatives; they can be obtained as follows. From Eq. (5.2), let

$$A = \frac{\partial}{\partial x} = \left(\frac{\partial}{\partial \xi}\right)\left(\frac{\partial \xi}{\partial x}\right) + \left(\frac{\partial}{\partial \eta}\right)\left(\frac{\partial \eta}{\partial x}\right)$$

Then,

$$\frac{\partial^2}{\partial x^2} = \frac{\partial A}{\partial x} = \frac{\partial}{\partial x}\left[\left(\frac{\partial}{\partial \xi}\right)\left(\frac{\partial \xi}{\partial x}\right) + \left(\frac{\partial}{\partial \eta}\right)\left(\frac{\partial \eta}{\partial x}\right)\right]$$

$$= \left(\frac{\partial}{\partial \xi}\right)\left(\frac{\partial^2 \xi}{\partial x^2}\right) + \underbrace{\left(\frac{\partial \xi}{\partial x}\right)\left(\frac{\partial^2}{\partial x\, \partial \xi}\right)}_{B} + \left(\frac{\partial}{\partial \eta}\right)\left(\frac{\partial^2 \eta}{\partial x^2}\right) + \underbrace{\left(\frac{\partial \eta}{\partial x}\right)\left(\frac{\partial^2}{\partial x\, \partial \eta}\right)}_{C} \quad (5.6)$$

The last step in Eq. (5.6) is obtained from the simple rule for differentiation of a product of two terms. The derivatives denoted by B and C in Eq. (5.6) are a "mixed bag"; they involve differentiation with respect to one variable in the (x, y, t) system and another variable in the (ξ, η, τ) system. This is not desirable, because we want to express $\partial^2/\partial x^2$ purely in terms of derivatives with respect to ξ, η, and τ. Therefore, we need to work further with the terms denoted by B and C. The mixed derivatives denoted by B and C can be reexpressed as follows.

$$B = \frac{\partial^2}{\partial x\, \partial \xi} = \frac{\partial}{\partial x}\left(\frac{\partial}{\partial \xi}\right)$$

Recalling the chain rule given by Eq. (5.2), we have

$$B = \left(\frac{\partial^2}{\partial \xi^2}\right)\left(\frac{\partial \xi}{\partial x}\right) + \left(\frac{\partial^2}{\partial \eta\, \partial \xi}\right)\left(\frac{\partial \eta}{\partial x}\right) \quad (5.7)$$

Similarly,

$$C = \frac{\partial^2}{\partial x\, \partial \eta} = \frac{\partial}{\partial x}\left(\frac{\partial}{\partial \eta}\right) = \left(\frac{\partial^2}{\partial \xi\, \partial \eta}\right)\left(\frac{\partial \xi}{\partial x}\right) + \left(\frac{\partial^2}{\partial \eta^2}\right)\left(\frac{\partial \eta}{\partial x}\right) \quad (5.8)$$

Substituting B and C from Eqs. (5.7) and (5.8) into Eq. (5.6) and rearranging the sequence of terms, we have

$$\boxed{\frac{\partial^2}{\partial x^2} = \left(\frac{\partial}{\partial \xi}\right)\left(\frac{\partial^2 \xi}{\partial x^2}\right) + \left(\frac{\partial}{\partial \eta}\right)\left(\frac{\partial^2 \eta}{\partial x^2}\right) + \left(\frac{\partial^2}{\partial \xi^2}\right)\left(\frac{\partial \xi}{\partial x}\right)^2 \\ + \left(\frac{\partial^2}{\partial \eta^2}\right)\left(\frac{\partial \eta}{\partial x}\right)^2 + 2\left(\frac{\partial^2}{\partial \eta \, \partial \xi}\right)\left(\frac{\partial \eta}{\partial x}\right)\left(\frac{\partial \xi}{\partial x}\right)} \qquad (5.9)$$

Equation (5.9) gives the second partial derivative with respect to x in terms of first, second, and mixed derivatives with respect to ξ and η, multiplied by various metric terms. Let us now continue to obtain the second partial with respect to y. From Eq. (5.3), let

$$D \equiv \frac{\partial}{\partial y} = \left(\frac{\partial}{\partial \xi}\right)\left(\frac{\partial \xi}{\partial y}\right) + \left(\frac{\partial}{\partial \eta}\right)\left(\frac{\partial \eta}{\partial y}\right)$$

Then,

$$\frac{\partial^2}{\partial y^2} = \frac{\partial D}{\partial y} = \frac{\partial}{\partial y}\left[\left(\frac{\partial}{\partial \xi}\right)\left(\frac{\partial \xi}{\partial y}\right) + \left(\frac{\partial}{\partial \eta}\right)\left(\frac{\partial \eta}{\partial y}\right)\right]$$

$$= \left(\frac{\partial}{\partial \xi}\right)\left(\frac{\partial^2 \xi}{\partial y^2}\right) + \underbrace{\left(\frac{\partial \xi}{\partial y}\right)\left(\frac{\partial^2}{\partial \xi \, \partial y}\right)}_{E} + \left(\frac{\partial}{\partial \eta}\right)\left(\frac{\partial^2 \eta}{\partial y^2}\right) + \underbrace{\left(\frac{\partial \eta}{\partial y}\right)\left(\frac{\partial^2}{\partial \eta \, \partial y}\right)}_{F} \qquad (5.10)$$

Using Eq. (5.3),

$$E = \frac{\partial}{\partial y}\left(\frac{\partial}{\partial \xi}\right) = \left(\frac{\partial^2}{\partial \xi^2}\right)\left(\frac{\partial \xi}{\partial y}\right) + \left(\frac{\partial^2}{\partial \eta \, \partial \xi}\right)\left(\frac{\partial \eta}{\partial y}\right) \qquad (5.11)$$

and

$$F = \frac{\partial}{\partial y}\left(\frac{\partial}{\partial \eta}\right) = \left(\frac{\partial^2}{\partial \eta \, \partial \xi}\right)\left(\frac{\partial \xi}{\partial y}\right) + \left(\frac{\partial^2}{\partial \eta^2}\right)\left(\frac{\partial \eta}{\partial y}\right) \qquad (5.12)$$

Substituting Eqs. (5.11) and (5.12) into Eq. (5.10), we have, after rearranging the sequence of terms,

$$\boxed{\frac{\partial^2}{\partial y^2} = \left(\frac{\partial}{\partial \xi}\right)\left(\frac{\partial^2 \xi}{\partial y^2}\right) + \left(\frac{\partial}{\partial \eta}\right)\left(\frac{\partial^2 \eta}{\partial y^2}\right) + \left(\frac{\partial^2}{\partial \xi^2}\right)\left(\frac{\partial \xi}{\partial y}\right)^2 \\ + \left(\frac{\partial^2}{\partial \eta^2}\right)\left(\frac{\partial \eta}{\partial y}\right)^2 + 2\left(\frac{\partial^2}{\partial \eta \, \partial \xi}\right)\left(\frac{\partial \eta}{\partial y}\right)\left(\frac{\partial \xi}{\partial y}\right)} \qquad (5.13)$$

Equation (5.13) gives the second partial derivative with respect to y in terms of first, second, and mixed derivatives with respect to ξ and η, multiplied by various metric terms. We now continue to obtain the second partial with respect to x and y.

$$\frac{\partial^2}{\partial x\,\partial y} = \frac{\partial}{\partial x}\left(\frac{\partial}{\partial y}\right) = \frac{\partial D}{\partial x} = \frac{\partial}{\partial x}\left[\left(\frac{\partial}{\partial \xi}\right)\left(\frac{\partial \xi}{\partial y}\right) + \left(\frac{\partial}{\partial \eta}\right)\left(\frac{\partial \eta}{\partial y}\right)\right]$$

$$= \underbrace{\left(\frac{\partial}{\partial \xi}\right)\left(\frac{\partial^2 \xi}{\partial x\,\partial y}\right) + \left(\frac{\partial \xi}{\partial y}\right)\left(\frac{\partial^2}{\partial \xi\,\partial x}\right)}_{B} + \underbrace{\left(\frac{\partial}{\partial \eta}\right)\left(\frac{\partial^2 \eta}{\partial x\,\partial y}\right) + \left(\frac{\partial \eta}{\partial y}\right)\left(\frac{\partial^2}{\partial \eta\,\partial x}\right)}_{C}$$

(5.14)

Substituting Eqs. (5.7) and (5.8) for B and C, respectively, into Eq. (5.14) and rearranging the sequence of terms, we have

$$\boxed{\begin{aligned}\frac{\partial^2}{\partial x\,\partial y} &= \left(\frac{\partial}{\partial \xi}\right)\left(\frac{\partial^2 \xi}{\partial x\,\partial y}\right) + \left(\frac{\partial}{\partial \eta}\right)\left(\frac{\partial^2 \eta}{\partial x\,\partial y}\right) + \left(\frac{\partial^2}{\partial \xi^2}\right)\left(\frac{\partial \xi}{\partial x}\right)\left(\frac{\partial \xi}{\partial y}\right) \\ &+ \left(\frac{\partial^2}{\partial \eta^2}\right)\left(\frac{\partial \eta}{\partial x}\right)\left(\frac{\partial \eta}{\partial y}\right) + \left(\frac{\partial^2}{\partial \xi\,\partial \eta}\right)\left[\left(\frac{\partial \eta}{\partial x}\right)\left(\frac{\partial \xi}{\partial y}\right) + \left(\frac{\partial \xi}{\partial x}\right)\left(\frac{\partial \eta}{\partial y}\right)\right]\end{aligned}}$$

(5.15)

Equation (5.15) gives the second partial derivative with respect to x and y in terms of first, second, and mixed derivatives with respect to ξ and η, multiplied by various metric terms.

Examine all the equations given in the boxes above. They represent all that is necessary to transform the governing flow equations obtained in Chap. 2 with x, y, and t as the independent variables to ξ, η, and τ as the *new* independent variables. Clearly, when this transformation is made, the governing equations in terms of ξ, η, and τ become rather lengthy. Let us consider a simple example, namely, that for inviscid, irrotational, steady, incompressible flow, for which Laplace's equation is the governing equation.

Example 5.1

Laplace's equation $\qquad \dfrac{\partial^2 \phi}{\partial x^2} + \dfrac{\partial^2 \phi}{\partial y^2} = 0 \qquad$ (5.16)

Transforming Eq. (5.16) from (x, y) to (ξ, η), where $\xi = \xi(x, y)$ and $\eta = \eta(x, y)$, we have from Eqs. (5.9) and (5.13),

$$\left(\frac{\partial^2 \phi}{\partial \xi^2}\right)\left(\frac{\partial \xi}{\partial x}\right)^2 + 2\left(\frac{\partial^2 \phi}{\partial \xi\,\partial \eta}\right)\left(\frac{\partial \eta}{\partial x}\right)\left(\frac{\partial \xi}{\partial x}\right) + \left(\frac{\partial^2 \phi}{\partial \eta^2}\right)\left(\frac{\partial \eta}{\partial x}\right)^2 + \left(\frac{\partial \phi}{\partial \xi}\right)\left(\frac{\partial^2 \xi}{\partial x^2}\right)$$

$$+ \left(\frac{\partial \phi}{\partial \eta}\right)\left(\frac{\partial^2 \eta}{\partial x^2}\right) + \left(\frac{\partial^2 \phi}{\partial \xi^2}\right)\left(\frac{\partial \xi}{\partial y}\right)^2 + 2\left(\frac{\partial^2 \phi}{\partial \eta\,\partial \xi}\right)\left(\frac{\partial \eta}{\partial y}\right)\left(\frac{\partial \xi}{\partial y}\right) + \left(\frac{\partial^2 \phi}{\partial \eta^2}\right)\left(\frac{\partial \eta}{\partial y}\right)^2$$

$$+ \left(\frac{\partial \phi}{\partial \xi}\right)\left(\frac{\partial^2 \xi}{\partial y^2}\right) + \left(\frac{\partial \phi}{\partial \eta}\right)\left(\frac{\partial^2 \eta}{\partial y^2}\right) = 0$$

Factoring some of the terms in the above equation, we have finally

$$\frac{\partial^2 \phi}{\partial \xi^2}\left[\left(\frac{\partial \xi}{\partial x}\right)^2 + \left(\frac{\partial \xi}{\partial y}\right)^2\right] + \frac{\partial^2 \phi}{\partial \eta^2}\left[\left(\frac{\partial \eta}{\partial x}\right)^2 + \left(\frac{\partial \eta}{\partial y}\right)^2\right]$$

$$+ 2\frac{\partial^2 \phi}{\partial \xi \, \partial \eta}\left[\left(\frac{\partial \eta}{\partial x}\right)\left(\frac{\partial \xi}{\partial x}\right) + \left(\frac{\partial \eta}{\partial y}\right)\left(\frac{\partial \xi}{\partial y}\right)\right]$$

$$+ \frac{\partial \phi}{\partial \xi}\left(\frac{\partial^2 \xi}{\partial x^2} + \frac{\partial^2 \xi}{\partial y^2}\right) + \frac{\partial \phi}{\partial \eta}\left(\frac{\partial^2 \eta}{\partial x^2} + \frac{\partial^2 \eta}{\partial y^2}\right) = 0 \quad (5.17)$$

Examine Eqs. (5.16) and (5.17); the former is Laplace's equation in the physical (x, y) space, and the latter is the transformed Laplace's equation in the computational (ξ, η) space. The transformed equation clearly contains many more terms. It is easy to mentally extrapolate what the governing continuity, momentum, and energy equations as derived in Chap. 2 would look like in the transformed space—lots and lots of terms.

Note: The need to apply the transformations for the second derivatives, namely, those given by Eqs. (5.9), (5.13), and (5.15), disappears when the governing flow equations are used in the strong conservation form expressed by Eq. (2.93). For a moment, return to Sec. 2.10 and examine Eq. (2.93), as well as the definitions of the column vectors given by Eqs. (2.94) to (2.98). Note that the viscous terms in F, G, and H, expressed by Eqs. (2.95) to (2.97), appear directly in the form τ_{xx}, τ_{xy}, $k\, \partial T/\partial x$, etc. These terms involve only first derivatives of the velocity (such as $\partial u/\partial x$, $\partial u/\partial y$) or first derivatives of the temperature. For the general transformation of these terms *inside* F, G, and H, only the transformation of the first derivatives, such as given by Eqs. (5.2) and (5.3), is needed. In turn, the first derivatives which appear in Eq. (2.93) are also transformed via Eqs. (5.2), (5.3), and (5.5). Therefore, when the governing flow equations are used in the form of Eq. (2.93), the transformation is carried out via a dual application of the first derivatives, i.e., a dual application of Eqs. (5.2), (5.3), and (5.5). In contrast, the governing equations expressed in the form of Eqs. (2.58a) to (2.58c), for example, have the viscous terms appearing directly as second derivatives. For a transformation of the governing equations in this form, both the first-derivative transformation, Eqs. (5.2), (5.3), and (5.5), and the second-derivative transformation, Eqs. (5.9), (5.13), and (5.15), are needed.

Once again we emphasize that Eqs. (5.1) to (5.3), (5.5), (5.9), (5.13), and (5.15) are used to transform the governing flow equations from the physical plane [(x, y) space] to the computational plane [(ξ, η) space], and that the purpose of the transformation in most CFD applications is to transform a nonuniform grid in physical space (such as shown in Fig. 5.2a) to a *uniform* grid in the computational space (such as shown in Fig. 5.2b). The transformed governing partial differential equations are then finite-differenced in the computational plane, where there exists a uniform $\Delta \xi$ and a uniform $\Delta \eta$, as shown in Fig. 5.2b. The flow-field variables are calculated at all grid points in the computational plane, such as points a, b, and c in Fig. 5.2b. These are the same flow-field variables which exist in the physical plane at the corresponding points a, b, and c in Fig. 5.2a. The transformation that accomplishes all this is given in general form by Eqs. (5.1a) to (5.1c). Of course, to carry out a solution for a given problem, the transformation given generically by Eqs. (5.1a) to (5.1c) must be explicitly specified. Examples of some specific transformations will be given in subsequent sections.

5.3 METRICS AND JACOBIANS

In Eqs. (5.2) to (5.15), the terms involving the geometry of the grid, such as $\partial \xi/\partial x$, $\partial \xi/\partial y$, $\partial \eta/\partial x$, $\partial \eta/\partial y$, are called *metrics*. If the transformation, Eqs. (5.1a) to (5.1c), is given analytically, then it is possible to obtain analytic values for the metric terms. However, in many CFD applications, the transformation, Eqs. (5.1a) to (5.1c), is given numerically, and hence the metric terms are calculated as finite differences.

Also, in many applications, the transformation may be more conveniently expressed as the *inverse* of Eqs. (5.1a) to (5.1c); that is, we may have available the *inverse transformation*

$$x = x(\xi, \eta, \tau) \tag{5.18a}$$

$$y = y(\xi, \eta, \tau) \tag{5.18b}$$

$$t = t(\tau) \tag{5.18c}$$

In Eqs. (5.18a) to (5.18c), ξ, η, and τ are the *independent* variables. However, in the derivative transformations given by Eqs. (5.2) to (5.15), the metric terms $\partial \xi/\partial x$, $\partial \eta/\partial y$, etc., are partial derivatives in terms of x, y, and t as the independent variables. Therefore, in order to calculate the metric terms in these equations from the inverse transformation in Eqs. (5.18a) to (5.18c), we need to relate $\partial \xi/\partial x$, $\partial \eta/\partial y$, etc., to the inverse forms $\partial x/\partial \xi$, $\partial y/\partial \eta$, etc. These inverse forms of the metrics are the values which can be directly obtained from the inverse transformation, Eqs. (5.18a) to (5.18c). Let us proceed to find such relations.

Consider a dependent variable in the governing flow equations, such as the x component of velocity, u. Let $u = u(x, y)$, where from Eqs. (5.18a) and (5.18b), $x = x(\xi, \eta)$ and $y = y(\xi, \eta)$. The total differential of u is given by

$$du = \frac{\partial u}{\partial x} dx + \frac{\partial u}{\partial y} dy \tag{5.19}$$

It follows from Eq. (5.19) that

$$\frac{\partial u}{\partial \xi} = \frac{\partial u}{\partial x}\frac{\partial x}{\partial \xi} + \frac{\partial u}{\partial y}\frac{\partial y}{\partial \xi} \tag{5.20}$$

and

$$\frac{\partial u}{\partial \eta} = \frac{\partial u}{\partial x}\frac{\partial x}{\partial \eta} + \frac{\partial u}{\partial y}\frac{\partial y}{\partial \eta} \tag{5.21}$$

Equations (5.20) and (5.21) can be viewed as two equations for the two unknowns $\partial u/\partial x$ and $\partial u/\partial y$. Solving the system of Eqs. (5.20) and (5.21) for $\partial u/\partial x$ using Cramer's rule, we have

$$\frac{\partial u}{\partial x} = \frac{\begin{vmatrix} \dfrac{\partial u}{\partial \xi} & \dfrac{\partial y}{\partial \xi} \\ \dfrac{\partial u}{\partial \eta} & \dfrac{\partial y}{\partial \eta} \end{vmatrix}}{\begin{vmatrix} \dfrac{\partial x}{\partial \xi} & \dfrac{\partial y}{\partial \xi} \\ \dfrac{\partial x}{\partial \eta} & \dfrac{\partial y}{\partial \eta} \end{vmatrix}} \tag{5.22}$$

In Eq. (5.22), the denominator determinant is identified as the *jacobian determinant*, denoted by

$$J \equiv \frac{\partial(x, y)}{\partial(\xi, \eta)} \equiv \begin{vmatrix} \dfrac{\partial x}{\partial \xi} & \dfrac{\partial y}{\partial \xi} \\ \dfrac{\partial x}{\partial \eta} & \dfrac{\partial y}{\partial \eta} \end{vmatrix} \tag{5.22a}$$

Therefore, Eq. (5.22) can be written in the following form, where the numerator determinant is displayed in its expanded form

$$\frac{\partial u}{\partial x} = \frac{1}{J}\left[\left(\frac{\partial u}{\partial \xi}\right)\left(\frac{\partial y}{\partial \eta}\right) - \left(\frac{\partial u}{\partial \eta}\right)\left(\frac{\partial y}{\partial \xi}\right)\right] \tag{5.23a}$$

Now let us return to Eqs. (5.20) and (5.21) and solve for $\partial u/\partial y$.

$$\frac{\partial u}{\partial y} = \frac{\begin{vmatrix} \dfrac{\partial x}{\partial \xi} & \dfrac{\partial u}{\partial \xi} \\ \dfrac{\partial x}{\partial \eta} & \dfrac{\partial u}{\partial \eta} \end{vmatrix}}{\begin{vmatrix} \dfrac{\partial x}{\partial \xi} & \dfrac{\partial y}{\partial \xi} \\ \dfrac{\partial x}{\partial \eta} & \dfrac{\partial y}{\partial \eta} \end{vmatrix}}$$

or

$$\frac{\partial u}{\partial y} = \frac{1}{J}\left[\left(\frac{\partial u}{\partial \eta}\right)\left(\frac{\partial x}{\partial \xi}\right) - \left(\frac{\partial u}{\partial \xi}\right)\left(\frac{\partial x}{\partial \eta}\right)\right] \tag{5.23b}$$

Examine Eqs. (5.23a) and (5.23b). They express the derivatives of the flow-field variables in physical space in terms of the derivatives of the flow-field variables in computational space. Equations (5.23a) and (5.23b) accomplish the same derivative transformations as given by Eqs. (5.2) and (5.3). However, unlike Eqs. (5.2) and (5.3) where the metric terms are $\partial \xi/\partial x$, $\partial \eta/\partial y$, etc., the new Eqs. (5.23a) and (5.23b) involve the inverse metrics $\partial x/\partial \xi$, $\partial y/\partial \eta$, etc. Also notice that Eqs. (5.23a) and (5.23b) include the jacobian of the transformation. Therefore, whenever you have the transformation given in the form of Eqs. (5.18a) to (5.18c), from which you can readily obtain the metrics in the form $\partial x/\partial \xi$, $\partial x/\partial \eta$, etc., the transformed governing flow equations can be expressed in terms of these inverse metrics and the jacobian J. To make this discussion more generic, let us write Eqs. (5.23a) and (5.23b) in a slightly more general form.

$$\boxed{\frac{\partial}{\partial x} = \frac{1}{J}\left[\left(\frac{\partial}{\partial \xi}\right)\left(\frac{\partial y}{\partial \eta}\right) - \left(\frac{\partial}{\partial \eta}\right)\left(\frac{\partial y}{\partial \xi}\right)\right]} \quad (5.24a)$$

and

$$\boxed{\frac{\partial}{\partial y} = \frac{1}{J}\left[\left(\frac{\partial}{\partial \eta}\right)\left(\frac{\partial x}{\partial \xi}\right) - \left(\frac{\partial}{\partial \xi}\right)\left(\frac{\partial x}{\partial \eta}\right)\right]} \quad (5.24b)$$

Since the dependent variable u was carried in Eqs. (5.23a) and (5.23b) just as an artifice to derive the inverse transformation, Eqs. (5.24a) and (5.24b) emphasize that the inverse derivative transformation can be applied to *any* dependent variable (not just u). Finally, we note that the second-derivative transformation can also be expressed in terms of the inverse metrics; i.e., there is the analog to Eqs. (5.9), (5.13), and (5.15) which contains the inverse metrics and the jacobian. We will not take the space to derive these analogous expressions here.

It is worthwhile to state the obvious. When in the literature you see the governing flow equations in the transformed coordinates and you see the jacobian J appearing in the transformed equations, you usually know that you are dealing with the inverse transformation and the inverse metrics in these equations. When you do *not* see J in the transformed equations, you are usually dealing with the direct transformation and the direct metrics as originally defined in Sec. 5.2. The only exception to these statements is the material to be discussed in Sec. 5.4. Once again, you are reminded that when you are given the direct transformation as represented by Eqs. (5.1a) to (5.1c), then the direct metrics such as $\partial \xi/\partial x$, $\partial \eta/\partial y$ are most easily obtained from this form of the transformation, and the derivative transformation embodied in Eqs. (5.2), (5.3), and (5.5) is the most straightforward. On the other hand, when you are given the inverse transformation as represented by (5.18a) to (5.18c), then the inverse metrics such as $\partial x/\partial \xi$, $\partial y/\partial \eta$ are most easily obtained, and the derivative transformation embodied in Eqs. (5.24a) and (5.24b) is the most straightforward.

You are reminded that in this chapter we have been treating two spatial variables, x and y. A similar but more lengthy set of results can be obtained for a three-dimensional spatial transformation from (x, y, z) to (ξ, η, ζ). Consult Ref. 13 for more details. Our discussion above has been intentionally limited to two dimensions in order to demonstrate the basic principles without cluttering the consideration with details.

Equations (5.24a) and (5.24b) can be obtained in a slightly more formal manner. Let us examine this more formal approach, because it leads to a general method for dealing directly with the different metrics—a general method which is a fairly direct way of extending the above results to three spatial dimensions, should you need to do so. Again, in the following we will deal with two spatial dimensions for simplicity. Consider the direct transformation for two dimensions, given as

$$\xi = \xi(x, y) \quad (5.25a)$$
$$\eta = \eta(x, y) \quad (5.25b)$$

(*Note*: Comparing Eqs. (5.25a) to (5.25b) with (5.1a) to (5.1c), you will observe that we have dropped $\tau = t$ from the present discussion. Since we are interested in only the spatial metrics in this discussion, the consideration of the transformation of time is not relevant.) From the expression for an exact differential, we have from Eqs. (5.25a) and (5.25b)

$$d\xi = \frac{\partial \xi}{\partial x} dx + \frac{\partial \xi}{\partial y} dy \quad (5.26a)$$

$$d\eta = \frac{\partial \eta}{\partial x} dx + \frac{\partial \eta}{\partial y} dy \quad (5.26b)$$

or, in matrix form,

$$\begin{bmatrix} d\xi \\ d\eta \end{bmatrix} = \begin{bmatrix} \frac{\partial \xi}{\partial x} & \frac{\partial \xi}{\partial y} \\ \frac{\partial \eta}{\partial x} & \frac{\partial \eta}{\partial y} \end{bmatrix} \begin{bmatrix} dx \\ dy \end{bmatrix} \quad (5.27)$$

Now consider the inverse transformation, given by

$$x = x(\xi, \eta) \quad (5.28a)$$
$$y = y(\xi, \eta) \quad (5.28b)$$

Taking the exact differentials, we have

$$dx = \frac{\partial x}{\partial \xi} d\xi + \frac{\partial x}{\partial \eta} d\eta \quad (5.29a)$$

$$dy = \frac{\partial y}{\partial \xi} d\xi + \frac{\partial y}{\partial \eta} d\eta \quad (5.29b)$$

or, in matrix form,

$$\begin{bmatrix} dx \\ dy \end{bmatrix} = \begin{bmatrix} \frac{\partial x}{\partial \xi} & \frac{\partial x}{\partial \eta} \\ \frac{\partial y}{\partial \xi} & \frac{\partial y}{\partial \eta} \end{bmatrix} \begin{bmatrix} d\xi \\ d\eta \end{bmatrix} \quad (5.30)$$

Solving Eq. (5.30) for the right-hand column matrix, i.e., multiplying by the inverse of the 2 × 2 coefficient matrix, we have

$$\begin{bmatrix} d\xi \\ d\eta \end{bmatrix} = \begin{bmatrix} \frac{\partial x}{\partial \xi} & \frac{\partial x}{\partial \eta} \\ \frac{\partial y}{\partial \xi} & \frac{\partial y}{\partial \eta} \end{bmatrix}^{-1} \begin{bmatrix} dx \\ dy \end{bmatrix} \quad (5.31)$$

GRIDS WITH APPROPRIATE TRANSFORMATIONS

Comparing Eqs. (5.27) and (5.31), we have

$$\begin{bmatrix} \dfrac{\partial \xi}{\partial x} & \dfrac{\partial \xi}{\partial y} \\ \dfrac{\partial \eta}{\partial x} & \dfrac{\partial \eta}{\partial y} \end{bmatrix} = \begin{bmatrix} \dfrac{\partial x}{\partial \xi} & \dfrac{\partial x}{\partial \eta} \\ \dfrac{\partial y}{\partial \xi} & \dfrac{\partial y}{\partial \eta} \end{bmatrix}^{-1} \tag{5.32}$$

Following the standard rules for creating the inverse of a matrix, Eq. (5.32) is written as

$$\begin{bmatrix} \dfrac{\partial \xi}{\partial x} & \dfrac{\partial \xi}{\partial y} \\ \dfrac{\partial \eta}{\partial x} & \dfrac{\partial \eta}{\partial y} \end{bmatrix} = \dfrac{\begin{bmatrix} \dfrac{\partial y}{\partial \eta} & -\dfrac{\partial x}{\partial \eta} \\ -\dfrac{\partial y}{\partial \xi} & \dfrac{\partial x}{\partial \xi} \end{bmatrix}}{\begin{vmatrix} \dfrac{\partial x}{\partial \xi} & \dfrac{\partial x}{\partial \eta} \\ \dfrac{\partial y}{\partial \xi} & \dfrac{\partial y}{\partial \eta} \end{vmatrix}} \tag{5.33}$$

Consider the determinant in the denominator of Eq. (5.33). Since the value of a determinant is unchanged by transposing its terms, we have

$$\begin{vmatrix} \dfrac{\partial x}{\partial \xi} & \dfrac{\partial x}{\partial \eta} \\ \dfrac{\partial y}{\partial \xi} & \dfrac{\partial y}{\partial \eta} \end{vmatrix} = \begin{vmatrix} \dfrac{\partial x}{\partial \xi} & \dfrac{\partial y}{\partial \xi} \\ \dfrac{\partial x}{\partial \eta} & \dfrac{\partial y}{\partial \eta} \end{vmatrix} \equiv J \tag{5.34}$$

Note that the right-hand determinant of Eq. (5.34) is precisely the jacobian J of the transformation, as can be seen from the definition of J given by Eq. (5.22a). Substituting Eq. (5.34) into (5.33), we have

$$\begin{bmatrix} \dfrac{\partial \xi}{\partial x} & \dfrac{\partial \xi}{\partial y} \\ \dfrac{\partial \eta}{\partial x} & \dfrac{\partial \eta}{\partial y} \end{bmatrix} = \dfrac{1}{J} \begin{bmatrix} \dfrac{\partial y}{\partial \eta} & -\dfrac{\partial x}{\partial \eta} \\ -\dfrac{\partial y}{\partial \xi} & \dfrac{\partial x}{\partial \xi} \end{bmatrix} \tag{5.35}$$

Comparing like elements of the two matrices in Eq. (5.35), we obtain the relationships for the direct metrics in terms of the inverse metrics, namely,

$$\frac{\partial \xi}{\partial x} = \frac{1}{J}\frac{\partial y}{\partial \eta} \qquad (5.36a)$$

$$\frac{\partial \eta}{\partial x} = -\frac{1}{J}\frac{\partial y}{\partial \xi} \qquad (5.36b)$$

$$\frac{\partial \xi}{\partial y} = -\frac{1}{J}\frac{\partial x}{\partial \eta} \qquad (5.36c)$$

$$\frac{\partial \eta}{\partial y} = \frac{1}{J}\frac{\partial x}{\partial \xi} \qquad (5.36d)$$

Hence, the above formalism leads directly to the relationship between the direct and inverse metrics. That the above results are consistent with our previous analyses can be seen by substituting Eqs. (5.36a) to (5.36d) into Eqs. (5.2) and (5.3), obtaining

$$\frac{\partial}{\partial x} = \frac{1}{J}\left[\left(\frac{\partial}{\partial \xi}\right)\left(\frac{\partial y}{\partial \eta}\right) - \left(\frac{\partial}{\partial \eta}\right)\left(\frac{\partial y}{\partial \xi}\right)\right]$$

$$\frac{\partial}{\partial y} = \frac{1}{J}\left[\left(\frac{\partial}{\partial \eta}\right)\left(\frac{\partial x}{\partial \xi}\right) - \left(\frac{\partial}{\partial \xi}\right)\left(\frac{\partial x}{\partial \eta}\right)\right]$$

The above two equations are identically Eqs. (5.24a) and (5.24b), which gives our derivative transformation expressed in terms of the inverse metrics. The extension of the above formalism to three spatial dimensions, leading to the three-dimensional counterpart of Eqs. (5.36a) to (5.36d), is straightforward.

5.4 FORM OF THE GOVERNING EQUATIONS PARTICULARLY SUITED FOR CFD REVISITED: THE TRANSFORMED VERSION

Return for a moment to Sec. 2.10, where we presented the *strong conservation form* of the governing flow equations, represented by Eq. (2.93). For the case of unsteady flow in two spatial dimensions, with no source terms, this equation reduces to

$$\frac{\partial U}{\partial t} + \frac{\partial F}{\partial x} + \frac{\partial G}{\partial y} = 0 \qquad (5.37)$$

(The treatment here of two spatial dimensions x and y rather than carrying all three dimensions x, y, and z is just for simplicity; the extension of the following analysis to three dimensions is straightforward.)

Question: When Eq. (5.37) is written in the transformed (ξ, η) space, can it be recast in strong conservation form; i.e., can it be written in a transformed form such that

$$\frac{\partial U_1}{\partial t} + \frac{\partial F_1}{\partial \xi} + \frac{\partial G_1}{\partial \eta} = 0 \qquad (5.38)$$

where F_1 and G_1 are suitable combinations of the original F and G flux vectors? If so, we will be able to retain in our transformed space all those advantages of the strong conservation form that were ascribed to some CFD calculations discussed in Sec. 2.10. The answer to the above question is *yes*. Let us see how and why.

First, transform the spatial variables in Eq. (5.37) according to the derivative transformation given by Eqs. (5.2) and (5.3).

$$\frac{\partial U}{\partial t} + \frac{\partial F}{\partial \xi}\left(\frac{\partial \xi}{\partial x}\right) + \frac{\partial F}{\partial \eta}\left(\frac{\partial \eta}{\partial x}\right) + \frac{\partial G}{\partial \xi}\left(\frac{\partial \xi}{\partial y}\right) + \frac{\partial G}{\partial \eta}\left(\frac{\partial \eta}{\partial y}\right) = 0 \qquad (5.39)$$

Multiply Eq. (5.39) by the jacobian J defined by Eq. (5.22a):

$$J\frac{\partial U}{\partial t} + J\left(\frac{\partial F}{\partial \xi}\right)\left(\frac{\partial \xi}{\partial x}\right) + J\left(\frac{\partial F}{\partial \eta}\right)\left(\frac{\partial \eta}{\partial x}\right) + J\left(\frac{\partial G}{\partial \xi}\right)\left(\frac{\partial \xi}{\partial y}\right) + J\left(\frac{\partial G}{\partial \eta}\right)\left(\frac{\partial \eta}{\partial y}\right) = 0 \qquad (5.40)$$

Putting Eq. (5.40) on the shelf for a moment, consider the simple derivative expansion of the term $JF(\partial \xi/\partial x)$, that is,

$$\frac{\partial[JF(\partial \xi/\partial x)]}{\partial \xi} = J\left(\frac{\partial \xi}{\partial x}\right)\frac{\partial F}{\partial \xi} + F\frac{\partial}{\partial \xi}\left(J\frac{\partial \xi}{\partial x}\right) \qquad (5.41)$$

Rearranging Eq. (5.41), we have

$$J\left(\frac{\partial F}{\partial \xi}\right)\left(\frac{\partial \xi}{\partial x}\right) = \frac{\partial[JF(\partial \xi/\partial x)]}{\partial \xi} - F\frac{\partial}{\partial \xi}\left(J\frac{\partial \xi}{\partial x}\right) \qquad (5.42)$$

Similarly, taking the η derivative of $JF\,(\partial \eta/\partial x)$ and rearranging, we have

$$J\left(\frac{\partial F}{\partial \eta}\right)\left(\frac{\partial \eta}{\partial x}\right) = \frac{\partial[JF(\partial \eta/\partial x)]}{\partial \eta} - F\frac{\partial}{\partial \eta}\left(J\frac{\partial \eta}{\partial x}\right) \qquad (5.43)$$

In a similar way, the terms $JG(\partial \xi/\partial y)$ and $JG(\partial \eta/\partial y)$ can be expanded and rearranged as

$$J\left(\frac{\partial G}{\partial \xi}\right)\left(\frac{\partial \xi}{\partial y}\right) = \frac{\partial[JG(\partial \xi/\partial y)]}{\partial \xi} - G\frac{\partial}{\partial \xi}\left(J\frac{\partial \xi}{\partial y}\right) \qquad (5.44)$$

and

$$J\left(\frac{\partial G}{\partial \eta}\right)\left(\frac{\partial \eta}{\partial y}\right) = \frac{\partial[JG(\partial \eta/\partial y)]}{\partial \eta} - G\frac{\partial}{\partial \eta}\left(J\frac{\partial \eta}{\partial y}\right) \qquad (5.45)$$

Substituting Eqs. (5.42) to (5.45) into Eq. (5.40) and factoring, we have

$$J\frac{\partial U}{\partial t} + \frac{\partial}{\partial \xi}\left(JF\frac{\partial \xi}{\partial x} + JG\frac{\partial \xi}{\partial y}\right) + \frac{\partial}{\partial \eta}\left(JF\frac{\partial \eta}{\partial x} + JG\frac{\partial \eta}{\partial y}\right)$$
$$- F\left[\frac{\partial}{\partial \xi}\left(J\frac{\partial \xi}{\partial x}\right) + \frac{\partial}{\partial \eta}\left(J\frac{\partial \eta}{\partial x}\right)\right]$$
$$- G\left[\frac{\partial}{\partial \xi}\left(J\frac{\partial \xi}{\partial y}\right) + \frac{\partial}{\partial \eta}\left(J\frac{\partial \eta}{\partial y}\right)\right] = 0 \quad (5.46)$$

The last two terms in Eq. (5.46), which appear in brackets, are zero, as follows. Substituting Eqs. (5.36a) to (5.36d) into these terms, we have

$$\frac{\partial}{\partial \xi}\left(J\frac{\partial \xi}{\partial x}\right) + \frac{\partial}{\partial \eta}\left(J\frac{\partial \eta}{\partial x}\right) = \frac{\partial}{\partial \xi}\left(\frac{\partial y}{\partial \eta}\right) - \frac{\partial}{\partial \eta}\left(\frac{\partial y}{\partial \xi}\right)$$
$$= \frac{\partial^2 y}{\partial \xi\, \partial \eta} - \frac{\partial^2 y}{\partial \eta\, \partial \xi} \equiv 0$$

and

$$\frac{\partial}{\partial \xi}\left(J\frac{\partial \xi}{\partial y}\right) + \frac{\partial}{\partial \eta}\left(J\frac{\partial \eta}{\partial y}\right) = \frac{\partial}{\partial \xi}\left(-\frac{\partial x}{\partial \eta}\right) + \frac{\partial}{\partial \eta}\left(\frac{\partial x}{\partial \xi}\right)$$
$$= -\frac{\partial^2 x}{\partial \xi\, \partial \eta} + \frac{\partial^2 x}{\partial \eta\, \partial \xi} \equiv 0$$

Thus, Eq. (5.46) can be written as

$$\boxed{\frac{\partial U_1}{\partial t} + \frac{\partial F_1}{\partial \xi} + \frac{\partial G_1}{\partial \eta} = 0} \quad (5.47)$$

where

$$U_1 = JU \quad (5.48a)$$
$$F_1 = JF\frac{\partial \xi}{\partial x} + JG\frac{\partial \xi}{\partial y} \quad (5.48b)$$
$$G_1 = JF\frac{\partial \eta}{\partial x} + JG\frac{\partial \eta}{\partial y} \quad (5.48c)$$

Equation (5.47) is the generic form of the governing flow equations written in strong conservation form in the transformed (ξ, η) space. Such a form was first obtained in 1974 by Viviand (Ref. 28) and Vinokur (Ref. 29).

Note in Eq. (5.47) that the newly defined flux vectors F_1 and G_1 are combinations of the physical flux vectors F and G, where the combinations involve the jacobian J and the *direct* metrics (not the inverse metrics, as defined in Sec. 5.3). Here is the exception to one of the statements made in Sec. 5.3; there it was stated that the appearance of the jacobian in the transformed equations signaled the use of the inverse metrics. This is not the case when the transformed equations are expressed in the strong conservation form given by Eq. (5.47). Indeed, if Eqs. (5.36a) and (5.36b) are substituted into the forms for F_1 and G_1, given by Eqs.

(5.48b) and (5.48c), we have

$$F_1 = JF\frac{\partial \xi}{\partial x} + JG\frac{\partial \xi}{\partial y} = F\frac{\partial y}{\partial \eta} - G\frac{\partial x}{\partial \eta} \quad (5.49a)$$

and

$$G_1 = JF\frac{\partial \eta}{\partial x} + JG\frac{\partial \eta}{\partial y} = -F\frac{\partial y}{\partial \xi} + G\frac{\partial x}{\partial \xi} \quad (5.49b)$$

Note that when F_1 and G_1 are expressed in terms of the inverse metrics as in Eqs. (5.49a) and (5.49b), the jacobian does not appear. These are not statements of critical importance—they just come under the heading of "interesting observations."

5.5 A COMMENT

Return to our road map given in Fig. 5.3. Up to this point in the present chapter we have dealt with the *concept* of a transformation from the physical (x, y) space to the computational (ξ, η) space, as reflected in the left column in Fig. 5.3. However, we have yet to examine any actual example of such a transformation; this is the subject of the center column in Fig. 5.3. In the previous sections, we have developed the transformation expressions in very general and generic terms. Keep in mind that such a transformation is consistent with the demands of *finite-difference* methods, where the finite-difference expressions are evaluated on a uniform grid. If such a uniform grid is compatible with the boundary geometry and the flow problem in the physical plane, then a transformed grid is not necessary, and all that we have discussed so far in this chapter is superfluous. However, for realistic problems with realistic geometries, this is generally not the case; either the nature of the flow problem itself (such as the viscous flow over a surface where a larger number of grid points should be packed closer to the surface) and/or the shape of the boundary (such as a curved surface that should be fitted with a curvilinear, boundary-fitted coordinate system) will usually demand a transformation which carries a nonuniform grid in the physical plane to a uniform grid in the computational plane. Such a transformation is inherently not required for finite-volume methods, which can deal directly with a nonuniform mesh in the physical plane.

In the remainder of this chapter, we will examine some actual transformations, i.e., some specific formulations represented by the generic form given by Eqs. (5.1a) to (5.1c). In the process, we will be dealing with specific aspects of *grid generation*. We will now be running down the center column of Fig. 5.3.

5.6 STRETCHED (COMPRESSED) GRIDS

Of all the grid generation techniques to be discussed, the simplest is treated in this section. It consists of stretching the grid in one or more coordinate directions.

Example 5.2. Consider the physical and computational planes shown in Fig. 5.4. Assume that we are dealing with the viscous flow over a flat surface, where the velocity varies rapidly near the surface as shown in the velocity profile sketched at the

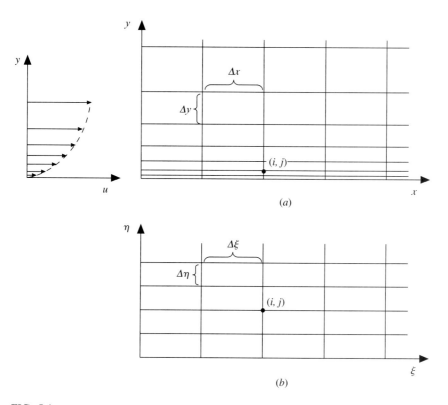

FIG. 5.4
Example of grid stretching. (*a*) Physical plane; (*b*) computational plane.

left of the physical plane. To calculate the details of this flow near the surface, a finely spaced grid in the *y* direction should be used, as sketched in the physical plane. However, far away from the surface, the grid can be coarser. Therefore, a proper grid should be one in which the horizontal coordinate lines become progressively more closely spaced in the vertical direction as the surface is approached. On the other hand, we wish to deal with a uniform grid in the computational plane, as also shown in Fig. 5.4. Examining Fig. 5.4, we see that the grid in the physical space is "stretched," as if a uniform grid were drawn on a piece of rubber and then the upper portion of the rubber were stretched upward in the *y* direction. A simple analytical transformation which can accomplish this grid stretching is

$$\xi = x \quad (5.50a)$$
$$\eta = \ln(y + 1) \quad (5.50b)$$

The *inverse* transformation is

$$x = \xi \tag{5.51a}$$
$$y = e^\eta - 1 \tag{5.51b}$$

Examine these equations more closely with Fig. 5.4 in mind. In both the physical and computational planes, the vertical grid lines are uniformly spaced in the x direction; this is reflected in both Eqs. (5.50a) and (5.51a). In the physical plane, Δx is the same throughout. In the computational plane, $\Delta \xi$ is the same throughout. Moreover, $\Delta x = \Delta \xi$. The grid is *not* stretched in the x direction. However, such is not the case for the horizontal grid lines. The horizontal lines are uniformly spaced in the *computational plane* by *intent*; we stipulate that $\Delta \eta$ be the same everywhere in the computational plane. In turn, what happens to the corresponding values of Δy in the physical plane? The answer is easily seen by differentiating Eq. (5.51b) with respect to η.

$$\frac{dy}{d\eta} = e^\eta$$

or
$$dy = e^\eta \, d\eta$$

Replacing dy and $d\eta$ with finite increments, we have approximately

$$\Delta y = e^\eta \, \Delta \eta \tag{5.52}$$

Note from Eq. (5.52) that as η becomes larger, i.e., as we move further *above* the plate in Fig. 5.4, the value of Δy becomes progressively larger for the same constant value of $\Delta \eta$. In other words, as we move in the vertical direction away from the plate, although we have a uniform grid in the computational plane, we encounter progressively larger values of Δy; that is, the grid in the physical plane appears to be *stretched* in the vertical direction. This is what is meant by a stretched grid. Moreover, the direct transformation given by Eqs. (5.50a) and (5.50b), or the inverse transformation given by Eqs. (5.51a) and (5.51b), is the *mechanics* by which the stretched grid is generated. This is the simplest essence of what is meant by grid generation.

Example 5.3. Let us examine what happens to the governing flow equations in both the physical and the computational planes. For simplicity, we will assume steady flow, and we will illustrate by means of the continuity equation. Taking the continuity equation in the form of Eq. (2.25), specializing it to a steady flow, and writing it in cartesian coordinates, we have

$$\frac{\partial(\rho u)}{\partial x} + \frac{\partial(\rho v)}{\partial y} = 0 \tag{5.53}$$

This is the continuity equation written in terms of the *physical plane*. This equation can be formally transformed to the computational plane using the generic derivative transformation given by Eqs. (5.2) and (5.3); the resulting form is

$$\frac{\partial(\rho u)}{\partial \xi}\left(\frac{\partial \xi}{\partial x}\right) + \frac{\partial(\rho u)}{\partial \eta}\left(\frac{\partial \eta}{\partial x}\right) + \frac{\partial(\rho v)}{\partial \xi}\left(\frac{\partial \xi}{\partial y}\right) + \frac{\partial(\rho v)}{\partial \eta}\left(\frac{\partial \eta}{\partial y}\right) = 0 \tag{5.54}$$

The metrics in Eq. (5.54) are obtained from the direct transformation given by Eqs. (5.50a) and (5.50b), namely,

$$\frac{\partial \xi}{\partial x} = 1 \qquad \frac{\partial \xi}{\partial y} = 0 \qquad \frac{\partial \eta}{\partial x} = 0 \qquad \frac{\partial \eta}{\partial y} = \frac{1}{y+1} \qquad (5.55)$$

Substituting the metrics from Eqs. (5.55) into (5.54), we obtain

$$\frac{\partial(\rho u)}{\partial \xi} + \frac{1}{y+1}\frac{\partial(\rho v)}{\partial \eta} = 0 \qquad (5.56)$$

However, from Eq. (5.50b), $y + 1 = e^{\eta}$. Therefore, Eq. (5.56) becomes

$$\frac{\partial(\rho u)}{\partial \xi} + \frac{1}{e^{\eta}}\frac{\partial(\rho v)}{\partial \eta} = 0$$

or

$$e^{\eta}\frac{\partial(\rho u)}{\partial \xi} + \frac{\partial(\rho v)}{\partial \eta} = 0 \qquad (5.57)$$

Equation (5.57) is the form of the continuity equation that holds in the computational plane. For the first time in this chapter, we have just witnessed an actual transformation of a governing flow equation from the physical plane to the computational plane; with this, it is hoped that some of the generic ideas presented in the earlier sections of this chapter are beginning to come more into focus.

Example 5.4. To illustrate further, let us repeat the above derivation but this time from the point of view of the *inverse* transformation defined by Eqs. (5.51a) and (5.51b). Returning to Eq. (5.53), and applying the generic inverse derivative transformation given by Eqs. (5.24a) and (5.24b),

$$\frac{1}{J}\left[\frac{\partial(\rho u)}{\partial \xi}\left(\frac{\partial y}{\partial \eta}\right) - \frac{\partial(\rho u)}{\partial \eta}\left(\frac{\partial y}{\partial \xi}\right)\right] + \frac{1}{J}\left[\frac{\partial(\rho v)}{\partial \eta}\left(\frac{\partial x}{\partial \xi}\right) - \frac{\partial(\rho v)}{\partial \xi}\left(\frac{\partial x}{\partial \eta}\right)\right] = 0 \qquad (5.58)$$

The inverse metrics in Eq. (5.58) are obtained from the inverse transformation given by Eqs. (5.51a) and (5.51b) as follows.

$$\frac{\partial x}{\partial \xi} = 1 \qquad \frac{\partial x}{\partial \eta} = 0 \qquad \frac{\partial y}{\partial \xi} = 0 \qquad \frac{\partial y}{\partial \eta} = e^{\eta} \qquad (5.59)$$

Substituting Eqs. (5.59) into (5.58), we have

$$e^{\eta}\frac{\partial(\rho u)}{\partial \xi} + \frac{\partial(\rho v)}{\partial \eta} = 0 \qquad (5.60)$$

This again is the transformed continuity equation. Indeed, Eq. (5.60) is identical to Eq. (5.57). All that we have done here is to demonstrate how the transformed equation can be obtained from either the direct transformation or the inverse transformation; the results are exactly the same.

Note in the above derivations that the continuity equation was first transformed by means of the derivative transformations; the results were Eq. (5.54) for the direct transformation and Eq. (5.58) for the inverse transformation. These are still generic transformations at this stage. The transformation only becomes specific when the specific metrics associated with the specific transformation are substituted into Eq. (5.54) or Eq. (5.58). We can now recognize the following important aspect of any transformation of the governing flow equations. *It is the metrics that carry all the specific information pertinent to a specific transformation.* Let us imagine the following division of effort. Assume you are responsible for numerically calculating a given flow field over a given body. Assume that the responsibility for the generation of the grid around the body rests with another person (or group) down the hall. When you are ready to make your calculations, you go to your friendly grid generation person, who will give you the *metrics* for the transformation. That is all the information about the transformation which you need in order to numerically solve the flow problem in the computational plane. On the other hand, you will also need to know the one-to-one correspondence of the location of each grid point in both the computational and physical planes in order to carry your solution back to the physical plane. For example, consider again the stretched grid discussed above. Solve the continuity equation in the form of Eq. (5.57), as well as the appropriate transformed versions of the momentum and energy equations (not shown here for simplicity) for the dependent flow-field variables in the *computational plane*. Among lots of other data from the solution, you will have the value of density at grid point (i, j), $\rho_{i,j}$, where the point (i, j) is located in the computational plane, as shown at the bottom of Fig. 5.4. However, from the one-to-one correspondence of the location of the same grid point (i, j) in the *physical* plane shown at the top of Fig. 5.4, you also know the value of the density at the point in the physical plane; namely, it is the same value $\rho_{i,j}$ obtained from the solution of the governing equations at grid point (i, j) in the computational plane.

Example 5.5. Let us consider a more elaborate version of grid stretching. The example is taken from Ref. 30 and 31, where the supersonic viscous flow over a blunt base is studied. Here, grid stretching is carried out in both the x and y directions. The physical and computational planes are illustrated in Fig. 5.5. The streamwise stretching in the x direction is accomplished through a transformation used by Holst (Ref. 32), given below:

$$x = \frac{\xi_0}{A} \{\sinh[(\xi - x_0)\beta_x] + A\} \tag{5.61}$$

where

$$A = \sinh(\beta_x x_0) \tag{5.62}$$

and

$$x_0 = \frac{1}{2\beta_x} \ln \frac{1 + (e^{\beta_x} - 1)\xi_0}{1 + (e^{-\beta_x} - 1)\xi_0} \tag{5.63}$$

In Eq. (5.61), ξ_0 is the location in the computational plane where the maximum clustering is to occur and β_x is a constant which controls the degree of clustering at ξ_0, with larger values of β_x providing a finer grid in the clustered region. The transverse stretching in the y direction is accomplished by dividing the physical plane into two sections: (1) the space directly behind the step and (2) the space above (both in front of and behind) the step. The transformation is based on that used by Roberts (Ref. 33) and is given by

$$y = \frac{(\beta_y + 1) - (\beta_y - 1)e^{-c(\eta - 1 - \alpha)/(1 - \alpha)}}{(2\alpha + 1)(1 + e^{-c(\eta - 1 - \alpha)/(1 - \alpha)})} \tag{5.64}$$

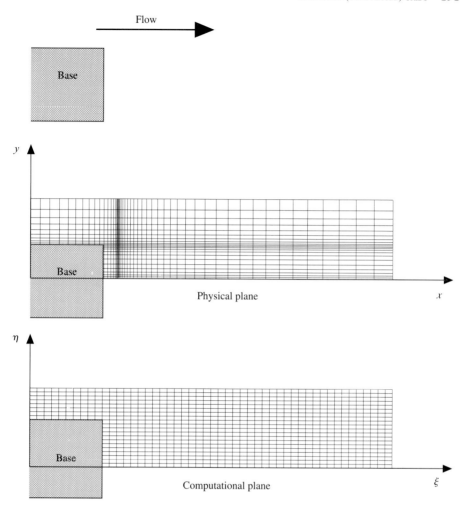

FIG. 5.5
Comparison of uniform and compressed grids. (*From Refs. 30 and 31.*)

where

$$c = \log \frac{\beta_y + 1}{\beta_y - 1}$$

In Eq. (5.64), β_y and α are appropriate constants and are different for the two sections identified above. The algebraic transformations given above result in the grid stretching shown in Fig. 5.5. Note in Fig. 5.5 that the blunt base itself has no active grid points inside it, for the obvious reason that no flow exists inside the solid base. The grid generation formulas given by Eqs. (5.61) to (5.64) are shown here strictly as an example of a more sophisticated stretched grid. This author is not necessarily recommending them above all others; the choice of a particular stretched grid is up to you as deemed most appropriate for your particular problem. Feel free to use whatever artistic license you feel comfortable with.

Referring to our chapter road map in Fig. 5.3, we have just finished the first item under grid generation, namely, stretched grids. We now move on to the next item, the all-important concept of boundary-fitted coordinate systems in general. In this context, it should be noted that the grids shown in Figs. 5.4 and 5.5 are effectively boundary-fitted coordinate systems in that the solid surfaces are coordinate lines in the grid. However, this is because the physical geometry of the flat plate (Fig. 5.4) and the blunt base (Fig. 5.5) conveniently fit into an already rectangular-configured grid. The next section treats the more general case of *curved* boundary surfaces, which obviously do not fit a rectangular grid in the physical plane.

5.7 BOUNDARY-FITTED COORDINATE SYSTEMS: ELLIPTIC GRID GENERATION

To introduce this section, let us examine a boundary-fitted coordinate system within the context of a straightforward problem. Consider the flow through the divergent duct shown in Fig. 5.6a. Curve *de* is the upper wall of the duct, and line *fg* is the centerline. For this flow, a simple rectangular grid in the physical plane is not appropriate, for the same reasons discussed in Sec. 5.1. Instead, we draw the curvilinear grid in Fig. 5.6a which allows both the upper boundary *de* and the centerline *fg* to be coordinate lines, exactly fitting these boundaries. In turn, the curvilinear grid in Fig. 5.6a must be transformed to a rectangular grid in the computational plane, Fig. 5.6b. This can be accomplished as follows. Let $y_s = f(x)$ be the ordinate of the upper surface *de* in Fig. 5.6a. Then the following transformation will result in a rectangular grid in (ξ, η) space:

$$\xi = x \tag{5.65}$$

$$\eta = \frac{y}{y_s} \quad \text{where } y_s = f(x) \tag{5.66}$$

For example, consider point *d* in the physical plane, where $y = y_d = y_s(x_d)$. When this coordinate is substituted into Eq. (5.66), we have

$$\eta_d = \frac{y_d}{y_s} = \frac{y_s(x_d)}{y_s(x_d)} = 1$$

Hence, in the computational plane, point *d* is located along $\eta = \eta_d = 1$. Now move to point *c* in the physical plane, where $y = y_c = y_s(x_c)$. The ordinate of point *c* is obviously different from that at point *d*; that is, $y_c > y_d$. However, when y_c is substituted into Eq. (5.66), we have

$$\eta_c = \frac{y_c}{y_s} = \frac{y_s(x_c)}{y_s(x_c)} = 1$$

Hence, in the computational plane, point *c* is located along $\eta = \eta_c = 1$. This is the same η coordinate as point *d* in the computational plane. From the above discussion, it is clear that *all* the points along the curved upper boundary in the physical plane fall, via the transformation given by Eq. (5.66), along the horizontal line $\eta = 1$ in

BOUNDARY-FITTED COORDINATE SYSTEMS: ELLIPTIC GRID GENERATION   **193**

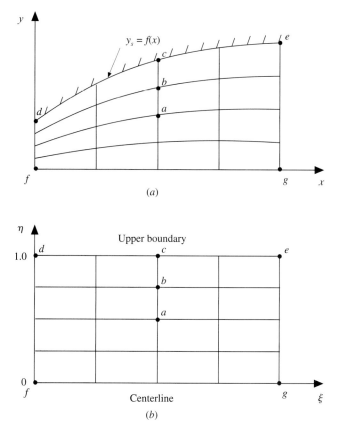

FIG. 5.6
A simple boundary-fitted coordinate system. (*a*) Physical plane; (*b*) computational plane.

the computational plane. This allows a uniform rectangular grid in the computational plane. Such is the essence of curvilinear boundary-fitted coordinate systems in the physical plane and their transformation to a uniform rectangular grid in the computational plane.

GUIDEPOST

The following discussion on elliptic grid generation, as well as the treatment of adaptive grids in Sec. 5.8, represents very important aspects of modern grid generation in CFD. This author strongly encourages you to study this material, at least from the point of view of understanding the basic ideas. However, because of the lack of sophistication, the applications in Part III will not deal with these ideas. Therefore, if you are looking for a shortcut at this

> stage, the following guidepost is suggested.
>
> Go directly to Sec. 5.9.

The above is a simple example of a boundary-fitted coordinate system. A more sophisticated example is shown in Fig. 5.7, which is an elaboration of the case illustrated in Fig. 5.2. Consider the airfoil shape given in Fig. 5.7a. A curvilinear system is wrapped around the airfoil, where one coordinate line $\eta = \eta_1 =$ constant is on the airfoil surface. This is the inner boundary of the grid, designated by Γ_1 in both the physical and computational planes in Fig. 5.7. The outer boundary of the grid is labeled Γ_2 in Fig. 5.7 and is given by $\eta = \eta_2 =$ constant. The shape and location of the inner boundary Γ_1 is fixed by the airfoil shape on which it is placed. The shape and location of the outer boundary Γ_2 is somewhat arbitrary—it is whatever you choose to draw. Examining this grid, we see that it clearly fits the boundary, and hence it is a *boundary-fitted coordinate system*. The lines which fan out from the inner boundary Γ_1 and which intersect the outer boundary Γ_2 are lines of constant ξ, such as line *ef* for which $\xi = \xi_1 =$ constant. The value of the constant is also your choice. That is, for each of the $\xi =$ constant curves, you designate a numerical value for ξ. For example, along curve *ef* you might designate $\xi = 0.1$. Along curve *gh* you might designate $\xi = 0.2$, and so forth. Also note that in Fig. 5.7a the lines of constant η totally enclose the airfoil, much like elongated circles; such a grid is called an *O-type grid* for airfoils. Another related curvilinear grid can have the $\eta =$ constant lines trailing downstream to the right, *not* totally enclosing the airfoil (except on the inner boundary Γ_1). Such a grid is called a *C-type grid*. We will see an example of a C-type grid shortly.

Question: What transformation will cast the curvilinear grid in Fig. 5.7a into a uniform grid in the computational plane as sketched in Fig. 5.7b? To answer this question, imagine that the curvilinear grid in the physical plane is drawn on top of a piece of graph paper ruled in cartesian (x, y) coordinates. Therefore, along the inner boundary Γ_1, the physical coordinates are known:

$$(x, y) \text{ known along } \Gamma_1$$

That is, for any given point on Γ_1, there is a set of two known numbers, namely, the x and y coordinates of that point. Similarly, the physical (x, y) coordinates of the outer boundary Γ_2 are also known, because Γ_2 is simply a rather arbitrarily drawn loop around the airfoil. Once this loop Γ_2 is specified, then the physical (x, y) coordinates along it are known:

$$(x, y) \text{ known along } \Gamma_2$$

This hints of a boundary-value problem where the boundary conditions (namely, the values of x and y) are known *everywhere* along the boundary. Recall from Sec. 3.4.3 that the solution of elliptic partial differential equations requires the specification of the boundary conditions *everywhere* along a boundary enclosing the domain. Therefore, let us consider the transformation in Fig. 5.7 to be defined by an *elliptic partial differential equation* [in contrast to the algebraic relations used in the case of the stretched grids, namely, Eqs. (5.51a) and (5.51b), and in the case of the simple

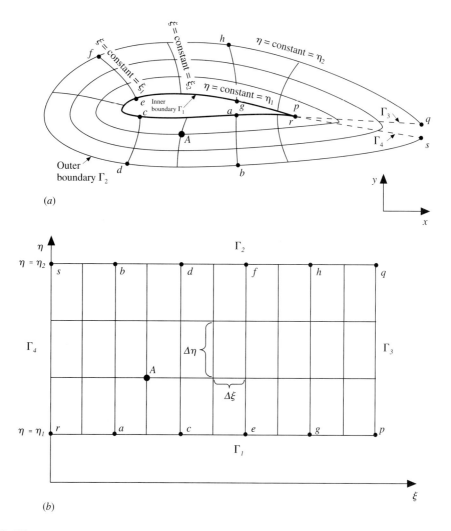

FIG. 5.7
Schematic of an elliptically generated, boundary-fitted grid. (*a*) Physical plane; (*b*) computational plane.

contoured duct in Fig. 5.6, namely, Eqs. (5.65) and (5.66)]. One of the simplest elliptic equations is Laplace's equation

$$\frac{\partial^2 \xi}{\partial x^2} + \frac{\partial^2 \xi}{\partial y^2} = 0 \tag{5.67}$$

$$\frac{\partial^2 \eta}{\partial x^2} + \frac{\partial^2 \eta}{\partial y^2} = 0 \tag{5.68}$$

In Eqs. (5.67) and (5.68), ξ and η are dependent variables and x and y are independent variables. Let us switch these roles and write the inverse, where x and y become the dependent variables. The result is

$$\alpha \frac{\partial^2 x}{\partial \xi^2} - 2\beta \frac{\partial^2 x}{\partial \xi \partial \eta} + \gamma \frac{\partial^2 x}{\partial \eta^2} = 0 \quad (5.69)$$

and

$$\alpha \frac{\partial^2 y}{\partial \xi^2} - 2\beta \frac{\partial^2 y}{\partial \xi \partial \eta} + \alpha \frac{\partial^2 y}{\partial \eta^2} = 0 \quad (5.70)$$

where

$$\alpha = \left(\frac{\partial x}{\partial \eta}\right)^2 + \left(\frac{\partial y}{\partial \eta}\right)^2$$

$$\beta = \left(\frac{\partial x}{\partial \xi}\right)\left(\frac{\partial x}{\partial \eta}\right) + \left(\frac{\partial y}{\partial \xi}\right)\left(\frac{\partial y}{\partial \eta}\right)$$

$$\gamma = \left(\frac{\partial x}{\partial \xi}\right)^2 + \left(\frac{\partial y}{\partial \xi}\right)^2$$

Equations (5.69) and (5.70) are elliptic partial differential equations with x, y as dependent variables and ξ, η as independent variables. Return to Fig. 5.7; a solution of Eqs. (5.69) and (5.70) allows the calculation of the (x, y) coordinates of grid points in the physical plane as a function of the (ξ, η) location of the same grid points in the computational plane. However, for a properly posed problem dealing with elliptic equations, we need to specify boundary conditions along the *entire* boundary of the domain, as stated in Sec. 3.4.3. Consider as our domain the computational plane shown in Fig. 5.7b, bounded above and below by Γ_2 and Γ_1, respectively, and on the side by Γ_3 and Γ_4. To this point in our discussion, we have specified the values of x and y along the boundaries Γ_1 and Γ_2 only; we need also to have some boundary conditions given along Γ_3 and Γ_4 to have a properly posed problem. To accomplish this, return to the physical plane in Fig. 5.7a. Imagine that we go to the extreme right of the O grid shown there, take a razor blade, and make a "cut" to the trailing edge of the airfoil. This cut now introduces two additional boundaries, namely, the curves *qp* and *sr*, denoted by Γ_3 and Γ_4, respectively. In Fig. 5.7a, the curves *qp* and *sr* are shown slightly separated; this is for clarity only. In reality, *qp* and *sr* are the *same* curve in the *xy* plane; *qp* simply denotes the upper surface of the cut and *sr* denotes the lower surface, but they lie on top of each other. In the physical plane, the points q and s lie on top of each other, and the points p and r also lie on top of each other; indeed, the entirety of Γ_3 lies on top of Γ_4. However, this is not the case in the $\xi\eta$ plane shown in Fig. 5.7b. Here, Γ_3 and Γ_4 are totally separated and form the right and left boundaries, respectively, of the domain in the computational space. It is almost as if the O grid in the physical plane, after the cut is made, is unwrapped, with Γ_4 being swung below and out to the left. In the computational plane, q and s are separate grid points and r and p are separate grid points. Return for a moment to the physical plane. The cut has been made rather arbitrarily, but once we make it, then we know the (x, y) coordinates along the cut.

That means we now have values of x and y specified along Γ_3 and Γ_4 in Fig. 5.7b. Reviewing the relationship between the physical and computational planes, we can state the following. The airfoil surface in the physical plane, curve *pgecar*, becomes the lower straight line denoted by Γ_1 in the computational plane. Similarly, the outer boundary in the physical plane, curve *qhfdbs*, becomes the upper straight line denoted by Γ_2 in the computational plane. The left and right sides of the rectangle in the computational plane are formed from the cut in the physical plane; the left side is line *rs* denoted by Γ_4 in Fig. 5.7b, and the right side is line *qp* denoted by Γ_3 in Fig. 5.7b.

The computational plane is sketched again in Fig. 5.8, just to emphasize what is happening. Here we emphasize that values of (x, y) are now *known along all four boundaries* $\Gamma_1, \Gamma_2, \Gamma_3$, and Γ_4. This is the essence of a properly posed boundary-value problem for the solution of elliptic partial differential equations. In turn, Eqs. (5.69) and (5.70) are such elliptic equations. For each grid point *inside* the domain shown in Fig. 5.7b, these equations can be solved numerically, along with the specified boundary values of (x, y) along $\Gamma_1, \Gamma_2, \Gamma_3$ and Γ_4, to give the corresponding (x, y) values of that same grid point in the physical plane. For example, consider the internal grid point labeled A in Fig. 5.8; this corresponds to the point labeled A in both the physical and computational planes in Fig. 5.7. At point A in the computational plane, Eqs. (5.69) and (5.70) are solved for its (x, y) coordinates. *This now locates point A in the physical plane.* In turn, for *all* the uniformly spaced grid points in the $\xi\eta$ plane, the solution of Eqs. (5.69) and (5.70) now locates these same points in a nonuniform manner in the xy plane. That is, a given grid point (ξ_i, η_j) in the computational plane corresponds to the *calculated* grid point (x_i, y_j) in physical space. The solution of Eqs. (5.69) and (5.70) is carried out by an appropriate finite-difference solution for elliptic equations; for example, relaxation techniques are popular for such equations. Because this transformation is being carried out via the solution of a system of elliptic partial differential equations, it is called *elliptic grid generation*.

Note that the above transformation, using an elliptic partial differential equation to generate the grid, does *not* involve closed-form analytic expressions; rather, it produces a set of *numbers* which locate a grid point (x_i, y_j) in physical space which corresponds to a given grid point (ξ_i, η_j) in computational space. In turn, the metrics in the governing flow equations (which are solved in the computational plane), such as $\partial\xi/\partial x$, $\partial\eta/\partial y$, are obtained from finite differences; central differences are frequently used for this purpose. For example, at any given grid point located at (i, j) in both the physical and computational planes, we can write for the metric at that point

$$\left(\frac{\partial \xi}{\partial x}\right)_{i,j} = \frac{\xi_{i+1,j} - \xi_{i-1,j}}{x_{i+1,j} - x_{i-1,j}}$$

and so forth. In turn, these values of the metrics are fed directly into the transformed governing flow equations that are being solved in the transformed plane, i.e., in the uniform grid in the $\xi\eta$ plane, thus allowing the flow field around the airfoil to be obtained.

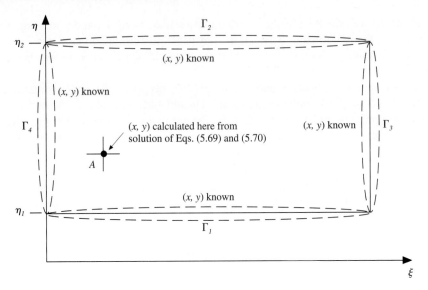

FIG. 5.8
Computational plane, illustrating the boundary conditions and an internal point.

Once again, it is important to keep in mind what we are doing here. Equations (5.69) and (5.70) have *nothing* to do with the physics of the flow field. They are simply elliptic partial differential equations *which we have chosen* to relate ξ and η to x and y and hence constitute a *transformation* (a one-to-one correspondence of grid points) from the physical plane to the computational plane. Because this transformation is governed by elliptic equations, it is an example of a general class of grid generation called *elliptic grid generation*, as stated earlier. Such elliptic grid generation was first used on a practical basis by Joe Thompson at Mississippi State University, and is described in detail in the pioneering article given in Ref. 34. This reference gives a great deal more detail, as well as more generalization, than presented in the present section; it is highly recommended that you examine this and other references before embarking on any elliptic grid generation of your own. The purpose of the present section is to help you understand the basic ideas.

The curvilinear, boundary-fitted coordinate system shown in Fig. 5.7a is illustrated in a qualitative sense in that figure for purposes of instruction. An actual grid generated about an airfoil using the above elliptic grid generation approach is shown in Fig. 5.9, which is a computer graphic taken from Ref. 6. Using Thompson's grid generation scheme (Ref. 34), Kothari and the present author (Ref. 6) have generated a boundary-fitted coordinate system around a Miley airfoil. (The Miley airfoil is an airfoil specially designed for low Reynolds number applications by Stan Miley at Mississippi State University.) In Fig. 5.9, the white speck in the middle of the figure is the airfoil, and the grid spreads away from the airfoil in all directions. In Ref. 6, low Reynolds number flows over airfoils were calculated by means of a time-marching finite-difference solution of the compressible Navier-Stokes equations. The free stream is subsonic; hence, the outer boundary must be placed far away from the airfoil because of the far-reaching propagation of disturbances in a subsonic flow. A detail of the grid in the near vicinity of the airfoil is shown in Fig. 5.10. Note from both Figs. 5.9 and 5.10 that the grid is a C-type grid in contrast to the O-type grid sketched in Fig. 5.7. The black areas in Figs. 5.9 and 5.10 are densely packed grid points that are not resolved by the computer graphics picture. The grid shown in Figs. 5.9 and 5.10 is precisely that used to obtain some of the low Reynolds number airfoil results discussed in Sec. 1.2, and in particular the results shown in Fig. 1.4. Also, Fig. 1.14 is an excellent example of a boundary-fitted grid wrapped outside *and* inside a gas turbine engine.

We end this section by emphasizing again that the elliptic grid generation, with its solution of elliptic partial differential equations to obtain the internal grid points, is *completely separate* from the finite-difference solution of the governing equations. The grid is generated first, before any solution of the governing equations is attempted. The use of Laplace's equation [Eqs. (5.67) and (5.68)] to obtain this

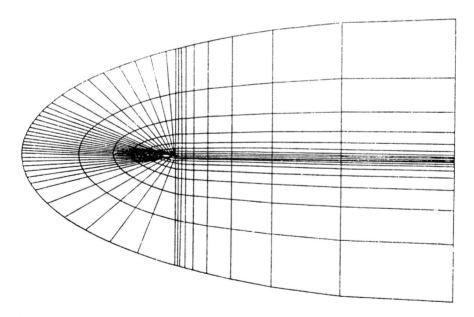

FIG. 5.9
Elliptically generated grid wrapped around a Miley airfoil, from the calculations by Kothari et al. The small, white speck at the focus of the grid is the airfoil. This reflects the necessity to place the far-field boundary a large distance from the body for a numerical solution of a subsonic flow.

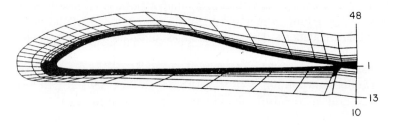

FIG. 5.10
A small, detailed section of the boundary-fitted grid of Fig. 5.9, in the near vicinity of the airfoil.

grid has nothing to do whatsoever with the physical aspects of the actual flow field. Here, Laplace's equation is simply used to generate the grid *only*.

5.8 ADAPTIVE GRIDS

The concept of a stretched grid as outlined in Sec. 5.6 is motivated by the desire to cluster a large number of closely spaced grid points in those regions of the flow where large gradients in the flow-field properties exist, hence improving the numerical accuracy of a given CFD calculation. This motivation is driven by more than just trying to minimize the truncation error with closely spaced points; it is also a matter of simply having enough grid points to properly capture the physics of the flow. A qualitative example of this is the viscous flow over a flat plate, sketched in Fig. 5.11. In the real physical flow, there will be a boundary layer that grows thicker with increasing downstream distance along the plate. Let x be the distance along the plate measured from the leading edge. The local thickness of the boundary layer is δ, where $\delta = \delta(x)$. Consider the grid shown in Fig. 5.11a; here we see a coarse grid where not a single grid point is placed in the real boundary layer. That is, $\Delta y > \delta$ for the first row of grid points above the plate. When a numerical calculation is made on this grid, and the no-slip condition of $u = 0$ is applied at the wall, a velocity profile is obtained like that sketched at the right of Fig. 5.11a. Some type of profile will be obtained, with u increasing in the y direction; it will be a boundary-layer-like profile but indicating a thickness far in excess of the real boundary-layer thickness. In contrast, consider the grid shown in Fig. 5.11b; this is also a coarse grid, with an equal number of points in the y direction as used in Fig. 5.11a. However, in Fig. 5.11b the grid is compressed such that at least some points are in the real boundary layer. That is, $\Delta y < \delta$ for the first row of grid points above the plate. When a numerical calculation is made on this grid, the resulting velocity profile shown at the right of Fig. 5.11b will be a more realistic representation of the real boundary layer. In essence, the coarse uniform grid shown in Fig. 5.11a misses the physical boundary layer altogether; the viscouslike velocity profile shown at the right is simply due to the application of the no-slip condition at the wall. In contrast,

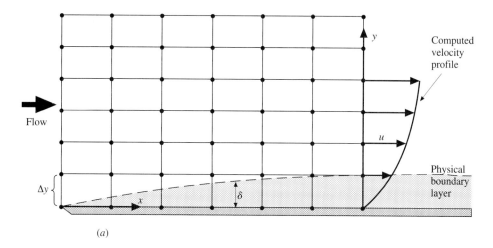

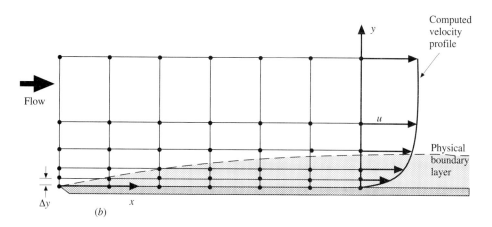

FIG. 5.11
Two sketches demonstrating the need to concentrate a number of grid points in the boundary layer. (*a*) No grid points in the boundary layer; (*b*) at least some points in the boundary layer.

the coarse but compressed grid shown in Fig. 5.11*b* at least captures some of the features of the real boundary layer.

Obviously, the purpose of the compressed (or stretched) grid is to put grid points in the flow field where the action is and to remove grid points from those regions where there is little or no action. However, as discussed in Sec. 5.6, a stretched grid is an algebraically generated grid which is set up *before* the solution of the flow is calculated. Moreover, once it is set up, it is locked in place for the entire flow-field solution. However, how do you know in advance where the major action is going to occur in the flow without actually solving the problem first? You may set up a stretched grid in advance, but you may miss completely the region where the real action is; i.e., you may not be so lucky as to set up your more closely spaced points so that they coincide with the regions of large gradients in the flow. Therein lies the motivation for an *adaptive grid*, which is the subject of this section.

An adaptive grid is a grid network that *automatically* clusters grid points in regions of high flow-field gradients; it uses the solution of the flow-field properties to locate the grid points in the physical plane. An adaptive grid can be visualized as one which evolves in steps of time in conjunction with a time-dependent solution of the governing flow-field equations, which computes the flow-field variables in steps of time. During the course of the solution, the grid points in the physical plane *move* in such a fashion to "adapt" to regions of large flow-field gradients as these gradients evolve with time. Hence, the actual grid points in the physical plane are constantly in motion during the solution of the flow field and become stationary only when the flow solution approaches a steady state. Therefore, unlike the stretched grid discussed in Sec. 5.6 and the elliptic grid generation discussed in Sec. 5.7, where the generation of the grid is completely separate from the flow-field solution, an adaptive grid is intimately linked to the flow-field solution and changes as the flow field changes. The hoped-for advantages of an adaptive grid are associated with the grid points being automatically clustered in regions where the "action" is occurring. These advantages are (1) increased accuracy for a fixed number of grid points or (2) for a given accuracy, fewer grid points are needed. Adaptive grids are still very new in CFD, and whether or not these advantages are always achieved is not well established.

An example of a simple adaptive grid is that used by Corda (Ref. 35) for the solution of viscous supersonic flow over a rearward-facing step. Here, the transformation is expressed in the form

$$\Delta x = \frac{B \, \Delta \xi}{1 + b(\partial g/\partial x)} \tag{5.71}$$

and

$$\Delta y = \frac{C \, \Delta \eta}{1 + c(\partial g/\partial y)} \tag{5.72}$$

where g is a primitive flow-field variable, such as p, ρ, or T. If $g = p$, then Eqs. (5.71) and (5.72) cluster the grid points in regions of large pressure gradients; if $g = T$, the grid points cluster in regions of large temperature gradients; and so forth. In Eqs. (5.71) and (5.72), $\Delta \xi$ and $\Delta \eta$ are fixed, uniform grid spacings in the computational $\xi \eta$ plane, b and c are constants chosen to increase or decrease the effect of the gradient in changing the grid spacing in the physical plane, B and C are scale factors, and Δx and Δy are the new grid spacing in the physical plane. Because $\partial g/\partial x$ and $\partial g/\partial y$ are changing with time during a time-dependent solution of the flow field, then clearly Δx and Δy change with time; i.e., the grid points move in the physical space. Clearly, in regions of the flow where $\partial g/\partial x$ and $\partial g/\partial y$ are large, Eqs. (5.71) and (5.72) yield small values of Δx and Δy for a given $\Delta \xi$ and $\Delta \eta$; this is the mechanism which clusters the grid points. This process is illustrated in Fig. 5.12, where the physical plane is shown in part (*a*) and the computational plane in part (b). Consider the specific grid point labeled N in Fig. 5.12b. This point is *fixed* in the $\xi\eta$ space; it does not move with time. So is its adjacent grid point, labeled $N + 1$. As usual, the distance between grid point N and $N + 1$ is $\Delta\xi$. Now examine the corresponding grid points in the physical plane, Fig. 5.12a. The location of points N and $N + 1$ in the physical plane at time level t are denoted by the black dots. The distance between these two points in the x direction is $(\Delta x)^t$, where the superscript t denotes the time level t. The x location of point N at time level t, denoted by x_N^t, depends on the various values of Δx between points 1 and 2, points 2 and 3, etc. That is,

$$x_N^t = \sum_{1}^{N}(\Delta x)_i^t \tag{5.73}$$

Now consider the situation at the next time level, $t + \Delta t$. Because $\partial g/\partial x$ will in general change from one time level to the next during the time-marching process,

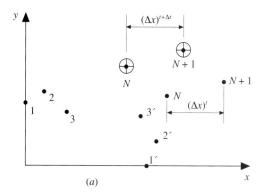

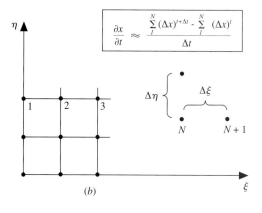

FIG. 5.12
Schematic of the mechanics of an adaptive grid. (a) Physical plane; (b) computational plane.

then Eq. (5.71) yields a new value of Δx, denoted by $(\Delta x)^{t+\Delta t}$ at time level $t + \Delta t$. Hence, the x location of point N shifts to a new value at time $t + \Delta t$, denoted by $x_N^{t+\Delta t}$. Of course, because of the simultaneous application of Eq. (5.72), the y location of point N shifts also. The new locations of points N and $N + 1$ at time level $t + \Delta t$ are shown by crossmarks in Fig. 5.12a; the new value of Δx, namely, $(\Delta x)^{t+\Delta t}$, is also shown. The new x location of point N at time level $t + \Delta t$ is given by

$$x_N^{t+\Delta t} = \sum_{1}^{N} (\Delta x)_i^{t+\Delta t} \qquad (5.74)$$

Equations analogous to (5.73) and (5.74) can be written for the y location.

You are reminded again that, in dealing with an adaptive grid, the computational plane consists of fixed points in the $\xi\eta$ space; these points are fixed in time; i.e., they do *not* move in the computational space. Moreover, $\Delta\xi$ is uniform and $\Delta\eta$ is uniform. Hence, the computational plane is the same as we have discussed in previous sections. The governing flow equations are solved in the computational plane, where the x, y, and t derivatives are transformed according to Eqs. (5.2), (5.3), and (5.5). In particular, examine the transformation given by Eq. (5.5) for the time derivative. In the case of stretched or boundary-fitted grids as discussed in Secs. 5.4 and 5.5, respectively, the metrics $\partial\xi/\partial t$ and $\partial\eta/\partial t$ were zero, and Eq. (5.5) yields $\partial/\partial t = \partial/\partial\tau$. However, for an adaptive grid,

$$\frac{\partial\xi}{\partial t} \equiv \left(\frac{\partial\xi}{\partial t}\right)_{x,y}$$

and

$$\frac{\partial\eta}{\partial t} \equiv \left(\frac{\partial\eta}{\partial t}\right)_{x,y}$$

are finite. Why? Because, although the grid points are fixed in the computational plane, the grid points in the physical plane are moving with time. The physical meaning of $(\partial\xi/\partial t)_{x,y}$ is the time rate of change of ξ at a *fixed* (x, y) location in the physical plane. Similarly, the physical meaning of $(\partial\eta/\partial t)_{x,y}$ is the time rate of change of η at a *fixed* (x, y) location in the physical plane. Imagine that you have your eyes locked to a fixed (x, y) location in the physical plane. As a function of time, the values of ξ and η associated with this *fixed* (x, y) location will change. This is why $\partial\xi/\partial t$ and $\partial\eta/\partial t$ are finite. In turn, when dealing with the transformed flow equations in the computational plane, all three terms on the right-hand side of Eq. (5.5) are finite and must be included in the transformed equations. In this fashion, the time metrics $\partial\xi/\partial t$ and $\partial\eta/\partial t$ automatically take into account the movement of the adaptive grid during the solution of the governing flow equations.

The values of the time metrics in the form shown in Eq. (5.5) are a bit cumbersome to evaluate; on the other hand, the related time metrics

$$\left(\frac{\partial x}{\partial t}\right)_{\xi,\eta} \quad \text{and} \quad \left(\frac{\partial y}{\partial t}\right)_{\xi,\eta}$$

are much easier to evaluate numerically; they stem directly from the forms of the adaptive grid transformation given by Eqs. (5.71) and (5.72). For example, return to Fig. 5.12. We can represent the time metric $(\partial x/\partial t)_{\xi, \eta}$ by taking the relative change in the x locations of points N and $N + 1$ and dividing by the time increment Δt. That is,

$$\left(\frac{\partial x}{\partial t}\right)_{\xi, \eta} = \frac{x_N^{t+\Delta t} - x_N^t}{\Delta t} \qquad (5.75a)$$

where $x_N^{t+\Delta t}$ and x_N^t are given by Eqs. (5.74) and (5.73), respectively. An analogous expression can be written for $(\partial y/\partial t)_{\xi, \eta}$ as follows:

$$\left(\frac{\partial y}{\partial t}\right)_{\xi, \eta} = \frac{y_M^{t+\Delta t} - y_M^t}{\Delta t} \qquad (5.75b)$$

where $y_M^{t+\Delta t}$ and y_M^t are given by expressions analogous to Eqs. (5.73) and (5.74), i.e.,

$$y_M^t = \sum_{1'}^{M} (\Delta y)_i^t$$

and

$$y_M^{t+\Delta t} = \sum_{1'}^{M} (\Delta y)_i^{t+\Delta t}$$

The meaning of M is as follows. Examine Fig. 5.12, where we have previously focused on the grid point labeled N; here, N is simply the value of the x index, namely, $i = N$. For the same grid point, M denotes the value of the corresponding y index, namely, $j = M$. The above summations are taken in the y direction, summed over points $1', 2', 3'$, etc., as shown in Fig. 5.12a. Since the time metrics $(\partial x/\partial t)_{\xi, \eta}$ and $(\partial y/\partial t)_{\xi, \eta}$ are the ones most directly obtained from the transformation given by Eqs. (5.71) and (5.72) and since the derivative transformation given by Eq. (5.5) involves the time metrics $(\partial \xi/\partial t)_{x, y}$ and $(\partial \eta/\partial t)_{x, y}$, we must find the relationship between these two sets of metrics. Let us proceed as follows.

Return to the general inverse transformation given by Eqs. (5.18a) to (5.18c). In particular, examine Eq. (5.18a), repeated below.

$$x = x(\xi, \eta, \tau) \qquad (5.18a)$$

Forming the exact differential, we have

$$dx = \left(\frac{\partial x}{\partial \xi}\right)_{\eta, \tau} d\xi + \left(\frac{\partial x}{\partial \eta}\right)_{\xi, \tau} d\eta + \left(\frac{\partial x}{\partial \tau}\right)_{\xi, \eta} d\tau \qquad (5.76)$$

In Eq. (5.76), the change in x, dx, is expressed in terms of changes in ξ, η, and τ, namely, $d\xi$, $d\eta$, and $d\tau$, respectively. If these changes are taking place with respect to time, holding x and y constant, Eq. (5.76) can be written as

$$\left(\frac{\partial x}{\partial t}\right)_{x,y}^{\!\!\!0} = \left(\frac{\partial x}{\partial \xi}\right)_{\eta, \tau} \left(\frac{\partial \xi}{\partial t}\right)_{x, y} + \left(\frac{\partial x}{\partial \eta}\right)_{\xi, \tau} \left(\frac{\partial \eta}{\partial t}\right)_{x, y} + \left(\frac{\partial x}{\partial \tau}\right)_{\xi, \eta} \left(\frac{\partial \tau}{\partial t}\right)_{x, y}^{\!\!\!1} \qquad (5.77)$$

In Eq. (5.77), $(\partial x/\partial t)_{x,y}$ is identically zero, because x is being held constant in this partial derivative. We are also stipulating that the generic Eq. (5.18c), where $t = t(\tau)$, is given by $t = \tau$. This is why $(\partial \tau/\partial t)_{x,y} = 1$ in Eq. (5.77). With these values, Eq. (5.77) becomes

$$-\left(\frac{\partial x}{\partial \tau}\right)_{\xi,\eta} = \left(\frac{\partial x}{\partial \xi}\right)_{\eta,\tau}\left(\frac{\partial \xi}{\partial t}\right)_{x,y} + \left(\frac{\partial x}{\partial \eta}\right)_{\xi,\tau}\left(\frac{\partial \eta}{\partial t}\right)_{x,y} \tag{5.78}$$

Note that we continue to carry the subscripts on the partial derivatives to avoid any confusion over what variables are held constant. Now consider Eq. (5.18b), repeated below,

$$y = y(\xi, \eta, \tau) \tag{5.18b}$$

Hence

$$dy = \left(\frac{\partial y}{\partial \xi}\right)_{\eta,\tau} d\xi + \left(\frac{\partial y}{\partial \eta}\right)_{\xi,\tau} d\eta + \left(\frac{\partial y}{\partial \tau}\right)_{\xi,\eta} d\tau \tag{5.79}$$

Thus, from this result we write

$$\cancel{\left(\frac{\partial y}{\partial t}\right)_{x,y}^{0}} = \left(\frac{\partial y}{\partial \xi}\right)_{\eta,\tau}\left(\frac{\partial \xi}{\partial t}\right)_{x,y} + \left(\frac{\partial y}{\partial \eta}\right)_{\xi,\tau}\left(\frac{\partial \eta}{\partial t}\right)_{x,y} + \left(\frac{\partial y}{\partial \tau}\right)_{\xi,\eta}\cancel{\left(\frac{\partial \tau}{\partial t}\right)_{x,y}^{1}} \tag{5.80}$$

or

$$-\left(\frac{\partial y}{\partial \tau}\right)_{\xi,\eta} = \left(\frac{\partial y}{\partial \xi}\right)_{\eta,\tau}\left(\frac{\partial \xi}{\partial t}\right)_{x,y} + \left(\frac{\partial y}{\partial \eta}\right)_{\xi,\tau}\left(\frac{\partial \eta}{\partial t}\right)_{x,y} \tag{5.81}$$

Examine Eqs. (5.78) and (5.81); they have in common the metrics $(\partial \xi/\partial t)_{x,y}$ and $(\partial \eta/\partial t)_{x,y}$. Using Cramer's rule, we solve Eqs. (5.78) and (5.81) first for $(\partial \xi/\partial t)_{x,y}$.

$$\left(\frac{\partial \xi}{\partial t}\right)_{x,y} = \frac{\begin{vmatrix} -\left(\frac{\partial x}{\partial \tau}\right)_{\xi,\eta} & \left(\frac{\partial x}{\partial \eta}\right)_{\xi,\tau} \\ -\left(\frac{\partial y}{\partial \tau}\right)_{\xi,\eta} & \left(\frac{\partial y}{\partial \eta}\right)_{\xi,\tau} \end{vmatrix}}{\begin{vmatrix} \left(\frac{\partial x}{\partial \xi}\right)_{\eta,\tau} & \left(\frac{\partial x}{\partial \eta}\right)_{\xi,\tau} \\ \left(\frac{\partial y}{\partial \xi}\right)_{\eta,\tau} & \left(\frac{\partial y}{\partial \eta}\right)_{\xi,\tau} \end{vmatrix}} \tag{5.82}$$

Recognizing that $\tau = t$, and that the denominator is the jacobian J, Eq. (5.82) becomes (dropping subscripts)

$$\frac{\partial \xi}{\partial t} = \frac{1}{J}\left[-\left(\frac{\partial x}{\partial t}\right)\left(\frac{\partial y}{\partial \eta}\right) + \left(\frac{\partial y}{\partial t}\right)\left(\frac{\partial x}{\partial \eta}\right)\right] \tag{5.83}$$

In a similar fashion, solving Eqs. (5.78) and (5.81) for $(\partial \eta/\partial t)_{x,y}$, we find that

$$\frac{\partial \eta}{\partial t} = \frac{1}{J}\left[-\left(\frac{\partial x}{\partial t}\right)\left(\frac{\partial y}{\partial \xi}\right) - \left(\frac{\partial y}{\partial t}\right)\left(\frac{\partial x}{\partial \xi}\right)\right] \quad (5.84)$$

Let us recapitulate. For an adaptive grid which is designed to evolve during the course of a time-marching solution, the governing flow equations, when transformed for solution in the computational $\xi \eta$ plane, must contain all the terms in the time transformation given by Eq. (5.5). We note that the time metrics in Eq. (5.5) are $\partial \xi/\partial t$ and $\partial \eta/\partial t$. These time metrics can be evaluated from Eqs. (5.83) and (5.84), respectively. In turn, in Eqs. (5.83) and (5.84), the terms $\partial x/\partial t$ and $\partial y/\partial t$ are calculated via Eqs. (5.75a) and (5.75b), respectively. The spatial metrics $\partial x/\partial \xi$, $\partial x/\partial \eta$, $\partial y/\partial \xi$, and $\partial y/\partial \eta$ which appear in Eqs. (5.83) and (5.84) as well as in the jacobian J can be replaced by central differences. For example,

$$\frac{\partial x}{\partial \xi} = \frac{x_{i+1,j} - x_{i-1,j}}{2\Delta \xi}$$

$$\frac{\partial x}{\partial \eta} = \frac{x_{i,j+1} - x_{i,j-1}}{2\Delta \eta}$$

$$\frac{\partial y}{\partial \xi} = \frac{y_{i+1,j} - y_{i-1,j}}{2\Delta \xi}$$

$$\frac{\partial y}{\partial \eta} = \frac{y_{i,j+1} - y_{i,j-1}}{2\Delta \eta}$$

where, in the above equations, $i = N$ and $j = M$.

An example of an adapted grid for the supersonic flow over a rearward-facing step is given in Fig. 5.13, taken from the work of Corda (Ref. 35). Flow is from left to right. The adapted grid shown in Fig. 5.13 is the final, steady-state grid obtained after the time-marching flow-field solution has reached its steady state at large time. Note that, as the steady state is approached, the time metrics $\partial \xi/\partial t$, $\partial \eta/\partial t$, $\partial x/\partial t$, and $\partial y/\partial t$ all approach zero; i.e., the grid points in the physical xy plane cease to move. Note in Fig. 5.13 that the grid points cluster around the expansion wave emanating from the top corner of the step and around the reattachment shock wave downstream of the step. It is interesting to note that the adapted grid itself is a type of *flow-field visualization method* that helps to identify the location of waves, shear layers, and other gradients in the flow. Returning to the original adaptive grid transformation

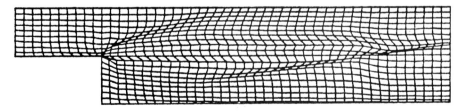

FIG. 5.13
Adapted grid for the rearward-facing step problem from Corda.

given by Eqs. (5.71) and (5.72), if $g = \rho$, then the grid points in the physical plane cluster in regions of large density gradients—this is the computational analog to a schlieren photograph taken in the laboratory. Note that the grid in Fig. 5.13 takes on the trappings of a type of "CFD schlieren" picture.

As a final note in this section, there are many different approaches for the generation of adaptive grids. The above discussion is just one; it is based on the ideas presented by Dwyer et al. in Ref. 36. Adaptive grids are in a current state of rapid development in modern CFD; you are encouraged to consult the modern literature on this extensive subject before embarking on any serious adaptive grid efforts of your own. The adaptive grid technique described in the present section was chosen for its simplicity, because our interest here is to give you just a feeling for the general idea.

5.9 SOME MODERN DEVELOPMENTS IN GRID GENERATION

As stated in Sec. 5.1, grid generation is a very active research and development activity within the general discipline of CFD. In this chapter, we have only introduced some of the basic ideas. However, let us take a quick look at two examples which reflect modern applications of grid generation within the practical world of aerodynamics.

The first example is the grid used to calculate the flow-field results over the Northrop F-20 airplane as presented in Figs. 1.6 and 1.7. Return to Chap. 1 for a moment and examine these figures. They were obtained by means of a numerical solution to the three-dimensional Euler equations, as described in Ref. 9. It is always a major challenge to construct a three-dimensional grid around a complex configuration such as the F-20. For the cases shown in Figs. 1.6 and 1.7, a three-dimensional boundary-fitted coordinate system is chosen with an elliptic grid generation following the ideas presented in Sec. 5.7, combined with an adaptive grid scheme following the ideas presented in Sec. 5.8. Sections of the grid are shown in Fig. 5.14, taken from Ref. 9. Here we see the grid coordinate lines in the surface of the body, the centerline plane, and the plane of the wing.

The fuselage is angled diagonally across the figure, with the nose at the lower left. The wing, tail, and rear portion of the fuselage appear solid white, due to the dense clustering of grid points in those regions where the grid has adapted to large flow-field gradients. Figure 5.14 represents a combination of grid generation ideas presented in this chapter, cast in the framework of a modern application of CFD. Figure 5.14 also sheds more light on how the results of Figs. 1.6 and 1.7 were obtained, thus helping to close the loop on some of the introductory discussion in Chap. 1.

A complete airplane is, in general, a complex geometric configuration, which sometimes requires grid generation even more elaborate than the single boundary-fitted grid discussed in Sec. 5.7 and as exemplified by Fig. 5.14. In many practical fluid dynamic applications, modern CFD solutions have employed a grid made up of two or more separate grids, with interfaces between each other. That is, the grid consists of two or more blocks, where each block is a separate grid different from

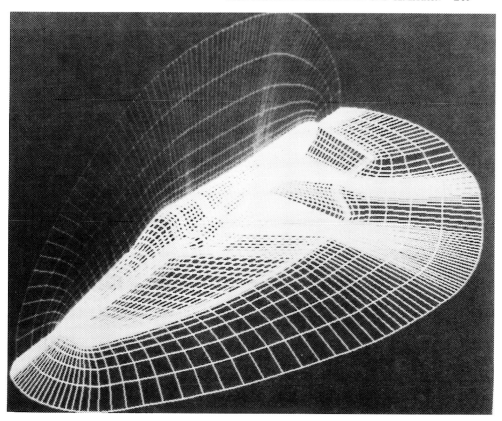

FIG. 5.14
An elliptically generated adaptive grid wrapped around an F-20 airplane configuration. The configuration surface, centerline plane, and wing plane are shown. (*From Ref. 9.*)

the others. These different blocks cover different zones of the flow field, and hence such grids are frequently called *zonal grids*. An example of a zonal grid is shown in Fig. 5.15, taken from Refs. 37 and 38. Here we see only part of a 20-block grid system for the computation of the flow over an F-16 fighter airplane. We see seven of the upper blocks in Fig. 5.15. The remaining blocks are used to help define the flow over the inlet, etc. One of the major problems encountered in the use of zonal block methods is the proper geometric interfacing across adjacent zones, a proper "connectivity" so that the accuracy of the CFD calculation is not compromised. Furthermore, each block can in principle be generated by a different scheme; i.e., one block might be an algebraic, stretched grid using cartesian coordinates (see Sec. 5.6); an adjacent block might be an algebraic grid using cylindrical coordinates; and yet another adjacent block might be an elliptically generated grid (see Sec. 5.7). This compounds the connectivity problems. A further discussion of these matters is beyond the scope of this book; for more details, see for example, Refs. 37 and 38.

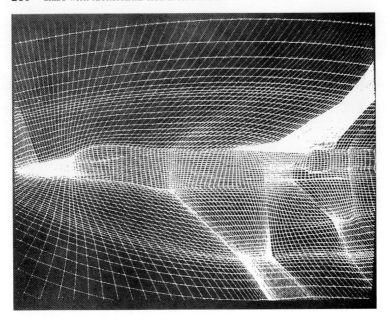

FIG. 5.15
A zonal grid wrapped around an F-16 airplane. Surface grid is shown as part of a 20-block grid. (*From Refs. 37 and 38. Copyright © 1990, AIAA. Reprinted with permission.*)

5.10 SOME MODERN DEVELOPMENTS IN FINITE-VOLUME MESH GENERATION: UNSTRUCTURED MESHES AND A RETURN TO CARTESIAN MESHES

To this point in Chap. 5, all the grids discussed and displayed have been couched in terms of finite-difference algorithm applications, with the understanding that whatever nonuniform grid exists in the physical space, there exists a transformation which will recast it as a uniform rectangular grid in the computational space. The finite-difference calculations are then made over this uniform grid in the computational space, after which the flow-field results are transferred directly back to the corresponding points in the physical space. Go back and look at some of the nonuniform grids in the physical space, such as Figs. 5.5, 5.9, 5.10, and 5.13 to 5.15. Although they are nonuniform, there is a certain "regularity" to them; the grid lines in physical space pertain to constant coordinate values ξ, η, and ζ in the transformed space. Moreover, a given family of coordinate lines do not intersect; i.e., lines of constant ξ do not cross, lines of constant η do not cross, etc. Hence, there is a certain "structure" to all these grids; such grids are called *structured grids*.

Here is another thought. The nonuniform grids you are looking at in physical space can also be visualized as a *mesh* of finite-volume cells. Since the finite-volume method does not demand a uniform, rectangular grid for computations (as does the finite-difference method), then such finite-volume calculations can be made directly in the physical plane on a nonuniform mesh. No transformations are necessary. Therefore, in the context of a finite-volume method, mesh generation simply involves the construction of the mesh in physical space. (Recall that our treatment of the finite volume method is introduced in this book via Problems 2.2 and 4.7.) Thus, if we wish, we can view Fig. 5.9 (for example) as a picture of a finite-volume *mesh*, on which the finite-volume calculations can be made directly. Moreover, in the same vein as described in the previous paragraph, the mesh represented by Fig. 5.9 is a *structured mesh*.

Here is yet another thought. There is nothing about the finite-volume method that demands a structured mesh; it can be applied to mesh cells of any arbitrary shape. This has given rise to the use of *unstructured meshes*. Perhaps the best way to describe what is meant by an unstructured mesh is to look at some. An unstructured mesh around a multielement airfoil is shown in Fig. 5.16, taken from Ref. 39. Another unstructured mesh for the calculation of the flow over a compression corner is shown in Fig. 5.17, taken from Ref. 40. Clearly, there is no regularity to these meshes. There are no coordinate lines that correspond to a constant ξ, η, and ζ. These grids are totally unstructured. This allows for maximum flexibility in matching mesh cells with the boundary surfaces and for putting cells where you want them. Constructing an unstructured mesh might be viewed in some sense as a work of art—you can shape the mesh cells as you like and put them wherever you want in the physical space. Of course, you have to develop the computer logic to automate the cell generator. Although unstructured meshes have been used for

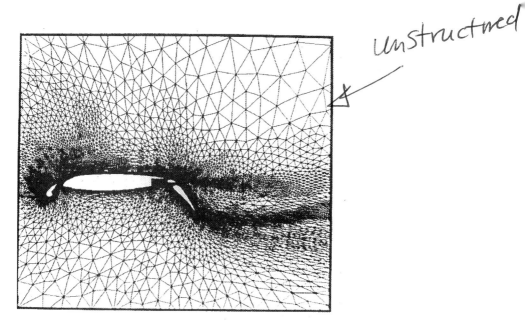

FIG. 5.16
An unstructured mesh around a multielement airfoil. (*From Ref. 39. Copyright © 1991, AIAA. Reprinted with permission.*)

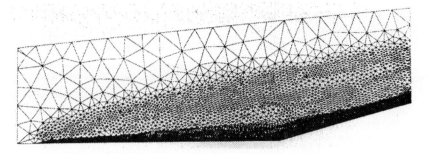

FIG. 5.17
An unstructured mesh around a compression corner. (*From Ref. 40. Copyright © 1991, AIAA. Reprinted with permission.*)

finite-element calculations in structural mechanics for a number of years, they are relatively new to the field of CFD. Indeed, in the field of grid or mesh generation for CFD, unstructured meshes are, at the time of writing, receiving a great deal of attention.

It is somewhat ironic that, at the same time that unstructured meshes have become popular, current advances are also being made in the extreme opposite direction: the use of cartesian meshes with the maximum degree of structure. At the beginning of this chapter, we momentarily considered a cartesian grid, such as shown in Fig. 5.1, and then immediately rejected it for general use because of the difficulty posed by grid points appearing inside the body, as well as the lack of grid points on the boundary surface. However, if we view Fig. 5.1 as a finite-volume mesh, then it takes on a new dimension, so to speak. The mesh cells away from the body can be rectangular, and those cells adjacent to the body can be modified in shape such that one side of each cell is along the body surface. This is shown schematically in Fig. 5.18. A cartesian mesh for the calculation of the flow over an airfoil (including flap deflection) is shown in Fig. 5.19, taken from Ref. 41. The generation of this mesh also incorporates some adaptation following the philosophy discussed in Sec. 5.8. A cartesian mesh around a double ellipsoid (a body shape somewhat like the space shuttle) is shown in Fig. 5.20, also taken from Ref. 41. In this case, the cartesian mesh is for the calculation of supersonic flow over the body, and the mesh adaptive procedure clusters the rectangularlike cells around the bow and canopy shock waves, as clearly seen in Fig. 5.20. Reference 41 represents one of the most recent investigations of cartesian meshes at the time of writing; for more details, consult this reference.

5.11 SUMMARY

This ends our general discussion of grid and mesh generation for the numerical computation of fluid-flow problems. Returning to the road map in Fig. 5.3, recall that our presentation has followed three general routes, as reflected by the three

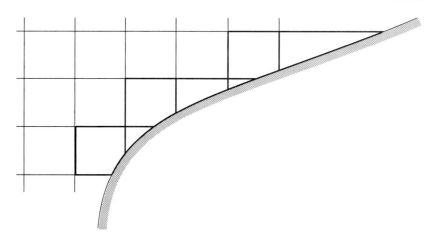

FIG. 5.18
Section of a cartesian mesh near a surface. Those mesh cells adjacent to the surface are highlighted by bold lines; they are modified so that one side of the cell lies along the surface.

vertical columns. Knowing that finite-difference solutions usually require that the computations be made in a transformed plane—the computational plane—the first column dealt with the general aspects of derivative transformations, along with the related aspects of metrics and jacobians. As part of this discussion, we demonstrated that the governing flow equations can be cast in a strong conservation form in the transformed space, analogous to their strong conservation form in the physical space presented in Chap. 2. Moving to the second column in Fig. 5.3, we discussed in some detail the various aspects of grid generation for use in finite-difference solutions, with examples given for stretched grids, elliptically generated boundary-fitted grids, adaptive grids, and zonal grids. Moving to the third column in Fig. 5.3, pertaining to mesh generation for finite-volume calculations, we note that the previously mentioned grids in physical space can also be viewed as finite-volume meshes. This is represented by the connecting lines from the various grid boxes to the main vertical trunk line under mesh generation in Fig. 5.3. However, in addition under the third column, we have the very modern considerations of unstructured meshes and cartesian meshes. The material presented in this chapter, as reflected in Fig. 5.3, is an important element of CFD; if you are uncertain or confused about any aspects represented by the various boxes in Fig. 5.3, make certain to review the relevant sections in this chapter before progressing further.

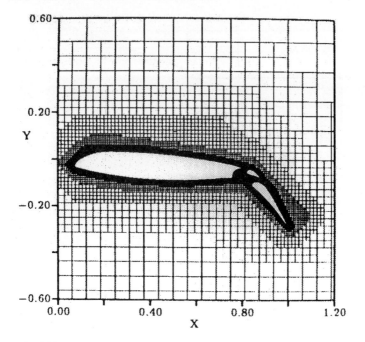

FIG. 5.19
A cartesian mesh for the calculation of the subsonic flow over a multielement airfoil. (*From Ref. 41. Copyright © 1991, AIAA. Reprinted with permission.*)

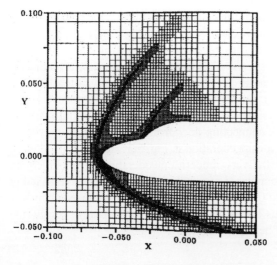

FIG. 5.20
A cartesian mesh for the calculation of the hypersonic flow over a double ellipsoid, a configuration somewhat like the space shuttle. (*From Ref. 41. Copyright © 1991, AIAA. Reprinted with permission.*)

PROBLEM

5.1. Consider a polar coordinate system drawn in the space about a circular cylinder. Discuss this system in relation to the general idea of a boundary-fitted coordinate system. Also, calculate the metrics for this system. *Note*: We will deal with such a coordinate system in Prob. 6.2, wherein the inviscid incompressible flow around a circular cylinder is calculated. The results obtained here are useful for Prob. 6.2.

CHAPTER 6

SOME SIMPLE CFD TECHNIQUES: A BEGINNING

Technique—The systematic procedure by which a complex or scientific task is accomplished.

The American Heritage Dictionary of the English Language, 1969

6.1 INTRODUCTION

This chapter is the last stepping-stone of Part II of this book; Part III deals with applications of CFD to various flow problems. To deal with any such applications, it is necessary to first understand the basic form and nature of the governing equations of fluid dynamics; this was the purpose of Part I. Second, it is necessary to understand the basics of various numerical discretizations that can be applied to these equations; this is the purpose of Part II. With the present chapter, we fulfill this purpose. Here we will take the basic numerical discretization approaches discussed earlier and mold them into various *techniques* that will allow the numerical solution of flow problems. It is in this chapter that we polish off the tools necessary for the various applications to be discussed in Part III.

Modern CFD is awash with different techniques—some old, some new, some quite simple and straightforward, and some very sophisticated and elaborate. They all have their strengths and weaknesses. In this light, let us set forth the philosophy of this chapter, and indeed the philosophy for the remainder of this book. This book is *not* intended to be an exposition of the latest state of the art in CFD. The state of the art can be found in the vast journal and technical report literature. This book is also *not* intended to be a source book for all the existing CFD techniques. (There are several advanced-level textbooks in CFD that provide a wide survey of many techniques, such as Refs. 13–18.) Rather, the purpose of this book is to provide the reader with a simple, uncluttered introduction to CFD and to establish a base from which the reader can move on to more advanced texts and courses in the field. Its role is somewhat analogous to a first undergraduate course in fluid dynamics, namely, to provide the student with some basic ideas, as well as the interest and motivation to go further with more advanced courses and state-of-the-art studies. Therefore, the CFD techniques discussed in this chapter are chosen for their simplicity as well as their usefulness. Our purpose here is to develop some CFD tools that are not overly sophisticated—tools that can be appreciated and understood at the introductory level adopted for this book but which are utilitarian enough to allow the solution of a variety of flows discussed in Part III. More sophisticated, state-of-the-art CFD techniques are discussed in Chap. 11, near the end of the book.

Finally, we note that any one particular CFD technique will not be appropriate for all problems; the diverse mathematical nature of different partial differential equations (such as described in Chap. 3) will ensure that some algorithms will work best for hyperbolic equations, others will work best for elliptic equations, etc. We will make this type of distinction as we progress.

Let us now begin to construct some techniques suitable for applications in CFD. We will do this in a generic way, leaving specific applications to specific problems for Part III. For simplicity, we will consider only two-dimensional flows; the extra work brought on by including a third dimension is not important for our purposes here. Wherever necessary, we will assume a calorically perfect gas (one with constant specific heats).

Finally, the road map for this chapter can be found in Fig. 6.35; note that it itemizes the techniques to be discussed here, along with the general nature of their applicability. Remember to consult this road map as you progress to each new section.

6.2 THE LAX-WENDROFF TECHNIQUE

The Lax-Wendroff technique is an explicit, finite-difference method particularly suited to marching solutions. The idea of numerical solutions obtained by marching in steps of time or space was discussed in Chap. 3; such marching solutions are associated with the solution of hyperbolic and parabolic partial differential equations. A good example of a flow-field problem governed by hyperbolic equations is the time-marching solution of an inviscid flow using the unsteady Euler equations. The behavior of such a time-marching solution is discussed in the Unsteady, Inviscid Flow Subsection of Sec. 3.4.1 and is sketched in Fig. 3.7. (It is recommended that you review this subsection before progressing further.)

For purposes of illustration, let us consider an unsteady, two-dimensional inviscid flow. The governing Euler equations are derived in Chap. 2 and itemized in Sec. 2.8.2. They are rearranged below in nonconservation form, obtained from Eqs. (2.82), (2.83a), (2.83b), and (2.85).

Continuity:
$$\frac{\partial \rho}{\partial t} = -\left(\rho \frac{\partial u}{\partial x} + u \frac{\partial \rho}{\partial x} + \rho \frac{\partial v}{\partial y} + v \frac{\partial \rho}{\partial y}\right) \quad (6.1)$$

x momentum:
$$\frac{\partial u}{\partial t} = -\left(u \frac{\partial u}{\partial x} + v \frac{\partial u}{\partial y} + \frac{1}{\rho} \frac{\partial p}{\partial x}\right) \quad (6.2)$$

y momentum:
$$\frac{\partial v}{\partial t} = -\left(u \frac{\partial v}{\partial x} + v \frac{\partial v}{\partial y} + \frac{1}{\rho} \frac{\partial p}{\partial y}\right) \quad (6.3)$$

Energy:
$$\frac{\partial e}{\partial t} = -\left(u \frac{\partial e}{\partial x} + v \frac{\partial e}{\partial y} + \frac{p}{\rho} \frac{\partial u}{\partial x} + \frac{p}{\rho} \frac{\partial v}{\partial y}\right) \quad (6.4)$$

In the above equations, we have assumed no body forces and no volumetric heat addition; that is, $\mathbf{f} = 0$ and $\dot{q} = 0$. Equation (6.4) is obtained from Eq. (2.85) by multiplying the momentum equation by velocity and then subtracting the result from Eq. (2.85)—the same type of derivation that generated Eq. (2.73) from Eq. (2.66). Equations (6.1) to (6.4) are hyperbolic with respect to time.

We now proceed to set up a numerical solution of Eqs. (6.1) to (6.4) using a time-marching approach; note that these equations are already arranged in a convenient form, with the time derivatives isolated on the left-hand side and the spatial derivatives on the right-hand side. The Lax-Wendroff method is predicated on a Taylor series expansion in time, as follows. Choose any dependent flow variable; for purposes of illustration, let us choose density ρ. Consider the two-dimensional grid shown in Fig. 6.1. Let $\rho_{i,j}^t$ denote the density at grid point (i, j) at time t. Then the density at the same grid point (i, j) at time $t + \Delta t$, denoted by $\rho_{i,j}^{t+\Delta t}$, is given by the Taylor series

$$\rho_{i,j}^{t+\Delta t} = \rho_{i,j}^t + \left(\frac{\partial \rho}{\partial t}\right)_{i,j}^t \Delta t + \left(\frac{\partial^2 \rho}{\partial t^2}\right)_{i,j}^t \frac{(\Delta t)^2}{2} + \cdots \quad (6.5)$$

When employing Eq. (6.5), we assume that the flow field at time t is known, and Eq. (6.5) gives the new flow field at time $t + \Delta t$. In Eq. (6.5), $\rho_{i,j}^t$ is known from the

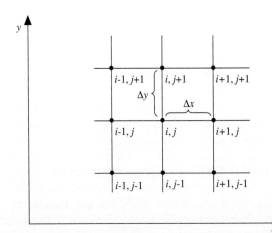

FIG. 6.1
Rectangular grid segment.

existing flow field at time t. If we can find numbers for $(\partial \rho/\partial t)^t_{i,j}$ and $(\partial^2 \rho/\partial t^2)^t_{i,j}$, then the value of density at the next step in time, $\rho^{t+\Delta t}_{i,j}$ can be calculated *explicitly* from Eq. (6.5). Analogous Taylor series are written for all the other dependent variables. For example,

$$u^{t+\Delta t}_{i,j} = u^t_{i,j} + \left(\frac{\partial u}{\partial t}\right)^t_{i,j} \Delta t + \left(\frac{\partial^2 u}{\partial t^2}\right)^t_{i,j} \frac{(\Delta t)^2}{2} + \cdots \qquad (6.6)$$

$$v^{t+\Delta t}_{i,j} = v^t_{i,j} + \left(\frac{\partial v}{\partial t}\right)^t_{i,j} \Delta t + \left(\frac{\partial^2 v}{\partial t^2}\right)^t_{i,j} \frac{(\Delta t)^2}{2} + \cdots \qquad (6.7)$$

$$e^{t+\Delta t}_{i,j} = e^t_{i,j} + \left(\frac{\partial e}{\partial t}\right)^t_{i,j} \Delta t + \left(\frac{\partial^2 e}{\partial t^2}\right)^t_{i,j} \frac{(\Delta t)^2}{2} + \cdots \qquad (6.8)$$

Equations (6.5) to (6.8) can be used to advance the flow-field variables at each grid point to the next step in time, based on known values of $\rho^t_{i,j}$, $u^t_{i,j}$, $v^t_{i,j}$, and $e^t_{i,j}$ at time t, as long as we can find numbers for the time derivatives evaluated at time t, that is, as long as we have numbers for $(\partial \rho/\partial t)^t_{i,j}$, $(\partial u/\partial t)^t_{i,j}$, $(\partial^2 u/\partial t^2)^t_{i,j}$, etc., which appear on the right side of Eqs. (6.5) to (6.8). Since Eqs. (6.5) to (6.8) are just mathematics, clearly the physics of the flow must enter the calculation somehow. Physics is what determines the time derivatives $(\partial \rho/\partial t)^t_{i,j}$, $(\partial^2 \rho/\partial t^2)^t_{i,j}$, etc., where the physics is embodied in the governing flow equations given by Eqs. (6.1) to (6.4). To be more specific, let us concentrate on the calculation of density at time $t + \Delta t$ as stipulated by Eq. (6.5). In this equation, a number for $(\partial \rho/\partial t)^t_{i,j}$ is obtained from the continuity equation, Eq. (6.1), where the spatial derivatives are given by second-order central differences. That is, from Eq. (6.1),

$$\left(\frac{\partial \rho}{\partial t}\right)^t_{i,j} = -\left(\rho^t_{i,j}\frac{u^t_{i+1,j} - u^t_{i-1,j}}{2\Delta x} + u^t_{i,j}\frac{\rho^t_{i+1,j} - \rho^t_{i-1,j}}{2\Delta x}\right.$$

$$\left. + \rho^t_{i,j}\frac{v^t_{i,j+1} - v^t_{i,j-1}}{2\Delta y} + v^t_{i,j}\frac{\rho^t_{i,j+1} - \rho^t_{i,j-1}}{2\Delta y}\right) \qquad (6.9)$$

In Eq. (6.9), all quantities on the right-hand side are known because the flow field at time t is known. Hence, Eq. (6.9) provides a number for $(\partial \rho/\partial t)^t_{i,j}$, which is inserted into Eq. (6.5). This takes care of the second term on the right side of Eq. (6.5). The third term, $(\partial^2 \rho/\partial t^2)^t_{i,j}$, is obtained in a similar fashion but requires more effort. Specifically, differentiate Eq. (6.1) with respect to time.

$$\frac{\partial^2 \rho}{\partial t^2} = -\rho\frac{\partial^2 u}{\partial x\, \partial t} + \frac{\partial u}{\partial x}\frac{\partial \rho}{\partial t} + u\frac{\partial^2 \rho}{\partial x\, \partial t} + \frac{\partial \rho}{\partial x}\frac{\partial u}{\partial t} + \rho\frac{\partial^2 v}{\partial y\, \partial t}$$

$$+ \frac{\partial v}{\partial y}\frac{\partial \rho}{\partial t} + v\frac{\partial^2 \rho}{\partial y\, \partial t} + \frac{\partial \rho}{\partial y}\frac{\partial v}{\partial t} \qquad (6.10)$$

The mixed second derivatives in Eq. (6.10), such as $\partial^2 u/(\partial x\, \partial t)$, are obtained by differentiating Eqs. (6.1) to (6.4) with respect to the proper spatial variable. For example, $\partial^2 u/(\partial x\, \partial t)$ is obtained by differentiating Eq. (6.2) with respect to x.

$$\frac{\partial^2 u}{\partial x\, \partial t} = -u\frac{\partial^2 u}{\partial x^2} + \left(\frac{\partial u}{\partial x}\right)^2 + v\frac{\partial^2 u}{\partial x\, \partial y} + \frac{\partial u}{\partial y}\frac{\partial v}{\partial x} + \frac{1}{\rho}\frac{\partial^2 p}{\partial x^2} - \frac{1}{\rho^2}\frac{\partial p}{\partial x}\frac{\partial \rho}{\partial x} \quad (6.11)$$

In Eq. (6.11), all terms on the right side are expressed as second-order, centered finite-difference quotients at time t; that is,

$$\left(\frac{\partial^2 u}{\partial x\, \partial t}\right)^t_{i,j} = -u^t_{i,j}\frac{u^t_{i+1,j} - 2u^t_{i,j} + u^t_{i-1,j}}{(\Delta x)^2}$$

$$+ \left(\frac{u^t_{i+1,j} - u^t_{i-1,j}}{2\Delta x}\right)^2 + v^t_{i,j}\frac{u^t_{i+1,j+1} + u^t_{i-1,j-1} - u^t_{i-1,j+1} - u^t_{i+1,j-1}}{4(\Delta x)(\Delta y)}$$

$$+ \frac{u^t_{i,j+1} - u^t_{i,j-1}}{2\Delta y}\frac{v^t_{i+1,j} - v^t_{i-1,j}}{2\Delta x} + \frac{1}{\rho^t_{i,j}}\frac{p^t_{i+1,j} - 2p^t_{i,j} + p^t_{i-1,j}}{(\Delta x)^2}$$

$$- \frac{1}{(\rho^t_{i,j})^2}\frac{p^t_{i+1,j} - p^t_{i-1,j}}{2\Delta x}\frac{\rho^t_{i+1,j} - \rho^t_{i-1,j}}{2\Delta x} \quad (6.12)$$

Examine Eq. (6.12); all terms on the right-hand side are known from the known flow field at time t. This provides a number for the left-hand side, i.e., a number for $(\partial^2 u/\partial x\, \partial t)^t_{i,j}$. In turn, this number is substituted for the term $\partial^2 u/(\partial x\, \partial t)$ which appears in Eq. (6.10). Continuing with the evaluation of Eq. (6.10), a number for $\partial^2 \rho/(\partial x\, \partial t)$ is found by differentiating Eq. (6.1) with respect to x and replacing all derivatives on the right side with second-order central differences, analogous to the form of Eq. (6.12). To conserve space, we will not write out the full result here. Continuing further with Eq. (6.10), a number for $\partial^2 v/(\partial y\, \partial t)$ is found by differentiating Eq. (6.3) with respect to y and replacing all derivatives on the right side with second-order central differences. The last mixed derivative in Eq. (6.10), $\partial^2 \rho/(\partial y\, \partial t)$, is found by differentiating Eq. (6.1) with respect to y and replacing all derivatives on the right side with second-order central differences. The only remaining derivatives on the right side of Eq. (6.10) are the first spatial derivatives, namely, $\partial u/\partial x$, $\partial v/\partial y$, $\partial \rho/\partial x$, and $\partial \rho/\partial y$, replaced by second-order central differences

$$\left(\frac{\partial u}{\partial x}\right)^t_{i,j} = \frac{u_{i+1,j} - u_{i-1,j}}{2\Delta x}$$

and so forth, as well as the first time derivatives $\partial \rho/\partial t$, $\partial u/\partial t$, and $\partial v/\partial t$. A number for $\partial \rho/\partial t$ has already been obtained from Eq. (6.9). Numbers for $\partial u/\partial t$ and $\partial v/\partial t$ are obtained in like fashion by inserting second-order central differences into the right-hand side of Eqs. (6.2) and (6.3), respectively. With all this, we finally obtain a number for $\partial^2 \rho/\partial t^2$ from Eq. (6.10). In turn, this is substituted into Eq. (6.5). Since $\partial \rho/\partial t$ was obtained earlier from Eq. (6.9), we now have known values at time t for all three terms on the right side of Eq. (6.5), namely, $\rho^t_{i,j}$, $(\partial \rho/\partial t)^t_{i,j}$, and $(\partial^2 \rho/\partial t^2)^t_{i,j}$. This allows the calculation of density at time $t + \Delta t$, namely, $\rho^{t+\Delta t}_{i,j}$, obtained from Eq. (6.5).

To find the remaining flow-field variables at grid point (i, j) at time $t + \Delta t$, we simply repeat the above procedure. For example, to find the value of the x component of velocity at time $t + \Delta t$, $u_{i,j}^{t+\Delta t}$, go to Eq. (6.6) and insert values for $(\partial u/\partial t)^t$ and $(\partial^2 u/\partial t^2)^t$ obtained from Eq. (6.2) in a like fashion as described above for the density. As you can see, the algebra marches on, but the idea is the same. To obtain the y component of velocity at time $t + \Delta t$, $v_{i,j}^{t+\Delta t}$, use Eq. (6.7), where values for $(\partial v/\partial t)^t$ and $(\partial^2 v/\partial t^2)^t$ are obtained from Eq. (6.3). To obtain the internal energy at time $t + \Delta t$, $e_{i,j}^{t+\Delta t}$, use Eq. (6.8), where values for $(\partial e/\partial t)^t$ and $(\partial^2 e/\partial t^2)^t$ are obtained from Eq. (6.4). With this, the flow-field variables at grid point (i, j) are now known at time $t + \Delta t$. This is illustrated schematically in Fig. 6.2, where the spatial grid in two time planes t and $t + \Delta t$ is shown. Examining this figure, we clearly see that the Lax-Wendroff method allows us to obtain *explicitly* the flow-field variables at grid point (i, j) at time $t + \Delta t$ from the *known* flow-field variables at grid points (i, j), $(i + 1, j)$, $(i - 1, j)$, $(i, j - 1)$, and $(i, j + 1)$ at time t. The flow-field variables at all other grid points at time $t + \Delta t$ are obtained in like fashion.

This is the essence and the details of the Lax-Wendroff method. It has second-order accuracy in both space and time. The idea is straightforward, but the algebra is lengthy; as you can see, most of the lengthy algebra is associated with the *second* time derivatives in Eqs. (6.6) to (6.8). Fortunately, there is a shortcut around much of this algebra—this is the subject of the next section.

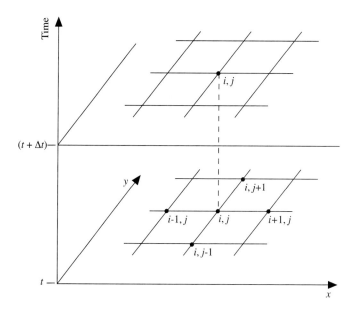

FIG. 6.2
A schematic of the grid for time marching.

6.3 MACCORMACK'S TECHNIQUE

MacCormack's technique is a variant of the Lax-Wendroff approach but is much simpler in its application. Like the Lax-Wendroff method, the MacCormack method is also an explicit finite-difference technique which is second-order-accurate in both space and time. First introduced in 1969 (Ref. 43), it became the most popular explicit finite-difference method for solving fluid flows for the next 15 years. Today, the MacCormack method has been mostly supplanted by more sophisticated approaches, some of which will be discussed in Chap. 11. However, the MacCormack method is very "student friendly;" it is among the easiest to understand and program. Moreover, the results obtained by using MacCormack's method are perfectly satisfactory for many fluid flow applications. For these reasons, MacCormack's method is highlighted here and will be used for some of the applications in Part III. It is an excellent method for introducing the fresh learner to the joys of CFD.

Consider again the two-dimensional grid shown in Fig. 6.1. For purposes of illustration, let us address again the solution of the Euler equations itemized in Eqs. (6.1) to (6.4). In Sec. 6.2 we discussed a time-marching solution using the Lax-Wendroff technique. Here, we will address a similar time-marching solution but using MacCormack's technique. As before, we assume that the flow field at each grid point in Fig. 6.1 is known at time t, and we proceed to calculate the flow-field variables at the same grid points at time $t + \Delta t$, as illustrated in Fig. 6.2. First, consider the density at grid point (i, j) at time $t + \Delta t$. In MacCormack's method, this is obtained from

$$\rho_{i,j}^{t+\Delta t} = \rho_{i,j}^t + \left(\frac{\partial \rho}{\partial t}\right)_{av} \Delta t \qquad (6.13)$$

where $(\partial \rho / \partial t)_{av}$ is a representative mean value of $\partial \rho / \partial t$ between times t and $t + \Delta t$. Compare Eq. (6.13) with its counterpart for the Lax-Wendroff method, Eq. (6.5). In Eq. (6.5), the time derivatives are evaluated at time t, and the carrying of the second derivative $(\partial^2 \rho / \partial t^2)_{i,j}^t$ is necessary to obtain second-order accuracy. In contrast, in Eq. (6.13), the value of $(\partial \rho / \partial t)_{av}$ is calculated so as to preserve second-order accuracy *without* the need to calculate values of the second time derivative $(\partial^2 \rho / \partial t^2)_{i,j}^t$, which is the term which involves a lot of algebra. With MacCormack's technique, this algebra is circumvented.

Similar relations are written for the other flow-field variables.

$$u_{i,j}^{t+\Delta t} = u_{i,j}^t + \left(\frac{\partial u}{\partial t}\right)_{av} \Delta t \qquad (6.14)$$

$$v_{i,j}^{t+\Delta t} = v_{i,j}^t + \left(\frac{\partial v}{\partial t}\right)_{av} \Delta t \qquad (6.15)$$

$$e_{i,j}^{t+\Delta t} = e_{i,j}^t + \left(\frac{\partial e}{\partial t}\right)_{av} \Delta t \qquad (6.16)$$

Let us illustrate by using the calculation of density as an example. Return to Eq. (6.13). The average time derivative, $(\partial \rho / \partial t)_{av}$, is obtained from a predictor-corrector philosophy as follows.

Predictor step. In the continuity equation, Eq. (6.1), replace the spatial derivatives on the right-hand side with *forward* differences.

$$\left(\frac{\partial \rho}{\partial t}\right)^t_{i,j} = -\left(\rho^t_{i,j}\frac{u^t_{i+1,j} - u^t_{i,j}}{\Delta x} + u^t_{i,j}\frac{\rho^t_{i+1,j} - \rho^t_{i,j}}{\Delta x}\right.$$

$$\left. + \rho^t_{i,j}\frac{v^t_{i,j+1} - v^t_{i,j}}{\Delta y} + v^t_{i,j}\frac{\rho^t_{i,j+1} - \rho^t_{i,j}}{\Delta y}\right) \quad (6.17)$$

In Eq. (6.17), all flow variables at time t are known values; i.e., the right-hand side is known. Now, obtain a *predicted* value of density, $(\bar{\rho})^{t+\Delta t}$, from the first two terms of a Taylor series, as follows.

$$(\bar{\rho})^{t+\Delta t}_{i,j} = \rho^t_{i,j} + \left(\frac{\partial \rho}{\partial t}\right)^t_{i,j}\Delta t \quad (6.18)$$

In Eq. (6.18), $\rho^t_{i,j}$ is known, and $(\partial \rho/\partial t)^t_{i,j}$ is a known number from Eq. (6.17); hence $(\bar{\rho})^{t+\Delta t}_{i,j}$ is readily obtained. The value of $(\bar{\rho})^{t+\Delta t}_{i,j}$ is only a *predicted* value of density; it is only first-order-accurate since Eq. (6.18) contains only the first-order terms in the Taylor series.

In a similar fashion, predicted values for u, v, and e can be obtained, i.e.,

$$(\bar{u})^{t+\Delta t}_{i,j} = u^t_{i,j} + \left(\frac{\partial u}{\partial t}\right)^t_{i,j}\Delta t \quad (6.19)$$

$$(\bar{v})^{t+\Delta t}_{i,j} = v^t_{i,j} + \left(\frac{\partial v}{\partial t}\right)^t_{i,j}\Delta t \quad (6.20)$$

$$(\bar{e})^{t+\Delta t}_{i,j} = e^t_{i,j} + \left(\frac{\partial e}{\partial t}\right)^t_{i,j}\Delta t \quad (6.20a)$$

In Eqs. (6.19) to (6.20a), numbers for the time derivatives on the right-hand side are obtained from Eqs. (6.2) to (6.4), respectively, with *forward* differences used for the spatial derivatives, similar to those shown in Eq. (6.17) for the continuity equation.

Corrector step. In the corrector step, we first obtain a *predicted* value of the *time derivative* at time $t + \Delta t$, $(\overline{\partial \rho/\partial t})^{t+\Delta t}_{i,j}$, by substituting the *predicted* values of ρ, u, and v into the right side of the continuity equation, replacing the spatial derivatives with *rearward* differences.

$$\overline{\left(\frac{\partial \rho}{\partial t}\right)}^{t+\Delta t}_{i,j} = -\left[(\bar{\rho})^{t+\Delta t}_{i,j}\frac{(\bar{u})^{t+\Delta t}_{i,j} - (\bar{u})^{t+\Delta t}_{i-1,j}}{\Delta x}\right.$$

$$+ (\bar{u})^{t+\Delta t}_{i,j}\frac{(\bar{\rho})^{t+\Delta t}_{i,j} - (\bar{\rho})^{t+\Delta t}_{i-1,j}}{\Delta x} + (\bar{\rho})^{t+\Delta t}_{i,j}\frac{(\bar{v})^{t+\Delta t}_{i,j} - (\bar{v})^{t+\Delta t}_{i,j-1}}{\Delta y}$$

$$\left. + (\bar{v})^{t+\Delta t}_{i,j}\frac{(\bar{\rho})^{t+\Delta t}_{i,j} - (\bar{\rho})^{t+\Delta t}_{i,j-1}}{\Delta y}\right] \quad (6.21)$$

The average value of the time derivative of density which appears in Eq. (6.13) is obtained from the arithmetic mean of $(\partial\rho/\partial t)^t_{i,j}$, obtained from Eq. (6.17), and $\overline{(\partial\rho/\partial t)}^{t+\Delta t}_{i,j}$, obtained from Eq. (6.21).

$$\left(\frac{\partial\rho}{\partial t}\right)_{\text{av}} = \frac{1}{2}\left[\underbrace{\left(\frac{\partial\rho}{\partial t}\right)^t_{i,j}}_{\text{From Eq. (6.17)}} + \underbrace{\left(\overline{\frac{\partial\rho}{\partial t}}\right)^{t+\Delta t}_{i,j}}_{\text{From Eq. (6.21)}}\right] \quad (6.22)$$

This allows us to obtain the final, "corrected" value of density at time $t + \Delta t$ from Eq. (6.13), repeated below:

$$\rho^{t+\Delta t}_{i,j} = \rho^t_{i,j} + \left(\frac{\partial\rho}{\partial t}\right)_{\text{av}} \Delta t \quad (6.13)$$

The predictor-corrector sequence described above yields the value of density at grid point (i, j) at time $t + \Delta t$, as illustrated in Fig. 6.2. This sequence is repeated at all grid points to obtain the density throughout the flow field at time $t + \Delta t$. To calculate u, v, and e at time $t + \Delta t$, the same technique is used, starting with Eqs. (6.14) to (6.16) and utilizing the momentum and energy equations in the form of Eqs. (6.2) to (6.4) to obtain the average time derivatives via the predictor-corrector sequence, using forward differences on the predictor and rearward differences on the corrector.

MacCormack's technique as described above, because a two-step predictor-corrector sequence is used with forward differences on the predictor and with rearward differences on the corrector, is a second-order-accurate method. Therefore, it has the same accuracy as the Lax–Wendroff method described in Sec. 6.2. However, the MacCormack method is much easier to apply, because there is no need to evaluate the second time derivatives as was the case for the Lax-Wendroff method. To see this more clearly, recall Eqs. (6.10) and (6.11), which are required for the Lax-Wendroff method. These equations represent a large number of additional calculations. Moreover, for a more complex fluid dynamic problem such as the flow of a viscous fluid, the differentiation of the continuity, momentum, and energy equations to obtain the second derivatives, first with respect to time, and then the mixed derivatives with respect to time and space, can be very tedious and provides an extra source for human error. MacCormack's method does *not* require such second derivatives and hence does *not* deal with equations such as (6.10) and (6.11).

In MacCormack's technique, the use of forward differences on the predictor and rearward differences on the corrector is not sacrosanct; the same order of accuracy is obtained by using rearward diffences on the predictor and forward differences on the corrector. Indeed, a time-marching solution can be carried out by alternating between these two sequences at every other time step, if you so choose.

GUIDEPOST

If you are anxious to start a computer project using MacCormack's technique, you can follow this guidepost now and return to Chap. 6 at a later time.

| Go to Sec. 6.6, | → | Then go to *all* |
| artificial viscosity. | | of Chap. 7, nozzle flows. |

On the other hand, if you want a broader perspective on various CFD techniques *before* you start the applications in Part III, simply continue to read the remaining sections in the present chapter.

6.4 SOME COMMENTS: VISCOUS FLOWS, CONSERVATION FORM, AND SPACE MARCHING

We have chosen to illustrate the Lax-Wendroff (Sec. 6.2) and MacCormack (Sec. 6.3) techniques by assuming an inviscid flow, using the nonconservation form of the Euler equations, and discussing a computational time-marching step. None of these have to be the case; these techniques can be applied just as well to viscous flows, to the conservation form of the governing flow equations, and to space marching. Let us examine each of these comments in turn.

6.4.1 Viscous Flows

Viscous flows are governed by the Navier-Stokes equations, summarized in Sec. 2.8.1. Written in the form for steady flow, these equations have a mathematical behavior which is partially elliptic. The Lax-Wendroff and MacCormack techniques are not appropriate for the solution of elliptic partial differential equations. However, the *unsteady* Navier-Stokes equations have a mixed parabolic and elliptic behavior, and therefore the Lax-Wendroff and MacCormack techniques are suitable. Indeed, the MacCormack technique has been used extensively for solutions of the unsteady Navier-Stokes equations by means of time-marching solutions. The idea is the same as discussed in Sec. 6.3; the Navier-Stokes equations are written with the time derivatives on the left side and spatial derivatives on the right side of the equations. The spatial derivatives are replaced in turn by forward and rearward differences on the predictor and corrector steps, respectively.* The approach is exactly the same as discussed in Sec. 6.3; the only difference is the larger number of spatial derivatives that are present in the Navier-Stokes equations compared to the Euler equations.

6.4.2 Conservation Form

For simplicity, we will continue to use the Euler equations in our discussion. The conservation form of the Euler equations suitable for CFD calculations was

---

* This statement is true for the convective terms. However, it has been the author's experience, as well as that of many others, that the *viscous* terms should be *centrally differenced* on both the predictor and corrector steps.

discussed in Sec. 2.10; this form was embodied in the generic equation given by Eq. (2.93). Rearranging this equation, and considering a two-dimensional flow, we have

$$\frac{\partial U}{\partial t} = -\frac{\partial F}{\partial x} - \frac{\partial G}{\partial y} + J \qquad (6.23)$$

where the elements of the column vectors U, F, G, and J are given by Eqs. (2.105) to (2.109), respectively. Clearly, values for the elements of U, namely, ρ, ρu, ρv, and $\rho(e + V^2/2)$, can be calculated in steps of time using either the Lax-Wendroff or the MacCormack technique. The approach is exactly the same as discussed in Secs. 6.2 and 6.3. Keep in mind that since the dependent variables in Eq. (6.23) are flux variables, the primitive variables have to be decoded at the end of each time step in the fashion given in Eqs. (2.100) to (2.104). At this stage, please return to Sec. 2.10 where such matters associated with the conservation form of the equations are discussed and review that material before progressing further. You will find that, by now, with the technical maturity you have obtained in the ensuing chapters, Sec. 2.10 will have renewed significance for you and your understanding will be enhanced. There is another reason for reviewing Sec. 2.10 right now—it leads directly into the material in the next subsection.

6.4.3 Space Marching

To illustrate the space-marching idea, let us apply MacCormack's technique to the two-dimensional flow shown in Fig. 6.3. The general flow direction is from left to right in the xy plane. For simplicity, assume the flow is inviscid; hence the governing flow equations are the Euler equations. In the generic, conservation form, this system of equations is given by Eq. (2.110), reduced to a two-dimensional form as

$$\frac{\partial F}{\partial x} = J - \frac{\partial G}{\partial y} \qquad (6.24)$$

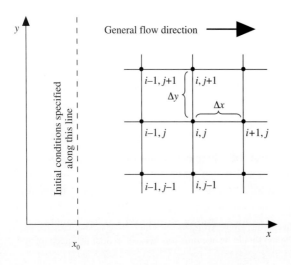

FIG. 6.3
A schematic of the grid for space marching.

For a subsonic flow, Eq. (6.24) is elliptic, and MacCormack's technique does not apply; indeed, *any* space-marching technique will not apply. However, as mentioned in Chap. 3, for a flow that is locally supersonic everywhere, Eq. (6.24) is hyperbolic. In this case, space marching is appropriate, and MacCormack's technique is applicable. With this in mind, notice that Eq. (6.24) is written with the x derivative isolated on the left-hand side and the source term and the y derivative on the right-hand side. Return to Fig. 6.3. Assume that the flow-field variables are known along the vertical line in the xy plane; this line is the *initial data line*. Also assume that the flow is locally supersonic everywhere. Then a solution can be obtained, starting with the initial data line and marching downstream in the x direction. We will illustrate the process for a single spatial step using MacCormack's technique. The ideas are the same as discussed in Sec. 6.3, except that here the spatial variable x performs the same role as the time variable t in Sec. 6.3. For example, in Fig. 6.3 assume the flow variables are known along a vertical line at a given x location. (The calculation was started using the initial data along the vertical line $x = x_0$.) Let this vertical line run through the grid points $(i, j+1)$, (i, j) and $(i, j-1)$ in Fig. 6.3. That is, the flow variables at these three grid points are considered known. MacCormack's technique allows the calculation of the flow variables at grid point $(i+1, j)$ from the known values at $(i, j+1)$, (i, j) and $(i, j-1)$, as follows. The value of the solution vector F in Eq. (6.24) at grid point $(i+1, j)$ can be found from

$$F_j^{i+1} = F_j^i + \left(\frac{\partial F}{\partial x}\right)_{av} \Delta x \tag{6.25}$$

Note that, in keeping with our previous notation, the index for the marching variable, in this case i, is used as a superscript. In Eq. (6.25), $(\partial F/\partial x)_{av}$ is a representative average value of the x derivative of F evaluated between x and $x + \Delta x$. It is found from Eq. (6.24) by means of a predictor-corrector approach, as follows.

Predictor step. In Eq. (6.24), replace the y derivative with a forward difference:

$$\left(\frac{\partial F}{\partial x}\right)_j^i = J_j^i - \frac{G_{j+1}^i - G_j^i}{\Delta y} \tag{6.26}$$

In Eq. (6.26), all terms on the right side are known numbers, because the flow is known along the vertical line through point (i, j). Calculate a predicted value for F at point $(i+1, j)$ from a Taylor series:

$$\bar{F}_j^{i+1} = F_j^i + \left(\frac{\partial F}{\partial x}\right)_j^i \Delta x \tag{6.27}$$

where, as in Sec. 6.3, the barred quantity represents a predicted quantity. Keep in mind that the shorthand vector notation shown in Eqs. (6.26) and (6.27) represents these operations on the individual continuity, momentum, and energy equations, where the elements of F and G are given by Eqs. (2.106) and (2.107), respectively. That is, $\bar{F}_j^{i+1}$ represents the *predicted* values of its individual elements, given for the present two-dimensional case by

$$\bar{F}_j^{i+1} = \begin{Bmatrix} \overline{(\rho u)}_j^{i+1} \\ \overline{(\rho u^2 + p)}_j^{i+1} \\ \overline{(\rho u v)}_j^{i+1} \\ \overline{\left[\rho u\left(e + \dfrac{u^2+v^2}{2}\right) + pu\right]}_j^{i+1} \end{Bmatrix} \qquad (6.28)$$

Before progressing further, the calculated values on the right side of Eq. (6.28) must be *decoded* to obtain predicted values of the primitive variables, as discussed in that part of Sec. 2.10 associated with Eqs. (2.111a) to (2.111e). These primitive variables are needed to form the numbers for the flux vector G in the corrector step, as follows.

Corrector step. Calculate a predicted value of $(\partial F/\partial x)_j^{i+1}$ at location $x + \Delta x$, denoted by $\overline{(\partial F/\partial x)}_j^{i+1}$, by inserting the predicted values for J and G into Eq. (6.24), using rearward differences. That is,

$$\overline{\left(\dfrac{\partial F}{\partial x}\right)}_j^{i+1} = \bar{J}_j^{i+1} - \dfrac{\bar{G}_j^{i+1} - \bar{G}_{j-1}^{i+1}}{\Delta y} \qquad (6.29)$$

In Eq. (6.29), the values of $\bar{G}_j^{i+1}$ and $\bar{G}_{j-1}^{i+1}$ are constructed from the predicted primitive variables which had been decoded earlier in the predictor step. The average value, $(\partial F/\partial x)_{av}$, is now formed as an arithmetic mean

$$\left(\dfrac{\partial F}{\partial x}\right)_{av} = \dfrac{1}{2}\left[\underbrace{\left(\dfrac{\partial F}{\partial x}\right)_j^i}_{\text{From Eq. (6.26)}} + \underbrace{\overline{\left(\dfrac{\partial F}{\partial x}\right)}_j^{i+1}}_{\text{From Eq. (6.29)}}\right] \qquad (6.30)$$

In turn, the final, corrected value of F_j^{i+1} is obtained from Eq. (6.25), repeated below:

$$F_j^{i+1} = F_j^i + \left(\dfrac{\partial F}{\partial x}\right)_{av} \Delta x \qquad (6.25)$$

Clearly, this spatial, downstream marching solution using MacCormack's technique is a direct analog of the time-marching solution discussed in Sec. 6.3, with the marching variable x playing the role of the earlier marching variable t.

There are two noteworthy differences associated with the downstream marching approach compared to the time-marching approach. The first has already been mentioned; it is associated with the need to decode the primitive variables from the flux variables. This decoding is simple when a time-marching solution of the conservation form of the equations is employed, as reflected in Eqs. (2.100) to (2.104), but it is more elaborate when a spatial-marching solution of the conservation form equations is used, as reflected in Eqs. (2.111a) to (2.111e). Of course, for a time-marching solution using the nonconservation form of the equations, no decoding is needed at all; the dependent variables are the primitive variables themselves, as we have seen in Secs. 6.2 and 6.3. The second difference between the two marching procedures, at least for explicit solutions, is that the downstream marching procedure demands the use of the conservation form of the governing equations so that the x derivative can be isolated as a single term, as displayed in Eq. (6.24). This can not be done with the nonconservation form of the equations, as a quick examination of Eqs. (6.1) to (6.4) will show. Here, with the time derivatives set to zero, three out of the four equations have two terms each involving x derivatives, and therefore a single x derivative can not be isolated on the left-hand side without another x derivative still appearing on the right-hand side. This, of course, destroys the explicit nature of the downstream marching approach as discussed here.

6.5 THE RELAXATION TECHNIQUE AND ITS USE WITH LOW-SPEED INVISCID FLOW

The relaxation technique is a finite-difference method particularly suited for the solution of elliptic partial differential equations. Low-speed, subsonic inviscid flow is governed by elliptic partial differential equations, as discussed in Sec. 3.4.3. Therefore, the relaxation technique is frequently applied to the solution of low-speed subsonic flow. Relaxation techniques can be either explicit or implicit. See Ref. 13 for an in-depth discussion of various relaxation techniques as applied to CFD problems. In the present section, we will describe an explicit relaxation technique, sometimes called a *point-iterative* method.

For purposes of illustration, let us consider an inviscid, incompressible, two-dimensional irrotational flow. For such a flow, the governing flow equations reduce to a single partial differential equation, namely, Laplace's equation, in terms of the scalar velocity potential Φ, where Φ is defined such that $\mathbf{V} = \nabla \Phi$. We will not provide the details here but rather make the assumption that you have some familiarity with such matters. If not, or if you simply need a review of the derivation, see, for example, Sec. 3.7 of Ref. 8. We will simply state here that the governing equation is

$$\frac{\partial^2 \Phi}{\partial x^2} + \frac{\partial^2 \Phi}{\partial y^2} = 0 \quad (6.31)$$

We wish to solve Eq. (6.31) numerically on the grid shown in Fig. 6.4. Replace the partial derivatives in Eq. (6.31) with second-order, central second differences, given by Eqs. (4.12) and (4.13).

$$\frac{\Phi_{i+1,j} - 2\Phi_{i,j} + \Phi_{i-1,j}}{(\Delta x)^2} + \frac{\Phi_{i,j+1} - 2\Phi_{i,j} + \Phi_{i,j-1}}{(\Delta y)^2} = 0 \quad (6.32)$$

Examining the grid in Fig. 6.4, note that grid points 1 through 20 constitute the

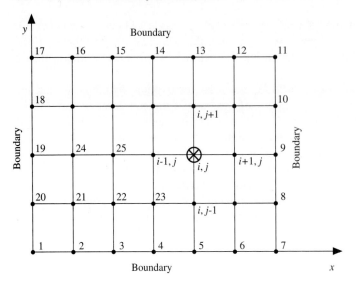

FIG. 6.4
Schematic for relaxation technique.

boundary of the domain. As discussed in Sec. 3.4.3, boundary conditions must be stipulated over the *entire* boundary enclosing the domain in order for the solution of an elliptic equation to be well-posed. In terms of the grid shown in Fig. 6.4, this means that Φ_1 through Φ_{20} are known values, equal to the given boundary conditions at points 1 through 20. The values of Φ at all other grid points—the internal grid points—are unknown. Equation (6.32), centered around grid point (i, j), contains five of these unknowns, namely, $\Phi_{i-1, j}$, $\Phi_{i, j}$, $\Phi_{i+1, j}$, $\Phi_{i, j+1}$, $\Phi_{i, j-1}$. In principle, Eq. (6.32) can be written around each of the internal grid points (there are 15 such points in Fig. 6.4), leading to a system of 15 linear algebraic equations with a total of 15 unknowns. There are several direct methods for solving these simultaneous equations. One is the standard Cramer's rule; however, the number of calculations required for the implementation of Cramer's rule is very large, due to the need to evaluate determinants of the size 15×15 for the present example. For any real calculation, hundreds or even thousands of grid points may be employed. Clearly, the use of Cramer's rule is out of the question for such applications. Another, and much more reasonable, direct solution is gaussian elimination (see, for example, Ref. 13). However, the simplest approach is to use a relaxation technique, as described below.

The relaxation technique is an iterative method, wherein values of four of the quantities in Eq. (6.32) are assumed to be the *known* values at iteration step n and only one of the quantities is treated as an unknown at iteration step $n + 1$. In Eq. (6.32), let us choose $\Phi_{i, j}$ as that unknown. Solving Eq. (6.32) for $\Phi_{i, j}$, we have

$$\Phi_{i,j}^{n+1} = \frac{(\Delta x)^2 (\Delta y)^2}{2(\Delta y)^2 + 2(\Delta x)^2} \left[\frac{\Phi_{i+1, j}^n + \Phi_{i-1, j}^n}{(\Delta x)^2} + \frac{\Phi_{i, j+1}^n + \Phi_{i, j-1}^n}{(\Delta y)^2} \right] \quad (6.33)$$

In Eq. (6.33), the superscripts n and $n + 1$ indicates the *iteration* step; it has nothing to do with our previous use of the superscript to designate a time- or space-marching step. Indeed, as we know, such marching is *not* appropriate for the solution of elliptic equations. Rather, in Eq. (6.33), $\Phi_{i,j}^{n+1}$ represents the unknown to be calculated at the next iteration step, $n + 1$, in terms of the known quantities $\Phi_{i+1,j}^{n}$, $\Phi_{i-1,j}^{n}$, $\Phi_{i,j+1}^{n}$, and $\Phi_{i,j-1}^{n}$ from the previous step n. (This approach is called the *Jacobi method*.) To get the whole process started, we first *assume* values for Φ at all grid points except one, at which Φ is treated as the unknown. Equation (6.33) is used to calculate that unknown. After repeated application of Eq. (6.33) to all the grid points, we have finished the first iteration, $n = 1$, and we go on to the next step, $n = 2$. This whole process is repeated for as many iterations as are necessary to converge to a solution. To be more specific, consider Eq. (6.33) applied at grid point 21 in Fig. 6.4. Assume that we have already carried out n iterations. Then, for the $n + 1$ iteration, Eq. (6.33) yields

$$\Phi_{21}^{n+1} = \frac{(\Delta x)^2 (\Delta y)^2}{2(\Delta y)^2 + 2(\Delta x)^2} \left[\frac{\Phi_{22}^n + \Phi_{20}}{(\Delta x)^2} + \frac{\Phi_{24}^n + \Phi_2}{(\Delta y)^2} \right] \quad (6.34)$$

In Eq. (6.34), Φ_{21}^{n+1} is the unknown; Φ_{22}^n and Φ_{24}^n are known from the previous iteration, and Φ_{20} and Φ_2 are known from the stipulated boundary conditions.

It is suggested that updated values of Φ be used as soon as possible on the right-hand side of Eq. (6.33). For example, after we have calculated Φ_{21}^{n+1} from Eq. (6.34), we move on to grid point 22, where an application of Eq. (6.33) yields

$$\Phi_{22}^{n+1} = \frac{(\Delta x)^2 (\Delta y)^2}{2(\Delta y)^2 + 2(\Delta x)^2} \left[\frac{\Phi_{23}^n + \Phi_{21}^{n+1}}{(\Delta x)^2} + \frac{\Phi_{25}^n + \Phi_3}{(\Delta y)^2} \right] \quad (6.35)$$

In Eq. (6.35), Φ_{22}^{n+1} is the unknown; Φ_{23}^n and Φ_{25}^n are known from the previous iteration, Φ_3 is known from the stipulated boundary condition, and Φ_{21}^{n+1} is known from Eq. (6.34), which was the immediately preceding calculation. In this fashion, the unknown Φ's at iteration $n + 1$ are progressively calculated along a given horizontal line, sweeping from left to right. (This approach is called the *Gauss-Seidel method*.) There is nothing magic about this sweeping direction. During the progressive solution of Eq. (6.33), we could just as well set up sequences that sweep from right to left, from top to bottom, or from bottom to top.

The above procedure is repeated for a number of iterations; *convergence* is achieved when $\Phi_{i,j}^{n+1} - \Phi_{i,j}^n$ becomes less than some prescribed value at all grid points. The degree to which you wish convergence to be achieved is up to you; the more iterations you take, the greater will be the accuracy.

Frequently, the convergence to a solution sometimes can be enhanced by a technique called *successive overrelaxation*. This is an extrapolation procedure based on the following idea. We interpret Eq. (6.33) as yielding an intermediate value of $\Phi_{i,j}$, denoted by $\overline{\Phi_{i,j}^{n+1}}$, where

$$\overline{\Phi_{i,j}^{n+1}} = \frac{(\Delta x)^2 (\Delta y)^2}{2(\Delta y)^2 + 2(\Delta x)^2} \left[\frac{\Phi_{i+1,j}^n + \Phi_{i-1,j}^{n+1}}{(\Delta x)^2} + \frac{\Phi_{i,j+1}^n + \Phi_{i,j-1}^{n+1}}{(\Delta y)^2} \right] \quad (6.36)$$

Note that we have chosen to write the value of $\Phi_{i-1,j}^{n+1}$ in Eq. (6.36) at iteration level $n + 1$ with the assumption that we are sweeping from left to right as discussed earlier, and hence the value of $\Phi_{i-1,j}^{n+1}$ is known at this stage. Similarly, $\Phi_{i,j-1}^{n+1}$ is known at this stage because we are starting our sweeping procedure at the bottom of the grid, and sequentially stepping to the next higher row of grid points. Then we use the value of $\Phi_{i,j}^n$ obtained at the end of the previous iteration, and $\Phi_{i,j}^{n+1}$ obtained from Eq. (6.36), to *extrapolate* a value for $\Phi_{i,j}^{n+1}$ as follows:

$$\Phi_{i,j}^{n+1} = \Phi_{i,j}^n + \omega(\overline{\Phi_{i,j}^{n+1}} - \Phi_{i,j}^n) \tag{6.37}$$

In Eq. (6.37), ω is a relaxation factor whose value is usually found by trial-and-error experimentation for a given problem. If $\omega > 1$, the above process is called *successive overrelaxation*. If $\omega < 1$, the process is called *underrelaxation* and is usually used when the convergence behavior is oscillating back and forth between some value. For overrelaxation, generally the value of ω is bounded by $1 < \omega < 2$ (see Ref. 13). In any event, the use of Eq. (6.37) with an appropriate value for ω can reduce the number of iterations necessary to achieve convergence and therefore reduce the computational time—in some problems by a factor of 30 according to Ref. 13.

6.6 ASPECTS OF NUMERICAL DISSIPATION AND DISPERSION; ARTIFICIAL VISCOSITY

Many aspects of life are never quite what they appear to be at first impression—CFD is no different. For example, in the present chapter we have discussed several techniques for the numerical solution of the governing flow equations. We have approached these discussions, as well as those in previous chapters, from the point of view that numerical solutions of the Euler or Navier-Stokes equations are being obtained within an accuracy determined by the truncation and round-off errors. The focus has been on the fact that we are solving some *specific partial differential equations* but that the numerical solutions are always somewhat in error.

There is a different perspective that we can take on this matter, one with a shade of difference compared to our previous discussions. For simplicity, let us consider a model equation, namely, the one-dimensional wave equation given by

$$\frac{\partial u}{\partial t} + a\frac{\partial u}{\partial x} = 0 \tag{6.38}$$

with $a > 0$. We consider (6.38) to be the *specific* partial differential equation that we want to solve numerically. Let us choose to discretize this equation by using a first-order forward difference in time and a first-order rearward difference in space. Then Eq. (6.38) is represented by the following difference equation:

$$\frac{u_i^{t+\Delta t} - u_i^t}{\Delta t} + a\frac{u_i^t - u_{i-1}^t}{\Delta x} = 0 \tag{6.39}$$

From our previous perspective, a solution of Eq. (6.39) represents a numerical solution of Eq. (6.38) within a certain accuracy as determined by the truncation and round-off errors. From our discussions in Chap. 4, we know that the accuracy of Eq. (6.39) is given by $O(\Delta t, \Delta x)$. Let us now take a slightly different point of view. To help establish this view, we replace $u_i^{t+\Delta t}$ and u_{i-1}^t in Eq. (6.39) with Taylor series expansions as follows:

$$u_i^{t+\Delta t} = u_i^t + \left(\frac{\partial u}{\partial t}\right)_i^t \Delta t + \left(\frac{\partial^2 u}{\partial t^2}\right)_i^t \frac{(\Delta t)^2}{2} + \left(\frac{\partial^3 u}{\partial t^3}\right)_i^t \frac{(\Delta t)^3}{6} + \cdots \qquad (6.40)$$

$$u_{i-1}^t = u_i^t - \left(\frac{\partial u}{\partial x}\right)_i^t \Delta x + \left(\frac{\partial^2 u}{\partial x^2}\right)_i^t \frac{(\Delta x)^2}{2} - \left(\frac{\partial^3 u}{\partial x^3}\right)_i^t \frac{(\Delta x)^3}{6} + \cdots \qquad (6.41)$$

Substituting Eqs. (6.40) and (6.41) into (6.39), we have

$$\left[\left(\frac{\partial u}{\partial t}\right)_i^t + \left(\frac{\partial^2 u}{\partial t^2}\right)_i^t \frac{\Delta t}{2} + \left(\frac{\partial^3 u}{\partial t^3}\right)_i^t \frac{(\Delta t)^2}{6} + \cdots\right]$$

$$+ a\left[\left(\frac{\partial u}{\partial x}\right)_i^t - \left(\frac{\partial^2 u}{\partial x^2}\right)_i^t \frac{\Delta x}{2} + \left(\frac{\partial^3 u}{\partial x^3}\right)_i^t \frac{(\Delta x)^2}{6} + \cdots\right] = 0 \qquad (6.42)$$

Rearranging Eq. (6.42), we obtain

$$\left(\frac{\partial u}{\partial t}\right)_i^t + a\left(\frac{\partial u}{\partial x}\right)_i^t = -\left(\frac{\partial^2 u}{\partial t^2}\right)_i^t \frac{\Delta t}{2} - \left(\frac{\partial^3 u}{\partial t^3}\right)_i^t \frac{(\Delta t)^2}{6}$$

$$+ \left(\frac{\partial^2 u}{\partial x^2}\right)_i^t \frac{a\,\Delta x}{2} - \left(\frac{\partial^3 u}{\partial x^3}\right)_i^t \frac{a(\Delta x)^2}{6} + \cdots \qquad (6.43)$$

Pause for a moment and examine Eq. (6.43). The left-hand side is exactly the left-hand side of the original partial differential equation given by Eq. (6.38); the right-hand side of Eq. (6.43) is the truncation error associated with the difference equation given by Eq. (6.39). Clearly, this truncation error is $O(\Delta t, \Delta x)$. Let us now replace the time derivatives on the right-hand side of Eq. (6.43) with x derivatives as follows. First, differentiate Eq. (6.43) with respect to t. (We will drop the subscript i and superscript t, since we know that all derivatives are being evaluated at point i and at time t.)

$$\frac{\partial^2 u}{\partial t^2} + a\frac{\partial^2 u}{\partial x\,\partial t} = -\frac{\partial^3 u}{\partial t^3}\frac{\Delta t}{2} - \frac{\partial^4 u}{\partial t^4}\frac{(\Delta t)^2}{6}$$

$$+ \frac{\partial^3 u}{\partial x^2\,\partial t}\frac{a\,\Delta x}{2} - \frac{\partial^4 u}{\partial x^3\,\partial t}\frac{a(\Delta x)^2}{6} + \cdots \qquad (6.44)$$

Also, differentiate Eq. (6.43) with respect to x and multiply by a.

$$a\frac{\partial^2 u}{\partial t \partial x} + a^2 \frac{\partial^2 u}{\partial x^2} = -\frac{\partial^3 u}{\partial t^2 \partial x}\frac{a\,\Delta t}{2} - \frac{\partial^4 u}{\partial t^3 \partial x}\frac{a(\Delta t)^2}{6}$$

$$+ \frac{\partial^3 u}{\partial x^3}\frac{a^2\,\Delta x}{2} - \frac{\partial^4 u}{\partial x^4}\frac{a^2(\Delta x)^2}{6} + \cdots \quad (6.45)$$

Subtracting Eq. (6.45) from (6.44), we have

$$\frac{\partial^2 u}{\partial t^2} = a^2\frac{\partial^2 u}{\partial x^2} - \frac{\partial^3 u}{\partial t^3}\frac{\Delta t}{2} - \frac{\partial^4 u}{\partial t^4}\frac{(\Delta t)^2}{6} + \frac{\partial^3 u}{\partial x^2 \partial t}\frac{a\,\Delta x}{2}$$

$$- \frac{\partial^4 u}{\partial x^3 \partial t}\frac{a(\Delta x)^2}{6} + \frac{\partial^3 u}{\partial t^2 \partial x}\frac{a\,\Delta t}{2} + \frac{\partial^4 u}{\partial t^3 \partial x}\frac{a(\Delta t)^2}{6}$$

$$- \frac{\partial^3 u}{\partial x^3}\frac{a^2\,\Delta x}{2} + \frac{\partial^4 u}{\partial x^4}\frac{a^2(\Delta x)^2}{6} + \cdots \quad (6.46)$$

We can express Eq. (6.46) in a more compact form by displaying only the first-order terms, i.e.,

$$\frac{\partial^2 u}{\partial t^2} = a^2\frac{\partial^2 u}{\partial x^2} + \frac{\Delta t}{2}\left[-\frac{\partial^3 u}{\partial t^3} + a\frac{\partial^3 u}{\partial t^2 \partial x} + O(\Delta t)\right]$$

$$+ \frac{\Delta x}{2}\left[a\frac{\partial^3 u}{\partial x^2 \partial t} - a^2\frac{\partial^3 u}{\partial x^3} + O(\Delta x)\right] \quad (6.47)$$

Equation (6.47) provides the expression for $\partial^2 u/\partial t^2$ which is to be substituted for the first term on the right-hand side of Eq. (6.43). Before carrying out this substitution, however, let us treat the second term on the right-hand side of Eq. (6.43), namely, the third time derivative. We do this by differentiating Eq. (6.47) with respect to time, yielding

$$\frac{\partial^3 u}{\partial t^3} = a^2\frac{\partial^3 u}{\partial x^2 \partial t} + O(\Delta t, \Delta x) \quad (6.48)$$

Differentiating Eq. (6.45) with respect to x and multiplying by a, we have

$$a^2\frac{\partial^3 u}{\partial x^2 \partial t} + a^3\frac{\partial^3 u}{\partial x^3} = O(\Delta t, \Delta x) \quad (6.49)$$

Adding Eqs. (6.48) and (6.49), we have

$$\frac{\partial^3 u}{\partial t^3} = -a^3\frac{\partial^3 u}{\partial x^3} + O(\Delta t, \Delta x) \quad (6.50)$$

Equation (6.50) provides an expression for the third time derivative to be inserted into both Eqs. (6.47) and (6.43). Returning to Eq. (6.47), we see two mixed derivatives with respect to t and x that must be treated. Differentiating Eq. (6.47) with respect to x, we have

$$\frac{\partial^3 u}{\partial t^3 \partial x} = a^2\frac{\partial^3 u}{\partial x^3} + O(\Delta t, \Delta x) \quad (6.51)$$

Also, rearranging Eq. (6.48), we have

$$\frac{\partial^3 u}{\partial x^2 \, \partial t} = \frac{1}{a^2} \frac{\partial^3 u}{\partial t^3} + O(\Delta t, \Delta x) \tag{6.52}$$

Substituting Eq. (6.50) into (6.52), we have

$$\frac{\partial^3 u}{\partial x^2 \, \partial t} = -a \frac{\partial^3 u}{\partial x^3} + O(\Delta t, \Delta x) \tag{6.53}$$

Substituting Eqs. (6.50), (6.51) and (6.53) into (6.47), we obtain

$$\frac{\partial^2 u}{\partial t^2} = a^2 \frac{\partial^2 u}{\partial x^2} + \frac{\Delta t}{2} \left[a^3 \frac{\partial^3 u}{\partial x^3} + a^3 \frac{\partial^3 u}{\partial x^3} + O(\Delta t, \Delta x) \right]$$

$$+ \frac{\Delta x}{2} \left[-a^2 \frac{\partial^3 u}{\partial x^3} - a^2 \frac{\partial^3 u}{\partial x^3} + O(\Delta t, \Delta x) \right] \tag{6.54}$$

Substituting Eqs. (6.54) and (6.50) into (6.43), we have

$$\frac{\partial u}{\partial t} + a \frac{\partial u}{\partial x} = -\frac{\partial^2 u}{\partial x^2} \frac{a^2 \, \Delta t}{2} - \frac{\partial^3 u}{\partial x^3} \frac{a^3 (\Delta t)^2}{2} + \frac{\partial^3 u}{\partial x^3} \frac{a^2 (\Delta x)(\Delta t)}{2}$$

$$+ \frac{\partial^3 u}{\partial x^3} \frac{a^3 (\Delta t)^2}{6} + \frac{\partial^2 u}{\partial x^2} \frac{a \, \Delta x}{2} - \frac{\partial^3 u}{\partial x^3} \frac{a \, (\Delta x)^2}{6}$$

$$+ O[(\Delta t)^3, (\Delta t)^2 (\Delta x), (\Delta t)(\Delta x)^2, (\Delta x)^3] \tag{6.55}$$

A rearrangement of Eq. (6.55), along with the definition of v as $v = a \, \Delta t / \Delta x$, yields

$$\boxed{\begin{aligned}\frac{\partial u}{\partial t} + a \frac{\partial u}{\partial x} &= \frac{a \, \Delta x}{2} (1 - v) \frac{\partial^2 u}{\partial x^2} + \frac{a (\Delta x)^2}{6} (3v - 2v^2 - 1) \frac{\partial^3 u}{\partial x^3} \\ &\quad + O[(\Delta t)^3, (\Delta t)^2 (\Delta x), (\Delta t)(\Delta x)^2, (\Delta x)^3] \end{aligned}} \tag{6.56}$$

Note that Eq. (6.56) is a *partial differential equation* in its own right, containing the terms $\partial u/\partial t$, $\partial u/\partial x$, $\partial^2 u/\partial x^2$, $\partial^3 u/\partial x^3$, etc. Finally, with Eq. (6.56) in mind, we are ready to emphasize the different perspective mentioned at the beginning of this long paragraph. Previously, we viewed an exact solution (no round-off error) of the difference equation, Eq. (6.39), as constituting a numerical solution of the *original* partial differential equation given by Eq. (6.38) but with an error given by the truncation error. However, there is another way of looking at this matter. In reality, the exact solution (no round-off error) of the difference equation, Eq. (6.39), constitutes an *exact solution* (no truncation error) of a *different* partial differential equation, namely, Eq. (6.56). Eq. (6.56) is called the *modified equation*. To repeat, when the difference equation, Eq. (6.39), is used to obtain a numerical solution of the original partial differential equation, Eq. (6.38), in reality this difference equation is solving quite a different partial differential equation—it is solving Eq. (6.56) instead of Eq. (6.38).

The derivation and display of the modified equation, as obtained above, is of more importance than just establishing a different perspective on the meaning of the exact solution of a difference equation. Equation (6.56) also gives us some information on the *behavior* to be expected of the numerical solution of the difference equation. For example, examine Eq. (6.56) closely. On the right-hand side there is a term involving $\partial^2 u/\partial x^2$. For a moment, shut out all other considerations from your mind and just visualize the governing equations for a viscous flow, namely, the Navier-Stokes equations given by Eqs. (2.58a) to (2.58c). These equations have terms such as $\partial^2 u/\partial x^2$ multiplied by the viscosity coefficient μ. These terms represent the dissipative aspect of the physical viscosity on the flow. Now return to Eq. (6.56). The term $\partial^2 u/\partial x^2$ appearing here acts as a dissipative term, much like the viscous terms in the Navier-Stokes equations. However, in Eq. (6.56), this term is a consequence of the numerical discretization embodied in the difference equation, Eq. (6.39), and is therefore purely of *numerical origin*, with no physical significance. For this reason, the appearance of this term (and those like it) within the framework of a numerical solution is called *numerical dissipation*. In turn, the coefficient in this term, such as $(a \, \Delta x/2)(1 - \nu)$ in Eq. (6.56), acts much like the physical viscosity and is therefore called the *artificial viscosity*. In CFD, the terms "numerical dissipation" and "artificial viscosity" are frequently used interchangeably and generally connote the diffusive behavior of a numerical solution—a behavior that is purely numerical in origin. For example, the original partial differential equation with which we began this section, Eq. (6.38), describes the propagation of a wave through an inviscid fluid in one dimension. In reality, if we start at time zero with an exact discontinuous wave as sketched in Fig. 6.5, then during the course of the solution the effect of numerical dissipation will be to spread out this wave in much the same way that real physical viscosity would spread the wave. Of course, the reason why the wave will spread in our numerical solution has nothing to do with physical viscosity; rather, it has everything to do with the fact that the exact numerical solution of the difference equation, Eq. (6.39), is a solution of Eq. (6.56) instead of the original partial differential equation given by Eq. (6.38), and Eq. (6.56) has some terms on the right-hand side that play the role of

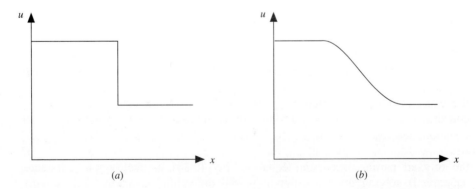

FIG. 6.5
Effect of numerical dissipation. (*a*) Initial wave at time $t = 0$. (*b*) Shape of the wave at some time $t > 0$ from the numerical solution as affected by numerical dissipation.

dissipation. Many algorithms used in CFD contain this effect of artificial viscosity implicitly in their procedure.

Somewhat related to the above concepts is the effect of *numerical dispersion*, which creates a numerical behavior different from that of numerical dissipation. Dispersion results in a distortion of the propagation of different phases of a wave, which shows up as "wiggles" in front of and behind the wave. This is illustrated schematically in Fig. 6.6. One of the values of deriving and displaying the modified equation associated with a given difference equation is that the relative behavior of diffusion and dispersion can be assessed. Numerical dissipation is the direct result of the even-order derivatives on the right-hand side of the modified equation ($\partial^2 u/\partial x^2$, $\partial^4 u/\partial x^4$, etc.), and numerical dispersion is the direct result of the odd-order derivatives ($\partial^3 u/\partial x^3$, etc.). Since the right-hand side of the modified equation is the truncation error, we can state that generally when the leading term of the truncation error is an even-order derivative, the numerical solution will display mainly dissipative behavior, and when the leading term is an odd-order derivative, the solution will display mainly dispersive behavior.

We come now to the bottom line of the discussion in this section. We have shown that artificial viscosity can appear within a given algorithm simply because of the form of the modified equation—such artificial viscosity is said to be present implicitly in the numerical solution. Although such artificial viscosity compromises the accuracy of a solution (which is a bad thing), it always serves to *improve* the *stability* of a solution (which is a good thing.) Indeed, for many applications in CFD, the solution does not have enough artificial viscosity implicitly in the algorithm, and the solution will go unstable unless more artificial viscosity is added *explicitly* to the calculation. This raises one of the most perplexing aspects of CFD. As you intentionally add more artificial viscosity to a numerical solution, you are increasing the probability of making the solution more inaccurate. On the other hand, by adding this artificial viscosity, you are at least able to obtain a stable solution, whereas without it, in some cases no solution would be attainable. (Flow problems with very strong gradients, such as shock waves, wherein such shock waves are

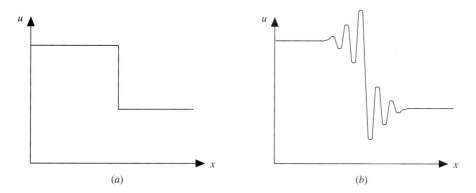

FIG. 6.6
Effect of numerical dispersion. (*a*) Initial wave at $t = 0$. (*b*) Shape of the wave at some time $t > 0$ from the numerical solution as affected by numerical dispersion.

captured within the flow by using a shock-capturing approach, are particularly sensitive and usually require the explicit addition of artificial viscosity for a stable and smooth solution.) Is any solution, no matter how inaccurate, better than no solution at all? The answer to this question for any given problem is a matter of circumstance and judgment. It is this author's opinion, backed up by the collective experience of the CFD community, that in those applications where the use of artificial viscosity has been necessary, the *judicious* use of this quantity has led, for the most part, to reasonable and sometimes very accurate numerical solutions. However, you have to know what you are doing in this regard.

Let us examine a specific form of artificial viscosity which has been reasonably successful in many applications and which has been used frequently in connection with MacCormack's technique described in Sec. 6.3. For purposes of illustration, assume that we are dealing with the governing flow equations in the form of Eq. (2.93), written below for an unsteady, two-dimensional flow.

$$\frac{\partial U}{\partial t} = -\frac{\partial F}{\partial x} - \frac{\partial G}{\partial y} + J \tag{6.57}$$

where U is the solution vector, $U = [\rho, \rho u, \rho v, \rho(e + V^2/2)]$. At each step of the time-marching solution, a small amount of artificial viscosity can be added in the following form:

$$S_{i,j}^t = \frac{C_x |p_{i+1,j}^t - 2p_{i,j}^t + p_{i-1,j}^t|}{p_{i+1,j}^t + 2p_{i,j}^t + p_{i-1,j}^t} (U_{i+1,j}^t - 2U_{i,j}^t + U_{i-1,j}^t)$$
$$+ \frac{C_y |p_{i,j+1}^t - 2p_{i,j}^t + p_{i,j-1}^t|}{p_{i,j+1}^t + 2p_{i,j}^t + p_{i,j-1}^t} (U_{i,j+1}^t - 2U_{i,j}^t + U_{i,j-1}^t) \tag{6.58}$$

Equation (6.58) is a fourth-order numerical dissipation expression; it is designed to "tweak" the calculations by a magnitude equivalent to a fourth-order term in the truncation error; i.e., it is equivalent to adding an extra fourth-order term to the right-hand side of the modified equations for the system of difference equations which are being solved. The fourth-order nature of Eq. (6.58) can be seen in the numerators, which are products of two second-order central difference expressions for second derivatives. In Eq. (6.58), C_x and C_y are two arbitrarily specified parameters; typical values of C_x and C_y range from 0.01 to 0.3. The choice is up to you and is usually determined after some experimentation with different values, assessing their effect on the particular calculation. In Eq. (6.58), U denotes the individual elements of the solutions vector, taken separately. To see this more clearly, assume that we are using MacCormack's technique. On the predictor step, $S_{i,j}^t$ is evaluated based on the known quantities at time t; on the corrector step, the values on the right-hand side of Eq. (6.58) are the predicted (barred) quantities, with the corresponding value of $S_{i,j}^t$ so obtained denoted by $\bar{S}_{i,j}^{t+\Delta t}$.

$$\bar{S}_{i,j}^{t+\Delta t} = \frac{C_x|\bar{p}_{i+1,j}^{t+\Delta t} - 2\bar{p}_{i,j}^{t+\Delta t} + \bar{p}_{i-1,j}^{t+\Delta t}|}{\bar{p}_{i+1,j}^{t+\Delta t} + 2\bar{p}_{i,j}^{t+\Delta t} + \bar{p}_{i-1,j}^{t+\Delta t}}(\bar{U}_{i+1,j}^{t+\Delta t} - 2\bar{U}_{i,j}^{t+\Delta t} + \bar{U}_{i-1,j}^{t+\Delta t})$$
$$+ \frac{C_y|\bar{p}_{i,j+1}^{t+\Delta t} - 2\bar{p}_{i,j}^{t+\Delta t} + \bar{p}_{i,j-1}^{t+\Delta t}|}{\bar{p}_{i,j+1}^{t+\Delta t} + 2\bar{p}_{i,j}^{t+\Delta t} + \bar{p}_{i,j-1}^{t+\Delta t}}(\bar{U}_{i,j+1}^{t+\Delta t} - 2\bar{U}_{i,j}^{t+\Delta t} + \bar{U}_{i-1,j}^{t+\Delta t}) \quad (6.59)$$

The values of $S_{i,j}^t$ and $\bar{S}_{i,j}^{t+\Delta t}$ are added to MacCormack's technique at the following stages of the calculation. Using the calculation of density from the continuity equation as an example, calculate $S_{i,j}^t$ from Eq. (6.58) with $U = \rho$. Then add the artificial viscosity term to Eq. (6.58), which now becomes

$$\bar{\rho}_{i,j}^{t+\Delta t} = \rho_{i,j}^t + \left(\frac{\partial \rho}{\partial t}\right)_{i,j}^t \Delta t + S_{i,j}^t \quad (6.60)$$

On the corrector step, the corrected value of density at time $t + \Delta t$ is obtained from Eq. (6.13) with the artificial viscosity $\bar{S}_{i,j}^{t+\Delta t}$ calculated from Eq. (6.59) added as an extra term, that is,

$$\rho_{i,j}^{t+\Delta t} = \rho_{i,j}^t + \left(\frac{\partial \rho}{\partial t}\right)_{av} \Delta t + \bar{S}_{i,j}^{t+\Delta t} \quad (6.61)$$

Note: There is nothing sacrosanct about the form for artificial viscosity expressed by Eqs. (6.58) and (6.59). It happens to be an empirically based expression which is given here just for the sake of discussion.

To what extent does the addition of artificial viscosity affect the accuracy of a problem? There is no pat answer to this question; it depends in a large part on the nature of the flow problem itself. However, some feel for the extent to which artificial viscosity can impact the solution of a flow problem can be obtained from Ref. 44; there, a series of numerical experiments are reported wherein the value of artificial viscosity was progressively varied and the resulting effects on the flow-field variables were examined. Some of the results are reviewed here so that you can obtain some of this feel. The flow problem is that of the supersonic viscous flow over a rearward-facing step, as shown in Fig. 6.7a. The finite-difference grid used for this study is shown in Fig. 6.7b. The flow field is calculated by means of a time-marching numerical solution of the Navier-Stokes equations using the MacCormack technique described in Sec. 6.3. The expression for artificial viscosity is given by Eqs. (6.58) and (6.59), and various calculations are made with values of C_x and C_y ranging from 0 to 0.3. The calculations are made for a freestream Mach number of 4.08 and a Reynolds number (based on step height) of 849. The step height is 0.51 cm, and the calculations are made for a surface which extends 12.5 cm downstream of the step and 2.04 cm upstream of the step. A calorically perfect gas with the ratio of specific heats equal to 1.31 is used (this is to partially simulate the "effective gamma" for partially dissociated air in a supersonic combustion ramjet environment). Figure 6.8 shows the computed pressure contours for the flow, using MacCormack's technique. Here, four different contour pictures are shown, one each

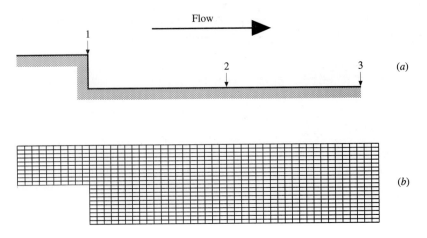

FIG. 6.7
(a) Rearward-facing step geometry. (b) 51 × 21 grid used for the calculations.

for a different value of C_x and C_y ranging from 0 to 0.3. At the top of Fig. 6.8 we see the computed flow field using zero artificial viscosity. The frame immediately below it gives the results wherein $C_x = C_y = 0.1$. The next frame corresponds to $C_x = C_y = 0.2$, and the final frame has $C_x = C_y = 0.3$. The expansion wave from the top corner and the recompression shock wave downstream of the step can be seen in all frames. However, careful examination of Fig. 6.8 shows that as C_x and C_y are progressively increased (the magnitude of the artificial viscosity is increased), the quantitative and qualitative aspects of the flow are perturbed. In Fig. 6.8a, where zero artificial viscosity is used, the recompression shock wave is fairly sharp and distinct, but there are wiggles ahead of and behind the shock. It is not easy to obtain a stable, converged solution in this case; the calculations are sensitive, and some "nursing" of the program is required. As the magnitude of the artificial viscosity is progressively increased, as shown in Fig. 6.8b to d, the solution behaves in a more stable fashion, but the structures of the resulting steady-state flows are somewhat different. This can be seen by comparing Fig. 6.8a and d; in Fig. 6.8d with heavy artificial viscosity, the recompression shock has been smoothed by the increased numerical dissipation. In contrast to Fig. 6.8a, we see no wiggles in Fig. 6.8d, and the shock wave is much more diffuse, while at the same time its location has translated upward. In Fig. 6.7a, three different axial locations are denoted by the numbers 1, 2, and 3. The velocity profiles (velocity versus vertical location y) for these three locations are shown in Fig. 6.9a to c. In each figure, the profiles are given for four different values of the artificial viscosity. Note that the velocity profiles are affected by artificial viscosity. Finally, the wall pressure distribution—the variation of pressure on the wall versus x location measured along the surface—is given in Fig. 6.10. Here, $x = 1$ cm is the location of the step, and the pressure distributions shown are those downstream of the step. The pressure at $x = 1$ cm is essentially the base pressure, i.e., the pressure on the vertical step itself. Four different curves are shown in Fig. 6.10, each one corresponding to a different value

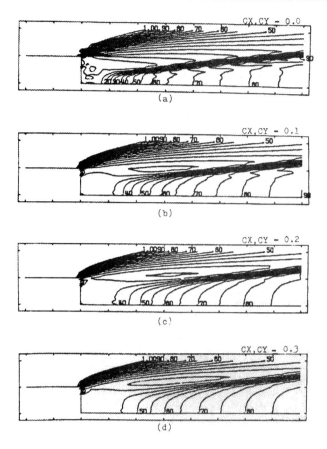

FIG. 6.8
Numerical experiment on the effects of artificial viscosity. Pressure contours calculated with values of the dissipation factors C_x and C_y ranging from 0 to 0.3. The freestream conditions are M_∞ = 4.08, T_∞ = 1046 K, ratio of specific heats γ = 1.31, and Reynolds number = 849 (based on a step height of 0.51 cm). The wall temperature $T_W = 0.2957 T_\infty$.

of artificial viscosity. Although the pressure distribution farther downstream of the step is relatively insensitive to the amount of artificial viscosity, the base pressure itself is quite sensitive to the artificial viscosity.

Note: The impact of artificial viscosity on the qualitative aspects of a flow solution is like that of the physical viscosity μ. By increasing the artificial viscosity, shock waves are thickened and smoothed, just like an increased physical coefficient of viscosity would cause. The details of separated flow regions are affected by artificial viscosity, just like an increase in physical viscosity would cause. By adding artificial viscosity, we are changing the overall entropy level of the flow field, just as physical viscosity would cause. Finally, by increasing the artificial viscosity in a numerical solution, we are in effect reducing the effective Reynolds number of the flow, just as an increase in μ would cause.

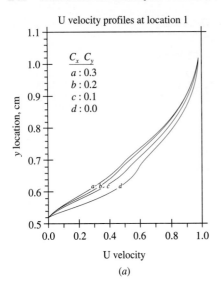

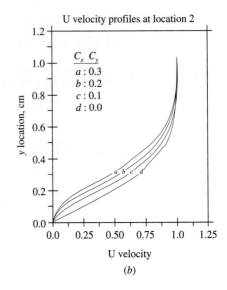

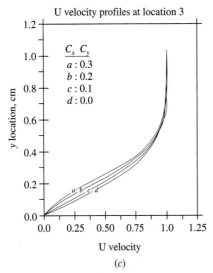

FIG. 6.9
Numerical experiment on the effects of artificial viscosity. Velocity profiles at the three locations marked in Fig. 6.7a. Same freestream conditions as listed in Fig. 6.8. The velocity given here is the nondimensional value referenced to freestream velocity.

The purpose of this section has been to introduce you to the concepts of numerical dissipation and the use of artificial viscosity for the stabilization and smoothing of some numerical solutions. Many applications in CFD do not require the addition of artificial viscosity. On the other hand, artificial viscosity, both implicit in an algorithm and explicitly added as needed, is a fact of life in many other CFD solutions. Such matters still remain a highly empirical aspect of CFD

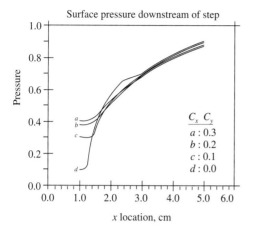

FIG. 6.10
Numerical experiment on the effects of artificial viscosity. Surface pressure distributions downstream of the step. Pressure given here is the nondimensional pressure referenced to freestream pressure. Same freestream conditions as listed in Fig. 6.8.

solutions; you will usually want to play around with various amounts of artificial viscosity until you are satisfied with the quality of the solution. Finally, this rather arbitrary, almost capricious aspect of CFD has been a thorn in the side of practioners for the past several decades. However, in the past few years, innovative methods in applied mathematics have addressed the problem of artificial viscosity in an intelligent fashion, leading to new algorithms which automatically use only the proper amount of artificial viscosity only in regions where it is needed. An example is the TVD (total-variation-diminishing) concept. Such aspects are discussed in Chap. 11. As you proceed further with your studies of CFD in the future, you will most likely reap the benefits of such mathematical advancements.

6.7 THE ALTERNATING-DIRECTION-IMPLICIT (ADI) TECHNIQUE

Let us return to the consideration of implicit solutions as exemplified by the Crank-Nicolson technique, introduced in Sec. 4.4. In this section, an example of a marching solution is given; Eq. (3.28) is used as a model equation with t as the marching variable. There exists only one other independent variable in the equation, namely, x. As long as we are dealing with linear equations, the implicit solutions using the Crank-Nicolson scheme are directly obtained from the use of Thomas' algorithm (see App. A). This is the case in Sec. 4.4, where a finite-difference representation of Eq. (3.28) is given in the tridiagonal form by Eq. (4.42). This tridiagonal form is readily solved by the use of Thomas' algorithm.

Note that the difference equation is linear. In Sec. 4.4, the original partial differential equation, Eq. (3.28), is linear, hence leading to a linear difference equation. In cases governed by nonlinear partial differential equations, a more general idea for obtaining linear difference equations is discussed in Sec. 11.3.1. When solving an inherently nonlinear problem by means of an implicit scheme, the matter of linearizing the difference equations is of utmost importance so that Thomas' algorithm (or some equivalent) can be used to expedite the calculations.

Such matters are discussed in Sec. 11.3.1; it is not necessary for us to elaborate further in this section.

The main thrust of the present section is concerned with the other aspect that destroys the tridiagonal nature of the difference equations, namely, multidimensionality involving more than one variable in addition to the marching variable. To see this more clearly, consider a model equation based on the unsteady, two-dimensional heat conduction equation, Eq. (3.27), written in two spatial dimensions as follows:

$$\frac{\partial T}{\partial t} = \alpha \left(\frac{\partial^2 T}{\partial x^2} + \frac{\partial^2 T}{\partial y^2} \right) \tag{6.62}$$

Paralleling the Crank-Nicolson development in Sec. 4.4, Eq. (6.62) can be written in finite-difference form as

$$\frac{T_{i,j}^{n+1} - T_{i,j}^n}{\Delta t} = \alpha \frac{\frac{1}{2}(T_{i+1,j}^{n+1} + T_{i+1,j}^n) + \frac{1}{2}(-2T_{i,j}^{n+1} - 2T_{i,j}^n) + \frac{1}{2}(T_{i-1,j}^{n+1} + T_{i-1,j}^n)}{(\Delta x)^2}$$

$$+ \alpha \frac{\frac{1}{2}(T_{i,j+1}^{n+1} + T_{i,j+1}^n) + \frac{1}{2}(-2T_{i,j}^{n+1} - 2T_{i,j}^n) + \frac{1}{2}(T_{i,j-1}^{n+1} + T_{i,j-1}^n)}{(\Delta y)^2} \tag{6.63}$$

Equation (6.63) is the equivalent in xy space to the one-dimensional form given by Eq. (4.40). However, unlike Eq. (4.40) which reduces to the tridiagonal form given by Eq. (4.42), Eq. (6.63) contains *five* unknowns, namely, $T_{i+1,j}^{n+1}$, $T_{i,j}^{n+1}$, $T_{i-1,j}^{n+1}$, $T_{i,j+1}^{n+1}$, and $T_{i,j-1}^{n+1}$, where the last two unknowns prevent a tridiagonal form. Hence, Thomas' algorithm can not be used. Although matrix methods exist which can solve Eq. (6.63), the computer time is much longer than that for a tridiagonal system. As a result, there is a distinct advantage in developing a scheme that will allow Eq. (6.62) to be solved by means of tridiagonal forms only. Such a scheme, namely, the alternating-direction-implicit (ADI) scheme, is the main subject of this section.

Recall that Eq. (6.62) is being solved by means of a marching technique; that is, $T(t + \Delta t)$ is being obtained in some fashion from the known values of $T(t)$. Let us achieve the solution of $T(t + \Delta t)$ in a *two-step process*, where intermediate values of T are found at an intermediate time, $t + \Delta t/2$, as follows. In the first step over a time interval $\Delta t/2$, replace the spatial derivatives in Eq. (6.62) with central differences, where only the x derivative is treated implicitly. That is, from Eq. (6.62),

$$\frac{T_{i,j}^{n+1/2} - T_{i,j}^n}{\Delta t/2} = \alpha \frac{T_{i+1,j}^{n+1/2} - 2T_{i,j}^{n+1/2} + T_{i-1,j}^{n+1/2}}{(\Delta x)^2} + \alpha \frac{T_{i,j+1}^n - 2T_{i,j}^n + T_{i,j-1}^n}{(\Delta y)^2} \tag{6.64}$$

Equation (6.64) reduces to the tridiagonal form

$$AT_{i-1,j}^{n+1/2} - BT_{i,j}^{n+1/2} + AT_{i+1,j}^{n+1/2} = K_i \tag{6.65}$$

where

$$A = \frac{\alpha \Delta t}{2(\Delta x)^2}$$

$$B = 1 + \frac{\alpha \Delta t}{(\Delta x)^2}$$

$$K_i = -T_{i,j}^n - \frac{\alpha \Delta t}{2(\Delta y)^2}(T_{i,j+1}^n - 2T_{i,j}^n + T_{i,j-1}^n)$$

Equation (6.65) yields a solution for $T_{i,j}^{n+1/2}$ for all i, keeping j fixed, using Thomas' algorithm. That is, examining Fig. 6.11, at a fixed value of j, we "sweep" in the x direction, using Eq. (6.65) to solve for $T_{i,j}^{n+1/2}$ for all values of i. If there are N grid points in the x direction, then we sweep from $i = 1$ to N. This sweep utilizes Thomas's algorithm once. This calculation is then repeated at the next row of grid points designated by $j + 1$. That is, replace j in Eq. (6.65) by $j + 1$ and solve for $T_{i,j+1}^{n+1/2}$ for all values of i from 1 to N, using Thomas' algorithm. If there are M grid points in the y direction, this process is repeated M times; i.e., there are M sweeps in the x direction, resulting in Thomas' algorithm being used M times. This sweeping

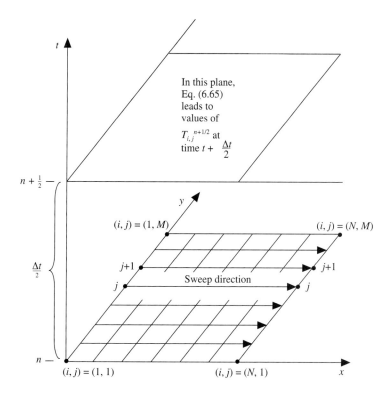

FIG. 6.11
First step in the ADI process. Sweeping in the x direction to obtain T at time $t + \Delta t/2$.

in the x direction is shown schematically in Fig. 6.11. At the end of this step, the values of T at the intermediate time $t + \Delta t/2$ are known at all grid points (i, j); that is, $T_{i,j}^{n+1/2}$ is known at all (i, j).

The second step of the ADI scheme takes the solution to time $t + \Delta t$, using the known values at time $t + \Delta t/2$. For this second step, the spatial derivatives in Eq. (6.62) are replaced with central differences, where the y derivative is treated implicitly. That is, from Eq. (6.62),

$$\frac{T_{i,j}^{n+1} - T_{i,j}^{n+1/2}}{\Delta t/2} = \alpha \frac{T_{i+1,j}^{n+1/2} - 2T_{i,j}^{n+1/2} + T_{i-1,j}^{n+1/2}}{(\Delta x)^2} + \alpha \frac{T_{i,j+1}^{n+1} - 2T_{i,j}^{n+1} + T_{i,j-1}^{n+1}}{(\Delta y)^2} \quad (6.66)$$

Equation (6.66) reduces to the tridiagonal form

$$CT_{i,j+1}^{n+1} - DT_{i,j}^{n+1} + CT_{i,j-1}^{n+1} = L_j \quad (6.67)$$

where

$$C = \frac{\alpha \, \Delta t}{2(\Delta y)^2}$$

$$D = 1 + \frac{\alpha \, \Delta t}{(\Delta y)^2}$$

$$L_j = -T_{i,j}^{n+1/2} - \frac{\alpha \, \Delta t}{2(\Delta x)^2} (T_{i+1,j}^{n+1/2} - 2T_{i,j}^{n+1/2} + T_{i-1,j}^{n+1/2})$$

Note that $T^{n+1/2}$ is known at all grid points from the first step. Equation (6.67) yields a solution for $T_{i,j}^{n+1}$ for all j, keeping i fixed, using Thomas' algorithm. That is, examining Fig. 6.12, at a fixed value of i, we sweep in the y direction, using Eq. (6.67) to solve for $T_{i,j}^{n+1}$ for all values of j, where j goes from 1 to M. This sweep utilizes Thomas's algorithm once. This calculation is then repeated at the next column of grid points designated by $i + 1$. That is, replace i in Eq. (6.67) by $i + 1$ and solve for $T_{i+1,j}^{n}$ for all values of j from 1 to M, using Thomas's algorithm. This process is repeated N times; i.e., there are N sweeps in the y direction, resulting in Thomas' algorithm being used N times. This sweeping in the y direction is shown schematically in Fig. 6.12. At the end of this step, the values of T at time $t + \Delta t$ are known at all grid points (i, j); that is, $T_{i,j}^{n+1}$ is known at all (i, j).

At the end of this two-step process, the dependent variable T has been marched a value Δt in the direction of t. Although there are two independent spatial variables x and y in addition to the marching variable t, this marching scheme involves only tridiagonal forms, and the solution has been achieved by the repeated application of Thomas' algorithm. Because the scheme involves two steps, one in which the difference equation is implicit in x and the other in which the difference equation is implicit in y, the source of the name of the scheme—*alternating-direction-implicit*—is obvious.

The ADI scheme is second-order-accurate in t, x, and y; that is, the truncation error is of $O[(\Delta t)^2, (\Delta x)^2, (\Delta y)^2]$. See Refs. 13 to 17 for details.

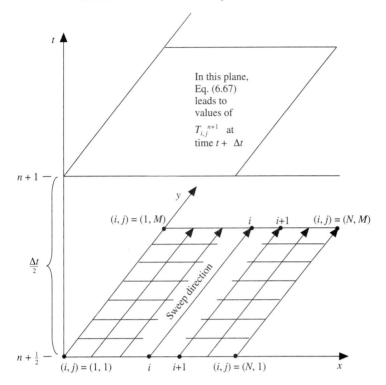

FIG. 6.12
Second step in the ADI process. Sweeping in the y direction to obtain T at time $t + \Delta t$.

This scheme has found application in many fluid flow problems. In the form described above, it is particularly useful for the solution of problems described by parabolic partial differential equations. Also, the scheme described above is a special form of a *general class* of schemes involving a splitting of two or more directions in an implicit solution of the governing flow equations so as to obtain tridiagonal forms. Hence, ADI can represent a general descriptor of a whole class of schemes, one of which has been described in this section. Another popular version of an ADI scheme is called *approximate factorization*; this is a more advanced topic which is discussed in Sec. 11.3.2.

6.8 THE PRESSURE CORRECTION TECHNIQUE: APPLICATION TO INCOMPRESSIBLE VISCOUS FLOW

A numerical technique for the solution of inviscid, incompressible flow was discussed in Sec. 6.5, namely, the relaxation technique. Inviscid, incompressible flow is governed by elliptic partial differential equations, and the relaxation technique, which is essentially an *iterative* process, is a classical numerical method for solving elliptic problems. In contrast, *viscous*, incompressible flow is governed by the incompressible Navier-Stokes equations, which exhibit a mixed elliptic-parabolic behavior, and hence the standard relaxation technique as described in Sec. 6.5 is not particularly helpful. The purpose of the present section is to describe an iterative process called the *pressure correction technique*, which has found widespread application in the numerical solution of the incompressible Navier-Stokes equations. The pressure correction technique has been developed for practical engineering solutions by Patankar and Spalding (Ref. 67) and is discussed at length in Ref. 68. The technique is embodied in an algorithm called SIMPLE (semi-implicit method for pressure-linked equations), pioneered by Patankar and Spalding, which has found widespread application over the past 20 years for both compressible and incompressible flows. However, in the present section we will focus on the use of the pressure correction method to solve incompressible, viscous flow.

Before describing the pressure correction method, there are two considerations associated with an incompressible flow solution that need to be addressed. They are the subject of the next two subsections.

6.8.1 Some Comments on the Incompressible Navier-Stokes Equations

The compressible Navier-Stokes equations are derived in Chap. 2 and summarized in Sec. 2.8.1. The incompressible Navier-Stokes equations can be obtained from the compressible form simply by setting density equal to a constant. That is, with ρ = constant, Eq. (2.29) becomes

$$\nabla \cdot \mathbf{V} = 0 \tag{6.68}$$

With the further assumption that μ is constant throughout the flow, Eqs. (2.50a) to (2.50c) combined with Eqs. (2.57a) to (2.57f) become

$$\rho \frac{Du}{Dt} = -\frac{\partial p}{\partial x} + 2\mu \frac{\partial^2 u}{\partial x^2} + \mu \frac{\partial}{\partial y}\left(\frac{\partial v}{\partial x} + \frac{\partial u}{\partial y}\right) + \mu \frac{\partial}{\partial z}\left(\frac{\partial u}{\partial z} + \frac{\partial w}{\partial x}\right) + \rho f_x \tag{6.69}$$

$$\rho \frac{Dv}{Dt} = -\frac{\partial p}{\partial y} + \mu \frac{\partial}{\partial x}\left(\frac{\partial v}{\partial x} + \frac{\partial u}{\partial y}\right) + 2\mu \frac{\partial^2 v}{\partial y^2} + \mu \frac{\partial}{\partial z}\left(\frac{\partial w}{\partial y} + \frac{\partial v}{\partial z}\right) + \rho f_y \tag{6.70}$$

$$\rho \frac{Dw}{Dt} = -\frac{\partial p}{\partial z} + \mu \frac{\partial}{\partial x}\left(\frac{\partial u}{\partial z} + \frac{\partial w}{\partial x}\right) + \mu \frac{\partial}{\partial y}\left(\frac{\partial w}{\partial y} + \frac{\partial v}{\partial z}\right) + 2\mu \frac{\partial^2 w}{\partial z^2} + \rho f_z \tag{6.71}$$

Note that in writing Eqs. (6.69) to (6.71), the terms in Eqs. (2.57a) to (2.57f) explicitly involving $\nabla \cdot \mathbf{V}$ have been set to zero due to Eq. (6.68). The fact that $\nabla \cdot \mathbf{V} = 0$ for incompressible flow allows a further reduction of Eqs. (6.69) to (6.71), as follows.

$$\nabla \cdot \mathbf{V} = \frac{\partial u}{\partial x} + \frac{\partial v}{\partial y} + \frac{\partial w}{\partial z} = 0 \tag{6.72}$$

Rearranging Eq. (6.72), we have

$$\frac{\partial u}{\partial x} = -\frac{\partial v}{\partial y} - \frac{\partial w}{\partial z} \qquad (6.72a)$$

Differentiating Eq. (6.72a) with respect to x, we obtain

$$\frac{\partial^2 u}{\partial x^2} = -\frac{\partial^2 v}{\partial x\, \partial y} - \frac{\partial^2 w}{\partial x\, \partial y} \qquad (6.73)$$

Adding $\partial^2 u/\partial x^2$ to both sides of Eq. (6.73) and multiplying by μ, we obtain

$$2\mu \frac{\partial^2 u}{\partial x^2} = \mu \frac{\partial^2 u}{\partial x^2} - \mu \frac{\partial^2 v}{\partial x\, \partial y} - \mu \frac{\partial^2 w}{\partial x\, \partial y} \qquad (6.74)$$

Substituting Eq. (6.74) for the second term on the right side of Eq. (6.69) and expanding other terms in Eq. (6.69), we obtain

$$\rho \frac{Du}{Dt} = -\frac{\partial p}{\partial x} + \mu \frac{\partial^2 u}{\partial x^2} - \mu \frac{\partial^2 v}{\partial x\, \partial y} - \mu \frac{\partial^2 w}{\partial x\, \partial y} + \mu \frac{\partial^2 v}{\partial x\, \partial y}$$
$$+ \mu \frac{\partial^2 u}{\partial y^2} + \mu \frac{\partial^2 u}{\partial z^2} + \mu \frac{\partial^2 w}{\partial x\, \partial y} + \rho f_x \qquad (6.75)$$

Canceling terms in Eq. (6.75), we obtain a convenient form of the x-momentum equation for a viscous, incompressible flow as

$$\rho \frac{Du}{Dt} = -\frac{\partial p}{\partial x} + \mu \left(\frac{\partial^2 u}{\partial x^2} + \frac{\partial^2 u}{\partial y^2} + \frac{\partial^2 u}{\partial z^2} \right) + \rho f_x$$

or
$$\rho \frac{Du}{Dt} = -\frac{\partial p}{\partial x} + \mu \nabla^2 u + \rho f_x \qquad (6.76)$$

where $\nabla^2 u$ is the laplacian of the x component of velocity, u. Equations (6.70) and (6.71) can be treated in a similar fashion. The resulting system of equations is the *incompressible Navier-Stokes equations*, summarized below.

$$\text{Continuity:} \quad \nabla \cdot \mathbf{V} = 0 \qquad (6.77)$$

$$x \text{ momentum:} \quad \rho \frac{Du}{Dt} = -\frac{\partial p}{\partial x} + \mu \nabla^2 u + \rho f_x \qquad (6.78)$$

$$y \text{ momentum:} \quad \rho \frac{Dv}{Dt} = -\frac{\partial p}{\partial y} + \mu \nabla^2 v + \rho f_y \qquad (6.79)$$

$$z \text{ momentum:} \quad \rho \frac{Dw}{Dt} = -\frac{\partial p}{\partial z} + \mu \nabla^2 w + \rho f_z \qquad (6.80)$$

Note that Eqs. (6.77) to (6.80) are self-contained; they are four equations for the four dependent variables u, v, w, and p. Through the assumptions of ρ = constant and μ = constant, the energy equation has been completely decoupled from the analysis. The implication here is that the continuity and momentum equations are all that are necessary to solve for the velocity and pressure fields in an incompressible flow, and that *if* a given problem involves heat transfer, and hence temperature gradients exist in the flow, the temperature field can be obtained directly from the energy equation *after* the velocity and pressure fields are obtained. In this section, we will not deal with a temperature field; rather, we will assume that T = constant, which is compatible with our earlier assumption that μ = constant [because $\mu = f(T)$]. Hence, Eqs. (6.77) to (6.80) are sufficient for our discussion here.

Clearly, from the above discussion we see that the incompressible Navier-Stokes equations are derived in a straightforward fashion from the compressible Navier-Stokes equations. In turn, this might lead us to think that a *numerical* solution of the incompressible equations might be obtained in a straightforward fashion from a numerical technique fashioned for the compressible equations. Unfortunately, this is not the case. For example, if we write a computer code to solve the compressible Navier-Stokes equations using a time-marching MacCormack's technique as described in Sec. 6.3, the explicit time step Δt is restricted by stability conditions. An approximate stability condition for an explicit Navier-Stokes solution is given in Ref. 13 as

$$\Delta t \leq \frac{1}{|u|/\Delta x + |v|/\Delta y + a\sqrt{1/(\Delta x)^2 + 1/(\Delta y)^2}} \tag{6.81}$$

For a compressible flow, the speed of sound a is finite, and Eq. (6.81) will yield a finite value of Δt for the numerical solution. However, for an incompressible flow, the speed of sound is theoretically infinite, and hence Eq. (6.81) would yield $\Delta t = 0$ for such a case. Clearly, for the numerical solution of an incompressible flow, something else must be done. This phenomenon is further reinforced by the observation that a compressible-flow CFD solution technique, when applied to a flow field where the Mach number is progressively reduced toward zero, takes progressively more time steps to converge; it is the author's experience that a compressible-flow code run for a flow which is everywhere at a local Mach number of about 0.2 or less takes a prohibitive amount of time to converge, and indeed has a tendency to be unstable at such a low Mach number.

For such reasons, in CFD, solution techniques for the incompressible Navier-Stokes equations are usually different from those used for the solution of the compressible Navier-Stokes equations. The pressure correction method, to be described shortly, transcends this difficulty; it has been used with reasonable success for compressible flow but with even more success for incompressible flow. It is an accepted and widely used technique for incompressible, viscous, CFD applications. Therefore, we focus on this method in the present section.

6.8.2 Some Comments on Central Differencing of the Incompressible Navier-Stokes Equations: The Need for a Staggered Grid

The incompressible continuity equation is given by Eq. (6.77), which in two dimensions is

$$\frac{\partial u}{\partial x} + \frac{\partial v}{\partial y} = 0 \tag{6.82}$$

A central difference equation representing Eq. (6.82) is

$$\frac{u_{i+1,j} - u_{i-1,j}}{2\Delta x} + \frac{v_{i,j+1} - v_{i,j-1}}{2\Delta y} = 0 \tag{6.83}$$

This difference equation numerically allows the checkerboard velocity distribution given in Fig. 6.13. Illustrated there is a zigzag type of distribution of both the x component and y component of the velocity, u and v, respectively. In the x direction, u varies as 20, 40, 20, 40, etc., at successive grid points, and in the y direction, v varies as 5, 2, 5, 2, etc., at successive grid points. If these numbers are substituted into Eq. (6.83), both terms are zero at every grid point; i.e., the discrete velocity distribution shown in Fig. 6.13 satisfies the central difference form of the continuity equation. On the other hand, the checkerboard velocity distribution in Fig. 6.13 is basically nonsense in terms of any real, physical flow field.

The problem described above does not occur for compressible flow, where the inclusion of the density variation in the continuity equation would generally wipe out the checkerboard pattern illustrated in Fig. 6.13 after the first time step.

A related problem is encountered in regard to central differences in the momentum equations, Eqs. (6.78) to (6.80). Imagine a two-dimensional discrete, checkerboard pressure pattern as illustrated in Fig. 6.14. In particular, consider the central difference formulation for the pressure gradients:

$$\frac{\partial p}{\partial x} = \frac{p_{i+1,j} - p_{i-1,j}}{2\Delta x} \tag{6.84a}$$

$$\frac{\partial p}{\partial y} = \frac{p_{i,j+1} - p_{i,j-1}}{2\Delta y} \tag{6.84b}$$

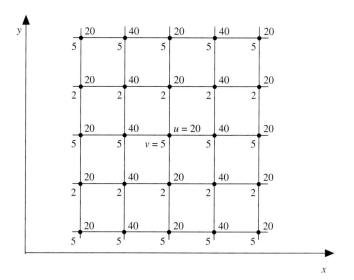

FIG. 6.13
Discrete checkerboard velocity distribution at each grid point; the number at the upper right is u and that at the lower left is v.

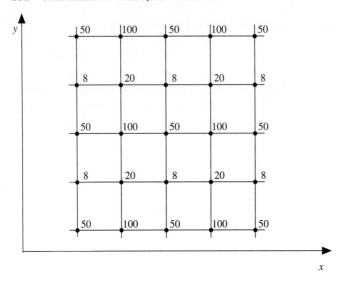

FIG. 6.14
Discrete checkerboard pressure distribution.

For the checkerboard pressure distribution illustrated in Fig. 6.14, Eqs. (6.84a) and (6.84b) give zero pressure gradients in the x and y directions, respectively. Clearly, the pressure field discretized in Fig. 6.14 would not be felt by the Navier-Stokes equations; rather, the numerical solution would effectively see only a uniform pressure in x and y.

In short, when central differences are used for the incompressible Navier-Stokes equations, the resulting difference equations are of a form that, when presented with the nonsensical velocity and pressure distributions shown in Figs. 6.13 and 6.14, will tend to perpetuate these distributions. Admittedly, some early central difference algorithms for incompressible viscous flow ignored this problem, and successful solutions were still obtained, presumably because of special treatment of the boundary conditions or by some other fortuitous aspect of the numerical procedure. However, given the weakness of the central difference formulation described above, we should justifiably feel uncomfortable, and we should look for some "fix" before embarking on the solution of a given problem.

Two such fixes are suggested. If upwind differences are used instead of central differences, the problem immediately goes away. A discussion of upwind differences is given in Sec. 11.4. However, another fix is to maintain central differencing but *stagger* the grid, as described below.

A staggered grid is illustrated in Fig. 6.15. Here, the pressures are calculated at the solid grid points, labeled $(i-1, j)$, (i, j), $(i+1, j)$, $(i, j+1)$, $(i, j-1)$, etc., and the velocities are calculated at the open grid points, labeled $(i-\frac{1}{2}, j)$, $(i+\frac{1}{2}, j)$, $(i, j+\frac{1}{2})$, $(i, j-\frac{1}{2})$, etc. Specifically, u is calculated at points $(i-\frac{1}{2}, j)$, $(i+\frac{1}{2}, j)$,

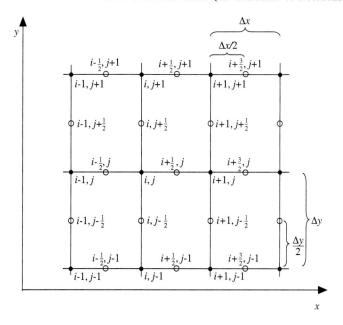

FIG. 6.15
Staggered grid.

etc., and v is calculated at different points $(i, j + \frac{1}{2})$, $(i, j - \frac{1}{2})$, etc. The key feature here is that pressures and velocities are calculated at *different* grid points. In Fig. 6.15, the open grid points are shown equidistant between the solid grid points, but this is not a necessity. An advantage of this staggered grid is, for example, that when $u_{i+1/2, j}$ is calculated, a central difference for $\partial p/\partial x$ yields $(p_{i+1, j} - p_{i, j})/\Delta x$; that is, the pressure gradient is based on *adjacent* pressure points, which eliminates the possibility of a checkerboard pressure pattern as sketched in Fig. 6.14. Also, a central difference expression for the continuity equation, Eq. (6.82), centered around point (i, j) becomes

$$\frac{u_{i+1/2, j} - u_{i-1/2, j}}{\Delta x} + \frac{v_{i, j+1/2} - v_{i, j-1/2}}{\Delta y} = 0 \qquad (6.85)$$

Because Eq. (6.85) is based on *adjacent* velocity points, the possibility of a checkerboard velocity pattern as sketched in Fig. 6.13 is eliminated.

6.8.3 The Philosophy of the Pressure Correction Method

The pressure correction technique is basically an iterative approach, where some innovative physical reasoning is used to construct the next iteration from the results of the previous iteration. The thought process is as follows:

1. Start the iterative process by guessing the pressure field. Denote the guessed pressures by p^*.
2. Use the values of p^* to solve for u, v, and w from the momentum equations. Since these velocities are those associated with the values of p^*, denote them by u^*, v^*, and w^*.
3. Since they were obtained from guessed values of p^*, the values u^*, v^*, and w^*, when substituted into the continuity equation, will not necessarily satisfy that equation. Hence, using the continuity equation, construct a pressure correction p' which when added to p^* will bring the velocity field more into agreement with the continuity equation. That is, the "corrected" pressure p is

$$p = p^* + p' \tag{6.86}$$

Corresponding velocity corrections u', v', and w' can be obtained from p' such that

$$u = u^* + u' \tag{6.87a}$$
$$v = v^* + v' \tag{6.87b}$$
$$w = w^* + w' \tag{6.87c}$$

4. In Eq. (6.86), designate the new value p on the left-hand side as the new value of p^*. Return to step 2, and repeat the process until a velocity field is found that *does* satisfy the continuity equation. When this is achieved, the correct flow field is at hand.

6.8.4 The Pressure Correction Formula

The pressure correction p' was introduced in Eq. (6.86). The calculation of the value of p' is the subject of this subsection. For simplicity, we will consider a two-dimensional flow; the additional terms associated with the third dimension are treated in a like manner. Also, we will neglect body forces.

The x- and y-momentum equations for an incompressible viscous flow are given by Eqs. (6.78) and (6.79), respectively. These equations are in nonconservation form. In conservation form, they are (see Sec. 2.8)

$$\frac{\partial(\rho u)}{\partial t} + \frac{\partial(\rho u^2)}{\partial x} + \frac{\partial(\rho uv)}{\partial y} = -\frac{\partial p}{\partial x} + \mu\left(\frac{\partial^2 u}{\partial x^2} + \frac{\partial^2 u}{\partial y^2}\right) \tag{6.88}$$

and

$$\frac{\partial(\rho v)}{\partial t} + \frac{\partial(\rho vu)}{\partial x} + \frac{\partial(\rho v^2)}{\partial y} = -\frac{\partial p}{\partial y} + \mu\left(\frac{\partial^2 v}{\partial x^2} + \frac{\partial^2 v}{\partial y^2}\right) \tag{6.89}$$

As discussed in Chap. 2, the conservation form follows directly from the model of an infinitely small volume fixed in space. Because of this model, a finite-difference form of Eqs. (6.88) and (6.89) will be somewhat akin to the discretized equations obtained from a finite-volume approach. The original formulation of the pressure correction method by Patankar and Spalding (Refs. 67 and 68) involved a finite-volume approach. In the present section, we will continue with a finite-difference approach; by using the conservation form of the governing partial differential equations, this finite-difference approach gives essentially the same discretized equations as would be obtained in a finite-volume method. We proceed to develop the discretized equations which are the basic tools of the pressure correction method. We choose to use a forward difference in time, and central differences for the spatial derivatives. Note that the pressure correction method is really a certain philosophy, i.e., a certain approach, as explained in Sec. 6.8.3, and the choice of any particular differencing scheme within this philosophy is generally satisfactory. That is, the scheme developed below is not the only approach; it is just a reasonable choice out of several.

Consider a region in a staggered grid as illustrated in Fig. 6.16. Recall that the pressures are evaluated at the solid grid points and the velocities at the open grid points. We will difference Eq. (6.88) centered around the point $(i + \frac{1}{2}, j)$ in Fig. 6.16. (For reference purposes, an equivalent finite-volume approach would deal with the shaded cell in Fig. 6.16.) We will need average values of v at the points a and b on the top and bottom, respectively, of the shaded cell. These are defined by linear interpolation between the two adjacent points; i.e., *define*

At point a : $\quad \bar{v}_{j+1/2} \equiv \frac{1}{2}\left(v_{i,j+1/2} + v_{i+1,j+1/2}\right)$ (6.90a)

At point b : $\quad \bar{v}_{j-1/2} \equiv \frac{1}{2}\left(v_{i,j-1/2} + v_{i+1,j-1/2}\right)$ (6.90b)

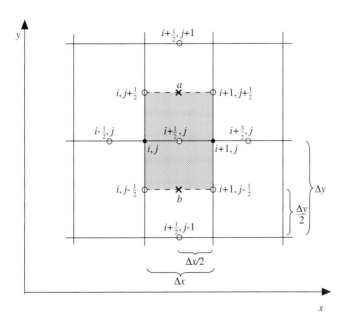

FIG. 6.16
Computational module for the x-momentum equation. The filled-in area is an effective control volume.

Centered around point $(i + \frac{1}{2}, j)$ a difference representation of Eq. (6.88) is

$$\frac{(\rho u)^{n+1}_{i+1/2,j} - (\rho u)^{n}_{i+1/2,j}}{\Delta t} = -\left[\frac{(\rho u^2)^{n}_{i+3/2,j} - (\rho u^2)^{n}_{i-1/2,j}}{2\Delta x}\right.$$

$$\left. + \frac{(\rho u \bar{v})^{n}_{i+1/2,j+1} - (\rho u v)^{n}_{i+1/2,j-1}}{2\Delta y}\right] - \frac{p^{n}_{i+1,j} - p^{n}_{i,j}}{\Delta x}$$

$$+ \mu \left[\frac{u^{n}_{i+3/2,j} - 2u^{n}_{i+1/2,j} + u^{n}_{i-1/2,j}}{(\Delta x)^2} + \frac{u^{n}_{i+1/2,j+1} - 2u^{n}_{i+1/2,j} + u_{i+1/2,j-1}}{(\Delta y)^2}\right] \quad (6.91)$$

or

$$\boxed{(\rho u)^{n+1}_{i+1/2,j} = (\rho u)^{n}_{i+1/2,j} + A\,\Delta t - \frac{\Delta t}{\Delta x}(p^{n}_{i+1,j} - p^{n}_{i,j})} \quad (6.92)$$

where, from Eq. (6.91),

$$A = -\left[\frac{(\rho u^2)^{n}_{i+3/2,j} - (\rho u^2)^{n}_{i-1/2,j}}{2\Delta x} + \frac{(\rho u \bar{v})^{n}_{i+1/2,j+1} - (\rho u v)^{n}_{i+1/2,j-1}}{2\Delta y}\right]$$

$$+ \mu \left[\frac{u^{n}_{i+3/2,j} - 2u^{n}_{i+1/2,j} + u^{n}_{i-1/2,j}}{(\Delta x)^2} + \frac{u^{n}_{i+1/2,j+1} - 2u^{n}_{i+1/2,j} + u_{i+1/2,j-1}}{(\Delta y)^2}\right]$$

Equation (6.92) is a difference equation representing the x-momentum equation. Note that $\bar{v}$ and v in Eqs. (6.91) and (6.92) are those values defined by Eqs. (6.90a and b), i.e., $\bar{v}$ and v use different grid points than those for u.

In like manner, a difference equation for the y-momentum equation is obtained. Here, we will difference Eq. (6.89) centered around point $(i, j + \frac{1}{2})$ as shown in Fig. 6.17. We define average values of u at the points c and d on the left and right sides of the shaded cell in Fig. 6.17 as follows:

At point c: $\quad u = \frac{1}{2}(u_{i-1/2,j} + u_{i-1/2,j+1})$
At point d: $\quad \bar{u} = \frac{1}{2}(u_{i+1/2,j} + u_{i+1/2,j+1})$

Using a forward difference in time and central differences in space, Eq. (6.89) becomes

$$\boxed{(\rho v)^{n+1}_{i,j+1/2} = (\rho v)^{n}_{i,j+1/2} + B\,\Delta t - \frac{\Delta t}{\Delta x}(p^{n}_{i,j+1} - p^{n}_{i,j})} \quad (6.93)$$

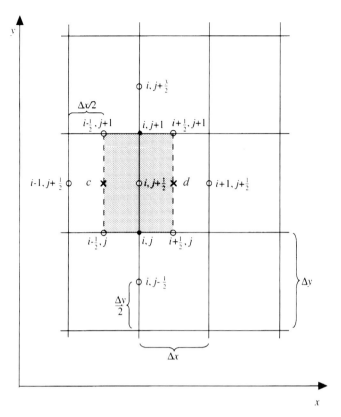

FIG. 6.17
Computational module for the y-momentum equation. The filled-in area is an effective control volume.

where

$$B = -\left[\frac{(\rho v \bar{u})^n_{i+1,j+1/2} - (\rho v u)^n_{i-1,j+1/2}}{2\Delta x} + \frac{(\rho v^2)^n_{i,j+3/2} - (\rho v^2)_{i,j-1/2}}{2\Delta y}\right]$$

$$+ \mu \left[\frac{v^n_{i+1,j+1/2} - 2v^n_{i,j+1/2} + v_{i-1,j+1/2}}{(\Delta x)^2} + \frac{v^n_{i,j+3/2} - 2v^n_{i,j+1/2} + v^n_{i,j-1/2}}{(\Delta x)^2}\right]$$

Note that u and $\bar{u}$ in Eq. (6.93) are those values defined by the average values at points c and d, i.e., u and $\bar{u}$ use different grid points than those for v.

As outlined in Sec. 6.8.3, at the beginning of each new iteration, $p = p^*$. For this situation, Eqs. (6.92) and (6.93) become, respectively,

$$(\rho u^*)^{n+1}_{i+1/2,j} = (\rho u^*)^n_{i+1/2,j} + A^* \, \Delta t - \frac{\Delta t}{\Delta x}(p^*_{i+1,j} - p^*_{i,j}) \qquad (6.94)$$

and
$$(\rho v^*)_{i,j+1/2}^{n+1} = (\rho v^*)_{i,j+1/2}^n + B^* \Delta t - \frac{\Delta t}{\Delta y}(p_{i,j+1}^* - p_{i,j}^*) \quad (6.95)$$

Subtracting Eq. (6.94) from Eq. (6.92), we have

$$(\rho u')_{i+1/2,j}^{n+1} = (\rho u')_{i+1/2,j}^n + A' \Delta t - \frac{\Delta t}{\Delta x}(p_{i+1,j}' - p_{i,j}')^n \quad (6.96)$$

where
$$(\rho u')_{i+1/2,j}^{n+1} = (\rho u)_{i+1/2,j}^{n+1} - (\rho u^*)_{i+1/2,j}^{n+1}$$
$$(\rho u')_{i+1/2,j}^n = (\rho u)_{i+1/2,j}^n - (\rho u^*)_{i+1/2,j}^n$$
$$A' = A - A^*$$
$$p_{i+1,j}' = p_{i+1,j} - p_{i+1,j}^*$$
$$p_{i,j}' = p_{i,j} - p_{i,j}^*$$

Subtracting Eq. (6.95) from Eq. (6.93), we obtain

$$(\rho v')_{i,j+1/2}^{n+1} = (\rho v')_{i,j+1/2}^n + B' \Delta t - \frac{\Delta t}{\Delta y}(p_{i,j+1}' - p_{i,j}') \quad (6.97)$$

where
$$(\rho v')_{i,j+1/2}^{n+1} = (\rho v)_{i,j+1/2}^{n+1} - (\rho v^*)_{i,j+1/2}^{n+1}$$
$$(\rho v')_{i,j+1/2}^n = (\rho v)_{i,j+1/2}^n - (\rho v^*)_{i,j+1/2}^n$$
$$B' = B - B^*$$
$$p_{i,j+1}' = p_{i,j+1} - p_{i,j+1}^*$$
$$p_{i,j}' = p_{i,j} - p_{i,j}^*$$

Eqs. (6.96) and (6.97) are the x- and y-momentum equations expressed in terms of the pressure and velocity corrections p', u', and v' defined by Eqs. (6.86), (6.87a), and (6.87b), respectively.

We are now in a position to obtain a formula for the pressure correction p' by insisting that the velocity field must satisfy the continuity equation. However, we are reminded that the pressure correction method is an iterative approach, and therefore there is no inherent reason why the formula designed to predict p' from one iteration to the next be physically correct; rather, we are concerned with only two aspects: (1) the formula for p' must yield the values that ultimately lead to the proper, converged solution, and (2) in the limit of the converged solution, the formula for p' must reduce to the physically correct continuity equation. That is, we are allowed to construct a formula for p' which is simply a *numerical artifice* designed to expedite the convergence of the velocity field to a solution that satisfies the continuity equation. When this convergence is achieved, $p' \to 0$, and the formula for p' reduces to the physically correct continuity equation.

With the above aspects in mind, let us proceed to obtain the pressure correction formula. Following Patankar (Ref. 68), let us arbitrarily set A', B',

$(\rho u')^n$, and $(\rho v')^n$ equal to zero in Eqs. (6.96) and (6.97), obtaining

$$(\rho u')^{n+1}_{i+1/2,j} = -\frac{\Delta t}{\Delta x}(p'_{i+1,j} - p'_{i,j})^n \tag{6.98}$$

and
$$(\rho v')^{n+1}_{i,j+1/2} = -\frac{\Delta t}{\Delta x}(p'_{i,j+1} - p'_{i,j})^n \tag{6.99}$$

Considering that we are simply constructing a numerical artifice which will provide some guidance in the iterative procedure, the above should not make you totally uncomfortable. Returning to the definition of $(\rho u')^{n+1}_{i+1/2,j}$ given just below Eq. (6.96), namely,

$$(\rho u')^{n+1}_{i+1/2,j} = (\rho u)^{n+1}_{i+1/2,j} - (\rho u^*)^{n+1}_{i+1/2,j}$$

we can write Eq. (6.98) as

$$(\rho u)^{n+1}_{i+1/2,j} = (\rho u^*)^{n+1}_{i+1/2,j} - \frac{\Delta t}{\Delta x}(p'_{i+1,j} - p'_{i,j})^n \tag{6.100}$$

Returning to the definition of $(\rho v')^{n+1}_{i,j+1/2}$ given just below Eq. (6.97), namely,

$$(\rho v')^{n+1}_{i,j+1/2} = (\rho v)^{n+1}_{i,j+1/2} - (\rho v^*)^{n+1}_{i,j+1/2}$$

we can write Eq. (6.99) as

$$(\rho v)^{n+1}_{i,j+1/2} = (\rho v^*)^{n+1}_{i,j+1/2} - \frac{\Delta t}{\Delta y}(p'_{i,j+1} - p'_{i,j})^n \tag{6.101}$$

Returning to the continuity equation

$$\frac{\partial(\rho u)}{\partial x} + \frac{\partial(\rho v)}{\partial y} = 0$$

and writing the corresponding central difference equations centered around point (i, j), we have

$$\frac{(\rho u)_{i+1/2,j} - (\rho u)_{i-1/2,j}}{\Delta x} + \frac{(\rho v)_{i,j+1/2} - (\rho v)_{i,j-1/2}}{\Delta y} = 0 \tag{6.102}$$

Substituting Eqs. (6.100) and (6.101) into (6.102) and dropping the superscripts, we have

$$\frac{(\rho u^*)_{i+1/2,j} - \Delta t/\Delta x(p'_{i+1,j} - p'_{i,j}) - (\rho u^*)_{i-1/2,j} + \Delta t/\Delta x(p'_{i,j} - p'_{i-1,j})}{\Delta x}$$
$$+ \frac{(\rho v^*)_{i,j+1/2} - \Delta t/\Delta y(p'_{i,j+1} - p'_{i,j}) - (\rho v^*)_{i,j-1/2} + \Delta t/\Delta x(p'_{i,j} - p'_{i,j-1})}{\Delta y} = 0$$

(6.103)

Rearranging Eq. (6.103), we obtain

$$\boxed{ap'_{i,j} + bp'_{i+1,j} + bp'_{i-1,j} + cp'_{i,j+1} + cp'_{i,j-1} + d = 0} \qquad (6.104)$$

where

$$a = 2\left[\frac{\Delta t}{(\Delta x)^2} + \frac{\Delta t}{(\Delta y)^2}\right]$$

$$b = -\frac{\Delta t}{(\Delta x)^2}$$

$$c = -\frac{\Delta t}{(\Delta y)^2}$$

$$d = \frac{1}{\Delta x}[(\rho u^*)_{i+1/2,j} - (\rho u^*)_{i-1/2,j}] + \frac{1}{\Delta y}[(\rho v^*)_{i,j+1/2} - (\rho v^*)_{i,j-1/2}]$$

Equation (6.104) is the *pressure correction formula*. It has an elliptic behavior, consistent with the fact that a pressure disturbance will propagate everywhere throughout an incompressible flow. Thus, Eq. (6.104) can be solved for p' by means of a numerical relaxation technique, such as described in Sec. 6.5.

Note that d in Eq. (6.104) is the central difference formulation of the left-hand side of the continuity equation expressed in terms of u^* and v^*. During the course of the iterative process, u^* and v^* define a velocity field that does *not* satisfy the continuity equation; hence in Eq. (6.104), $d \neq 0$ for all but the last iteration. In this sense, d is a *mass source* term. By definition, in the last iteration, the velocity field has converged to a field that satisfies the continuity equation, and hence, theoretically, $d = 0$ for this last iteration. In this sense, although a mathematical artifice was used to obtain Eq. (6.104), in the last iterative step we can construe Eq. (6.104) as being a proper physical statement of the conservation of mass.

It is interesting to note that the pressure correction formula, Eq. (6.104), is a central difference formulation of the *Poisson equation* in terms of the pressure correction p'.

$$\frac{\partial^2 p'}{\partial x^2} + \frac{\partial^2 p'}{\partial y^2} = Q \qquad (6.105)$$

If the second partial derivatives in Eq. (6.105) are replaced by central differences and if $Q = d/(\Delta t\, \Delta x)$, then Eq. (6.104) is obtained. (This short derivation is left as

Prob. 6.1.) Poisson's equation is one of the well-known equations from classical physics and mathematics, and it is worthwhile to observe that the pressure correction formula is nothing more than a difference equation representation of the Poisson equation for p'. We also note that the Poisson equation is an *elliptic* equation, which mathematically verifies the elliptic behavior of the pressure correction formula.

6.8.5 The Numerical Procedure: The SIMPLE Algorithm

To bring all the above discussion into perspective, we now summarize the numerical steps for the pressure correction method. The following description is the essence of the SIMPLE algorithm as set forth in Patankar (Ref. 68). The acronym SIMPLE stems from semi-implicit method for pressure-linked equations. The semi-implicit terminology refers to our arbitrary setting of A', B', $(\rho u')^n$, and $(\rho v')^n$ equal to zero in Eqs. (6.96) and (6.97), thus allowing the pressure correction formula, Eq. (6.104), to have p' appearing at only four grid points. If this artifice had not been used, the resulting pressure correction formula would have included velocities at neighboring grid points. These velocities are in turn influenced by pressure corrections in their neighborhood, and the resulting pressure correction formula would have reached much further into the flow field, essentially coupling the entire pressure correction field in one equation. This would have represented a "fully implicit" equation. Instead, because of the above artifice, Eq. (6.104) contains pressure corrections at only four grid points, and hence it is termed as only *semi-implicit* by Patankar (Ref. 68).

The step-by-step procedure for the SIMPLE algorithm is as follows:

1. Keeping in mind the staggered grid as sketched in Fig. 6.15, guess values of $(p^*)^n$ at all the "pressure" grid points (the filled points in Fig. 6.15). Also, arbitrarily set values of $(\rho u^*)^n$ and $(\rho v^*)^n$ at the proper "velocity" grid points (the open points in Fig. 6.15). Here, we are considering the grid points internal to the flow field; the treatment of points on the boundaries will be discussed later.
2. Solve for $(\rho u^*)^{n+1}$ from Eq. (6.94) and $(\rho v^*)^{n+1}$ from Eq. (6.95) at all appropriate internal grid points.
3. Substitute these values of $(\rho u^*)^{n+1}$ and $(\rho v^*)^{n+1}$ into Eq. (6.104), and solve for p' at all interior grid points. (This solution can be carried out by a relaxation procedure such as described in Sec. 6.5.)
4. Calculate p^{n+1} at all internal grid points from Eq. (6.86), i.e.,

$$p^{n+1} = (p^*)^n + p'$$

5. The values of p^{n+1} obtained in step 4 are used to solve the momentum equations again. For this, we designate p^{n+1} obtained above as the *new* values of $(p^*)^n$ to be inserted into Eqs. (6.94) and (6.95). With this interpretation, return to step 2 and repeat steps 2 to 5 until convergence is achieved. A reasonable criterion to use for a measure of convergence is when the mass source term d approaches zero.

When convergence is achieved, the velocity distribution has been obtained which satisfies the continuity equation. The whole function of the pressure correction formula, Eq. (6.104), is to aim the iteration process in such a direction that, when the velocity distribution is calculated from the momentum equations, it will eventually converge to the correct distribution which satisfies the continuity equation.

Something needs to be said in regard to the superscripts n and $n + 1$ used in the above equations. Equations (6.88) and (6.89) are the *unsteady* momentum equations, and hence the corresponding difference equations, Eqs. (6.92) and (6.93), utilize the standard superscript notation, n for a given time level and $n + 1$ for the next time level. On the other hand, the terms that were neglected in the derivation of the pressure correction formula, Eq. (6.104), result in a stepwise iteration process (the process described by steps 2–5 above) which in no way is timewise-accurate. However, this is no problem, because the pressure correction method is designed to solve for a *steady* flow, and we obtain this steady flow via an iterative process. From this point of view, it is best to interpret the superscripts n and $n + 1$ in the above equations as simply designating *sequential iteration steps*, with no significance to any *real* transient variation. Also in this sense, the value of Δt that appears in the above equations can be viewed simply as a parameter which has some effect on the speed at which convergence is achieved.

On a related matter, Eq. (6.104) may exhibit a divergent (rather than a convergent) behavior for some applications. Patankar suggests using some underrelaxation in such cases; i.e., instead of using Eq. (6.86) in step 4, use the equation

$$p^{n+1} = (p^*)^n + \alpha_p p' \tag{6.106}$$

where α_p is an underrelaxation factor; a value of about 0.8 is suggested. It may also be helpful in some cases to underrelax the values of u^* and v^* obtained from Eqs. (6.94) and (6.95).

6.8.6 Boundary Conditions for the Pressure Correction Method

How are boundary conditions specified consistent with the philosophy of the pressure correction method? This question is addressed here. For geometric simplicity, consider the constant-area duct sketched in Fig. 6.18; a staggered grid is distributed inside the duct. For an incompressible viscous flow, the physical problem is uniquely specified if:

1. At the inflow boundary, p and v are specified and u is allowed to float. If p is specified, then p' is zero at the inflow boundary. Hence, in Fig. 6.18,

$$p'_1 = p'_3 = p'_5 = p'_7 = 0$$

   v_2, v_4, v_6 are specified and held fixed.

2. At the outflow boundary, p is specified and u and v are allowed to float. Hence

$$p'_8 = p'_{10} = p'_{12} = p'_{14} = 0$$

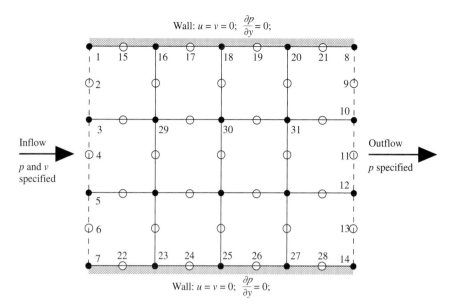

FIG. 6.18
Grid schematic for the discussion of boundary conditions for the pressure correction method.

3. *At the walls*, the viscous, no-slip condition holds at the wall. Hence, the velocity at the wall is zero.

$$u_{15} = u_{17} = u_{19} = u_{21} = u_{22} = u_{24} = u_{26} = u_{28} = 0$$

For the numerical solution, we need one more boundary condition at the wall. Since Eq. (6.104) has elliptic behavior and is solved by a relaxation technique, a boundary condition associated with p' must be specified over the *complete* boundary containing the computational domain. From items 1 and 2 above, we have $p' = 0$ at the inflow and outflow boundaries. A condition associated with p' at the walls can be derived as follows. Evaluate the y-momentum equation at the wall, where $u = v = 0$. With these velocity values inserted into Eq. (6.79), we have at the wall (neglecting body forces)

$$\left(\frac{\partial p}{\partial y}\right)_w = \mu \left(\frac{\partial^2 v}{\partial x^2} + \frac{\partial^2 v}{\partial y^2}\right)_w \tag{6.107}$$

Since $v_w = 0$, then in Eq. (6.107), $(\partial^2 v/\partial x^2)_w = 0$. Also, in the near vicinity of the wall, v is small; hence, in Eq. (6.107) we can reasonably assume that $(\partial^2 v/\partial y^2)_w$ is small. Thus, from Eq. (6.107) we can comfortably state the approximate (but reasonable) pressure boundary condition at the wall to be given by

$$\left(\frac{\partial p}{\partial y}\right)_w = 0 \tag{6.108}$$

Discretizing Eq. (6.108), we have (referring to Fig. 6.18)

$$p_1 = p_3 \qquad p_{16} = p_{29} \qquad p_5 = p_7 \qquad \text{etc.}$$

With this, a pressure boundary condition is numerically specified over the *complete* boundary enclosing the computational domain.

> **GUIDEPOST**
>
> The previous discussion on the pressure correction method is all that you need to know to tackle the incompressible viscous flow problem in Sec. 9.4, namely, the solution of Couette flow by means of an iterative solution of the two-dimensional incompressible Navier-Stokes equations. Therefore, if you are anxious to set up a computer project using the pressure correction method, you can
>
> Go directly to Sec. 9.4.
>
> However, if you choose this route, make certain to afterward return to the present location in the book and resume your general reading with the next section on computer graphics.

6.9 SOME COMPUTER GRAPHIC TECHNIQUES USED IN CFD

We end this chapter on simple CFD techniques with a discussion of some computer graphic "techniques" that are frequently employed in the presentation of CFD data. This section is different from the preceding sections in that we are not going to present any specific numerical technique for the solution of a flow problem; rather, in the present section we discuss how computer graphics is used as an essential *tool* by the computational fluid dynamicist to display the *results* of a CFD calculation. There are various graphical techniques used for the presentation of data, and hence it seems appropriate to include in this chapter on techniques an overview of such graphical techniques that are most frequently encountered in CFD.

We can classify the ways that CFD data are usually presented under six categories, to be discussed below. The computational fluid dynamicist usually implements these various modes of graphical representation via the use of existing computer graphic software rather than developing the details of new computer graphic programs himself or herself. It is generally not the purview of CFD to be involved with the development details of computer graphic software but rather to simply use this software as a *tool*. We will reflect this attitude in the present section. There are many existing software packages used by computational fluid dynamicists today. In the case of this author's students, TECPLOT, a software package provided by Amtec Engineering, is used. For this reason, many of the various computer graphic figures presented in this section were generated with TECPLOT; this is not to be construed as an endorsement of a specific product but rather simply as an example of a standard graphics software approach. New techniques and software for computer graphics are evolving as rapidly as those for CFD itself, so when your time comes, you will want to make your own choice of an appropriate graphics software package.

The majority of ways that CFD results are presented graphically can be classified under six general categories. Illustrations of these categories constitute the remainder of this section.

6.9.1 *xy* Plots

You are perhaps most familiar with *xy* plots; you have been dealing with them at least since your first course in algebra. On a two-dimensional graph, they represent the variation of one dependent variable versus another independent variable. Return for a moment to Fig. 1.6*b* to *f*. These are good examples of *xy* plots. In this case they are plots of pressure coefficient versus nondimensional chordwise distance; each different plot, from Fig. 1.6*b* to *f*, corresponds to a different spanwise station. Such *xy* plots are the simplest and most straightforward category of computer graphical representation of CFD results. Although such graphs are not particularly sophisticated, they still remain the most *precise* quantitative way to present numerical data on a graph; that is, another person can readily read quantitative data from curves on an *xy* plot without making any mental or arithmetic interpolation.

6.9.2 Contour Plots

A disadvantage of *xy* plots as described above is that they usually do not illustrate the *global* nature of a set of CFD results all in one view. On the other hand, contour plots do provide such a global view.

A contour line is a line along which some property is constant. We have already seen some contour plots. For example, return to Fig. 1.6*a*. This is a contour plot for pressure coefficient on the surface of an F-20 fighter airplane. Each line corresponds to a constant value of pressure coefficient. Generally, contours are plotted such that the *difference* between the quantitative value of the dependent variable from one contour line to an adjacent contour line is held constant. In this fashion, in regions where the dependent variable is rapidly changing in space, the adjacent contour lines are closely spaced together; in contrast, in regions where the dependent variable is slowly changing in space, the adjacent contour lines are widely spaced. In Fig. 1.6*a*, the regions where the contour lines are bunched together indicate regions of large pressure gradients on the surface—in this case pinpointing regions where shock waves are present on the surface of the airplane. Another example of contour plots is given in Fig. 6.8*a* to *d*. Here, pressure contours are shown for the two-dimensional, viscous, supersonic flow over a rearward-facing step. The regions of large gradients in the flow—the expansion wave from the top edge of the step and the recompression shock wave further downstream—are clearly seen in these contour plots.

It is clear from examining these contour plots that the *global* nature of the flow is seen in one single view; to obtain the same global feeling for the results from *xy* plots, say to ascertain the locations of the shock and expansion waves, we would have to examine a number of *xy* plots. Contour plots are clearly a superior graphical representation from this point of view. On the other hand, it requires more effort to read precise quantitative data from a contour plot as compared to a curve in an *xy* plot. Although each contour may be labeled as to the constant numerical value of the property it represents, the obtaining of numerical values *between* contour lines requires some mental and/or numerical interpolation in space, an imprecise process to say the least.

The plotting of a contour diagram by hand is a long, laborious process, although such plots were made (very infrequently) by some intrepid souls before the advent of the computer. This is in contrast to xy plots, which have been made by hand with aplomb since the days of Rene Descartes in the seventeenth century. Therefore, the proliferation of contour plots with the advent of the computer is understandable; in CFD, contour plots are one of the most commonly found graphical representations of data.

Let us examine a few more examples of contour plots from some modern CFD applications, pointing out various nuances and subcategories. For example, consider Fig. 6.19a and b. These are contour plots of the transverse velocity (the y component of velocity, v) in the flow field behind a detonation wave propagating through a combustible mixture of H_2, O_2, and argon. The detonation wave is propagating from left to right; the front of the wave is seen as the almost perpendicular cluster of contours at the right of the figures. The detonation wave is propagating into a uniform gas, which is the region to the right of the front; by definition there are no contours in this uniform, constant-property region—it appears as a totally clear region at the extreme right in the figures. Combustion of the hydrogen and oxygen occurs behind the detonation front. Because of the physical presence of slight disturbances in the flow behind the front, the flow field becomes two-dimensional, with transverse waves, along with various slip lines, as can be seen in the contour plot. The purpose of including Fig. 6.19a and b in this discussion is to point out the effect of the *number* of contour lines chosen for a given graph. Figure 6.19a contains 15 different contour levels; each contour is labeled with a number or letter, and the value of the transverse velocity in centimeters per second is given in the table of contour values at the right of the graph. Now examine Fig. 6.19b; this is the same set of data but plotted with 35 contour levels. Clearly, Fig. 6.19b gives a sharper, clearer picture of the flow field than Fig. 6.19a. This comparison clearly illustrates the value of including a sufficiently large number of contour lines in your plot.

Figure 6.19a and b is an example of *line* contour plots. Another type of contour plot is a *flooded contour*, illustrated in Fig. 6.20. This figure shows the same data for the transverse velocity as presented in Fig. 6.19b, but instead of using lines, a constant property is denoted by a constant *intensity of color shading*. In this case, gray is the color, and Fig. 6.20 is called a *gray-scale color map*. Therefore, instead of illustrating the flow with a discrete number of contour lines, the regions between these lines are simply filled with a color intensity that denotes the value of the flow-field property—the regions between the lines are "flooded" with color intensity. The color-coded velocity scale is shown at the right of Fig. 6.20.

The author wishes to thank James Weber, one of his graduate students at the University of Maryland, for providing these figures obtained as part of his doctoral research. These calculations were made using a finite-volume scheme called the *flux-corrected transport* (FCT) method, as described in Ref. 69.

Let us examine contour plots for another type of flow-field situation, in this case the shock-shock interaction problem sketched in Fig. 6.21. Here, the straight oblique shock from a wedge in a Mach 8 flow impinges on the bow shock from a cylinder placed above the wedge. The interaction of the shock waves from the

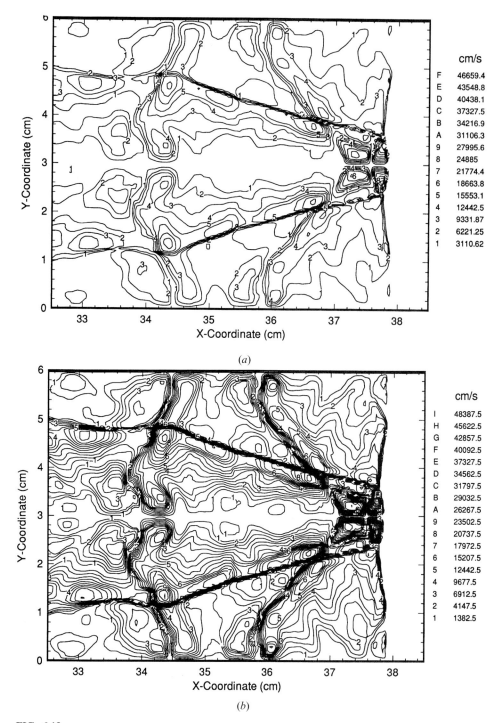

FIG. 6.19
Transverse velocity line contours in the flow behind a detonation wave propagating into a 20% H_2, 10% O_2, and 70% A mixture: (*a*) 15 contour levels; (*b*) 35 contour levels. The comparison between (*a*) and (*b*) shows the clarity obtained by using an increased number of contour lines. (*Calculations made by, and figure obtained from, James Weber, University of Maryland.*)

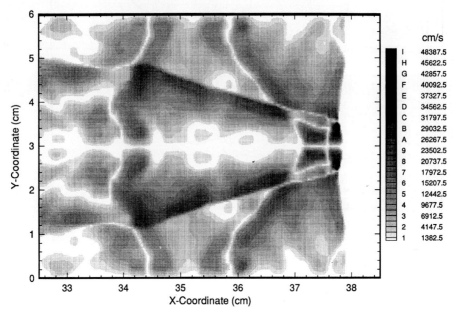

FIG. 6.20
Flooded contours on a gray-scale color map. Same data for transverse velocity as given in Fig. 6.19b. (*Calculations by, and figure obtained from, James Weber, University of Maryland.*)

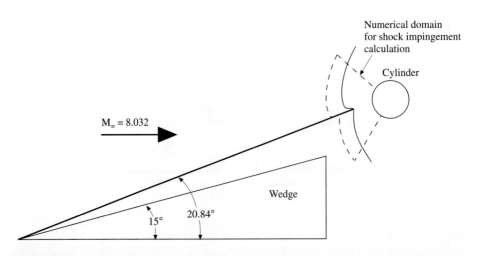

FIG. 6.21
Schematic of the shock–shock interaction from a wedge shock impinging on a bow shock from a cylinder mounted above the wedge. (*Obtained from Charles Lind, University of Maryland.*)

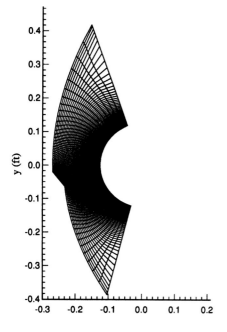

FIG. 6.22
Boundary-fitted grid for the shock interaction calculation. (*Obtained from Charles Lind, University of Maryland.*)

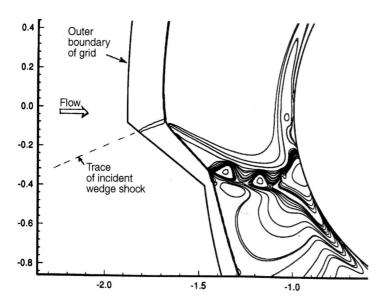

FIG. 6.23
Density contour plot of the type IV shock–shock interaction; $M_\infty = 5.04$, Reynolds number (based on cylinder diameter) $= 3.1 \times 10^5$. (*Calculations by, and figure obtained from, Charles Lind, University of Maryland.*)

wedge and the cylinder creates a complex flow field in the region of interaction. Notice in Fig. 6.21 that the curved shock wave from the cylinder becomes kinked in the region where the wedge shock impinges. The boundary-fitted coordinate system shown in Fig. 6.22 is created to solve the Navier-Stokes equations in this interaction region, using a finite-difference method. Density contours for this flow field (designated a *type IV shock interaction* because of the special geometric features of the angles made by the two intersecting shocks) are shown in Fig. 6.23. The incident wedge shock enters the computational grid from the lower left, as shown. The bow shock from the cylinder is identified by the very sharp clustering of contour lines bordering the left of the interaction region. The flow downstream of the bow shock is a complex region of refracted shocks and slip lines. However, the details of this type of flow can be made clearer by constructing a contour plot not of density but of the density *gradient*. Such a plot is shown in Fig. 6.24. This is a flooded contour plot on a gray-scale color map of *density gradient*. It is interesting to note that, in a physical laboratory situation, actual photographs of shock waves can be made by means of a special optical system called a *schlieren system*. In a schlieren photograph, shock and expansion waves are made visible by the refraction of light waves through the flow, which creates a pattern of various dark and light intensities proportional to the magnitude of the local *gradient* of density in the flow. Hence, the flooded contour plot in Fig. 6.24 is really a *CFD-generated schlieren picture* of the flow-field, analogous in every sense to a schlieren photograph that would be obtained in the laboratory. This illustrates another subcategory of contour plots and shows the tremendous versatility of the whole concept of contour plotting.

The author wishes to thank Charles Lind, a graduate student at the University of Maryland, for providing these figures obtained as part of his doctoral research.

The contour plots shown in Figs. 6.8 and 6.19 to 6.24 are made from two-dimensional flow-field calculations; the representation of these contour lines in the plane of the paper is therefore sufficient to give a global picture covering the whole geometric extent of the flow. But what happens when you have a three-dimensional flow? One answer is a multizone three-dimensional contour plot, such as shown in Fig. 6.25, taken from Ref. 70. Here we see pressure contours drawn for the transonic flow over an airplane wing, where the graphic shows the three-dimensional flow in perspective. Contours in three vertical planes at different spanwise stations are shown, along with contours on the upper surface of the wing. Such a plot gives a reasonable global picture of the three-dimensional flow over the wing, including the location of a shock wave on the upper surface near the leading edge, evidenced by the bunching up of some of the contour lines. An improvization of such a three-dimensional perspective plot is the "straight-on" composite view shown in Fig. 6.26, taken from Ref. 71. Here we see helicity density contours shown directly (not in perspective) in four different cross-flow planes for the flow over an ogive-cylinder in low-speed, subsonic flow at a 40° angle of attack. The side view of the body is also shown, with the axial locations of the four cross-flow planes clearly marked. In addition, some of the streamlines in the separated flow over the top of the body are shown in the side view.

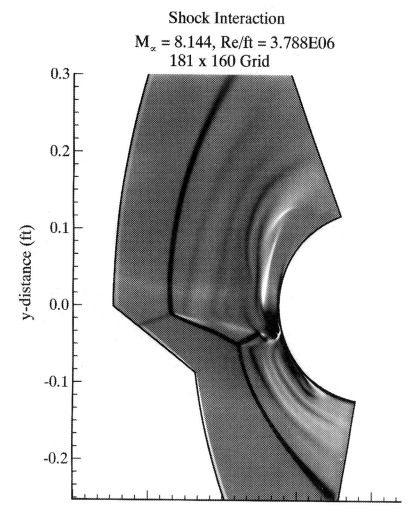

FIG. 6.24
Computer-produced schlieren of the type IV interaction; a flooded contour plot of density gradient from a CFD calculation. (*Calculations by, and figure obtained from, Charles Lind, University of Maryland.*)

6.9.3 Vector and Streamline Plots

A vector plot is a display of a vector quantity (in CFD, usually velocity) at discrete grid points, showing both magnitude and direction, where the base of each vector is located at the respective grid point. We have already seen examples of vector plots for two-dimensional flows in Figs. 1.13, 1.15, 1.19, 1.23, and 1.25, and for a three-dimensional flow in Fig. 1.21. Return to these figures, and examine them from the point of view of examples of a computer graphic technique. For convenience, a vector plot for the compressible subsonic flow over a forward-facing step is shown

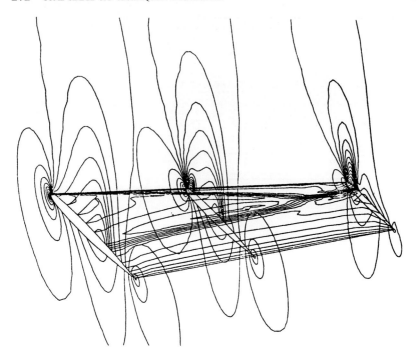

FIG. 6.25
Pressure contours in the three-dimensional transonic flow over an ONERA M6 wing. Euler solution. $M_\infty = 0.835$, angle of attack = 3.06°. (*From Ref. 70. Courtesy of Elsevier Science Publishers.*)

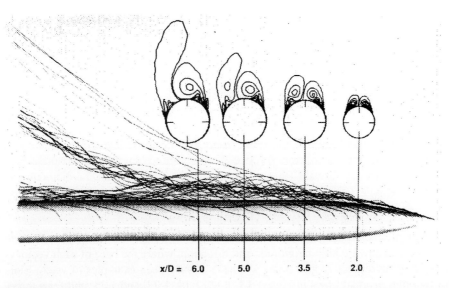

FIG. 6.26
Off-surface streamlines and helicity density contours around an ogive cylinder. $M_\infty = 0.28$, angle of attack = 40°, Reynolds number based on body diameter = 3×10^6. (*From Ref. 71.* Copyright © 1991, AIAA. Reprinted with permission.

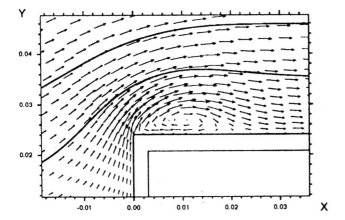

FIG. 6.27
Two-dimensional vector plot and streamlines for the compressible subsonic flow over a forward-facing step. (*After Ref. 72. Courtesy of Amtec Engineering, Bellevue, Washington.*)

in Fig. 6.27, obtained from Ref. 72. Two streamlines are also shown in Fig. 6.27—an example of a composite plot illustrating two variables on the same graph.

In CFD, as in all aspects of fluid dynamics in general, illustrations of streamlines are excellent tools for examining the nature of a flow. We have already seen two-dimensional streamline plots in Figs. 1.3, 1.10, 1.17, and 1.27. An example of a three-dimensional streamline plot is seen in Fig. 1.7. The three-dimensional particle tracks shown in Fig. 1.8 are essentially in this same category. Return to these figures, and examine them in light of our discussion here. For further illustration, Fig. 6.28 shows a composite of both streamlines and velocity vectors for the inviscid flow over the surface of a three-dimensional hypersonic body shape, obtained from Ref. 72.

6.9.4 Scatter Plots

In a scatter plot, a symbol (square, circle, etc.) is drawn at discrete grid points in the flow, where the magnitude of some scalar quantity (pressure, temperature, etc.) is indicated by either the size of the symbol, its shading, its color, or some combination thereof. For example, Fig. 6.29 is a scatter plot for the compressible subsonic flow over a forward-facing step, obtained from Ref. 72. The diameter of each circle indicates the magnitude of the y component of velocity, and the shading of each circle indicates the magnitude of the density.

6.9.5 Mesh Plots

Mesh plots consist of lines connecting grid points in either a two- or three-dimensional grid. We have already seen examples of two-dimensional mesh plots in Figs. 1.9, 1.11, 5.9, 5.10, 5.13 to 5.17, 5.19, and 5.20. A three-dimensional mesh plot is shown in Fig. 1.26. Examine again these figures, this time as examples of a computer graphic technique. A computer graphic display of a mesh for a three-

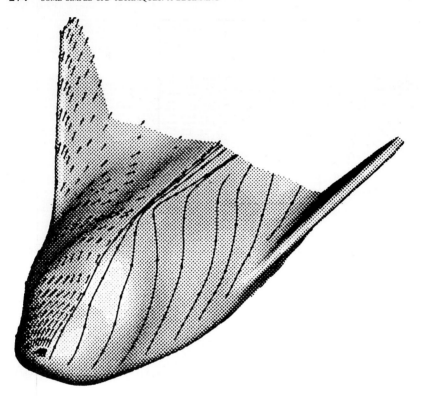

FIG. 6.28
Three-dimensional vectors and streamlines on the surface of a hypersonic body. (*From Ref. 72. Courtesy of Amtec Engineering, Bellevue, Washington.*)

dimensional flow calculation over a wing is shown in Fig. 6.30, obtained from Ref. 70. This mesh was used for the calculation of the transonic flow results shown in Fig. 6.25. Another type of mesh plot is that showing only the mesh on the surface of a body, such as illustrated in Fig. 6.31. Here, the mesh covers the entire body, including both top and bottom surfaces, but the computer graphic display is designed to remove the hidden lines, therefore obtaining a clearer picture. Finally, another improvization is shown in Fig. 6.32, which illustrates a three-dimensional mesh with the body shape shown as a light-source-shaded surface. This figure is a dramatic example of the quality and sophistication of modern computer graphics.

6.9.6 Composite Plots

Many of the categories of different plots described above can be combined into a single plot, called a *composite plot*. Figure 6.27 is a simple example of a composite plot, where two quantities are overlaid in the same graph. Figure 6.33 illustrates a composite plot showing four different zones on a body surface, where different graphical results are shown in each zone.

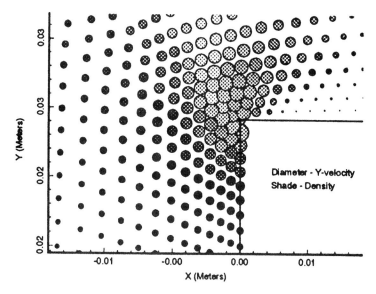

FIG. 6.29
A two-dimensional scatter plot showing the magnitude of the y component of velocity (diameter of circles) and the value of density (shading of circles) for the compressible subsonic flow over a forward-facing step. (*From Ref. 72. Courtesy of Amtec Engineering, Bellevue, Washington.*)

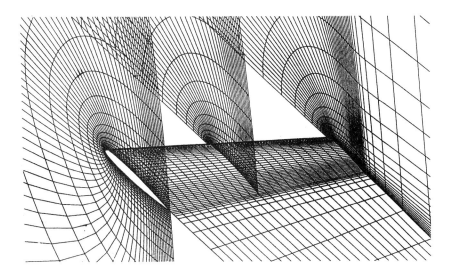

FIG. 6.30
A three-dimensional mesh plot for the calculation of the flow over an airplane wing used to obtain the flow field results shown in Fig. 6.25. (*From Ref. 70. Courtesy of Elsevier Science Publishers.*)

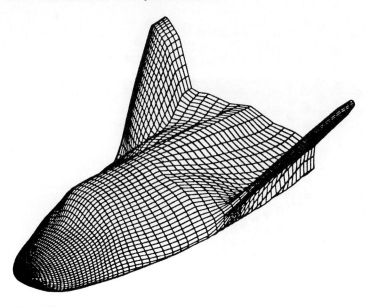

FIG. 6.31
A three-dimensional mesh plot on the surface of a body. The mesh is wrapped completely over the body, but the computer graphic shown here is designed to remove the hidden lines, for clarity. (*From Ref. 72. Courtesy of Amtec Engineering, Bellevue, Washington.*)

6.9.7 Summary on Computer Graphics

Computer graphics is a dynamically evolving discipline, and the CFD community is constantly taking advantage of any new techniques for the display of its data. The graphical display of three-dimensional CFD results, a dream in the minds of researchers just 20 years ago, is commonplace today. As one final example, we offer Fig. 6.34, obtained from Ref. 73. Here we see a complete airplane displayed in three-dimensional perspective, with contours of pressure coefficient displayed over its surface. Such a computer graphics display is not only a technical record of quantitative results, it is also an aesthetic work of art.

Yes, a work of art—that is part of modern computer graphics. Perhaps no better example of this can be given than the Engineering Research Center for Computational Field Simulation—a National Science Foundation Center for engineering research established at Mississippi State University for the purpose of enhancing the methods of grid generation, CFD, and computer graphics. Under the direction of Dr. Joe Thompson at Mississippi State, this interdisciplinary center has become one of the world's leading sources of new advancements in CFD and in graphical displays. An important and unique aspect of this center is that on its staff are faculty from the university's department of art—a true testimonial that computer graphics today is a work of art.

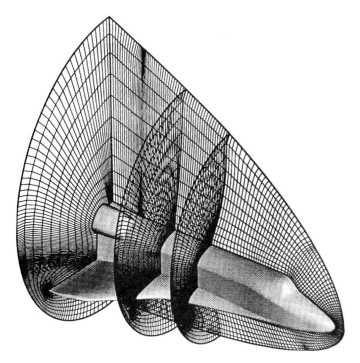

FIG. 6.32
A three-dimensional mesh plot for the calculation of the flow over a body, where the body is shown as a light-source-shaded surface. (*From Ref. 72. Courtesy of Amtec Engineering, Bellevue, Washington.*)

Finally, we note that Chap. 12 contains many additional examples of CFD results displayed by computer graphic techniques; you may want to turn to Chap. 12 at this point in your reading and just flip through those results to further enhance your appreciation of the roll of computer graphics in CFD.

6.10 SUMMARY

In this chapter, we have taken the final steps in our discussion of the basics of the numerics necessary for the numerical solution of the governing flow equations. In particular, we have tied together the fundamental aspects of numerical discretization discussed in Chap. 4 and shown how they can be put together to form various *techniques* for the numerical solution of the continuity, momentum, and energy equations. We have seen that the choice of an appropriate numerical technique is closely related to the mathematical behavior of the original partial differential equations. (Is the problem driven by elliptic, parabolic, or hyperbolic behavior, or some combination thereof?) The techniques discussed in this chapter have been well-established over the past two decades (and longer for some cases). They are intentionally chosen for their relative simplicity and straightforward aspects—they establish a certain foundation that will enable you to better appreciate the more

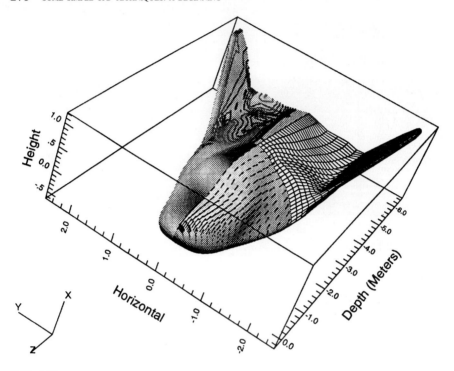

FIG. 6.33
A composite plot, illustrating four different zones on a body surface, where different graphical results are shown in each zone. (*From Ref. 72. Courtesy of Amtec Engineering, Bellevue, Washington.*)

modern, more sophisticated techniques which you will find in more advanced studies of CFD and in the modern applications of CFD in industry and research laboratories.

We did not begin this chapter with a road map, but it is fitting that we end with one. Figure 6.35 gives the general path; look over this road map and make certain that you feel comfortable with the details associated with each box in the map. If you are not quite certain about any aspects, return to the appropriate section in this chapter and review the material again. After you have done this, you will be ready to forge ahead to Part III of this book.

Remember the techniques in this chapter were chosen on the basis of their relative simplicity, while at the same time being sufficient for the applications to be treated in Part III. These techniques are essentially "student-friendly," and they have a great deal to offer in opening up the joys and power of CFD to the beginning student. Indeed, the material in Parts I and II of this book is intended to introduce the reader to some of the philosophy, the definitions, and concepts of the discipline in a (hopefully) easy-to-understand manner. The intent is to make you feel comfortable with the material, not to overwhelm you with some of the more

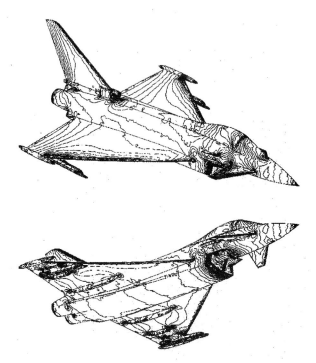

FIG. 6.34
Three-dimensional pressure coefficient contours over the surface of a generic fighter aircraft. $M_\infty = 0.85$, angle of attack = $10°$, angle of yaw = $30°$. (*From Ref. 73. Courtesy of Elsevier Science Publishers.*)

mathematically sophisticated new techniques that represent the current state-of-the-art of CFD. However, you will not be left totally in the dark about the modern CFD; Part IV will introduce you to some aspects of this modern state-of-the-art, but at a proper stage in your learning process so that it hopefully will make sense to you. So press on to Parts III and IV; Part III will serve to reinforce what you have already learned, and Part IV will inform you about some of the new techniques that characterize CFD today.

PROBLEMS

6.1. Show that the pressure correction formula, Eq. (6.104), is a central difference formulation of Poisson's equation for the pressure correction, namely, Eq. (6.105).

6.2. The velocity potential for an incompressible, inviscid, irrotational flow over a circular cylinder is governed by Laplace's equation, as described in Sec. 6.5. Write Laplace's equation in polar coordinates. Write a computer program that numerically solves this equation for the velocity potential in the flow field around the cylinder. Plot the velocity

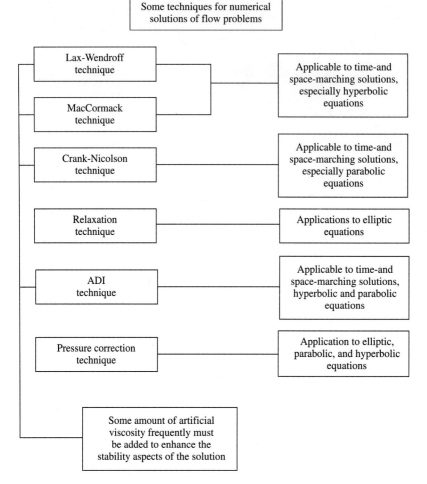

FIG. 6.35
Road map for Chap. 6.

potential as a function of radial distance away from the cylinder, at several angular positions. Calculate and plot the pressure coefficient distribution along the surface of the cylinder. Finally, compare all these numerical results with exact results from the classical analytical solution for the flow over a cylinder.

PART III

SOME APPLICATIONS

We are now ready to examine *precisely* how CFD can be used to solve various flow problems. We have had to wait to do this until we completed our study of the governing flow equations and their mathematical behavior in Part I. We have also had to wait until we completed our study of the basic numerical aspects of discretization of partial differential equations (finite differences) or of integral equations (finite volumes), our discussion of grid generation and transformations, and our development of various techniques in Part II. We have had to wait for all these aspects to fall into place before we could address some applications, because the application of CFD generally requires the simultaneous knowledge of all the above aspects. However, we are now ready to launch into the world of applications—the subject of Part III of this book. The applications chosen in Part III all have a common theme; they are flow problems which are relatively basic and straightforward and which, for the most part, have exact analytical or semianalytical solutions obtained from an independent, theoretical study. These choices are made for three reasons: (1) they allow the clear illustration of the details of application of CFD to flow problems without muddying these applications with complex fluid dynamic details; (2) they are flow problems which you are most likely to be familiar with from previous studies, and therefore feel somewhat comfortable with; and (3) the known, exact solutions allow a direct comparison with our CFD results, thus allowing us to obtain a feeling for how accurate CFD can be and what it takes to achieve this accuracy. Therefore, each of the chapters in Part III treats a specific flow problem and highlights the application of CFD to that specific problem. The applications will involve one or more of the techniques discussed in Chap. 6. In the process, you will have the opportunity to see the detailed implementation of these techniques, to see their strengths and weaknesses, to obtain a better understanding of what these techniques really mean, and to obtain a feeling for the real "nitty-gritty" aspects of working through CFD solutions for various types of flows. No effort is made to deal with the complex three-dimensional flows which constitute the bulk of modern CFD attention today—this is left to more advanced studies and workplace applications well beyond the scope of this book. Some of these modern applications were discussed in Chap. 1 and serve as an incentive for you to pursue further studies of CFD after finishing this book (further studies in the form of advanced books, courses, and applications in the workplace).

CHAPTER 7

NUMERICAL SOLUTIONS OF QUASI-ONE-DIMENSIONAL NOZZLE FLOWS

> *When you measure what you are speaking about, and express it in numbers, you know something about it; but when you cannot measure it, when you cannot express it in numbers, your knowledge is of a meager and unsatisfactory kind: it may be the beginning of knowledge, but you have scarcely, in your thoughts, advanced to the stage of science.*
>
> William Thomson, Lord Kelvin, from Popular Lectures and Addresses, 1891–1894

7.1 INTRODUCTION: THE FORMAT FOR CHAPTERS IN PART III

For the next four chapters, which constitute Part III, the following format will be followed. Each chapter will deal with a specific flow field; for example, the present chapter deals with the quasi-one-dimensional flow through a convergent-divergent nozzle. Each chapter will be subdivided into three main parts:

1. *Physical description of the flow.* The physical aspects of the flow will be described, and pertinent equations and relationships obtained from the analytical solution will be reviewed. If experimental data are appropriate, they will be

discussed. The purpose here is to give you a physical understanding of the flow field, to be calculated subsequently with our CFD techniques.
2. *CFD solution: intermediate steps.* A specific CFD technique (one of those discussed in Chap. 6) will be chosen for the numerical solution of the flow problem. The pertinent partial differential equations or the integral form of the equations, as the case may be, most suited for the specific CFD technique as applied to the specific flow problem will be set forth. The solution will be set up, step by step, and the numerical operations will be carried through in detail for the first few steps. Numbers will be given for all stages of the calculation during these intermediate steps so that you can compare them directly with your own calculations. Such matters as the calculations at internal points, at the boundaries, the numerical implementation of boundary conditions, and the determination of step size (if appropriate) will be covered in detail.
3. *CFD solution: final results.* Tabulations and graphs of the final numerical solution to the flow field will be given. These final results will be compared with the exact analytical (and/or experimental) results, and an evaluation of the accuracy of the CFD solution will be made.

Note: You have an option at this stage. You can decide simply to read these chapters, obtain a detailed understanding of the implementation of various CFD techniques to various problems, and get a feeling for the results. Or, you can also decide to write your own computer programs to calculate the answers yourself. It is for those of you who make the latter decision that some of the intermediate numbers obtained on the way toward a solution will be given. These numbers will be boxed and easy to follow so that you can check on the early calculational aspects of the problem. Also, the final answers will be given in some detail so that you can check the final results obtained from your computer program. You are strongly encouraged to make this latter decision: to write your own computer programs for the various solutions as we progress through the next four chapters. Simply reading the material is certainly worthwhile, but it is analogous to sitting on the sidelines watching a football game. By writing your own programs and calculating along with the steps given in the book, you will be playing the game yourself and getting your hands dirty. To really learn CFD, you must get your hands dirty; i.e., you must wade into the calculations and do them yourself. The flow problems and their CFD solutions given in the next four chapters are suitable for personal computers; you do not need a powerful mainframe or even a major workstation for their solution. Indeed, the present author has used his own Macintosh computer for the solutions described herein.

In some cases, more than one CFD technique will be used to solve the same flow problem. This is done to give you some comparison of the strengths and weaknesses of one technique versus another and a feeling for the relative difficulty of setting up one technique on the computer compared to another.

We are finally ready to go. This author wishes you happy computing!

7.2 INTRODUCTION TO THE PHYSICAL PROBLEM: SUBSONIC-SUPERSONIC ISENTROPIC FLOW

The flow problem discussed here can be found in any gas dynamic textbook; for example, it is covered in detail in Chap. 10 of the author's book *Fundamentals of Aerodynamics*, 2d ed. (Ref. 8), as well as in Chap. 5 of the author's book *Modern Compressible Flow*, 2d ed. (Ref. 21). In the present section, we will review some of the important physical and analytical aspects of this flow.

We consider the steady, isentropic flow through a convergent-divergent nozzle as sketched in Fig. 7.1. The flow at the inlet to the nozzle comes from a reservoir where the pressure and temperature are denoted by p_0, and T_0, respectively. The cross-sectional area of the reservoir is large (theoretically, $A \to \infty$), and hence the velocity is very small ($V \to 0$). Thus, p_0 and T_0 are the stagnation values, or *total* pressure and *total* temperature, respectively. The flow expands isentropically to supersonic speeds at the nozzle exit, where the exit pressure, temperature, velocity, and Mach number are denoted by p_e, T_e, V_e, and M_e, respectively. The flow is locally subsonic in the convergent section of the nozzle, sonic at the throat (minimum area), and supersonic at the divergent section. The sonic flow ($M = 1$) at the throat means that the local velocity at this location is equal to

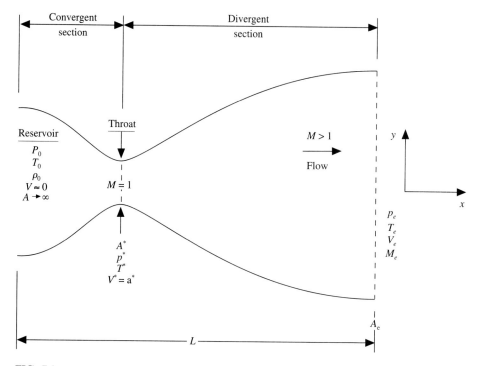

FIG. 7.1
Schematic for subsonic-supersonic isentropic nozzle flow.

the local speed of sound. Using an asterisk to denote sonic flow values, we have at the throat $V = V^* = a^*$. Similarly, the sonic flow values of pressure and temperature are denoted by p^* and T^*, respectively. The area of the sonic throat is denoted by A^*. We assume that at a given section, where the cross-sectional area is A, the flow properties are uniform across that section. Hence, although the area of the nozzle changes as a function of distance along the nozzle, x, and therefore in reality the flow field is two-dimensional (the flow varies in the two-dimensional xy space), we make the *assumption* that the flow properties vary only with x; this is tantamount to assuming uniform flow properties across any given cross section. Such flow is defined as *quasi-one-dimensional* flow.

The governing continuity, momentum, and energy equations for this quasi-one-dimensional, steady, isentropic flow can be expressed, respectively, as

Continuity :
$$\rho_1 V_1 A_1 = \rho_2 A_2 V_2 \tag{7.1}$$

Momentum :
$$p_1 A_1 + \rho_1 V_1^2 A_1 + \int_{A_1}^{A_2} p \, dA = p_2 A_2 + \rho_2 V_2^2 A_2 \tag{7.2}$$

Energy :
$$h_1 + \frac{V_1^2}{2} = h_2 + \frac{V_2^2}{2} \tag{7.3}$$

where subscripts 1 and 2 denote different locations along the nozzle. In addition, we have the perfect gas equation of state,

$$p = \rho R T \tag{7.4}$$

as well as the relation for a calorically perfect gas,

$$h = c_p T \tag{7.5}$$

Equations (7.1) to (7.5) can be solved analytically for the flow through the nozzle. Some results are as follows. The Mach number variation through the nozzle is governed exclusively by the area ratio A/A^* through the relation

$$\left(\frac{A}{A^*}\right)^2 = \frac{1}{M^2}\left[\frac{2}{\gamma+1}\left(1 + \frac{\gamma-1}{2}M^2\right)\right]^{(\gamma+1)/(\gamma-1)} \tag{7.6}$$

where γ = ratio of specific heats = c_p/c_v. For air at standard conditions, $\gamma = 1.4$. For a nozzle where A is specified as a function of x, hence A/A^* is known as a function of x, then Eq. (7.6) allows the (implicit) calculation of M as a function of x. This is sketched in Fig. 7.2b. In turn, the variation of pressure, density, and temperature as a function of Mach number (and hence as a function of A/A^*, thus x) is given,

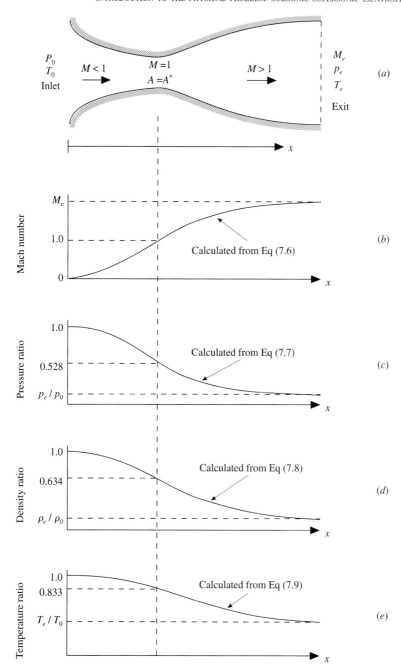

FIG. 7.2
Qualitative aspects of quasi-one-dimensional nozzle flow: isentropic subsonic-supersonic solution.

respectively, by

$$\frac{p}{p_0} = \left(1 + \frac{\gamma - 1}{2} M^2\right)^{-\gamma/(\gamma - 1)} \tag{7.7}$$

$$\frac{\rho}{\rho_0} = \left(1 + \frac{\gamma - 1}{2} M^2\right)^{-1/(\gamma - 1)} \tag{7.8}$$

$$\frac{T}{T_0} = \left(1 + \frac{\gamma - 1}{2} M^2\right)^{-1} \tag{7.9}$$

These variations are sketched in Fig. 7.2c to e.

The nozzle flow described above just does not "happen" by itself. That is, if you take the nozzle sketched in Fig. 7.2a and place it on a desk in front of you, the air does not just start flowing through the nozzle by itself. As with all mechanical systems, it takes a force to accelerate a given mass; the nozzle flow is no different. In this case, the force exerted on the gas to accelerate it through the nozzle is supplied by the pressure ratio across the nozzle, p_0/p_e. For a nozzle with a specified area ratio A_e/A^*, the pressure ratio required to establish the subsonic-supersonic isentropic flow sketched in Fig. 7.2 must be a very specific value, namely, that value shown in Fig. 7.2c. This pressure ratio is a boundary condition applied to the flow; in the laboratory, it is provided by a high-pressure air reservoir at the inlet and/or a vacuum source at the exit.

7.3 CFD SOLUTION OF SUBSONIC-SUPERSONIC ISENTROPIC NOZZLE FLOW: MACCORMACK'S TECHNIQUE

At this point, you are reminded that *any* numerical solution of the steady, isentropic quasi-one-dimensional nozzle flow is overkill; we have a closed-form analytical solution as described in Sec. 7.2, and therefore in general a numerical solution is not needed. However, that is not the point. What we want to accomplish here is to illustrate the application of various CFD techniques, and we are intentionally choosing a flow problem with a known analytic solution for this illustration. That is, we are following the philosophy as set forth in Sec. 7.1.

In this section we choose to illustrate the application of MacCormack's technique as described in Sec. 6.3. In particular, we will set up a time-marching, finite-difference solution for the quasi-one-dimensional nozzle flow. Before progressing further, pause at this point, return to Sec. 6.3, and read it again, carefully. In the present section, we will assume that you fully understand MacCormack's technique to the extent described in Sec. 6.3. Also, reexamine Fig. 1.32b, which illustrates the major ideas that feed into this application.

7.3.1 The Setup

In this section, we will set up three eschelons of equations as follows:

1. The governing flow equations will be couched in terms of *partial differential equations* suitable for the time-marching solution of quasi-one-dimensional flow (the closed-form algebraic equations discussed in Sec. 7.2 are for a steady flow and are not suitable for the present purpose).

2. The finite-difference expressions pertaining to MacCormack's technique as applied to this problem will be set up.

3. Other details for the numerical solution (such as the calculation of the time step and the treatment of boundary conditions) will be formulated.

THE GOVERNING FLOW EQUATIONS. Beginning with step 1 above, recall that we have derived the governing partial differential equations for inviscid flow (the Euler equations) in Chap. 2; these are summarized in Eqs. (2.82) to (2.86). Since we are dealing with a one-dimensional inviscid flow for our nozzle problem, it would seem appropriate to take Eqs. (2.82) to (2.86), simply write them down for one-dimensional flow, and proceed ahead. After all, these equations have been derived in Chap. 2 in the most general sense, and we should be able to make use of them. *However, such is not the case with quasi-one-dimensional nozzle flow.* Why? The answer lies with the simplifying assumption we have made with quasi-one-dimensional flow as described in Sec. 7.2, namely, we assume that the flow properties are uniform across any given cross section of the nozzle. In so doing, we have somewhat twisted the physics of the flow.* Return to Fig. 7.1 for a moment. Note that, in reality, the real nozzle flow is a two-dimensional flow because, with the area changing as a function of x, in actuality there will be flow-field variations in both the x and y directions. This is the real physics of the flow, and Eqs. (2.82) to (2.86) properly describe such a two-dimensional flow. On the other hand, the *assumption* of quasi-one-dimensional flow dictates that the flow properties are functions of x only. Since this assumption twists the real physics of the flow, then Eqs. (2.82) to (2.86) are not necessarily appropriate for quasi-one-dimensional flow. On the other hand, for the equations that *are* appropriate for quasi-one-dimensional flow, we would at least like for the overall physical principles of (1) mass conservation, (2) Newton's second law, and (3) energy conservation to hold exactly, in spite of our twisted physics due to the quasi-one-dimensional assumption. To ensure that these physical principles are satisfied, we must return to the *integral forms* of the governing equations derived in Chap. 2 and apply these integral forms to a control volume consistent with the quasi-one-dimensional assumption. Let us proceed.

---

* To say that we are "twisting" the physics of the flow is a rather strong statement in order to emphasize a point. What we are really doing with our quasi-one-dimensional assumptions is constructing a *simplified engineering model* of the flow. Such *modeling* to simplify more complicated problems is done very frequently in engineering and physical science. Of course, the price we pay for such modeling is usually some compromise with the real physics of the flow.

We start with the integral form of the continuity equation given by Eq. (2.19), repeated below:

$$\frac{\partial}{\partial t}\iiint_{\mathscr{V}} \rho\, d\mathscr{V} + \iint_{S} \rho \mathbf{V}\cdot d\mathbf{S} = 0 \qquad (2.19)$$

We apply this equation to the shaded control volume shown in Fig. 7.3. This control volume is a slice of the nozzle flow, where the infinitesimal thickness of the slice is dx. On the left side of the control volume, consistent with the quasi-one-dimensional assumption, the density, velocity, pressure, and internal energy, denoted by ρ, V, p, and e, respectively, are uniform over the area A. Similarly, on the right side of the control volume, the density, velocity, pressure, and internal energy, denoted by $\rho + d\rho$, $V + dV$, $p + dp$, and $e + de$, respectively, are uniform over the area $A + dA$. Applied to the control volume in Fig. 7.3, the volume integral in Eq. (2.19) becomes, in the limit as dx becomes very small,

$$\frac{\partial}{\partial t}\iiint_{\mathscr{V}} \rho\, d\mathscr{V} = \frac{\partial}{\partial t}(\rho A\, dx) \qquad (7.10)$$

where $A\, dx$ is the volume of the control volume in the limit of dx becoming vanishingly small. The surface integral in Eq. (2.19) becomes

$$\iint_{S} \rho \mathbf{V}\cdot d\mathbf{S} = -\rho V A + (\rho + d\rho)(V + dV)(A + dA) \qquad (7.11)$$

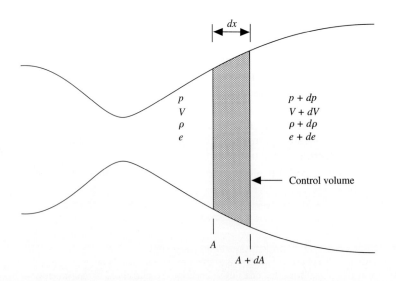

FIG. 7.3
Control volume for deriving the partial differential equations for unsteady, quasi-one-dimensional flow.

where the minus sign on the leading term on the right-hand side is due to the vectors **V** and **dS** pointing in opposite directions over the left face of the control volume, and hence the dot product is negative. (Recall from Chap. 2 that **dS** always points *out* of the control volume, by convention.) Expanding the triple product term in Eq. (7.11), we have

$$\iint_S \rho \mathbf{V} \cdot \mathbf{dS} = -\rho V A + \rho V A + \rho V \, dA + \rho A \, dV + \rho \, dV \, dA$$

$$+ AV \, d\rho + V \, dA \, d\rho + A \, dV \, d\rho + d\rho \, dV \, dA \quad (7.12)$$

In the limit as dx becomes very small, the terms involving *products* of differentials in Eq. (7.12), such as $\rho \, dV \, dA$, $d\rho \, dV \, dA$, go to zero much faster than those terms involving only one differential. Hence, in Eq. (7.12), all terms involving *products* of differentials can be dropped, yielding in the limit as dx becomes very small

$$\iint_S \rho \mathbf{V} \cdot \mathbf{dS} = \rho V \, dA + \rho A \, dV + AV \, d\rho = d(\rho AV) \quad (7.13)$$

Substituting Eqs. (7.10) and (7.13) into (2.19), we have

$$\frac{\partial}{\partial t}(\rho A \, dx) + d(\rho AV) = 0 \quad (7.14)$$

Dividing Eq. (7.14) by dx and noting that $d(\rho AV)/dx$ is, in the limit as dx goes to zero, the definition of the partial derivative with respect to x, we have

$$\boxed{\frac{\partial(\rho A)}{\partial t} + \frac{\partial(\rho AV)}{\partial x} = 0} \quad (7.15)$$

Equation (7.15) is the partial differential equation form of the continuity equation *suitable for unsteady, quasi-one-dimensional flow*. It ensures that mass is conserved for this model of the flow.

It is interesting to pause for a moment and compare this with the general continuity equation for three-dimensional flow, Eq. (2.82*b*), specialized for one-dimensional flow. For such a case, Eq. (2.82*b*) becomes

$$\frac{\partial \rho}{\partial t} + \frac{\partial(\rho u)}{\partial x} = 0 \quad (7.16)$$

where u is the x component of velocity. Clearly, Eq. (7.16) is different from Eq. (7.15). Equation (7.16) applies to a *truly* one-dimensional flow, where A is constant with respect to x. It does *not* represent a proper statement of the conservation of mass for our *model* of quasi-one-dimensional flow, where $A = A(x)$; instead, Eq. (7.15) is a proper statement of mass conservation for our model. Of course, note that for the special case of constant-area flow, Eq. (7.15) reduces to Eq. (7.16).

We now turn to the integral form of the x component of the momentum equation, (from Prob. 2.2) written below for an inviscid flow (neglecting the viscous stress terms) with no body forces,

$$\frac{\partial}{\partial t}\iiint_{\mathcal{V}}(\rho u)\,d\mathcal{V} + \iint_{S}(\rho u\mathbf{V})\cdot d\mathbf{S} = -\iint_{S}(p\,d\mathbf{S})_x \tag{7.17}$$

where the term $(p\,d\mathbf{S})_x$ denotes the x component of the vector $p\,d\mathbf{S}$. We apply Eq. (7.17) to the shaded control volume in Fig. 7.3. In Eq. (7.17), the integrals on the left side are evaluated in the same manner as discussed above in regard to the continuity equation. That is,

$$\frac{\partial}{\partial t}\iiint_{\mathcal{V}}(\rho u)\,d\mathcal{V} = \frac{\partial}{\partial t}(\rho V A\,dx) \tag{7.18}$$

and

$$\iint_{S}(\rho u\mathbf{V})\cdot d\mathbf{S} = -\rho V^2 A + (\rho + d\rho)(V + dV)^2(A + dA) \tag{7.19}$$

The evaluation of the pressure force term on the right-hand side of Eq. (7.17) is best carried out with the aid of Fig. 7.4. Here, the x components of the vector $p\,d\mathbf{S}$ are shown on all four sides of the control volume. Remember that $d\mathbf{S}$ always points *away* from the control volume; hence any x component $(p\,d\mathbf{S})_x$ that acts toward the left (in the negative x direction) is a negative quantity, and any x component $(p\,d\mathbf{S})_x$ that acts toward the right (in the positive x direction) is a positive quantity. Also note that the x component of $p\,d\mathbf{S}$ acting on the top and bottom inclined faces of the control volume in Fig. 7.4 can be expressed as the pressure p acting on the *component* of the inclined area projected perpendicular to the x direction, $(dA)/2$; hence, the contribution of each inclined face (top or bottom) to the pressure integral in Eq. (7.17) is $-p(dA/2)$. All together, the right-hand side of Eq. (7.17) is expressed as follows:

$$\iint(p\,d\mathbf{S})_x = -pA + (p + dp)(A + dA) - 2p\left(\frac{dA}{2}\right) \tag{7.20}$$

Substituting Eqs. (7.18) to (7.20) into (7.17), we have

$$\frac{\partial}{\partial t}(\rho V A\,dx) - \rho V^2 A + (\rho + d\rho)(V + dV)^2(A + dA)$$
$$= pA - (p + dp)(A + dA) + p\,dA \tag{7.21}$$

Canceling like terms and ignoring products of differentials, Eq. (7.21) becomes in the limit of dx becoming very small

$$\frac{\partial}{\partial t}(\rho V A\,dx) + d(\rho V^2 A) = -A\,dp \tag{7.22}$$

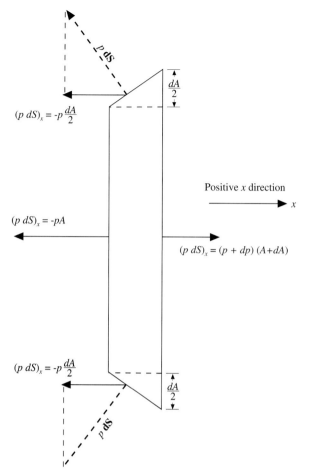

FIG. 7.4
The forces in the x direction acting on the control volume.

Dividing Eq. (7.22) by dx and taking the limit as dx goes to zero, we obtain the partial differential equation

$$\frac{\partial(\rho V A)}{\partial t} + \frac{\partial(\rho V^2 A)}{\partial x} = -A\frac{\partial p}{\partial x} \tag{7.23}$$

We could live with Eq. (7.23) as it stands—it represents the *conservation* form of the momentum equation for quasi-one-dimensional flow. However, let us obtain the equivalent nonconservation form. This is done by multiplying the continuity equation, Eq. (7.15), by V, obtaining

$$V\frac{\partial(\rho A)}{\partial t} + V\frac{\partial(\rho V A)}{\partial x} = 0 \tag{7.24}$$

and then subtracting Eq. (7.24) from Eq. (7.23).

$$\frac{\partial(\rho V A)}{\partial t} - V\frac{\partial(\rho A)}{\partial t} + \frac{\partial(\rho V^2 A)}{\partial x} - V\frac{\partial(\rho V A)}{\partial x} = -A\frac{\partial p}{\partial x} \tag{7.25}$$

Expanding the derivatives on the left-hand side of Eq. (7.25) and canceling like terms, we have

$$\rho A \frac{\partial V}{\partial t} + \rho A V \frac{\partial V}{\partial x} = -A \frac{\partial p}{\partial x} \qquad (7.26)$$

Dividing Eq. (7.26) by A, we finally obtain

$$\boxed{\rho \frac{\partial V}{\partial t} + \rho V \frac{\partial V}{\partial x} = -\frac{\partial p}{\partial x}} \qquad (7.27)$$

Equation (7.27) is the momentum equation appropriate for quasi-one-dimensional flow, written in nonconservation form.

One of the reasons for obtaining the nonconservation form of the momentum equation is to compare it with the general result expressed by Eq. (2.83a). For one-dimensional flow with no body forces, Eq. (2.83a) is written as

$$\rho \frac{\partial u}{\partial t} + \rho u \frac{\partial u}{\partial x} = -\frac{\partial p}{\partial x} \qquad (7.28)$$

This is stylistically the same form as Eq. (7.27) for quasi-one-dimensional flow. Equations (7.27) and (7.28) simply demonstrate that the classic form of Euler's equation, generically written as

$$dp = -\rho V \, dV$$

holds for both types of flow.

Finally, let us consider the integral form of the energy equation, as obtained in Prob. 2.2. For an adiabatic flow ($\dot{q} = 0$) with no body forces and no viscous effects, the integral form of the energy equation is

$$\frac{\partial}{\partial t} \iiint_{\mathcal{V}} \rho\left(e + \frac{V^2}{2}\right) d\mathcal{V} + \iint_S \rho\left(e + \frac{V^2}{2}\right) \mathbf{V} \cdot \mathbf{dS} = -\iint_S (p\mathbf{V}) \cdot \mathbf{dS} \qquad (7.29)$$

Applied to the shaded control volume in Fig. 7.3, and keeping in mind the pressure forces shown in Fig. 7.4, Eq. (7.29) becomes

$$\frac{\partial}{\partial t}\left[\rho\left(e + \frac{V^2}{2}\right)A\,dx\right] - \rho\left(e + \frac{V^2}{2}\right)VA$$

$$+ (\rho + d\rho)\left[e + de + \frac{(V+dV)^2}{2}\right](V+dV)(A+dA)$$

$$= -\left[-pVA + (p+dp)(V+dV)(A+dA) - 2\left(pV\frac{dA}{2}\right)\right] \qquad (7.30)$$

Neglecting products of differentials and canceling like terms, Eq. (7.30) becomes

$$\frac{\partial}{\partial t}\left[\rho\left(e + \frac{V^2}{2}\right)A\,dx\right] + d(\rho e V A) + \frac{d(\rho V^3 A)}{2} = -d(pAV) \qquad (7.31)$$

or

$$\frac{\partial}{\partial t}\left[\rho\left(e+\frac{V^2}{2}\right)A\,dx\right] + d\left[\rho\left(e+\frac{V^2}{2}\right)VA\right] = -d(pAV) \qquad (7.32)$$

Taking the limit as dx approaches zero, Eq. (7.32), becomes the following partial differential equation:

$$\frac{\partial[\rho(e+V^2/2)A]}{\partial t} + \frac{\partial[\rho(e+V^2/2)VA]}{\partial x} = -\frac{\partial(pAV)}{\partial x} \qquad (7.33)$$

Equation (7.33) is the conservation form of the energy equation expressed in terms of the total energy $e + V^2/2$, appropriate for unsteady, quasi-one-dimensional flow. Let us obtain from Eq. (7.33) the nonconservation form expressed in terms of internal energy by itself. The latter can be achieved by multiplying Eq. (7.23) by V, obtaining

$$\frac{\partial[\rho(V^2/2)A]}{\partial t} + \frac{\partial[\rho(V^3/2)A]}{\partial x} = -AV\frac{\partial p}{\partial x} \qquad (7.34)$$

and subtracting Eq. (7.34) from (7.33), yielding

$$\frac{\partial(\rho eA)}{\partial t} + \frac{\partial(\rho eVA)}{\partial x} = -p\frac{\partial(AV)}{\partial x} \qquad (7.35)$$

Equation (7.35) is the conservation form of the energy equation expressed in terms of internal energy e, suitable for quasi-one-dimensional flow. The nonconservation form is then obtained by multiplying the continuity equation, Eq. (7.15), by e,

$$e\frac{\partial(\rho A)}{\partial t} + e\frac{\partial(\rho AV)}{\partial x} = 0 \qquad (7.36)$$

and subtracting Eq. (7.36) from Eq. (7.35), yielding

$$\rho A\frac{\partial e}{\partial t} + \rho AV\frac{\partial e}{\partial x} = -p\frac{\partial(AV)}{\partial x} \qquad (7.37)$$

Expanding the right-hand side and dividing by A, Eq. (7.37) becomes

$$\rho\frac{\partial e}{\partial t} + \rho V\frac{\partial e}{\partial x} = -p\frac{\partial V}{\partial x} - p\frac{V}{A}\frac{\partial A}{\partial x}$$

or

$$\boxed{\rho\frac{\partial e}{\partial t} + \rho V\frac{\partial e}{\partial x} = -p\frac{\partial V}{\partial x} - pV\frac{\partial(\ln A)}{\partial x}} \qquad (7.38)$$

Equation (7.38) is the nonconservation form of the energy equation expressed in terms of internal energy, appropriate to unsteady, quasi-one-dimensional flow.

The reason for obtaining the energy equation in the form of Eq. (7.38) is that, for a calorically perfect gas, it leads directly to a form of the energy equation in terms of temperature T. For our solution of the quasi-one-dimensional nozzle flow of a calorically perfect gas, this is a fundamental variable, and therefore it is convenient to deal with it as the primary dependent variable in the energy equation. For a calorically perfect gas

$$e = c_v T$$

Hence, Eq. (7.38) becomes

$$\rho c_v \frac{\partial T}{\partial t} + \rho V c_v \frac{\partial T}{\partial x} = -p \frac{\partial V}{\partial x} - pV \frac{\partial (\ln A)}{\partial x} \quad (7.39)$$

As an interim summary, our continuity, momentum, and energy equations for unsteady, quasi-one-dimensional flow are given by Eqs. (7.15), (7.27), and (7.39), respectively. Take the time to look at these equations; you see three equations with four unknown variables ρ, V, p, and T. The pressure can be eliminated from these equations by using the equation of state

$$p = \rho R T \quad (7.40)$$

along with its derivative

$$\frac{\partial p}{\partial x} = R \left(\rho \frac{\partial T}{\partial x} + T \frac{\partial \rho}{\partial x} \right) \quad (7.41)$$

With this, we expand Eq. (7.15) and rewrite Eqs. (7.27) and (7.39), respectively, as

Continuity: $$\frac{\partial (\rho A)}{\partial t} + \rho A \frac{\partial V}{\partial x} + \rho V \frac{\partial A}{\partial x} + V A \frac{\partial \rho}{\partial x} = 0 \quad (7.42)$$

Momentum: $$\rho \frac{\partial V}{\partial t} + \rho V \frac{\partial V}{\partial x} = -R \left(\rho \frac{\partial T}{\partial x} + T \frac{\partial \rho}{\partial x} \right) \quad (7.43)$$

Energy: $$\rho c_v \frac{\partial T}{\partial t} + \rho V c_v \frac{\partial T}{\partial x} = -\rho R T \left[\frac{\partial V}{\partial x} + V \frac{\partial (\ln A)}{\partial x} \right] \quad (7.44)$$

At this stage, we could readily proceed to set up our numerical solution of Eqs. (7.42) to (7.44). Note that these are written in terms of dimensional variables. This is fine, and many CFD solutions are carried out directly in terms of such dimensional variables. Indeed, this has an added engineering advantage because it gives you a feeling for the magnitudes of the real physical quantities as the solution progresses. However, for nozzle flows, the flow-field variables are frequently expressed in terms of nondimensional variables, such as those sketched in Fig. 7.2, where the flow variables are referenced to their reservoir values. The nondimensional variables p/p_0, ρ/ρ_0, and T/T_0 vary between 0 and 1, which is an "aesthetic" advantage when presenting the results. Because fluid dynamicists dealing with nozzle flows so frequently use these nondimensional terms, we will follow suit here. (A number of CFD practitioners prefer to always deal with nondimensional variables, whereas others prefer dimensional variables; as far as the numerics are concerned, there should be no real difference, and the choice is really a matter of your personal preference.) Therefore, returning to Fig. 7.1, where the reservoir temperature and density are denoted by T_0 and ρ_0, respectively, we define the nondimensional temperature and density, respectively, as

$$T' = \frac{T}{T_0} \qquad \rho' = \frac{\rho}{\rho_0}$$

where (for the time being) the prime denotes a dimensionless variable. Moreover, letting L denote the length of the nozzle, we define a dimensionless length as

$$x' = \frac{x}{L}$$

Denoting the speed of sound in the reservoir as a_0, where

$$a_0 = \sqrt{\gamma R T_0}$$

we define a dimensionless velocity as

$$V' = \frac{V}{a_0}$$

Also, the quantity L/a_0 has the dimension of time, and we define a dimensionless time as

$$t' = \frac{t}{L/a_0}$$

Finally, we ratio the local area A to the sonic throat area A^* and define a dimensionless area as

$$A' = \frac{A}{A^*}$$

Returning to Eq. (7.42) and introducing the nondimensional variables, we have

$$\frac{\partial(\rho' A')}{\partial t'}\left(\frac{\rho_0 A^*}{L/a_0}\right) + \rho' A' \frac{\partial V'}{\partial x'}\left(\frac{\rho_0 A^* a_0}{L}\right) + \rho' V' \frac{\partial A'}{\partial x'}\left(\frac{\rho_0 a_0 A^*}{L}\right)$$
$$+ V' A' \frac{\partial \rho'}{\partial x'}\left(\frac{a_0 A^* \rho_0}{L}\right) = 0 \quad (7.45)$$

Note that A' is a function of x' only; it is *not* a function of time (the nozzle geometry is fixed, invariant with time). Hence, in Eq. (7.45) the time derivative can be written as

$$\frac{\partial(\rho' A')}{\partial t'} = A' \frac{\partial \rho'}{\partial t'}$$

With this, Eq. (7.45) becomes

Continuity:
$$\frac{\partial \rho'}{\partial t'} = -\rho' \frac{\partial V'}{\partial x'} - \rho' V' \frac{\partial (\ln A')}{\partial x'} - V' \frac{\partial \rho'}{\partial x'} \qquad (7.46)$$

Returning to Eq. (7.43) and introducing the nondimensional variables, we have

$$\rho' \frac{\partial V'}{\partial t'} \left(\frac{\rho_0 a_0}{L/a_0}\right) + \rho' V' \frac{\partial V'}{\partial x'} \left(\frac{\rho_0 a_0^2}{L}\right) = -R \left(\rho' \frac{\partial T'}{\partial x'} + T' \frac{\partial \rho'}{\partial x'}\right) \left(\frac{\rho_0 T_0}{L}\right)$$

or
$$\rho' \frac{\partial V'}{\partial t'} = -\rho' V' \frac{\partial V'}{\partial x'} - \left(\rho' \frac{\partial T'}{\partial x'} + T' \frac{\partial \rho'}{\partial x'}\right) \frac{RT_0}{a_0^2} \qquad (7.47)$$

In Eq. (7.47), note that

$$\frac{RT_0}{a_0^2} = \frac{\gamma RT_0}{\gamma a_0^2} = \frac{a_0^2}{\gamma a_0^2} = \frac{1}{\gamma}$$

Hence, Eq. (7.47) becomes

Momentum:
$$\frac{\partial V'}{\partial t'} = -V' \frac{\partial V'}{\partial x'} - \frac{1}{\gamma} \left(\frac{\partial T'}{\partial x'} + \frac{T'}{\rho'} \frac{\partial \rho'}{\partial x'}\right) \qquad (7.48)$$

Returning to Eq. (7.44) and introducing the nondimensional variables, we have

$$\rho' c_v \frac{\partial T'}{\partial t'} \left(\frac{\rho_0 T_0}{L/a_0}\right) + \rho' V' c_v \frac{\partial T'}{\partial x'} \left(\frac{\rho_0 a_0 T_0}{L}\right)$$
$$= -\rho' RT' \left[\frac{\partial V'}{\partial x'} + V' \frac{\partial (\ln A')}{\partial x'}\right] \left(\frac{\rho_0 T_0 a_0}{L}\right) \qquad (7.49)$$

In Eq. (7.49), the factor R/c_v is given by

$$\frac{R}{c_v} = \frac{R}{R/(\gamma - 1)} = \gamma - 1$$

Hence, Eq. (7.49) becomes

Energy:
$$\frac{\partial T'}{\partial t'} = -V' \frac{\partial T'}{\partial x'} - (\gamma - 1)T' \left[\frac{\partial V'}{\partial x'} + V' \frac{\partial (\ln A')}{\partial x'}\right] \qquad (7.50)$$

That is it! We are finally finished with the first eschelon as itemized at the beginning of this subsection. After what may seem like an interminable manipulation of the governing equations, we have finally set up that particular form of the equations that will be most appropriate as well as convenient for the time-marching solution of quasi-one-dimensional nozzle flow, namely, Eqs. (7.46), (7.48), and (7.50).

THE FINITE-DIFFERENCE EQUATIONS. We now proceed to the next echelon, namely, the setting up of the finite-difference expressions using MacCormack's explicit technique for the numerical solution of Eqs. (7.46), (7.48), and (7.50). To implement a finite-difference solution, we divide the x axis along the nozzle into a number of discrete grid points, as shown in Fig. 7.5. (Recall that in our quasi-one-dimensional nozzle assumption, the flow variables *across* the nozzle cross section at any particular grid point, say point i, are uniform.) In Fig. 7.5, the first grid point, labeled point 1, is assumed to be in the reservoir. The points are evenly distributed along the x axis, with Δx denoting the spacing between grid points. The last point, namely, that at the nozzle exit, is denoted by N; we have a total number of N grid points distributed along the axis. Point i is simply an arbitrary grid point, with points $i-1$ and $i+1$ as the adjacent points. Recall from Sec. 6.3 that MacCormack's technique is a predictor-corrector method. In the time-marching approach, remember that we know the flow-field variables at time t, and we use the difference equations to solve explicitly for the variables at time $t + \Delta t$.

First, consider the predictor step. Following the discussion in Sec. 6.3, we set up the spatial derivatives as forward differences. Also, to reduce the complexity of the notation, we will drop the use of the prime to denote a dimensionless variable. In what follows, *all* variables are the nondimensional variables, denoted earlier by the prime notation. Analogous to Eq. (6.17), from Eq. (7.46) we have

$$\left(\frac{\partial \rho}{\partial t}\right)_i^t = -\rho_i^t \frac{V_{i+1}^t - V_i^t}{\Delta x} - \rho_i^t V_i^t \frac{\ln A_{i+1} - \ln A_i}{\Delta x} - V_i^t \frac{\rho_{i+1}^t - \rho_i^t}{\Delta x} \quad (7.51)$$

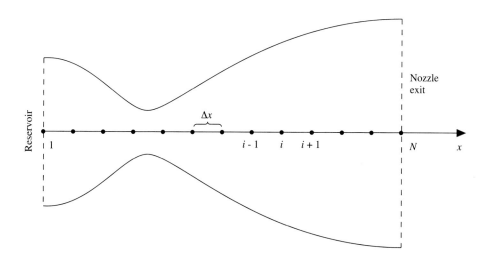

FIG. 7.5
Grid point distribution along the nozzle.

From Eq. (7.48), we have

$$\left(\frac{\partial V}{\partial t}\right)_i^t = -V_i^t \frac{V_{i+1}^t - V_i^t}{\Delta x} - \frac{1}{\gamma}\left(\frac{T_{i+1}^t - T_i^t}{\Delta x} + \frac{T_i^t}{\rho_i^t}\frac{\rho_{i+1}^t - \rho_i^t}{\Delta x}\right) \quad (7.52)$$

From Eq. (7.50), we have

$$\left(\frac{\partial T}{\partial t}\right)_i^t = -V_i^t \frac{T_{i+1}^t - T_i^t}{\Delta x} - (\gamma - 1)T_i^t\left(\frac{V_{i+1}^t - V_i^t}{\Delta x} + V_i^t \frac{\ln A_{i+1} - \ln A_i}{\Delta x}\right) \quad (7.53)$$

Analogous to Eqs. (6.18) to (6.21), we obtain predicted values of ρ, V, and T, denoted by barred quantities, from

$$\bar{\rho}_i^{t+\Delta t} = \rho_i^t + \left(\frac{\partial \rho}{\partial t}\right)_i^t \Delta t \quad (7.54)$$

$$\bar{V}_i^{t+\Delta t} = V_i^t + \left(\frac{\partial V}{\partial t}\right)_i^t \Delta t \quad (7.55)$$

$$\bar{T}_i^{t+\Delta t} = T_i^t + \left(\frac{\partial T}{\partial t}\right)_i^t \Delta t \quad (7.56)$$

In Eqs. (7.54) to (7.56), ρ_i^t, V_i^t, and T_i^t are known values at time t. Numbers for the time derivatives in Eqs. (7.54) to (7.56) are supplied directly by Eqs. (7.51) to (7.53).

Moving to the corrector step, we return to Eqs. (7.46), (7.48), and (7.50) and replace the spatial derivatives with rearward differences, using the predicted (barred) quantities. Analogous to Eq. (6.22), we have from Eq. (7.46)

$$\left(\frac{\overline{\partial \rho}}{\partial t}\right)_i^{t+\Delta t} = -\bar{\rho}_i^{t+\Delta t}\frac{\bar{V}_i^{t+\Delta t} - \bar{V}_{i-1}^{t+\Delta t}}{\Delta x} - \bar{\rho}_i^{t+\Delta t}\bar{V}_i^{t+\Delta t}\frac{\ln A_i - \ln A_{i-1}}{\Delta x}$$

$$- \bar{V}_i^{t+\Delta t}\frac{\bar{\rho}_i^{t+\Delta t} - \bar{\rho}_{i-1}^{t+\Delta t}}{\Delta x} \quad (7.57)$$

From Eq. (7.48), we have

$$\left(\frac{\overline{\partial V}}{\partial t}\right)_i^{t+\Delta t} = -\bar{V}_i^{t+\Delta t}\frac{\bar{V}_i^{t+\Delta t} - \bar{V}_{i-1}^{t+\Delta t}}{\Delta x} - \frac{1}{\gamma}\left(\frac{\bar{T}_i^{t+\Delta t} - \bar{T}_{i-1}^{t+\Delta t}}{\Delta x} + \frac{\bar{T}_i^{t+\Delta t}}{\bar{\rho}_i^{t+\Delta t}}\frac{\bar{\rho}_i^{t+\Delta t} - \bar{\rho}_{i-1}^{t+\Delta t}}{\Delta x}\right)$$

$$(7.58)$$

From Eq. (7.50), we have

$$\left(\frac{\overline{\partial T}}{\partial t}\right)_i^{t+\Delta t} = -\bar{V}_i^{t+\Delta t}\frac{\bar{T}_i^{t+\Delta t} - \bar{T}_{i-1}^{t+\Delta t}}{\Delta x} - (\gamma - 1)\bar{T}_i^{t+\Delta t}$$

$$\times \left(\frac{\bar{V}_i^{t+\Delta t} - \bar{V}_{i-1}^{t+\Delta t}}{\Delta x} + \bar{V}_i^{t+\Delta t}\frac{\ln A_i - \ln A_{i-1}}{\Delta x}\right) \quad (7.59)$$

Analogous to Eq. (6.22), the average time derivatives are given by

$$\left(\frac{\partial \rho}{\partial t}\right)_{av} = 0.5\left[\underbrace{\left(\frac{\partial \rho}{\partial t}\right)_i^t}_{\text{From Eq. (7.51)}} + \underbrace{\left(\overline{\frac{\partial \rho}{\partial t}}\right)_i^{t+\Delta t}}_{\text{From Eq. (7.57)}}\right] \quad (7.60)$$

$$\left(\frac{\partial V}{\partial t}\right)_{av} = 0.5\left[\underbrace{\left(\frac{\partial V}{\partial t}\right)_i^t}_{\text{From Eq. (7.52)}} + \underbrace{\left(\overline{\frac{\partial V}{\partial t}}\right)_i^{t+\Delta t}}_{\text{From Eq. (7.58)}}\right] \quad (7.61)$$

$$\left(\frac{\partial T}{\partial t}\right)_{av} = 0.5\left[\underbrace{\left(\frac{\partial T}{\partial t}\right)_i^t}_{\text{From Eq. (7.53)}} + \underbrace{\left(\overline{\frac{\partial T}{\partial t}}\right)_i^{t+\Delta t}}_{\text{From Eq. (7.59)}}\right] \quad (7.62)$$

Finally, analogous to Eqs. (6.13) to (6.16), we have for the corrected values of the flow-field variables at time $t + \Delta t$

$$\rho_i^{t+\Delta t} = \rho_i^t + \left(\frac{\partial \rho}{\partial t}\right)_{av} \Delta t \quad (7.63)$$

$$V_i^{t+\Delta t} = V_i^t + \left(\frac{\partial V}{\partial t}\right)_{av} \Delta t \quad (7.64)$$

$$T_i^{t+\Delta t} = T_i^t + \left(\frac{\partial T}{\partial t}\right)_{av} \Delta t \quad (7.65)$$

Keep in mind that all the variables in Eqs. (7.51) to (7.65) are the *nondimensional* values. Also, Eqs. (7.51) to (7.65) constitute our second eschelon of equations, namely, the finite-difference expressions of the governing equations in a form that pertains to MacCormack's technique.

CALCULATION OF TIME STEP. We now proceed to the third and final eschelon of equations mentioned at the beginning of this section, namely, the setting up of other details necessary for the numerical solution of the quasi-one-dimensional nozzle flow problem. First, we ask the question: What about the magnitude of Δt? The governing system of equations, Eqs. (7.42) to (7.44), is hyperbolic with respect to time. Recalling our discussion of stability considerations in Sec. 4.5, a stability constraint exists on this system analogous to that found in Eq. (4.84), namely,

$$\Delta t = C \frac{\Delta x}{a + V} \quad (7.66)$$

Recall from Sec. 4.5 that C is the *Courant number*; the simple stability analysis of a linear hyperbolic equation carried out in Sec. 4.5 gives the result that $C \leq 1$ for an explicit numerical solution to be stable. The present application to subsonic-supersonic isentropic nozzle flow is governed by *nonlinear* partial differential equations, namely, Eqs. (7.46), (7.48), and (7.50). In this case, the exact stability criterion for a linear equation, namely, that $C \leq 1$, can only be viewed as general guidance for our present nonlinear problem. However, it turns out to be quite good guidance, as we shall see. Also note that, in contrast to Eq. (4.84), Eq. (7.66) is written with the sum $a + V$ in the denominator. Equation (7.66) is the *Courant-Friedrichs-Lowry (CFL) criterion* for a one-dimensional flow, where V is the local flow velocity at a point in the flow and a is the local speed of sound. Equation (7.66), along with $C \leq 1$, simply states that Δt must be less than, or at best equal to, the time it takes a sound wave to move from one grid point to the next. Equation (7.66) is in dimensional form. However, when t, x, a, and V are nondimensionalized, the nondimensional form of Eq. (7.66) is exactly the same form as the dimensional case. (Prove this to yourself.) Hence, we will hereafter treat the variables in Eq. (7.66) as our nondimensional variables defined earlier. That is, in Eq. (7.66), Δt is the increment in nondimensional time and Δx is the increment in nondimensional space; Δt and Δx in Eq. (7.66) are precisely the same as appear in the nondimensional equations (7.51) to (7.65). Examining Eq. (7.66) more carefully, we note that, although Δx is the same throughout the flow, both V and a are variables. Hence, at a given grid point at a given time step, Eq. (7.66) is written as

$$(\Delta t)_i^t = C \frac{\Delta x}{a_i^t + V_i^t} \qquad (7.67)$$

At an adjacent grid point, we have from Eq. (7.66)

$$(\Delta t)_{i+1}^t = C \frac{\Delta x}{a_{i+1}^t + V_{i+1}^t} \qquad (7.68)$$

Clearly, $(\Delta t)_i^t$ and $(\Delta t)_{i+1}^t$ obtained from Eqs. (7.67) and (7.68), respectively are, in general, different values. Hence, in the implementation of the time-marching solution, we have two choices:

1. In utilizing Eqs. (7.54) to (7.56) and (7.63) to (7.65), we can, at each grid point i, employ the *local* values of $(\Delta t)_i^t$ determined from Eq. (7.67). In this fashion, the flow-field variables at each grid point in Fig. 7.5 will be advanced in time according to their own, local time step. Hence, the resulting flow field at time $t + \Delta t$ will be in a type of *artificial "time warp,"* with the flow-field variables at a given grid point corresponding to some nonphysical time different from that of the variables at an adjacent grid point. Clearly, such a *local time-stepping* approach does not realistically follow the *actual, physical transients* in the flow and hence cannot be used for an accurate solution of the *unsteady* flow. However, if the final steady-state flow field in the limit of large time is the only desired result, then the intermediate variation of the flow-field variables with time is irrelevant. Indeed, if such is the case, the *local* time stepping will frequently lead

to *faster* convergence to the steady state. This is why some practitioners use the local time-stepping approach. However, there is always a philosophical question that arises here, namely, does the *local* time-stepping method always lead to the *correct* steady state? Although the answer is usually yes, there is still some reason for a small feeling of discomfort in this regard.

2. The other choice is to calculate $(\Delta t)_i^t$ at all the grid points, $i = 1$ to $i = N$, and then choose the *minimum* value for use in Eqs. (7.54) to (7.56) and (7.63) to (7.65). That is,

$$\Delta t = \text{minimum}(\Delta t_1^t, \Delta t_2^t, \ldots, \Delta t_i^t, \ldots, \Delta t_N^t) \qquad (7.69)$$

The resulting Δt obtained from Eq. (7.69) is then used in Eqs. (7.54) to (7.56) and (7.63) to (7.65). In this fashion, the flow-field variables at all the grid points at time $t + \Delta t$ all correspond to the *same* physical time. Hence, the time-marching solution is following the actual unsteady flow variations that would exist in nature; i.e., the solution gives a time-accurate solution of the actual transient flow field, consistent with the unsteady continuity, momentum, and energy equations. This consistent time marching is the approach we will use in the present book. Although it may require more time steps to approach the steady state in comparison to the "local" time stepping described earlier, we can feel comfortable that the consistent time-marching approach is giving us the physically meaningful transient variations—which frequently are of intrinsic value by themselves. Thus, in our subsequent calculations, we will use Eq. (7.69) to determine the value of Δt.

BOUNDARY CONDITIONS. Another aspect of the numerical solution is that of *boundary conditions*—an all-important aspect, because without the physically proper implementation of boundary conditions and their numerically proper representation, we have no hope whatsoever in obtaining a proper numerical solution to our flow problem. First, let us examine the physical boundary conditions for the subsonic-supersonic isentropic flow shown in Fig. 7.2, which is the subject of this section. Returning to Fig. 7.5, we note that grid points 1 and N represent the two boundary points on the x axis. Point 1 is essentially in the reservoir; it represents an *inflow* boundary, with flow coming from the reservoir and entering the nozzle. In contrast, point N is an *outflow* boundary, with flow leaving the nozzle at the nozzle exit. Moreover, the flow velocity at point 1 is a very low, subsonic value. (The flow velocity at point 1, which corresponds to a finite area ratio A_1/A^*, cannot be precisely zero; if it were, there would be no mass flow entering the nozzle. Hence, point 1 does not correspond *exactly* to the reservoir, where by definition the flow velocity is zero. That is, the area for the reservoir is theoretically infinite, and we are clearly starting our own calculation at point 1 where the cross-sectional area is finite.) Hence, not only is point 1 an *inflow* boundary, it is a *subsonic* inflow boundary. *Question*: Which flow quantities should be specified at this subsonic inflow boundary and which should be calculated as part of the solution (i.e., allowed to "float" as a function of time)? A formal answer can be obtained by using the method of characteristics for an unsteady, one-dimensional flow, as introduced in

Chap. 3. We did not develop the method of characteristics in Chap. 3 to the extent necessary to precisely study this question about the boundary conditions; indeed, such a matter is beyond the scope of this book. However, we will mention the result of such a study, which you will find to be physically acceptable. In a subsection of Sec. 3.4.1, we indicated that unsteady, inviscid flow is governed by hyperbolic equations, and therefore for one-dimensional unsteady flow there exist two real characteristic lines through any point in the xt plane. This is illustrated in Fig. 3.6; return to this figure and examine it carefully before continuing on. Note that the two characteristic lines through point P in Fig. 3.6 are labeled left- and right-running characteristics, respectively. Physically, these two characteristics represent infinitely weak Mach waves which are propagating upstream and downstream, respectively. Both Mach waves are traveling at the speed of sound a. Now turn to Fig. 7.6, which shows our convergent-divergent nozzle (Fig. 7.6a) with an xt diagram sketched below it (Fig. 7.6b). Concentrate on grid point 1 in the xt plane in Fig. 7.6b. At point

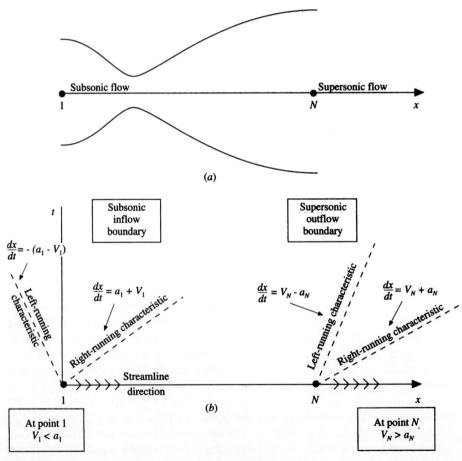

FIG. 7.6
Study of boundary conditions: subsonic inflow and supersonic outflow.

1, the local flow velocity is subsonic, $V_1 < a_1$. Hence, the left-running characteristic at point 1 travels *upstream*, to the left in Fig. 7.6; i.e., the left-running Mach wave, which is traveling toward the left (relative to a moving fluid element) at the speed of sound easily works its way *upstream* against the low-velocity subsonic flow, which is slowly moving from left to right. Hence, in Fig. 7.6b, we show the left-running characteristic running to the left with a combined speed $a_1 - V_1$ (relative to the fixed nozzle in Fig. 7.6a). Since the domain for the flow field to be calculated is contained between grid points 1 and N, then at point 1 we see that the left-running characteristic is propagating *out of* the domain; it is propagating to the left, away from the domain. In contrast, the right-running characteristic, which is a Mach wave propagating to the right at the speed of sound relative to a fluid element, is clearly moving toward the right in Fig. 7.6b. This is for two reasons: (1) the fluid element at point 1 is already moving toward the right, and (2) the right-running Mach wave (characteristic) is moving toward the right at the speed of sound relative to the fluid element. Hence, the right-running characteristic is propagating to the right (relative to the nozzle) at a combined velocity of $V_1 + a_1$. What we see here is that the right-running characteristic is propagating from point 1 *into* the domain of the calculation.

What does all this have to do with boundary conditions? The method of characteristics tells us that at a boundary where one characteristic propagates *into* the domain, then the value of one dependent flow-field variable must be *specified* at that boundary, and if one characteristic line propagates *out* of the domain, then the value of another dependent flow-field variable must be allowed to *float* at the boundary; i.e., it must be calculated in steps of time as a function of the timewise solution of the flow field. Also, note that at point 1 a streamline flows *into* the domain, across the inflow boundary. In terms of denoting what should and should not be specified at the boundary, the streamline *direction* plays the same role as the characteristic directions; i.e., the streamline moving *into* the domain at point 1 stipulates that the value of a second flow-field variable must be *specified* at the inflow boundary. *Conclusion*: At the *subonic inflow boundary*, we must *stipulate* the values of *two* dependent flow-field variables, whereas the value of *one* other variable must be allowed to *float*. (Please note that the above discussion has been intentionally hand-waving and somewhat intuitive; a rigorous mathematical development is deferred for your future studies, beyond the scope of this book.)

Let us apply the above ideas to the *outflow* boundary, located at grid point N in Fig. 7.6. As before, the left-running characteristic at point N propagates to the left at the speed of sound *a relative to a fluid element*. However, because the speed of the fluid element itself is supersonic, the left-running characteristic is carried *downstream* at the speed (relative to the nozzle) of $V_N - a_N$. The right-running characteristic at point N propagates to the right at the speed of sound a relative to the fluid element, and thus it is swept downstream at the speed (relative to the nozzle) of $V_N + a_N$. Hence, at the *supersonic outflow boundary*, we have both characteristics propagating *out* of the domain; so does the streamline at point N. Therefore, there are *no* flow-field variables which require their values to be stipulated at the supersonic outflow boundary; *all* variables must be allowed to *float* at this boundary.

The above discussion details how the inflow and outflow boundary conditions are to be handled on an *analytical* basis. The *numerical* implementation of this discussion is carried out as follows.

Subsonic inflow boundary (point 1). Here, we must allow one variable to float; we choose the velocity V_1, because on a physical basis we know the mass flow through the nozzle must be allowed to adjust to the proper steady state, and allowing V_1 to float makes the most sense as part of this adjustment. The value of V_1 changes with time and is calculated from information provided by the flow-field solution over the internal points. (The *internal* points are those *not* on a boundary, i.e., points 2 through $N-1$ in Fig. 7.5). We use linear extrapolation from points 2 and 3 to calculate V_1. This is illustrated in Fig. 7.7. Here, the slope of the linear extrapolation line is determined from points 2 and 3 as

$$\text{Slope} = \frac{V_3 - V_2}{\Delta x}$$

Using this slope to find V_1 by linear extrapolation, we have

$$V_1 = V_2 - \frac{V_3 - V_2}{\Delta x} \Delta x$$

or
$$V_1 = 2V_2 - V_3 \tag{7.70}$$

All other flow-field variables are specified. Since point 1 is viewed as essentially the reservoir, we stipulate the density and temperature at point 1 to be their respective stagnation values, ρ_0 and T_0, respectively. These are held *fixed*, independent of time. Hence, in terms of the *nondimensional* variables, we have

$$\left.\begin{array}{l} \rho_1 = 1 \\ T_1 = 1 \end{array}\right\} \text{fixed, independent of time} \tag{7.71}$$

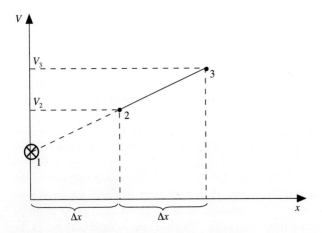

FIG. 7.7
Sketch for linear extrapolation.

Supersonic outflow boundary (point N). Here, we must allow *all* flow-field variables to float. We again choose to use linear extrapolation based on the flow-field values at the internal points. Specifically, we have, for the *nondimensional* variables,

$$V_N = 2V_{N-1} - V_{N-2} \qquad (7.72a)$$
$$\rho_N = 2\rho_{N-1} - \rho_{N-2} \qquad (7.72b)$$
$$T_N = 2T_{N-1} - T_{N-2} \qquad (7.72c)$$

NOZZLE SHAPE AND INITIAL CONDITIONS. The nozzle shape, $A = A(x)$, is specified and held fixed, independent of time. For the case illustrated in this section, we choose a parabolic area distribution given by

$$A = 1 + 2.2(x - 1.5)^2 \qquad 0 \le x \le 3 \qquad (7.73)$$

Note that $x = 1.5$ is the throat of the nozzle, that the convergent section occurs for $x < 1.5$, and that the divergent section occurs for $x > 1.5$. This nozzle shape is drawn to scale in Fig. 7.8.

To start the time-marching calculations, we must stipulate *initial* conditions for ρ, T, and V as a function of x; that is, we must set up values of ρ, T, and V at time $t = 0$. In *theory*, these initial conditions can be purely arbitrary. In practice, there are

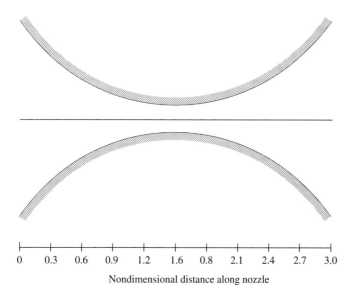

FIG. 7.8
Shape of the nozzle used for the present calculations. This geometric picture is not unique; for a calorically perfect gas, what is germane is the area ratio distribution along the nozzle. Hence, assuming a two-dimensional nozzle, the ordinates of the shape shown here can be ratioed by any constant factor, and the nozzle solution would be the same.

two reasons why you want to choose the initial conditions *intelligently*:

1. The closer the initial conditions are to the final steady-state answer, the faster the time-marching procedure will converge, and hence the shorter will be the computer execution time.
2. If the initial conditions are too far away from reality, the initial timewise gradients at early time steps can become huge; i.e., the *time derivatives* themselves are initially very large. For a given time step Δt and a given spatial resolution Δx, it has been the author's experience that *inordinately* large gradients during the early part of the time-stepping procedure can cause the program to go unstable. In a sense, you can visualize the behavior of a time-marching solution as a stretched rubber band. At early times, the rubber band is highly stretched, thus providing a *strong* potential to push the flow field *rapidly* toward the steady-state solution. As time progresses, the flow field gets closer to the steady-state solution, and the rubber band progressively relaxes, hence slowing down the rate of approach [i.e., at larger times, the values of the time derivatives calculated from Eqs. (7.60) to (7.62) become progressively smaller]. At the beginning of the calculation, it is wise not to pick initial conditions which are so far off that the rubber band is "stretched too far," and may even break.

Therefore, in your choice of initial conditions, you are encouraged to use *any* knowledge you may have about a given problem in order to intelligently pick some initial conditions. For example, in the present problem, we know that ρ and T *decrease* and *V increases* as the flow expands through the nozzle. Hence, we choose initial conditions that *qualitatively* behave in the same fashion. For simplicity, let us assume linear variations of the flow-field variables, as a function of x. For the present case, we assume the following values at time $t = 0$.

$$\rho = 1 - 0.3146x \qquad (7.74a)$$
$$T = 1 - 0.2314x \qquad \text{initial conditions at } t = 0 \qquad (7.74b)$$
$$V = (0.1 + 1.09x)T^{1/2} \qquad (7.74c)$$

7.3.2 Intermediate Numerical Results: The First Few Steps

In this section, we give a few numerical results which reflect the first stages of the calculation. This is to give you a more solid impression of what is going on and to provide some intermediate results for you to compare with when you write and run your own computer solution to this problem.

The first step is to feed the nozzle shape and the initial conditions into the program. These are given by Eqs. (7.73) and (7.74); the resulting numbers are tabulated in Table 7.1. The values of ρ, V, and T given in this table are for $t = 0$.

The next step is to put these initial conditions into Eqs. (7.51) to (7.53) to initiate calculations pertaining to the predictor step. For purposes of illustration, let us return to the sketch shown in Fig. 7.5 and focus on the calculations associated

TABLE 7.1
Nozzle shape and initial conditions

$\dfrac{x}{L}$	$\dfrac{A}{A^*}$	$\dfrac{\rho}{\rho_0}$	$\dfrac{V}{a_0}$	$\dfrac{T}{T_0}$
0	5.950	1.000	0.100	1.000
0.1	5.312	0.969	0.207	0.977
0.2	4.718	0.937	0.311	0.954
0.3	4.168	0.906	0.412	0.931
0.4	3.662	0.874	0.511	0.907
0.5	3.200	0.843	0.607	0.884
0.6	2.782	0.811	0.700	0.861
0.7	2.408	0.780	0.790	0.838
0.8	2.078	0.748	0.877	0.815
0.9	1.792	0.717	0.962	0.792
1.0	1.550	0.685	1.043	0.769
1.1	1.352	0.654	1.122	0.745
1.2	1.198	0.622	1.197	0.722
1.3	1.088	0.591	1.268	0.699
1.4	1.022	0.560	1.337	0.676
1.5	1.000	0.528	1.402	0.653
1.6	1.022	0.497	1.463	0.630
1.7	1.088	0.465	1.521	0.607
1.8	1.198	0.434	1.575	0.583
1.9	1.352	0.402	1.625	0.560
2.0	1.550	0.371	1.671	0.537
2.1	1.792	0.339	1.713	0.514
2.2	2.078	0.308	1.750	0.491
2.3	2.408	0.276	1.783	0.468
2.4	2.782	0.245	1.811	0.445
2.5	3.200	0.214	1.834	0.422
2.6	3.662	0.182	1.852	0.398
2.7	4.168	0.151	1.864	0.375
2.8	4.718	0.119	1.870	0.352
2.9	5.312	0.088	1.870	0.329
3.0	5.950	0.056	1.864	0.306

with grid point i. We will choose $i = 16$, which is the grid point at the throat of the nozzle drawn in Fig. 7.8. From the initial data given in Table 7.1, we have

$$\rho_i = \rho_{16} = 0.528$$
$$\rho_{i+1} = \rho_{17} = 0.497$$
$$V_i = V_{16} = 1.402$$
$$V_{i+1} = V_{17} = 1.463$$
$$T_i = T_{16} = 0.653$$
$$T_{i+1} = T_{17} = 0.630$$
$$\Delta x = 0.1$$
$$A_i = A_{16} = 1.0 \quad \ln A_{16} = 0$$
$$A_{i+1} = A_{17} = 1.022 \quad \ln A_{17} = 0.02176$$

Substitute these values into Eq. (7.51).

$$\left(\frac{\partial \rho}{\partial t}\right)^{t=0}_{16} = -0.528\left(\frac{1.463 - 1.402}{0.1}\right) - 0.528(1.402)\left(\frac{0.02176 - 0}{0.1}\right)$$

$$- 1.402\left(\frac{0.497 - 0.528}{0.1}\right)$$

$$= \boxed{-0.0445}$$

Substitute the above values into Eq. (7.52)

$$\left(\frac{\partial V}{\partial t}\right)^{t=0}_{16} = -1.402\left(\frac{1.463 - 1.402}{0.1}\right)$$

$$- \frac{1}{1.4}\left[\frac{0.630 - 0.653}{0.1} + \frac{0.653}{0.528}\left(\frac{0.497 - 0.528}{0.1}\right)\right]$$

$$= \boxed{-0.418}$$

Substitute the above values into Eq. (7.53).

$$\left(\frac{\partial T}{\partial t}\right)^{t=0}_{16} = -1.402\left(\frac{0.630 - 0.653}{0.1}\right) - (1.4 - 1)(0.653)$$

$$\times \left[\frac{1.463 - 1.402}{0.1} + 1.402\left(\frac{0.02176 - 0}{0.1}\right)\right]$$

$$= \boxed{0.0843}$$

Please note: The numbers shown in the *boxes* above are the precise numbers, rounded to three significant figures, that came out of the author's Macintosh computer. If you choose to run through the above calculations with your hand calculator using all the above entries, there will be slight differences because the numbers you feed into the calculator are *already* rounded to three significant figures, and hence the subsequent arithmetic operations on your calculator will lead to slight errors compared to the computer results. That is, your hand-calculator results may not always give you *precisely* the numbers you will find in the boxes, but they will certainly be close enough to check the results.

The next step is to calculate the *predicted* values (the "barred" quantities) from Eqs. (7.54) to (7.56). To do this, we first note that Δt is calculated from Eq. (7.69), which picks the minimum value of Δt_i from all those calculated from Eq. (7.67) evaluated for all internal points $i = 2, 3, \ldots, 30$. We do not have the space to show all these calculations here. As a sample calculation, let us calculate $(\Delta t)^{t=0}_{16}$ from Eq. (7.67). At present, we will assume a Courant number equal to 0.5; that is, $C = 0.5$. Also, in nondimensional terms, the speed of sound is given by

$$a = \sqrt{T} \tag{7.75}$$

where in Eq. (7.75) both a and T are the *nondimensional* values (a denotes the local speed of sound divided by a_0). Derive Eq. (7.75) for yourself. Thus, from Eq. (7.67), we have

$$(\Delta t)_{16}^{t=0} = C\left[\frac{\Delta x}{(T_{16})^{1/2} + V_{16}}\right] = 0.5\left[\frac{0.1}{(0.653)^{1/2} + 1.402}\right] = 0.0226$$

This type of calculation is made at all the interior grid points, and the minimum value is chosen. The resulting minimum value is

$$\Delta t = 0.0201$$

With this, we can calculate $\bar{\rho}$, $\bar{V}$, and $\bar{T}$ as follows. From Eq. (7.54), noting that $t = 0 + \Delta t = \Delta t$,

$$\bar{\rho}_{16}^{t=\Delta t} = \rho_{16}^{t=0} + \left(\frac{\partial \rho}{\partial t}\right)_{16}^{t=0} \Delta t = 0.528 + (-0.0445)(0.0201)$$

$$= \boxed{0.527}$$

From Eq. (7.55)

$$\bar{V}_{16}^{t=\Delta t} = V_{16}^{t=0} + \left(\frac{\partial V}{\partial t}\right)_{16}^{t=0} \Delta t = 1.402 + (-0.418)(0.0201)$$

$$= \boxed{1.39}$$

From Eq. (7.56)

$$\bar{T}_{16}^{t=\Delta t} = T_{16}^{t=0} + \left(\frac{\partial T}{\partial t}\right)_{16}^{t=0} \Delta t = 0.653 + (0.0843)(0.0201)$$

$$= \boxed{0.655}$$

At this stage, we note that the above calculations are carried out over *all* the internal grid points $i = 2$ to 30. The calculations are too repetitive to include here. Simply note that when the predictor step is completed, we have $\bar{\rho}$, $\bar{V}$, and $\bar{T}$ at all the internal grid points $i = 2$ to 30. This includes, of course, $\bar{\rho}_{15}^{t=\Delta t}$, $\bar{V}_{15}^{t=\Delta t}$, and $\bar{T}_{15}^{t=\Delta t}$. Focusing again on grid point 16, we now insert these *barred* quantities at grid points 15 and 16 into Eqs. (7.57) to (7.59). This is the beginning of the corrector step. From Eq. (7.57) we have

$$\overline{\left(\frac{\partial \rho}{\partial t}\right)}_{16}^{t=\Delta t} = -0.527(0.653) - 0.527(1.39)(-0.218) - 1.39(-0.368)$$

$$= \boxed{0.328}$$

From Eq. (7.58) we have

$$\overline{\left(\frac{\partial V}{\partial t}\right)}_{16}^{t=\Delta t} = -1.39(0.653) - \frac{1}{1.4}\left(-0.257 + \frac{0.655}{0.527}\right) = \boxed{-0.400}$$

From Eq. (7.59) we have

$$\left(\frac{\partial T}{\partial t}\right)^{t=\Delta t}_{16} = -1.39(-0.257) - (1.4-1)(0.655)[0.653 + 1.39(-0.218)]$$
$$= \boxed{0.267}$$

With these values, we form the *average* time derivatives using Eqs. (7.60) to (7.62). From Eq. (7.60), we have at grid point $i = 16$,

$$\left(\frac{\partial \rho}{\partial t}\right)_{av} = 0.5(-0.0445 + 0.328) = \boxed{0.142}$$

From Eq. (7.61), we have at grid point $i = 16$,

$$\left(\frac{\partial V}{\partial t}\right)_{av} = 0.5(-0.418 - 0.400) = \boxed{-0.409}$$

From Eq. (7.62), we have at grid point $i = 16$,

$$\left(\frac{\partial T}{\partial t}\right)_{av} = 0.5(0.0843 + 0.267) = \boxed{0.176}$$

We now complete the corrector step by using Eqs. (7.63) to (7.65). From Eq. (7.63), we have at $i = 16$,

$$\rho^{t=\Delta t}_{16} = 0.528 + 0.142(0.0201) = \boxed{0.531}$$

From Eq. (7.64), we have at $i = 16$,

$$V^{t=\Delta t}_{16} = 1.402 + (-0.409)(0.0201) = \boxed{1.394}$$

From Eq. (7.65), we have at $i = 16$,

$$T^{t=\Delta t}_{16} = 0.653 + 0.176(0.0201) = \boxed{0.656}$$

Defining a nondimensional pressure as the local static pressure divided by the reservoir pressure p_0, the equation of state is given by

$$p = \rho T$$

where p, ρ, and T are *nondimensional* values. Thus, at grid point $i = 16$, we have

$$p^{t=\Delta t}_{16} = \rho^{t=\Delta t}_{16} T^{t=\Delta t}_{16} = 0.531(0.656) = \boxed{0.349}$$

This now completes the corrector step for grid point $i = 16$. When the above corrector-step calculations are carried out for all grid points from $i = 2$ to 30, then we have completed the corrector step for all the *internal* grid points.

It remains to calculate the flow-field variables at the boundary points. At the subsonic inflow boundary ($i = 1$), V_1 is calculated by linear extrapolation from grid points 2 and 3. At the end of the corrector step, from a calculation identical to that given above, the values of V_2 and V_3 at time $t = \Delta t$ are $V_2 = 0.212$ and $V_3 = 0.312$. Thus, from Eq. (7.70), we have

$$V_1 = 2V_2 - V_3 = 2(0.212) - 0.312 = \boxed{0.111}$$

At the supersonic outflow boundary ($i = 31$) all the flow-field variables are calculated by linear extrapolation from Eqs. (7.72a) to (7.72c). At the end of the corrector step, from a calculation identical to that given above, $V_{29} = 1.884$, $V_{30} = 1.890$, $\rho_{29} = 0.125$, $\rho_{30} = 0.095$, $T_{29} = 0.354$, and $T_{30} = 0.332$. When these values are inserted into Eqs. (7.72a) to (7.72c), we have

$$V_{31} = 2V_{30} - V_{29} = 2(1.890) - 1.884 = \boxed{1.895}$$

$$\rho_{31} = 2\rho_{30} - \rho_{29} = 2(0.095) - 0.125 = \boxed{0.066}$$

$$T_{31} = 2T_{30} - T_{29} = 2(0.332) - 0.354 = \boxed{0.309}$$

With this, we have completed the calculation of all the flow-field variables at all the grid points after the first time step, i.e., at time $t = \Delta t$. A tabulation of these variables is given in Table 7.2. Note that the Mach number is included in this tabulation. In terms of the nondimensional velocity and temperature, the Mach number (which is already a dimensionless parameter defined as the local velocity divided by the local speed of sound) is given by

$$M = \frac{V}{\sqrt{T}} \tag{7.76}$$

Examine Table 7.2 closely. By reading across the line labeled $I = 16$, you will find the familiar numbers that we have generated for grid point $i = 16$ in the above discussion. Take the time to make this comparison. The entries for all other internal grid points are calculated in a like manner. Also note the values at the boundary points, labeled $I = 1$ and $I = 31$ in Table 7.2. You will find the numbers to be the same as discussed above.

7.3.3 Final Numerical Results: The Steady-State Solution

Compare the flow-field results obtained after one time step (Table 7.2) with the same quantities at the previous time (in this case the initial conditions given in Table 7.1). Comparing these two tables, we see that the flow-field variables *have changed*. For example, the nondimensional density at the throat (where $A = 1$) has changed from 0.528 to 0.531, a 0.57 percent change over one time step. This is the natural behavior of a time-marching solution—the flow-field variables change from one time step to the next. However, in the approach toward the steady-state solution, at larger values of time (after a large number of time steps), the *changes* in the flow-field variables from one time step to the next become smaller and approach zero in

TABLE 7.2
Flow-field variables after the first time step

I	$\dfrac{x}{L}$	$\dfrac{A}{A^*}$	$\dfrac{\rho}{\rho_0}$	$\dfrac{V}{a_0}$	$\dfrac{T}{T_0}$	$\dfrac{p}{p_0}$	M
1	0.000	5.950	1.000	0.111	1.000	1.000	0.111
2	0.100	5.312	0.955	0.212	0.972	0.928	0.215
3	0.200	4.718	0.927	0.312	0.950	0.881	0.320
4	0.300	4.168	0.900	0.411	0.929	0.836	0.427
5	0.400	3.662	0.872	0.508	0.908	0.791	0.534
6	0.500	3.200	0.844	0.603	0.886	0.748	0.640
7	0.600	2.782	0.817	0.695	0.865	0.706	0.747
8	0.700	2.408	0.789	0.784	0.843	0.665	0.854
9	0.800	2.078	0.760	0.870	0.822	0.625	0.960
10	0.900	1.792	0.731	0.954	0.800	0.585	1.067
11	1.000	1.550	0.701	1.035	0.778	0.545	1.174
12	1.100	1.352	0.670	1.113	0.755	0.506	1.281
13	1.200	1.198	0.637	1.188	0.731	0.466	1.389
14	1.300	1.088	0.603	1.260	0.707	0.426	1.498
15	1.400	1.022	0.567	1.328	0.682	0.387	1.609
16	1.500	1.000	0.531	1.394	0.656	0.349	1.720
17	1.600	1.022	0.494	1.455	0.631	0.312	1.833
18	1.700	1.088	0.459	1.514	0.605	0.278	1.945
19	1.800	1.198	0.425	1.568	0.581	0.247	2.058
20	1.900	1.352	0.392	1.619	0.556	0.218	2.171
21	2.000	1.550	0.361	1.666	0.533	0.192	2.282
22	2.100	1.792	0.330	1.709	0.510	0.168	2.393
23	2.200	2.078	0.301	1.748	0.487	0.146	2.504
24	2.300	2.408	0.271	1.782	0.465	0.126	2.614
25	2.400	2.782	0.242	1.813	0.443	0.107	2.724
26	2.500	3.200	0.213	1.838	0.421	0.090	2.834
27	2.600	3.662	0.184	1.858	0.398	0.073	2.944
28	2.700	4.168	0.154	1.874	0.376	0.058	3.055
29	2.800	4.718	0.125	1.884	0.354	0.044	3.167
30	2.900	5.312	0.095	1.890	0.332	0.032	3.281
31	3.000	5.950	0.066	1.895	0.309	0.020	3.406

the limit of large time. At this stage, the steady state (for all practical purposes) has been achieved, and the calculation can be stopped. This termination of the calculation can be done automatically by the computer program itself by having a test in the program to sense when the changes in the flow-field variables become smaller than some prescribed value (prescribed by you, depending on your desired accuracy of the final "steady-state" solution). Another option, and that preferred by the present author, is to simply stop the calculation after a prescribed number of time steps, look at the results, and see if they have approached the stage where the flow-field variables are not materially changing any more. If such is not the case, simply resume the calculations, and carry them out for the requisite number of time steps until you do see that the steady-state results have been reached.

What patterns do the timewise variations of the flow-field variables take? Some feeling for the answer is provided by Fig. 7.9, which shows the variation of ρ,

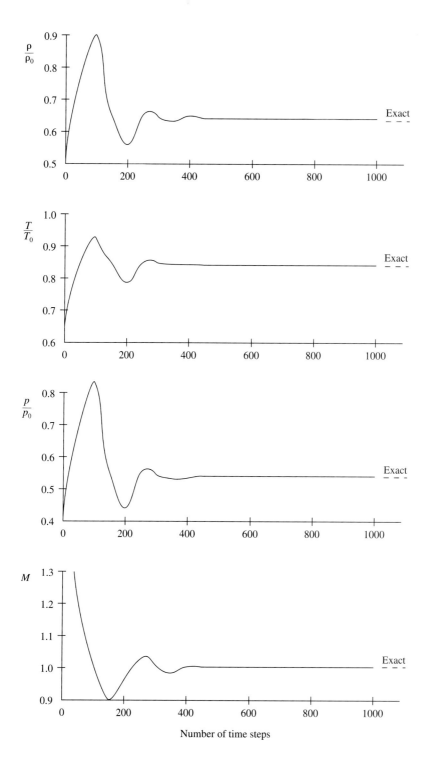

FIG. 7.9
Timewise variations of the density, temperature, pressure, and Mach number at the nozzle throat (at grid point $i = 15$, where $A = 1$).

T, p, and M at the nozzle throat plotted versus the number of time steps. The abscissa starts at zero, which represents the initial conditions, and ends at time step 1000. Hence, the abscissa is essentially a time axis, with time increasing to the right. Note that the largest changes take place at early times, after which the final, steady-state value is approached almost asymptotically. Here is the "rubber band effect" mentioned previously; at early times the rubber band is "stretched" tightly, and therefore the flow-field variables are driven by a stronger potential and hence change rapidly. At later times, as the steady state is approached, the rubber band is less stretched; it becomes more "relaxed", and the changes become much smaller with time. The dashed lines to the right of the curves shown in Fig. 7.9 represent the exact, analytical values as obtained from the equations discussed in Sec. 7.2. Note that the numerical time-marching procedure converges to the proper theoretical steady-state answer. We also note that no artificial viscosity has been explicitly added for these calculations; it is not needed.

It is interesting to examine the variation of the time derivatives as a function of time itself, or equivalently as a function of the number of time steps. Once again focusing on the nozzle throat (at grid point $i = 16$), Fig. 7.10 gives the variation of the time derivatives of nondimensional density and velocity as a function of the

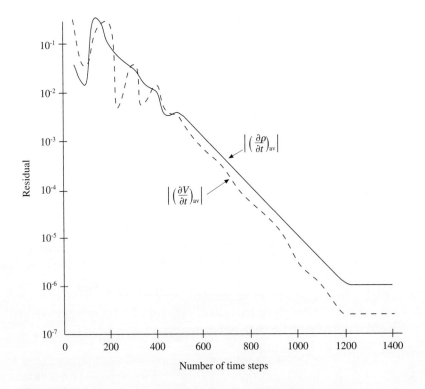

FIG. 7.10
Timewise variations of the absolute values of the time derivatives of nondimensional density and velocity at the nozzle throat (at grid point $i = 16$).

number of time steps. These are the *average* time derivatives calculated from Eqs. (7.60) and (7.61), respectively. The *absolute value* of these time derivatives is shown in Fig. 7.10. From these results, note two important aspects:

1. At early times, the time derivatives are large, and they oscillate in value. These oscillations are associated with various unsteady compression and expansion waves which propagate through the nozzle during the transient process. (See Chap. 7 of Ref. 21 for a discussion of unsteady wave motion in a duct.)

2. At later times, the time derivatives rapidly grow small, changing by six orders of magnitude over a span of 1000 time steps. This is, of course, what we want to see happen. In the theoretical limit of the steady state (which is achieved at infinite time), the time derivatives should go to zero. However, numerically this will never happen over a finite number of time steps. In fact, the results shown in Fig. 7.10 indicate that the values of the time derivatives plateau after 1200 time steps. This seems to be a characteristic of MacCormack's technique. However, the values of the time derivatives at these plateaus are so small that, for all practical purposes, the numerical solution has arrived at the steady-state solution. Indeed, in terms of the values of the flow-field variables themselves, the results of Fig. 7.9 indicate that the steady state is realistically achieved after 500 time steps, during which the time derivatives in Fig. 7.10 have decreased only by two orders of magnitude.

Return to Eqs. (7.46) and (7.48) for a moment; we might visualize that what is being plotted in Fig. 7.10 are the numerical values of the right-hand side of these equations. As time progresses and as the steady-state is approached, the right-hand side of these equations should approach zero. Since the *numerical* values of the right-hand side are not precisely zero, they are called *residuals*. This is why the ordinate in Fig. 7.10 is labeled as the residual. When CFD experts are comparing the relative merits of two or more different algorithms for a time-marching solution to the steady state, the magnitude of the residuals and their rate of decay are often used as figures of merit. That algorithm which gives the fastest decay of the residuals to the smallest value is usually looked upon most favorably.

Another insight to the mechanics of the timewise variation of the flow and its approach to the steady state is provided by the mass flow variations shown in Fig. 7.11. Here, the nondimensional mass flow ρVA (where ρ, V, and A are the nondimensional values) is plotted as a function of nondimensional distance through the nozzle. Six different curves are shown, each for a different time during the course of the time-marching procedure. The dashed curve is the variation of ρVA which pertains to the initial conditions, and hence it is labeled $0\Delta t$. The strange-looking, distorted sinelike variation of this dashed curve is simply the product of the assumed initial values for ρ and V combined with the specified parabolic variation of the nozzle area ratio A. After 50 time steps, the mass flow distribution through the nozzle has changed considerably; this is given by the curve labeled $50\Delta t$. After 100 time steps ($100\Delta t$), the mass flow distribution has changed radically; the mass flow variation is simply flopping around inside the nozzle due to

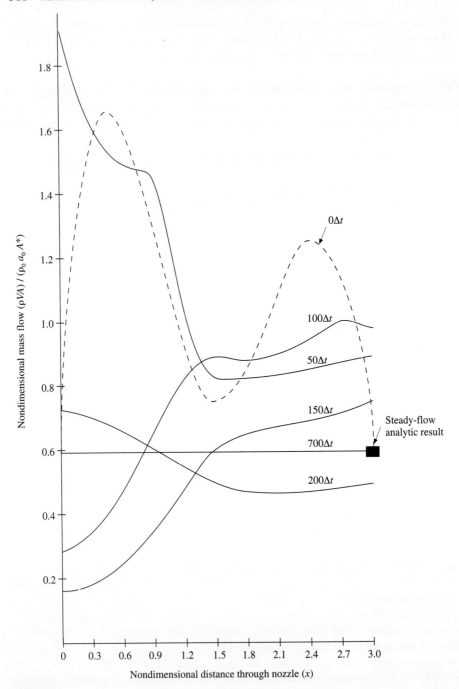

FIG. 7.11
Instantaneous distributions of the nondimensional mass flow as a function of distance through the nozzle at six different times during the time-marching approach to the steady state.

the transient variation of the flow-field variables. However, after 200 time steps ($200\Delta t$), the mass flow distribution is beginning to settle down, and after 700 time steps ($700\Delta t$), the mass flow distribution is a straight, horizontal line across the graph. This says that the mass flow has converged to a *constant*, steady-state value throughout the nozzle. This agrees with our basic knowledge of steady-state nozzle flows, namely, that

$$\rho V A = \text{constant}$$

Moreover, it has converged to essentially the *correct value* of the steady mass flow, which in terms of the *nondimensional* variables evaluated at the nozzle throat is given by

$$\rho V A = \rho^* \sqrt{T^*} \quad \text{(at throat)} \quad (7.77)$$

where ρ^* and T^* are the nondimensional density and temperature at the throat, and where $M = 1$. [Derive Eq. (7.77) yourself—it is easy.] From the analytical equations discussed in Sec. 7.2, when $M = 1$ and $\gamma = 1.4$, we have $\rho^* = 0.634$ and $T^* = 0.833$. With these numbers, Eq. (7.77) yields

$$\rho V A = \text{constant} = 0.579$$

This value is given by the dark square in Fig. 7.11; the mass flow result for $700\Delta t$ agrees reasonably well with the dark square.

Finally, let us examine the steady-state results. From the discussion above and from examining Fig. 7.9, the steady state is, for all practical purposes, reached after about 500 time steps. However, being very conservative, we will examine the results obtained after 1400 time steps; between 700 and 1400 time steps, there is no change in the results, at least to the three-decimal-place accuracy given in the tables herein.

A feeling for the graphical accuracy of the numerically obtained steady state is given by Fig. 7.12. Here, the steady-state nondimensional density and Mach number distributions through the nozzle are plotted as a function of nondimensional distance along the nozzle. The numerical results, obtained after 1400 time steps, are given by the solid curves, and the exact analytical results are given by the circles. The analytical results are obtained from the equations discussed in Sec. 7.2; they can readily be obtained from the tables at the back of most compressible flow texts, such as Ref. 21. They can also be obtained by writing your own short computer program to calculate numbers from the theoretically derived equations in Sec. 7.2. In any event, the comparison shown in Fig. 7.12 clearly demonstrates that the numerical results agree very well with the exact analytical values, certainly to within graphical accuracy.

The detailed numerical results, to three decimal places, are tabulated in Table 7.3. These are the results obtained after 1400 time steps. They are given here for you to compare numbers from your own computer program. It is interesting to note that the elapsed nondimensional time, starting at zero with the initial conditions, is, after 1400 time steps, a value of 28.952. Since time is nondimensionalized by the quantity L/a_0, let us assume a case where the length of the nozzle is 1 m and the reservoir temperature is the standard sea level value, $T = 288$ K. For this case, $L/a_0 = (1 \text{ m})/(340.2 \text{ m/s}) = 2.94 \times 10^{-3}$ s. Hence, the total *real* time that has

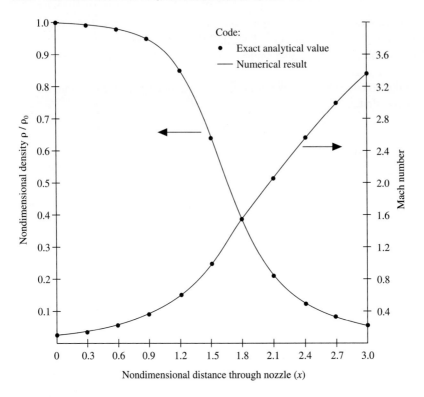

FIG. 7.12
Steady-state distributions of nondimensional density and Mach number as a function of nondimensional distance through the nozzle. Comparison between the exact analytical values (circles) and the numerical results (solid curves).

elapsed over the 1400 time steps is $(2.94 \times 10^{-3})(28.952) = 0.0851$ s. That is, the nozzle flow, starting from the assumed initial conditions, takes only 85.1 ms to reach steady-state conditions; in reality, since convergence is obtained for all practical purposes after about 500 time steps, the practical convergence time is more on the order of 30 ms.

A comparison between some of the numerical results and the corresponding exact analytical values is given in Table 7.4; this provides you with a more detailed comparison than is given in Fig. 7.12. Compared are the numerical and analytical results for the density ratio and Mach number. Note that the numerical results, to three decimal places, are not in precise agreement with the analytical values; there is a small percentage disagreement between the two sets of results, ranging from 0.3 to 3.29 percent. This amount of error is not discernable on the graphical display in Fig. 7.12. At first thought, there might be three reasons for these small numerical inaccuracies: (1) a small inflow boundary condition error, (2) truncation errors associated with the finite value of Δx, such as discussed in Sec. 4.3, and (3) possible effects of the Courant number being substantially less than unity (recall that in the

TABLE 7.3
Flow-field variables after 1400 time steps (nonconservation form of the governing equations)

I	$\dfrac{x}{L}$	$\dfrac{A}{A^*}$	$\dfrac{\rho}{\rho_0}$	$\dfrac{V}{a_0}$	$\dfrac{T}{T_0}$	$\dfrac{p}{p_0}$	M	$\dot{m}$
1	0.000	5.950	1.000	0.099	1.000	1.000	0.099	0.590
2	0.100	5.312	0.998	0.112	0.999	0.997	0.112	0.594
3	0.200	4.718	0.997	0.125	0.999	0.996	0.125	0.589
4	0.300	4.168	0.994	0.143	0.998	0.992	0.143	0.591
5	0.400	3.662	0.992	0.162	0.997	0.988	0.163	0.589
6	0.500	3.200	0.987	0.187	0.995	0.982	0.187	0.589
7	0.600	2.782	0.982	0.215	0.993	0.974	0.216	0.588
8	0.700	2.408	0.974	0.251	0.989	0.963	0.252	0.588
9	0.800	2.078	0.963	0.294	0.985	0.948	0.296	0.587
10	0.900	1.792	0.947	0.346	0.978	0.926	0.350	0.587
11	1.000	1.550	0.924	0.409	0.969	0.895	0.416	0.586
12	1.100	1.352	0.892	0.485	0.956	0.853	0.496	0.585
13	1.200	1.198	0.849	0.575	0.937	0.795	0.594	0.585
14	1.300	1.088	0.792	0.678	0.911	0.722	0.710	0.584
15	1.400	1.022	0.721	0.793	0.878	0.633	0.846	0.584
16	1.500	1.000	0.639	0.914	0.836	0.534	0.99	0.584
17	1.600	1.022	0.551	1.037	0.789	0.434	1.167	0.584
18	1.700	1.088	0.465	1.155	0.737	0.343	1.345	0.584
19	1.800	1.198	0.386	1.263	0.684	0.264	1.528	0.585
20	1.900	1.352	0.318	1.361	0.633	0.201	1.710	0.586
21	2.000	1.550	0.262	1.446	0.585	0.153	1.890	0.587
22	2.100	1.792	0.216	1.519	0.541	0.117	2.065	0.588
23	2.200	2.078	0.179	1.582	0.502	0.090	2.233	0.589
24	2.300	2.408	0.150	1.636	0.467	0.070	2.394	0.590
25	2.400	2.782	0.126	1.683	0.436	0.055	2.549	0.590
26	2.500	3.200	0.107	1.723	0.408	0.044	2.696	0.591
27	2.600	3.662	0.092	1.759	0.384	0.035	2.839	0.591
28	2.700	4.168	0.079	1.789	0.362	0.029	2.972	0.592
29	2.800	4.718	0.069	1.817	0.342	0.024	3.105	0.592
30	2.900	5.312	0.061	1.839	0.325	0.020	3.225	0.595
31	3.000	5.950	0.053	1.862	0.308	0.016	3.353	0.585

calculations discussed so far, the Courant number is chosen to be 0.5), such as discussed at the end of Sec. 4.5. Let us examine each of these reasons in turn.

INFLOW BOUNDARY CONDITION ERROR. There is a "built-in" error at the inflow boundary. At the first grid point, at $x = 0$, we *assume* that the density, pressure, and temperature are the reservoir properties ρ_0, p_0, and T_0, respectively. This is strictly true only if $M = 0$ at this point. In reality, there is a finite area ratio at $x = 0$, namely, $A/A^* = 5.95$, and hence a finite Mach number must exist at $x = 0$, both numerically and analytically (to allow a finite value of mass flow through the nozzle). Hence, in Table 7.4, the numerical value of ρ/ρ_0 at $x = 0$ is equal to 1.0— this is our prescribed boundary condition. On the other hand, the exact analytical

TABLE 7.4
Density ratio and Mach number distributions through the nozzle

$\dfrac{x}{L}$	$\dfrac{A}{A^*}$	$\dfrac{\rho}{\rho_0}$ (numerical results)	$\dfrac{\rho}{\rho_0}$ (exact analytical results)	Difference, %	M (numerical results)	M (exact analytical results)	Difference, %
0.000	5.950	1.000	0.995	0.50	0.099	0.098	1.01
0.100	5.312	0.998	0.994	0.40	0.112	0.110	1.79
0.200	4.718	0.997	0.992	0.30	0.125	0.124	0.08
0.300	4.168	0.994	0.990	0.40	0.143	0.140	2.10
0.400	3.662	0.992	0.987	0.50	0.163	0.160	1.84
0.500	3.200	0.987	0.983	0.40	0.187	0.185	1.07
0.600	2.782	0.982	0.978	0.41	0.216	0.214	0.93
0.700	2.408	0.974	0.970	0.41	0.252	0.249	1.19
0.800	2.078	0.963	0.958	0.52	0.296	0.293	1.01
0.900	1.792	0.947	0.942	0.53	0.350	0.347	0.86
1.000	1.550	0.924	0.920	0.43	0.416	0.413	0.72
1.100	1.352	0.892	0.888	0.45	0.496	0.494	0.40
1.200	1.198	0.849	0.844	0.59	0.594	0.592	0.34
1.300	1.088	0.792	0.787	0.63	0.710	0.709	0.14
1.400	1.022	0.721	0.716	0.69	0.846	0.845	0.12
1.500	1.000	0.639	0.634	0.78	0.999	1.000	0.10
1.600	1.022	0.551	0.547	0.73	1.167	1.169	0.17
1.700	1.088	0.465	0.461	0.87	1.345	1.348	0.22
1.800	1.198	0.386	0.382	1.04	1.528	1.531	0.20
1.900	1.352	0.318	0.315	0.94	1.710	1.715	0.29
2.000	1.550	0.262	0.258	1.53	1.890	1.896	0.32
2.100	1.792	0.216	0.213	1.39	2.065	2.071	0.29
2.200	2.078	0.179	0.176	1.68	2.233	2.240	0.31
2.300	2.408	0.150	0.147	2.00	2.394	2.402	0.33
2.400	2.782	0.126	0.124	2.38	2.549	2.557	0.31
2.500	3.200	0.107	0.105	1.87	2.696	2.706	0.37
2.600	3.662	0.092	0.090	2.17	2.839	2.848	0.32
2.700	4.168	0.079	0.078	1.28	2.972	2.983	0.37
2.800	4.718	0.069	0.068	1.45	3.105	3.114	0.29
2.900	5.312	0.061	0.059	3.29	3.225	3.239	0.43
3.000	5.950	0.053	0.052	1.89	3.353	3.359	0.18

this is our prescribed boundary condition. On the other hand, the exact analytical value of ρ/ρ_0 at $x = 0$ is 0.995, giving a 0.5 percent error. This built-in error is not viewed as serious, and we will not be concerned with it here.

TRUNCATION ERROR: THE MATTER OF GRID INDEPENDENCE. The matter of *grid independence* is a serious consideration in CFD, and this stage of our data analysis is a perfect time to introduce the concept. In general, when you solve a problem using CFD, you are employing a finite number of grid points (or a finite mesh) distributed over the flow field. Assume that you are using N grid points. If everything goes well during your solution, you will get some numbers out for the flow-field variables at these N grid points, and these numbers may look qualitatively good to you. However, assume that you rerun your solution, this time using twice as many grid points, $2N$, distributed over the same domain; i.e., you have decreased the value of the increment Δx (and also Δy in general, if you are dealing with a two-dimensional solution). You may find that the values of your flow-field variables are quite different for this second calculation. If this is the case, then your solution is a function of the number of grid points you are using—an untenable situation. You must, if at all practical, continue to increase the number of grid points until you reach a solution which is no longer sensitive to the number of points. When you reach this situation, then you have achieved *grid independence*.

Question: Do we have grid independence for the present calculation? Recall that we have used 31 grid points distributed evenly through the nozzle. To address this question, let us double the number of grid points; i.e., let us halve the value of Δx by using 61 grid points. Table 7.5 compares the steady-state results for density, temperature, and pressure ratios, as well as for Mach numbers, at the throat for both the cases using 31 and 61 grid points. Also tabulated in Table 7.5 are the exact analytical results. Note that although doubling the number of grid points did improve the numerical solution, it did so only marginally. The same is true for all locations within the nozzle. In other words, the two steady-state numerical solutions are essentially the same, and therefore we can conclude that our original calculations using 31 grid points is essentially *grid-independent*. This grid independent solution does not agree *exactly* with the analytical results, but it is certainly close enough for our purposes. The degree of grid independence that you need to achieve in a given problem depends on what you want out of the solution. Do you need extreme accuracy? If so, you need to press the matter of grid independence in a very detailed fashion. Can you tolerate answers that can be a little less precise numerically (such as the 1 or 2 percent accuracy shown in the present calculations)? If so, you can slightly relax the criterion for extreme grid independence and use fewer grid points, thus saving computer time (which frequently means saving money). The proper decision depends on the circumstances. However, you should always be conscious of the question of grid independence and resolve the matter to your satisfaction for any CFD problem you solve. For example, in the present problem, do you think you can drive the numerical results shown in Table 7.5 to agree exactly with the analytical results by using more and more grid points? If so, how many grid points

TABLE 7.5
Demonstration of grid independence

	Conditions at the nozzle throat			
	$\dfrac{\rho^*}{\rho_0}$	$\dfrac{T^*}{T_0}$	$\dfrac{p*}{p_0}$	M
Case 1: 31 points	0.639	0.836	0.534	0.999
Case 2: 61 points	0.638	0.835	0.533	1.000
Exact analytical solution	0.634	0.833	0.528	1.000

will you need? You might want to experiment with this question by running your own program and seeing what happens.

COURANT NUMBER EFFECTS. At the end of Sec. 4.5, we broached the possibility that if the Courant number were too small, and hence the analytical domain for a given grid point were much smaller than the numerical domain, there might be problems in regard to the accuracy of the solution, albeit the solution will be very stable. Do we have such a problem with the present calculations? We have employed $C = 0.5$ for the present calculations. Is this too small, considering that the stability criterion for *linear* hyperbolic equations (see Sec. 4.5) is $C \leq 1.0$? To examine this question, we can simply repeat the previous calculations but with progressively higher values of the Courant number. The resulting steady-state flow-field values at the nozzle throat are tabulated in Table 7.6; the tabulations are given for six different values of C, starting at $C = 0.5$ and ranging to 1.2. For values ranging to as high as $C = 1.1$, the results were only marginally different, as seen in Table 7.6. By increasing C to as high as 1.1., the numerical results do not agree any better with the exact analytical results (as shown in Table 7.6) than the results at lower values of C. Hence, all our previous results obtained by using $C = 0.5$ are not tainted by any noticeable error due to the smaller-than-necessary value of C. Indeed, if anything, the numerical results for $C = 0.5$ in Table 7.6 are marginally *closer* to the exact analytical solution than the results for higher Courant numbers. For the steady-state numerical results tabulated in Table 7.6, the number of time steps was adjusted each time C was changed so that the nondimensional time at the end of each run was essentially the same. This adjustment is necessary because the value of Δt calculated from Eqs. (7.66) and (7.69) will obviously be different for different values of C. For example, when $C = 0.5$ as in our previous results, we carried out the time-marching procedure to 1400 time steps, which corresponded to a nondimensional time of 28.952. When C is increased to 0.7, the number of time steps carried out was $1400(\frac{5}{7}) = 1000$. This corresponded to a nondimensional time of 28.961—essentially the same as for the previous run. In the same manner, all the numerical data compared in Table 7.6 pertain to the same nondimensional time.

TABLE 7.6
Courant number effects

Courant number	$\dfrac{\rho^*}{\rho_0}$	$\dfrac{T^*}{T_0}$	$\dfrac{p*}{p_0}$	M
0.5	0.639	0.836	0.534	0.999
0.7	0.639	0.837	0.535	0.999
0.9	0.639	0.837	0.535	0.999
1.0	0.640	0.837	0.535	0.999
1.1	0.640	0.837	0.535	0.999
1.2	Program went unstable and blew up			
Exact analytical solution	0.634	0.833	0.528	1.000

It is interesting to note that for the present application, the CFL criterion as first introduced by Eq. (4.84), namely, that $C \leq 1$, does not hold exactly. In Table 7.6, we show results where $C = 1.1$; a stable solution is obtained in spite of the fact that the CFL criterion is violated. However, as noted in Table 7.6, when the Courant number is increased to 1.2, instabilities do occur, and the program blows up. Therefore, for the flow problem we have been discussing in this chapter, which is governed by *nonlinear* hyperbolic partial differential equations, the CFL criterion (which is based on linear equations) does not hold exactly. However, from the above results, we can see that the CFL criterion is certainly a good *estimate* for the value of Δt; it is the most reliable estimate for Δt that we can use, even though the governing equations are nonlinear.

7.4 CFD SOLUTION OF PURELY SUBSONIC ISENTROPIC NOZZLE FLOW

In this section we treat the case of purely subsonic flow through a duct. The physical aspects of such a flow are described in detail in Refs. 8 and 21. They differ from the subsonic-supersonic isentropic solution described in Sec. 7.2 in the following ways:

1. For subsonic flow in the duct, there are an *infinite* number of possible isentropic flow solutions, each one corresponding to a specific pressure ratio p_e/p_0, between the exit and the reservoir. Two such solutions are sketched in Fig. 7.13. For one case, labeled case a, the exit pressure is $(p_e)_a$, where $(p_e)_a$ is only slightly smaller than the reservoir pressure p_0. This small pressure difference across the nozzle causes a "gentle wind" to blow through the duct, with the local Mach number increasing with distance in the convergent portion of the duct, reaching a peak value at the minimum area section (where this peak value of M is considerably less than 1), and then decreasing in the divergent section, resulting in the value of the Mach number at the exit $(M_e)_a$, being very small. If the exit pressure is reduced, hence creating a larger pressure difference across the nozzle, the flow through the nozzle will be faster. For example, for the case labeled b in Fig. 7.13, where $(p_e)_b < (p_e)_a$, the Mach number is larger through the nozzle, albeit still purely subsonic throughout. If the exit pressure is further reduced, there will be some value of p_e, say $(p_e)_c$, which results in the Mach number at the throat just barely grazing unity, such as sketched in Fig. 7.13. At the same time, the pressure at the minimum area section will equal $0.528 p_0$, which corresponds to local sonic conditions. Examining Fig. 7.13 carefully, we note that for exit pressures p_e such that $(p_e)_c < p_e < p_0$, the flow through the duct will be purely subsonic. There are an infinite number of such flows, corresponding to the infinite choice of p_e in the range from p_0 to $(p_e)_c$. Therefore, when the flow is totally subsonic throughout the duct, the local flow properties are dictated by *both* the local area ratio A/A_t (where A_t is the minimum area—the throat area) *and* the pressure ratio across the nozzle p_e/p_0. This is in contrast to the subsonic-supersonic case described in Sec. 7.2, where the local Mach number is strictly a function of the area ratio only [from Eq. (7.6)].

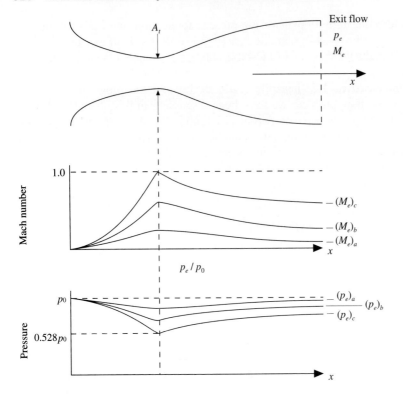

FIG. 7.13
Schematic of purely subsonic flow in a convergent-divergent nozzle.

2. In the subsonic case, the Mach number at the minimum area A_t is less than 1. Hence, A_t is *not* the same as A^*, which is defined in Sec. 7.2 as the *sonic* throat area; that is, A^* is the throat area which corresponds to sonic flow. Hence, in the purely subsonic flow case, A^* is simply a reference area; moreover, in this situation, $A^* < A_t$.

The exact analytical solution of the purely subsonic flow case proceeds as follows. The exit-to-reservoir pressure ratio must be specified; that is, p_e/p_0 is given. Since the total pressure is constant through the nozzle, the value of p_e/p_0 defines M_e through Eq. (7.7), i.e.,

$$\frac{p_e}{p_0} = \left(1 + \frac{\gamma - 1}{2} M_e^2\right)^{-\gamma/(\gamma - 1)} \tag{7.78}$$

Once M_e is known from a solution of Eq. (7.78), the value of A^* can be calculated from Eq. (7.6) as

$$\frac{A_e}{A^*} = \frac{1}{M_e^2} \left[\frac{2}{\gamma + 1}\left(1 + \frac{\gamma - 1}{2} M_e^2\right)\right]^{(\gamma + 1)/(\gamma - 1)} \tag{7.79}$$

where A^* is simply a reference value in this case; A^* is smaller than the throat area A_t. In turn, with A^* known, the local area divided by A^*, namely, A/A^*, determines the local Mach number M via Eq. (7.6). Finally, this local value of M determines the local values of p/p_0, ρ/ρ_0, and T/T_0 from Eqs. (7.7) to (7.9).

7.4.1 The Setup: Boundary and Initial Conditions

For this calculation, we will specify a nozzle with the following area distribution, where all symbols are in dimensional terms:

$$\frac{A}{A_t} = \begin{cases} 1 + 2.2\left(\frac{x}{L} - 1.5\right)^2 & \text{for } 0 \leq \frac{x}{L} \leq 1.5 & (7.80a) \\ 1 + 0.2223\left(\frac{x}{L} - 1.5\right)^2 & \text{for } 1.5 \leq \frac{x}{L} \leq 3.0 & (7.80b) \end{cases}$$

In the above equations, A_t denotes the area of the nozzle throat. Keep in mind that as long as the flow is subsonic at the throat, A_t is *not* equal to A^*; indeed, $A_t > A^*$. A plot of the area distribution given by Eqs. (7.80a) and (7.80b) is shown in Fig. 7.14.

The governing flow equations are the same as used for the subsonic-supersonic solutions discussed in Sec. 7.3, namely, Eqs. (7.46), (7.48), and (7.50).

The treatment of the boundary conditions for the present subsonic flow solution must reflect the need to *specify* the pressure ratio across the nozzle in order to have a unique solution, as discussed at the beginning of Sec. 7.4. Referring to Fig. 7.15, the subsonic inflow boundary, point 1, is treated exactly as discussed in the Boundary Conditions subsection to Sec. 7.3.1. However, in the present problem, the *outflow boundary* is also *subsonic*. In the context of the discussion of the Boundary Conditions subsection, we have at the subsonic outflow boundary one characteristic line (the right-running characteristic) propagating to the right and the other characteristic line (the left-running characteristic) propagating to the left. Also, the streamline at point N is moving toward the right. Examining Fig. 7.15, we see at point N one characteristic moving *out* of the domain, namely, the right-running characteristic, as well as the flow along the streamline moving *out* of the domain. Consistent with our discussion in the Boundary Condition subsection to

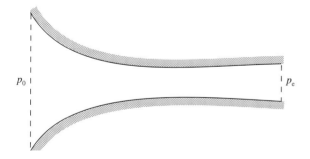

FIG. 7.14
Sketch of nozzle for the purely subsonic flow solution discussed in Sec. 7.4.

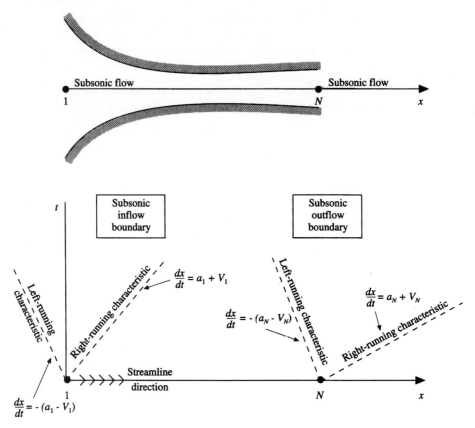

FIG. 7.15
Study of boundary conditions for subsonic inflow and outflow.

Sec. 7.3.1 this means that two flow variables should be allowed to *float* at the boundary point N. On the other hand, we also see at point N *one* characteristic moving *into* the domain, namely, the left-running characteristic. As discussed in the Boundary Conditions subsection, this means that the value of *one* flow variable must be *specified* at the boundary point N. Of course, this is consistent with our earlier physical discussion, namely, that to have a unique solution of the purely subsonic flow in the duct, we need to specify the pressure ratio across the nozzle, p_0/p_e; that is, for a fixed p_0, we need to *specify* the exit pressure p_e.

How do we implement the specification of p_e within the numerical solution? Returning to the governing equations, Eqs. (7.46), (7.48) and (7.50), we note that the dependent variables in these equations are density, velocity, and temperature—not pressure. However, through the equation of state,

$$p = \rho RT \tag{7.81}$$

Hence, specifying the value of p_e is the same as specifying the product $\rho_e R T_e$. In terms of the nondimensional variables in Eqs. (7.46), (7.48), and (7.50), we can express Eq. (7.81) evaluated at the duct exit as

$$p'_e = \rho'_e T'_e \tag{7.82}$$

The numerical implementation of the boundary conditions is carried out as follows. The subsonic inflow boundary is treated exactly as in the Boundary Conditions subsection to Sec. 7.3.1; that is, Eqs. (7.70) and (7.71) are used here. For the subsonic *outflow* boundary, we have

$$p'_N = \text{specified value} \tag{7.83}$$

Since ρ'_N and T'_N are the dependent variables appropriate to our governing equations, we must make certain that both ρ'_N and T'_N, which will vary as a function of time, are strongly coupled to the pressure boundary condition given by Eq. (7.83); that is, no matter how ρ'_N and T'_N vary from one time step to the next, at *each* time step they must satisfy the constraint that

$$\rho'_N T'_N = p'_N = \text{specified value} \tag{7.84}$$

One way to accomplish this strong coupling is as follows. Let us linearly extrapolate T'_N obtaining

$$T'_N = 2T'_{N-1} - T'_{N-2} \tag{7.85}$$

From this value of T'_N, calculate ρ'_N from the equation of state such that Eq. (7.83) is satisfied, i.e.,

$$\rho'_N = \frac{p'_N}{T'_N} = \frac{\text{specified value}}{T'_N} \tag{7.86}$$

The values of T'_N from Eq. (7.85) along with ρ'_N from Eq. (7.86) ensure that p'_N remains constant at the specified value. *Alternatively*, we could obtain ρ'_N by linear extrapolation,

$$\rho'_N = 2\rho'_{N-1} - \rho'_{N-2} \tag{7.87}$$

and calculate T'_N from the equation of state,

$$T'_N = \frac{p'_N}{\rho'_N} = \frac{\text{specified value}}{\rho'_N} \tag{7.88}$$

The values of ρ'_N and T'_N obtained from Eqs. (7.87) and (7.88), respectively, also ensure that p'_N remains constant at the specified value. (It has been the author's experience that either combination works equally as well, i.e., the combination of Eqs. (7.85) and (7.86) where temperature is extrapolated or the combination of Eqs. (7.87) and (7.88) where density is extrapolated.) Finally, as before, the velocity at the downstream boundary is extrapolated:

$$V'_N = 2V'_{N-1} - V'_{N-2} \tag{7.89}$$

Note: There is more than meets the eye to the way we have set up the boundary conditions to this problem. We will return to this point in Sec. 7.4.2.

Finally, for the initial conditions, let us somewhat arbitrarily set up the following variations:

$$\rho' = 1.0 - 0.023x' \qquad (7.90a)$$
$$T' = 1.0 - 0.009333x' \qquad (7.90b)$$
$$V' = 0.05 + 0.11x' \qquad (7.90c)$$

These specify the initial flow field at time $t = 0$.

We will carry out a time-marching solution of the purely subsonic flowfield using MacCormack's predictor-corrector explicit finite-difference method just as utilized for the previous subsonic-supersonic solution. The details are exactly the same. Indeed, to treat the subsonic flow described in this section, only a slight modification is needed to the computer program you might have written for the previous case—just the initial conditions, the nozzle shape, and the downstream boundary conditions need to be changed. Therefore, no further details are needed here.

7.4.2 Final Numerical Results: MacCormack's Technique

In Sec. 7.3.2 we discussed some intermediate results pertaining to the detailed calculations on the first time step. Since exactly the same method is being used here, there is no need to discuss the intermediate calculations. Let us proceed to the final numerical results.

A feeling for the timewise variation of the flow field as it marches toward the steady-state solution is provided by Fig. 7.16 and 7.17. These results pertain to a

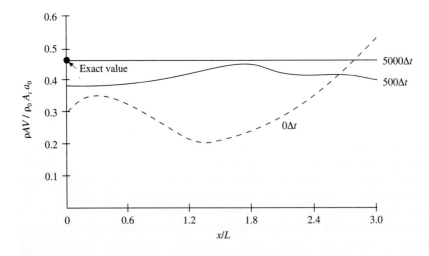

FIG. 7.16
Variation of mass flow through the nozzle at different times; purely subsonic flow case with $p_e/p_0 = 0.93$.

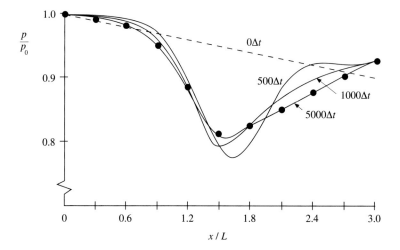

FIG. 7.17
Variation of the pressure distribution through the nozzle at different times; purely subsonic flow case with $p_e/p_0 = 0.93$. Circles indicate exact analytical values.

specified pressure ratio across the nozzle of $p_e/p_0 = 0.93$. The distribution of the nondimensional mass flow through the nozzle at three different times is shown in Fig. 7.16. The dashed curve labeled $0\Delta t$ corresponds to the initial condition. Note that after 500 time steps (the curve labeled $500\Delta t$) the mass flow is moving toward the steady-state value. After 5000 time steps, the mass flow has converged to a horizontal line; that is, $\rho AV = $ constant. The dark circle is the exact analytical value; note that good agreement is achieved between the numerical and the analytical results. The timewise variation of the pressure distribution through the nozzle at four different times is shown in Fig. 7.17. Again, the dashed line is the initial distribution. Note that the initial condition for the pressure ratio at the exit is slightly below the specified value of 0.93; however, after the first time step, the boundary condition imposed by Eq. (7.84) is in effect, with $p_e/p_0 = 0.93$. This is reflected in the fact that the pressure distributions for $500\Delta t$, $1000\Delta t$, and $5000\Delta t$ all meet at the same point at the nozzle exit. The dark circles in Fig. 7.17 give the exact analytical values.

The final, steady-state values of the flow-field variables, including the mass flow, as a function of distance through the nozzle are tabulated in Table 7.7. For these calculations, 31 grid points are distributed through the nozzle, and the Courant number is 0.5. These results are for time step 5000. This is a conservative number of time steps; in reality, convergence is obtained for all practical purposes after 2500 time steps. The convergence behavior of the solution is further indicated by the values of the residuals (the average nondimensional time derivatives), which are on the order of 10^{-2} after 500 time steps, 10^{-3} after 2500 time steps, and 10^{-5} after 5000 time steps.

A comparison of the numerical results after 5000 time steps with the exact analytical results is tabulated in Table 7.8. The accuracy of the numerical results for

TABLE 7.7
Flow-field variables after 5000 time steps—subsonic flow

I	$\dfrac{x}{L}$	$\dfrac{A}{A_t}$	$\dfrac{\rho}{\rho_0}$	$\dfrac{V}{a_0}$	$\dfrac{T}{T_0}$	$\dfrac{p}{p_0}$	M	$\dot{m}$
1	0.000	5.950	1.000	0.079	1.000	1.000	0.079	0.469
2	0.100	5.312	0.998	0.089	0.999	0.997	0.089	0.472
3	0.200	4.718	0.998	0.099	0.999	0.997	0.099	0.467
4	0.300	4.168	0.996	0.113	0.998	0.995	0.113	0.468
5	0.400	3.662	0.995	0.128	0.998	0.992	0.128	0.467
6	0.500	3.200	0.992	0.147	0.997	0.989	0.147	0.467
7	0.600	2.782	0.989	0.170	0.995	0.984	0.170	0.466
8	0.700	2.408	0.984	0.197	0.993	0.977	0.197	0.466
9	0.800	2.078	0.977	0.229	0.991	0.968	0.230	0.466
10	0.900	1.792	0.968	0.268	0.987	0.955	0.270	0.465
11	1.000	1.550	0.955	0.314	0.982	0.937	0.317	0.465
12	1.100	1.352	0.938	0.367	0.975	0.914	0.371	0.465
13	1.200	1.198	0.916	0.424	0.966	0.885	0.431	0.465
14	1.300	1.088	0.892	0.480	0.955	0.853	0.491	0.466
15	1.400	1.022	0.871	0.524	0.946	0.824	0.539	0.467
16	1.500	1.000	0.862	0.542	0.942	0.812	0.559	0.467
17	1.600	1.002	0.863	0.540	0.943	0.814	0.556	0.467
18	1.700	1.009	0.865	0.535	0.944	0.816	0.551	0.467
19	1.800	1.020	0.869	0.526	0.946	0.822	0.541	0.467
20	1.900	1.036	0.875	0.516	0.948	0.829	0.530	0.467
21	2.000	1.056	0.881	0.502	0.951	0.838	0.515	0.467
22	2.100	1.080	0.888	0.487	0.954	0.847	0.499	0.467
23	2.200	1.109	0.896	0.470	0.957	0.857	0.481	0.467
24	2.300	1.142	0.903	0.453	0.960	0.867	0.462	0.467
25	2.400	1.180	0.911	0.434	0.963	0.877	0.443	0.467
26	2.500	1.222	0.918	0.416	0.966	0.887	0.423	0.467
27	2.600	1.269	0.925	0.398	0.970	0.897	0.404	0.467
28	2.700	1.320	0.932	0.379	0.972	0.906	0.385	0.467
29	2.800	1.376	0.938	0.362	0.975	0.915	0.366	0.467
30	2.900	1.436	0.944	0.344	0.977	0.923	0.348	0.467
31	3.000	1.500	0.949	0.327	0.980	0.930	0.331	0.466

this purely subsonic case is about the same as that obtained for the subsonic-supersonic isentropic flow case (see Table 7.4).

It is interesting to note the values of time required to come to a reasonable steady state. For the present case, $t' = t/(L/a_0) = 84.3$. This is to be compared with the nondimensional time required for convergence in the subsonic-supersonic flow case calculated earlier, which after 500 time steps was 10.3. For the same nozzle length L and reservoir speed of sound a_0, the subsonic flow takes a much longer time to converge to the steady state. That is, in part, a reflection of the time it takes for a fluid element to travel through the nozzle, which we will call the *transit time*. For the steady state to be reached, there should be a time lapse of several transit times—this is required for the history of the initial conditions to "flush through" the nozzle. For a purely subsonic flow, the fluid elements have an average velocity much

TABLE 7.8
Comparison between numerical and analytical values

$\dfrac{x}{L}$	$\dfrac{A}{A_t}$	$\dfrac{\rho}{\rho_0}$ (numerical results)	$\dfrac{\rho}{\rho_0}$ (exact analytical results)	Difference, %	M (numerical results)	M (exact analytical results)	Difference, %
0.000	5.950	1.000	0.997	0.30	0.079	0.077	2.50
0.100	5.312	0.998	0.996	0.20	0.089	0.086	3.30
0.200	4.718	0.998	0.995	0.30	0.099	0.097	2.00
0.300	4.168	0.996	0.994	0.20	0.113	0.110	2.65
0.400	3.662	0.995	0.992	0.30	0.128	0.126	1.56
0.500	3.200	0.992	0.990	0.20	0.147	0.144	2.04
0.600	2.782	0.989	0.986	0.30	0.170	0.167	1.76
0.700	2.408	0.984	0.981	0.30	0.197	0.194	1.52
0.800	2.078	0.977	0.975	0.20	0.230	0.226	1.74
0.900	1.792	0.968	0.966	0.20	0.270	0.265	1.85
1.000	1.550	0.955	0.953	0.21	0.317	0.312	1.58
1.100	1.352	0.938	0.936	0.21	0.371	0.365	1.62
1.200	1.198	0.916	0.916	0.00	0.431	0.423	1.86
1.300	1.088	0.892	0.893	0.11	0.491	0.480	2.24
1.400	1.022	0.871	0.875	0.46	0.539	0.524	2.78
1.500	1.000	0.862	0.867	0.58	0.559	0.541	3.22
1.600	1.002	0.863	0.868	0.58	0.556	0.539	3.06
1.700	1.009	0.865	0.870	0.57	0.551	0.534	3.09
1.800	1.020	0.869	0.874	0.58	0.541	0.526	2.77
1.900	1.036	0.875	0.879	0.46	0.530	0.514	3.02
2.000	1.056	0.881	0.885	0.45	0.515	0.500	2.91
2.100	1.080	0.888	0.892	0.45	0.499	0.485	2.81
2.200	1.109	0.896	0.898	0.33	0.481	0.468	2.91
2.300	1.142	0.903	0.906	0.33	0.462	0.450	2.60
2.400	1.180	0.911	0.913	0.22	0.443	0.431	2.71
2.500	1.222	0.918	0.920	0.22	0.423	0.413	2.36
2.600	1.269	0.925	0.926	0.11	0.404	0.394	2.48
2.700	1.320	0.932	0.933	0.11	0.385	0.376	2.34
2.800	1.376	0.938	0.939	0.11	0.366	0.358	2.19
2.900	1.436	0.944	0.944	0.00	0.348	0.340	2.30
3.000	1.500	0.949	0.949	0.00	0.331	0.324	2.11

lower than for the subsonic-supersonic case; hence the transit time for the subsonic case is much larger. For this reason, nature simply takes a longer time to establish a steady subsonic flow compared to that for a steady supersonic flow. Such a trend is clearly evident in our results here.

7.4.3 The Anatomy of a Failed Solution

In our discussion of the way we have set up the boundary conditions in Sec. 7.4.1, we noted that there was more to it than meets the eye. Let us examine this comment further.

Consider a case where $p_e/p_0 = 0.9$; this is a stronger pressure ratio across the nozzle than the case discussed in Sec. 7.4.2, where $p_e/p_0 = 0.93$. Therefore, the flow Mach number inside the nozzle will be larger. However, according to the exact analytical solution, the steady flow through our nozzle with $p_e/p_0 = 0.9$ should still be subsonic everywhere; the highest Mach number, which occurs at the throat, is theoretically $M_t = 0.721$, and the theoretical Mach number at the exit is 0.391. Nevertheless, under the same conditions as those used in Sec. 7.4.2 (same initial conditions, Courant number, and boundary condition treatment) the case with $p_e/p_0 = 0.9$ goes unstable and blows up. It is instructional to investigate the behavior of the blowup and to surmise the reasons for this behavior.

The distribution of pressure through the nozzle at four different times is shown in Fig. 7.18. The dashed line labeled $0\Delta t$ is the initial distribution at time $t = 0$. After 400 time steps (the curve labeled $400\Delta t$), the flow appears to be moving toward a qualitatively proper solution. After 800 time steps, for the most part, the solution appears to be approaching a proper steady-state result; for example, at $800\Delta t$, the numerical results show the Mach number at the throat $M_t = 0.704$, very close to the theoretical value of 0.721. A further comparison is given by the solid circles in Fig. 7.18, which represent the exact analytical results for p/p_0. Note that in the convergent section of the nozzle ($x/L < 1.5$), the proper steady state is almost achieved. However, tracing through the curve for $800\Delta t$, we see a small but disturbing oscillation forming at the downstream boundary. At $1200\Delta t$, this oscillation has escalated enormously, and shortly thereafter the solution blows up. This behavior, which occurs when $p_e/p_0 = 0.9$, is completely different than the behavior shown in Fig. 7.17 for the case of $p_e/p_0 = 0.93$, where the steady state is successfully approached after about 2500 time steps.

Why do the oscillations shown in Fig. 7.18 develop? In short, they are due to finite waves that reflect off the downstream boundary, where the reflection is due to purely numerical reasons. Since we are holding the exit pressure p_e constant throughout the calculation, there is every reason to expect that finite, unsteady compression and expansion waves that are traveling to the right in the unsteady nozzle flow will reflect off this constant-pressure boundary. If these waves are strong enough, massive oscillations will occur near the downstream boundary. Given enough time, the oscillations eventually lead to a blowup of the calculations. Obviously, for a less severe pressure ratio across the nozzle, such as $p_e/p_0 = 0.93$, the weaker unsteady waves produced within the nozzle, when reflected from the downstream boundary, do not set up an oscillation.

Let us reexamine the downstream boundary condition on a physical basis. Our stipulation of a *fixed*, constant pressure at the exit is physically valid only in the *steady-state* case. During the unsteady flow, finite compression and expansion waves travel up and down the nozzle. As these waves travel out of the nozzle at the downstream boundary, *all* the flow variables, including the pressure, *fluctuate* with time. This is the real *physical* situation. (See Chap. 7 of Ref. 21 for a discussion of unsteady, one-dimensional, finite-wave properties.) In the numerical calculation discussed above, we are not allowing the pressure at the downstream boundary to fluctuate; we are stipulating that it is fixed, independent of time. This is the proper boundary condition as the flow approaches the steady state, but it is physically

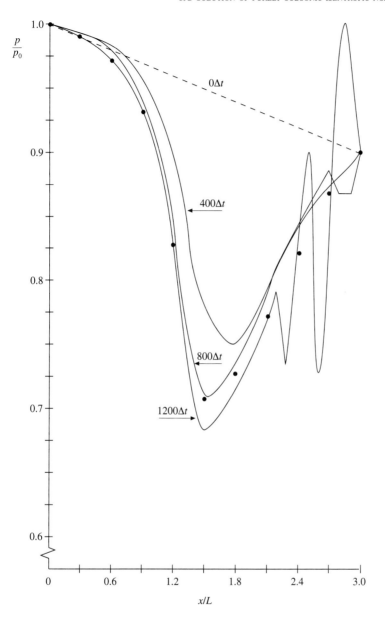

FIG. 7.18
Variation of the pressure distribution through the nozzle at different times; purely subsonic flow case with $p_e/p_0 = 0.90$. Note the oscillatory behavior after 1200 time steps. Circles indicate exact analytical values.

improper during the unsteady flow variations that take place during the time-marching process. As a result, with the numerically fixed pressure at the exit, we are in part "bottling up" the waves inside the nozzle. When the pressure ratio p_e/p_0 is strong enough (such as in the case of $p_e/p_0 = 0.9$), the unsteady finite waves produced at early times in the nozzle are sufficiently strong, and the nonphysical reflections from the constant-pressure boundary eventually grow into the type of oscillations shown in Fig. 7.18, with the result that the calculations blow up. On the other hand, if the pressure ratio p_e/p_0 is milder (such as in the case of $p_e/p_0 = 0.93$), the unsteady finite waves are weaker, and we are able to obtain a proper, steady state, as described earlier.

There are several "fixes" that we could try in order to improve the behavior of the attempted solution for $p_e/p_0 = 0.9$. First, we could simply try different initial conditions, ones that are closer to the steady-state answers. In this fashion, the unsteady finite waves set up during the transient approach to the steady state will be weaker, therefore diminishing the tendency for the oscillation buildup as reflected in Fig. 7.18. Second, we could add some artificial viscosity for the reasons discussed in Sec. 6.6. Note that so far in the our nozzle calculations, we have not explicitly added artificial viscosity. However, one of the purposes of artificial viscosity is to help damp the type of oscillations shown in Fig. 7.18. Such a ploy may be effective for the present case.

We will not pursue either of these possible fixes here because we need to turn our attention to other, more pressing matters. We will have the opportunity to explore the matter of adding artificial viscosity to our nozzle calculation in Sec. 7.6, which deals with a shock-capturing case.

7.5 THE SUBSONIC-SUPERSONIC ISENTROPIC NOZZLE SOLUTION REVISITED: THE USE OF THE GOVERNING EQUATIONS IN CONSERVATION FORM

In Chap. 2 we made a distinction between the nonconservation form and the conservation form of the governing flow equations. We made the point that, theoretically, either form of the equations is a suitable representation of the fundamental physical principles of mass conservation, Newton's second law, and energy conservation. However, in CFD, there are some good numerical reasons to use one form or the other for the solution of certain flow problems. An important example is the case of shock capturing (see Sec. 2.10), where we noted that the conservation form of the equations is the proper form to employ; the nonconservation form will lead to poor numerical results.

In the present section, we take the opportunity to examine the differences between results obtained from the nonconservation form of the equations and those obtained from the conservation form. We will first cast the governing equations for quasi-one-dimensional flow in conservation form. Then we will set up the numerical solution of these equations using MacCormack's technique, as applied to the subsonic-supersonic isentropic flow case. The matter of shock capturing within the nozzle will be deferred until Sec. 7.6. Finally, we will compare the numerical results obtained from the conservation form of the equations to those obtained from the nonconservation form.

7.5.1 The Basic Equations in Conservation Form

Returning to Eq. (7.15), repeated below.

$$\frac{\partial(\rho A)}{\partial t} + \frac{\partial(\rho A V)}{\partial x} = 0 \tag{7.15}$$

This is the continuity equation for quasi-one-dimensional flow. It is already in conservation form. Nondimensionalizing the variables according to the forms given in Sec. 7.2, we have

$$\frac{\partial\left(\frac{\rho}{\rho_0}\frac{A}{A^*}\right)}{\partial\left(\frac{t}{L/a_0}\right)} \left(\frac{\rho_0 A^* a_0}{L}\right) + \frac{\partial\left(\frac{\rho}{\rho_0}\frac{A}{A^*}\frac{V}{a_0}\right)}{\partial(x/L)} \left(\frac{\rho_0 A^* a_0}{L}\right) = 0$$

or

$$\boxed{\frac{\partial(\rho' A')}{\partial t'} + \frac{\partial(\rho' A' V')}{\partial x'}} \tag{7.91}$$

As before, the primes in Eq. (7.91) denote the nondimensional variables.

Return to Eq. (7.23), repeated below.

$$\frac{\partial(\rho A V)}{\partial t} + \frac{\partial(\rho A V^2)}{\partial x} = -A\frac{\partial p}{\partial x} \tag{7.23}$$

This is the momentum equation for quasi-one-dimensional flow. It is already in conservation form. Let us combine the two x derivatives in Eq. (7.23) as follows. Since

$$\frac{\partial(pA)}{\partial x} = p\frac{\partial A}{\partial x} + A\frac{\partial p}{\partial x} \tag{7.92}$$

we can add Eq. (7.92) to Eq. (7.23), obtaining

$$\frac{\partial(\rho A V)}{\partial t} + \frac{\partial(\rho A V^2 + pA)}{\partial x} = p\frac{\partial A}{\partial x} \tag{7.93}$$

Nondimensionalizing Eq. (7.93), we have

$$\frac{\partial\left(\frac{\rho}{\rho_0}\frac{A}{A^*}\frac{V}{a_0}\right)}{\partial\left(\frac{t}{L/a_0}\right)} \left(\frac{\rho_0 A^* a_0^2}{L}\right) + \frac{\partial\left[\frac{\rho}{\rho_0}\frac{A}{A^*}\frac{V^2}{a_0^2}(\rho_0 A^* a_0^2) + \frac{p}{p_0}\frac{A}{A^*}(p_0 A^*)\right]}{\partial\left(\frac{x}{L}\right)L}$$

$$= \frac{p}{p_0}\frac{\partial(A/A^*)}{\partial(x/L)}\left(\frac{p_0 A^*}{L}\right)$$

or

$$\frac{\partial(\rho' A' V')}{\partial t'} + \frac{\partial[\rho' A' V'^2 + p' A'(p_0/\rho_0 a_0^2)]}{\partial x'} = p'\frac{\partial A'}{\partial x'}\left(\frac{p_0}{\rho_0 a_0^2}\right) \tag{7.94}$$

However,

$$\frac{p_0}{\rho_0 a_0^2} = \frac{\rho_0 R T_0}{\rho_o a_0^2} = \frac{\rho_0 R T_0}{\rho_0 \gamma R T_0} = \frac{1}{\gamma}$$

Thus, Eq. (7.94) becomes

$$\frac{\partial(\rho'A'V')}{\partial t'} + \frac{\partial[\rho'A'V'^2 + (1/\gamma)p'A']}{\partial x'} = \frac{1}{\gamma}p'\frac{\partial A'}{\partial x'} \qquad (7.95)$$

Returning to Eq. (7.33), repeated below.

$$\frac{\partial[\rho(e+V^2/2)A]}{\partial t} + \frac{\partial[\rho(e+V^2/2)AV]}{\partial x} = -\frac{\partial(pAV)}{\partial x} \qquad (7.33)$$

This is the energy equation for quasi-one-dimensional flow. It is already in conservation form. Combining the x derivatives in Eq. (7.33), we have

$$\frac{\partial[\rho(e+V^2/2)A]}{\partial t} + \frac{\partial[\rho(e+V^2/2)AV + pAV]}{\partial x} = 0 \qquad (7.96)$$

Let us define a nondimensional internal energy as follows:

$$e' = \frac{e}{e_0} \qquad \text{where } e_0 = c_v T_0 = \frac{RT_0}{\gamma - 1}$$

With this, the nondimensional form of Eq. (7.96) is obtained as follows.

$$\frac{\partial\left\{\frac{\rho}{\rho_0}\left[\frac{e}{e_0}(e_0) + \frac{V^2}{2a_0^2}(a_0^2)\right]\frac{A}{A^*}\right\}}{\partial\left(\frac{t}{L/a_0}\right)}\left(\frac{\rho_0 A^* a_0}{L}\right)$$

$$+ \frac{\partial\left\{\frac{\rho}{\rho_0}\left[\frac{e}{e_0}(e_0) + \frac{V^2}{2a_0^2}(a_0^2)\right]\frac{V}{a_0}\frac{A}{A^*}(\rho_0 a_0 A^*) + \left(\frac{p}{p_0}\frac{A}{A^*}\frac{V}{a_0}\right)(p_0 A^* a_0)\right\}}{\partial\left(\frac{x}{L}\right)L} = 0$$

(7.97)

Since $e_0 = RT_0/(\gamma - 1)$, Eq. (7.97) becomes

$$\frac{\partial\left[\rho'\left(\frac{e'}{\gamma-1} + \frac{\gamma}{2}V'^2\right)A'\right]}{\partial t'}\left(\frac{\rho_0 A^* a_0 RT_0}{L}\right)$$

$$+ \frac{\partial\left[\rho'\left(\frac{e'}{\gamma-1} + \frac{\gamma}{2}V'^2\right)V'A'\left(\frac{\rho_0 a_0 A^* RT_0}{L}\right) + (p'A'V')\left(\frac{p_0 A^* a_0}{L}\right)\right]}{\partial x'} = 0 \qquad (7.98)$$

Divide Eq. (7.98) by $\rho_0 A^* a_0 R T_0 / L$.

$$\frac{\partial\left[\rho'\left(\frac{e'}{\gamma-1}+\frac{\gamma}{2}V'^2\right)A'\right]}{\partial t'} + \frac{\partial\left[\rho'\left(\frac{e'}{\gamma-1}+\frac{\gamma}{2}V'^2\right)V'A' + p'A'V'\left(\frac{p_0}{\rho_0 R T_0}\right)\right]}{\partial x'} = 0$$

(7.99)

However, in Eq. (7.99),

$$\frac{p_0}{\rho_0 R T_0} = \frac{\rho_0 R T_0}{\rho_0 R T_0} = 1$$

Thus, Eq. (7.99) becomes

$$\boxed{\frac{\partial\left[\rho'\left(\frac{e'}{\gamma-1}+\frac{\gamma}{2}V'^2\right)A'\right]}{\partial t'} + \frac{\partial\left[\rho'\left(\frac{e'}{\gamma-1}+\frac{\gamma}{2}V'^2\right)V'A' + p'A'V'\right]}{\partial x'} = 0} \quad (7.100)$$

Equations (7.91), (7.95), and (7.100) are the nondimensional conservation form of the continuity, momentum, and energy equations for quasi-one-dimensional flow, respectively. Return to Eq. (2.93), which is a generic form of the governing equations for unsteady, three-dimensional flow. The equations for quasi-one-dimensional flow can be expressed in a similar generic form. Let us define the elements of the solutions vector U, the flux vector F, and the source term J as follows.

$$U_1 = \rho' A'$$
$$U_2 = \rho' A' V'$$
$$U_3 = \rho'\left(\frac{e'}{\gamma-1}+\frac{\gamma}{2}V'^2\right)A'$$
$$F_1 = \rho' A' V'$$
$$F_2 = \rho' A' V'^2 + \frac{1}{\gamma} p' A'$$
$$F_3 = \rho'\left(\frac{e'}{\gamma-1}+\frac{\gamma}{2}V'^2\right)V'A' + p'A'V'$$
$$J_2 = \frac{1}{\gamma} p' \frac{\partial A'}{\partial x'}$$

With these elements, Eqs. (7.91), (7.95), and (7.100) can be written, respectively, as

$$\boxed{\frac{\partial U_1}{\partial t'} = -\frac{\partial F_1}{\partial x'}} \quad (7.101a)$$

$$\boxed{\frac{\partial U_2}{\partial t'} = -\frac{\partial F_2}{\partial x'} + J_2} \quad (7.101b)$$

$$\boxed{\frac{\partial U_3}{\partial t'} = -\frac{\partial F_3}{\partial x'}} \quad (7.101c)$$

We are now finished with the governing equations for quasi-one-dimensional flow. Equations (7.101a) to (7.101c) represent the continuity, momentum, and energy equations for quasi-one-dimensional flow, in conservation form. These are the equations we wish to numerically solve using MacCormack's technique.

Before setting up this numerical solution, keep in mind from our discussions in Chap. 2 that in the conservation form of the equations the dependent variables (the variables for which we directly obtain numbers) are *not* the primitive variables. For example, in Eqs. (7.101a) to (7.101c), our numerical solution will give us numbers directly for U_1, U_2, and U_3 in steps of time; this is why U is called the *solutions vector*. To obtain the primitive variables (ρ, V, T, p, etc.), we must *decode* the elements U_1, U_2, and U_3 as follows. From the definitions of U_1, U_2, and U_3 given above, we have

$$\rho' = \frac{U_1}{A'} \tag{7.102}$$

$$V' = \frac{U_2}{U_1} \tag{7.103}$$

$$T' = e' = (\gamma - 1)\left(\frac{U_3}{U_1} - \frac{\gamma}{2} V'^2\right) \tag{7.104}$$

$$p' = \rho' T' \tag{7.105}$$

Note in Eq. (7.104) that we have recognized the fact that $e' = T'$, or

$$e' \equiv \frac{e}{e_0} = \frac{c_v T}{c_v T_0} = \frac{T}{T_0} = T'$$

Therefore, after we obtain U_1, U_2, and U_3 at each time step from the numerical solution of Eqs. (7.101a) to (7.101c), we can immediately calculate the corresponding primitive variables at each time step, ρ', V', T', and p', from Eqs. (7.102) to (7.105).

7.5.2 The Setup

Return to Eqs. (7.101a) to (7.101c) for a moment; we note that the flux vector elements F_1, F_2, and F_3 are couched in terms of the primitive variables [see the relations for F_1, F_2, and F_3 immediately preceding Eqs. (7.101a) to (7.101c). It has been the author's experience that when the computer program is written with F_1, F_2, and F_3 expressed directly in terms of ρ', V', p', and e', instabilities develop during the course of the time-marching solution. For example, in the present example of quasi-one-dimensional, subsonic-supersonic, isentropic nozzle flow, instabilities develop in the subsonic section which finally cause the program to blow up after about 300 time steps. This behavior is an example of a lack of "purity" in the formulation of the governing equations in conservation form, a lack which eventually causes numerical problems. If we were to write a computer program to implement the equations exactly as written in Sec. 7.5.1, we would set up the numerical solution of Eqs. (7.101a) to (7.101c) for U_1, U_2, and U_3 at each time step. We would then *decode* these elements of the solutions vector to obtain the primitive variables at each time step, as shown in Eqs. (7.102) to (7.105). These primitive variables ρ', V', e', and p' would, in turn, be used to construct F_1, F_2, and F_3 for use in the solution of Eqs. (7.101a) to (7.101c) for the next time step, and so forth. As stated above, in the author's experience, when the primitive variables are used to construct F_1, F_2, and F_3, numerical difficulties occasionally arise. This is somehow connected to the fact that the dependent variables which appear *explicitly* in Eqs. (7.101a) to (7.101c) are U_1, U_2, and U_3—not the primitive variables. For this reason, it is best to couch F_1, F_2, and F_3 *directly* in terms of the dependent variables U_1, U_2, and U_3 and avoid the use of the primitive variables in Eqs. (7.101a) to (7.101c). That is, in Eqs. (7.101a) to (7.101c), we will write

$$F_1 = F_1(U_1, U_2, U_3) \qquad (7.106a)$$

$$F_2 = F_2(U_1, U_2, U_3) \qquad (7.106b)$$

$$F_3 = F_3(U_1, U_2, U_3) \qquad (7.106c)$$

$$J_2 = J(U_1, U_2, U_3) \qquad (7.106d)$$

such that the governing equations are "purely" in terms of the elements of the solution vector, i.e., in terms of U_1, U_2, and U_3 only. Let us proceed to obtain the specific forms indicated by Eqs. (7.106a) to (7.106d).

"PURE" FORM OF THE FLUX TERMS. Consider the flux term F_1, given in Sec. 7.5.1 by

$$F_1 = \rho' A' V' \qquad (7.107)$$

Substituting Eqs. (7.102) and (7.103) for ρ' and V', respectively, into Eq. (7.107), we have

$$\boxed{F_1 = U_2} \qquad (7.108)$$

Consider the flux term F_2, given in Sec. 7.5.1 by

$$F_2 = \rho' A' V'^2 + \frac{1}{\gamma} p' A' \qquad (7.109)$$

From Eq. (7.105), the pressure in Eq. (7.109) can be replaced by the product $\rho' T'$. In turn, ρ', V', and T' can be expressed in terms of U_1, U_2, and U_3 via Eqs. (7.102) to (7.104). Hence, Eq. (7.109) becomes

$$F_2 = \frac{U_2^2}{U_1} + \frac{1}{\gamma} U_1 (\gamma - 1) \left[\frac{U_3}{U_1} - \frac{\gamma}{2} \left(\frac{U_2}{U_1} \right)^2 \right]$$

or

$$\boxed{F_2 = \frac{U_2^2}{U_1} + \frac{\gamma - 1}{\gamma} \left(U_3 - \frac{\gamma}{2} \frac{U_2^2}{U_1} \right)} \qquad (7.110)$$

Consider the flux term F_3, given in Sec. 7.5.1 by

$$F_3 = \rho'\left(\frac{e'}{\gamma - 1} + \frac{\gamma}{2}V'^2\right)V'A' + p'A'V' \tag{7.111}$$

Substituting Eqs. (7.102) to (7.105) into Eq. (7.111), we have

$$F_3 = U_2\left(\frac{U_3}{U_1} - \frac{\gamma}{2}V'^2 + \frac{\gamma}{2}V'^2\right) + U_2 T'$$

$$= \frac{U_2 U_3}{U_1} + (\gamma - 1)U_2\left[\frac{U_3}{U_1} - \frac{\gamma}{2}\left(\frac{U_2}{U_1}\right)^2\right]$$

or
$$F_3 = \gamma\frac{U_2 U_3}{U_1} - \frac{\gamma(\gamma - 1)}{2}\frac{U_2^3}{U_1^2} \tag{7.112}$$

Finally, the source term J_2 was given in Sec. 7.5.1 as

$$J_2 = \frac{1}{\gamma}p'\frac{\partial A'}{\partial x'} \tag{7.113}$$

From Eq. (7.105), this becomes

$$J_2 = \frac{1}{\gamma}\rho' T'\frac{\partial A'}{\partial x'} \tag{7.114}$$

Substituting Eqs. (7.102) and (7.104) into (7.114), we have

$$J_2 = \frac{1}{\gamma}\frac{U_1}{A'}(\gamma - 1)\left[\frac{U_3}{U_1} - \frac{\gamma}{2}\left(\frac{U_2}{U_1}\right)^2\right]\frac{\partial A'}{\partial x'}$$

or
$$J_2 = \frac{\gamma - 1}{\gamma}\left(U_3 - \frac{\gamma}{2}\frac{U_2^2}{U_1}\right)\frac{\partial(\ln A')}{\partial x'} \tag{7.115}$$

We now return to our governing flow equations in conservation form as given by Eqs. (7.101a) to (7.101c). With F_1, F_2, F_3, and J_2 given by Eqs. (7.108), (7.110), (7.112), and (7.115), respectively, then Eqs. (7.101a) to (7.101c) are expressed in terms of U_1, U_2, and U_3 only—the primitive variables are nowhere to be found. This is the "pure" form of the governing equations in conservation form; it is the form which we will use in the following sections. When a computer program is written to solve the equations in this pure form, the solution is stable and convergence to a steady state is achieved.

A comment: The behavior discussed above, namely, that instabilities are sometimes encountered when F_1, F_2, and F_3 are constructed in terms of the primitive variables, whereas a stable solution is obtained when F_1, F_2, and F_3 are constructed in terms of U_1, U_2, and U_3, is one of those nonintuitive peculiarities of CFD. So what if F_1, F_2, and F_3 are written in terms of ρ', V', T', and p' instead of

U_1, U_2, U_3? On a *theoretical* basis, there is no difference. However, on a numerical basis, there is a big difference—the difference between instability and stability. This author has no simple mathematical explanation for this behavior. Let us simply consider it as part of the "art" of CFD. On the other hand, we have here an example of the advantages to be obtained by writing our CFD programs using the most consistent, or pure, form of the equations and by treating all steps in the computer program in a consistent fashion, i.e., by not changing horses in midstream.

BOUNDARY CONDITIONS. The boundary conditions for the subsonic-supersonic isentropic flow solution using the conservation form of the governing equations are theoretically the same as discussed in the Boundary Conditions subsection in Sec. 7.3.1; i.e., at the subsonic inflow boundary two properties are held fixed and one is allowed to float, and at the supersonic outflow boundary all properties are allowed to float. In the present formulation, as before, we hold ρ' and T' fixed at the inflow boundary, both equal to 1.0, and allow V' to float. By holding ρ' fixed, then U_1 at grid point $i = 1$ is fixed, independent of time, via $U_1 = \rho' A'$. That is,

$$U_{1\,(i=1)} = (\rho' A')_{i=1} = A'_{i=1} = \text{fixed value}$$

The floating value of V' at the inflow boundary is calculated at the end of each time step by linearly extrapolating U_2 from the known values at the internal grid points $i = 2$ and 3, that is,

$$U_{2\,(i=1)} = 2U_{2\,(i=2)} - U_{2\,(i=3)} \qquad (7.116)$$

and then obtaining V' at $i = 1$ from Eq. (7.103). Since V' floats at the inflow boundary, so does the value of U_3, which is given by

$$U_3 = \rho'\left(\frac{e'}{\gamma - 1} + \frac{\gamma}{2} V'^2\right) A' \qquad (7.117)$$

Since $\rho' A' = U_1$ and $e' = T'$, Eq. (7.117) is written as

$$U_3 = U_1\left(\frac{T'}{\gamma - 1} + \frac{\gamma}{2} V'^2\right) \qquad (7.118)$$

The value of U_3 ($i = 1$) is found by inserting the value of V' at $i = 1$, calculated above, as well as the *fixed* value $T' = 1$, into Eq. (7.118). Note that the values of U_1, U_2, and U_3 calculated at grid point $i = 1$ are used in turn to obtain the values of the flux terms F_1, F_2, and F_3 at grid point $i = 1$. These values of the flux terms at the inflow boundary are needed to form the rearward differences that appear in Eqs. (7.101a) to (7.101c) during the corrector step of MacCormack's technique. The values of F_1, F_2, and F_3 at the inflow boundary are calculated from Eqs. (7.108), (7.110), and (7.112), respectively, using U_1, U_2, and U_3 at grid point $i = 1$.

The flow properties at the downstream, supersonic outflow boundary are obtained by linear extrapolation from the two adjacent internal points. If N denotes the grid point at the outflow boundary, then

$$(U_1)_N = 2(U_1)_{N-1} - (U_1)_{N-2} \tag{7.119a}$$
$$(U_2)_N = 2(U_2)_{N-1} - (U_2)_{N-2} \tag{7.119b}$$
$$(U_3)_N = 2(U_3)_{N-1} - (U_3)_{N-2} \tag{7.119c}$$

The values of F_1, F_2, and F_3 at grid point $i = N$ are obtained from the values of U_1, U_2, and U_3 at point $i = N$, using Eqs. (7.108), (7.110), and (7.112), respectively. These flux values are needed to form the forward differences that appear in Eqs. (7.101a) to (7.101c) during the predictor step of MacCormack's technique. Of course, the primitive variables at the downstream outflow boundary are obtained from Eqs. (7.102) to (7.105).

INITIAL CONDITIONS. Since the dependent variables being solved in Eqs. (7.101a) to (7.101c) are U_1, U_2, and U_3, we need initial conditions for these same variables at time $t = 0$ in order to start the finite-difference solution. The initial conditions for U_1, U_2, and U_3 also allow initial conditions for F_1, F_2, and F_3 to be obtained from Eqs. (7.108), (7.110), and (7.112), respectively. Such initial conditions for F_1, F_2, and F_3 are needed to form the x derivatives on the right-hand sides of Eqs. (7.101a) to (7.101c) at the first time step.

For the present calculations the same nozzle shape as given by Eq. (7.73) is used. The initial conditions for U_1, U_2, and U_3 were synthesized by assuming the following variations of ρ' and T':

$$\left. \begin{array}{l} \rho' = 1.0 \\ T' = 1.0 \end{array} \right\} \text{ for } 0 \leq x' \leq 0.5 \qquad \begin{array}{l}(7.120a)\\(7.120b)\end{array}$$

$$\left. \begin{array}{l} \rho' = 1.0 - 0.366(x' - 0.5) \\ T' = 1.0 - 0.167(x' - 0.5) \end{array} \right\} \text{ for } 0.5 \leq x' \leq 1.5 \qquad \begin{array}{l}(7.120c)\\(7.120d)\end{array}$$

$$\left. \begin{array}{l} \rho' = 0.634 - 0.3879(x' - 1.5) \\ T' = 0.833 - 0.3507(x' - 1.5) \end{array} \right\} \text{ for } 1.5 \leq x' \leq 3.5 \qquad \begin{array}{l}(7.120e)\\(7.120f)\end{array}$$

These variations are slightly more realistic than those assumed in the Nozzle Shape and Initial Conditions subsection of Sec. 7.3.1; this is in anticipation that the stability behavior of the finite-difference formulation using the conservation form of the governing equations might be slightly more sensitive, and therefore it is useful to start with more improved initial conditions than those given in Sec. 7.3.1 by Eqs. (7.74a) to (7.74c). The initial condition for the variation of V' is synthesized by taking advantage of the fact that one of the dependent variables in our governing equations, namely, U_2, is physically the local mass flow; that is, $U_2 = \rho' A' V'$. Therefore, *for the initial conditions only,* let us assume a constant mass flow through the nozzle and calculate V' as

$$V' = \frac{U_2}{\rho' A'} = \frac{0.59}{\rho' A'} \tag{7.121}$$

The value 0.59 is chosen for U_2 because it is close to the exact analytical value of the steady-state mass flow (which for this case is 0.579). Therefore, the initial condition for V' as a function of x' is obtained by substituting the ρ' variation given by Eqs. (7.120a), (7.120c), and (7.120e) into Eq. (7.121). Finally, the initial conditions for U_1, U_2, U_3 are obtained by substituting the above variations for ρ', T', and V' into the definitions given in Sec. 7.5.1, namely,

$$U_1 = \rho' A' \tag{7.122a}$$

$$U_2 = \rho' A' V' \tag{122b}$$

$$U_3 = \rho' \left(\frac{e'}{\gamma - 1} + \frac{\gamma}{2} V'^2 \right) A' \tag{7.122c}$$

where $e' = T'$. Of course, for the initial conditions described above, V' is calculated such that $U_2 = \rho' A' V' = 0.59$.

TIME STEP CALCULATION. The governing equations for unsteady, quasi-one-dimensional flow in *conservation* form are hyperbolic partial differential equations, just as are the governing equations in *nonconservation* form which are employed in Sec. 7.3. Therefore, for an explicit finite-difference solution, the stability criterion for the time step increment Δt is specified by the CFL criterion. In turn, for the calculations in the present section, the value of Δt is obtained precisely as described in Sec. 7.3.1 and given by Eqs. (7.67) to (7.69). Hence, no further elaboration is given here.

7.5.3 Intermediate Calculations: The First Time Step

In the same spirit as Sec. 7.3.2, which gave some intermediate calculations using the nonconservation form of the governing equations, we carry out the same idea in the present section for the conservation form. Since the sequence of calculations is somewhat modified when the conservation form is used, it will be useful to go through some of the details of the computation for the first time step. As explained earlier, the presentation of these intermediate results will not only be instructional but they will also allow you to check the accuracy of your computer program, should you choose to write one for the solution of the present problem.

The nozzle shape and initial conditions for the present calculations are given in Table 7.9. The nozzle shape is the same as used for the calculations in Sec. 7.3 and is sketched in Fig. 7.8. The current initial conditions are different from those used in Sec. 7.3, principally to take advantage of the fact that U_2 is the local mass flow, and following the adage that we should choose initial conditions as intelligently as possible, we assume an initial constant mass flow distribution through the nozzle. This is obvious from the column labeled $\dot{m}$ in Table 7.9. Here, $\dot{m}$ is nondimensional, denoted by $\dot{m} = \rho A V / \rho_0 A^* a_0$. The values of ρ', T', and V' in Table 7.9 are obtained from Eqs. (7.120a) to (7.120f) and (7.121), respectively. Also shown in Table 7.9 are the corresponding initial conditions for U_1, U_2, and U_3, obtained from Eqs. (7.122a) to (7.122c), respectively.

TABLE 7.9
Initial conditions for the case using the conservation form

$\dfrac{x}{L}$	$\dfrac{A}{A*}$	$\dfrac{\rho}{\rho_0}$	$\dfrac{V}{a_0}$	$\dfrac{T}{T_0}$	$\dot{m}$	U_1	U_2	U_3
0.000	5.950	1.000	0.099	1.000	0.590	5.950	0.590	14.916
0.100	5.312	1.000	0.111	1.000	0.590	5.312	0.590	13.326
0.200	4.718	1.000	0.125	1.000	0.590	4.718	0.590	11.847
0.300	4.168	1.000	0.142	1.000	0.590	4.168	0.590	10.478
0.400	3.662	1.000	0.161	1.000	0.590	3.662	0.590	9.222
0.500	3.200	1.000	0.184	1.000	0.590	3.200	0.590	8.076
0.600	2.782	0.963	0.220	0.983	0.590	2.680	0.590	6.679
0.700	2.408	0.927	0.264	0.967	0.590	2.232	0.590	5.502
0.800	2.078	0.890	0.319	0.950	0.590	1.850	0.590	4.525
0.900	1.792	0.854	0.386	0.933	0.590	1.530	0.590	3.728
1.000	1.550	0.817	0.466	0.916	0.590	1.266	0.590	3.094
1.100	1.352	0.780	0.559	0.900	0.590	1.055	0.590	2.604
1.200	1.198	0.744	0.662	0.883	0.590	0.891	0.590	2.241
1.300	1.088	0.707	0.767	0.866	0.590	0.769	0.590	1.983
1.400	1.022	0.671	0.861	0.850	0.590	0.685	0.590	1.811
1.500	1.000	0.634	0.931	0.833	0.590	0.634	0.590	1.705
1.600	1.022	0.595	0.970	0.798	0.590	0.608	0.590	1.614
1.700	1.088	0.556	0.975	0.763	0.590	0.605	0.590	1.557
1.800	1.198	0.518	0.951	0.728	0.590	0.620	0.590	1.521
1.900	1.352	0.479	0.911	0.693	0.590	0.647	0.590	1.498
2.000	1.550	0.440	0.865	0.658	0.590	0.682	0.590	1.479
2.100	1.792	0.401	0.821	0.623	0.590	0.719	0.590	1.458
2.200	2.078	0.362	0.783	0.588	0.590	0.753	0.590	1.430
2.300	2.408	0.324	0.757	0.552	0.590	0.779	0.590	1.389
2.400	2.782	0.285	0.744	0.517	0.590	0.793	0.590	1.333
2.500	3.200	0.246	0.749	0.482	0.590	0.788	0.590	1.259
2.600	3.662	0.207	0.777	0.477	0.590	0.759	0.590	1.170
2.700	4.168	0.169	0.840	0.412	0.590	0.702	0.590	1.071
2.800	4.718	0.130	0.964	0.377	0.590	0.612	0.590	0.975
2.900	5.312	0.091	1.221	0.342	0.590	0.483	0.590	0.917
3.000	5.950	0.052	1.901	0.307	0.590	0.310	0.590	1.023

To illustrate the intermediate calculations, let us focus on grid point $i = 16$, which, as seen in Fig. 7.8, is at the throat of the nozzle. We will follow MacCormack's explicit predictor-corrector technique, described at length in previous sections.

Predictor step. To start the calculation, we use the initial conditions for U_1, U_2, and U_3 to calculate the initial values of F_1, F_2, and F_3 at grid points $i = 16$ and 17. From Table 7.9, the initial values for the U's are

$$(U_1)_{i=16} = 0.634 \quad (U_2)_{i=16} = 0.590 \quad (U_3)_{i=16} = 1.705$$
$$(U_1)_{i=17} = 0.608 \quad (U_2)_{i=17} = 0.590 \quad (U_3)_{i=17} = 1.614$$

From Eq. (7.108)

$$(F_1)_{i=16} = (U_2)_{i=16} = \boxed{0.590}$$

$$(F_1)_{i=17} = (U_2)_{i=17} = \boxed{0.590}$$

From Eq. (7.110)

$$(F_2)_{i=16} = \left[\frac{U_2^2}{U_1} + \frac{\gamma-1}{\gamma}\left(U_3 - \frac{\gamma}{2}\frac{U_2^2}{U_1}\right)\right]_{i=16}$$

$$= \frac{(0.590)^2}{0.634} + \frac{0.4}{1.4}\left[1.705 - 0.7\frac{(0.590)^2}{0.634}\right]$$

$$= \boxed{0.926}$$

$$(F_2)_{i=17} = \frac{(0.590)^2}{0.608} + \frac{0.4}{1.4}\left[1.614 - 0.7\frac{(0.590)^2}{0.608}\right]$$

$$= \boxed{0.919}$$

From Eq. (7.112)

$$(F_3)_{i=16} = \left[\frac{\gamma U_2 U_3}{U_1} - \frac{\gamma(\gamma-1)}{2}\frac{U_2^3}{U_1^2}\right]_{i=16}$$

$$= \frac{1.4(0.590)(1.705)}{0.634} - \frac{1.4(0.4)(0.590)^3}{2(0.634)^2}$$

$$= \boxed{2.078}$$

$$(F_3)_{i=17} = \frac{1.4(0.590)(1.614)}{0.608} - \frac{1.4(0.4)(0.590)^3}{2(0.608)^2}$$

$$= \boxed{2.036}$$

From Eq. (7.113) we have

$$J_2 = \frac{1}{\gamma}p'\frac{\partial A'}{\partial x'} = \frac{1}{\gamma}\rho'T'\frac{\partial A'}{\partial x'}$$

Hence

$$(J_2)_{i=16} = \frac{1}{1.4}(0.634)(0.833)\left(\frac{1.022-1.0}{0.1}\right) = 0.083$$

Note that by using Eq. (7.113) for J_2 rather than the expression given by Eq. (7.115), we are breaking slightly with the purity of the governing equations as described in the first subsection of Sec. 7.5.2. This is being done for simplicity [Eq. (7.113) is much shorter than Eq. (7.115)]. The results are not compromised. The value of $\Delta x'$ is L/N, where L is the length of the nozzle and N is the number of increments along the nozzle, which for the present case is 30. Hence

$$\Delta x' = \frac{L}{N} = \frac{3.0}{30} = 0.1$$

From Eq. (7.101a), using forward differences for the x derivatives, we have

$$\left(\frac{\partial U_1}{\partial t'}\right)^{t'}_{i=16} = -\frac{(F_1)_{i=17} - (F_1)_{i=16}}{\Delta x'} = -\frac{0.590 - 0.590}{0.1} = \boxed{0}$$

From Eq. (7.101b), we have

$$\left(\frac{\partial U_2}{\partial t'}\right)^{t'}_{i=16} = -\frac{(F_2)_{i=17} - (F_2)_{i=16}}{\Delta x'} + J_2$$

$$= -\frac{0.919 - 0.926}{0.1} + 0.083 = \boxed{0.156}$$

(*Please note:* Once again, remember that since we are giving the numbers in the present section to three decimal places, if you are following along with a hand calculator using these three-place figures, some small numerical errors may result in your hand calculations. In the above, and throughout this section, the numbers that appear in boxes are the exact numbers that came from the author's Macintosh computer.) Finally, from Eq. (7.101c), we have

$$\left(\frac{\partial U_3}{\partial t'}\right)^{t'}_{i=16} = -\frac{(F_3)_{i=17} - (F_3)_{i=16}}{\Delta x'} = -\frac{2.036 - 2.078}{0.1} = \boxed{0.416}$$

To obtain the predicted values of the flow quantities, we must first obtain the value of the time step $\Delta t'$. This is carried out as mentioned in the last subsection of Sec. 7.5.2 and as given by Eqs. (7.67) to (7.69) in Sec. 7.3.1. After scanning all the grid points from $i = 1$ to $i = 31$, the minimum value of $\Delta t'$ is found to be, using a Courant number of $C = 0.5$,

$$\Delta t' = 0.0267$$

We proceed to find the predicted values of U_1, U_2, and U_3, denoted by the barred quantities.

$$(\bar{U}_1)^{t'+\Delta t'}_{i=16} = (U_1)^{t'}_{i=16} + \left(\frac{\partial U_1}{\partial t'}\right)^{t'}_{i=16} \Delta t'$$
$$= 0.634 + 0\Delta t' = \boxed{0.634}$$

$$(\bar{U}_2)^{t'+\Delta t'}_{i=16} = (U_2)^{t'}_{i=16} + \left(\frac{\partial U_2}{\partial t'}\right)^{t'}_{i=16} \Delta t'$$
$$= 0.590 + 0.156(0.0267) = \boxed{0.594}$$

$$(\bar{U}_3)^{t'+\Delta t'}_{i=16} = (U_3)^{t'}_{i=16} + \left(\frac{\partial U_3}{\partial t'}\right)^{t'}_{i=16} \Delta t'$$
$$= 1.705 + 0.416(0.0267) = \boxed{1.716}$$

At this stage, the *predicted* values of the primitive variables can be decoded from $\bar{U}_1$, $\bar{U}_2$, and $\bar{U}_3$, using Eqs. (7.102) to (7.105). For example, from Eq. (7.102),

$$(\bar{\rho}')_{i=16}^{t'+\Delta t'} = \frac{(\bar{U}_1)_{i=16}^{t'+\Delta t'}}{(A')_{i=16}} = \frac{0.634}{1.0} = \boxed{0.634}$$

and from Eqs. (7.103) and (7.104),

$$(\bar{T}')_{i=16}^{t'+\Delta t'} = (\gamma - 1) \left\{ \frac{(\bar{U}_3)_{i=16}^{t'+\Delta t'}}{(\bar{U}_1)_{i=16}^{t'+\Delta t'}} - \frac{\gamma}{2} \left[\frac{(\bar{U}_2)_{i=16}^{t'+\Delta t'}}{(\bar{U}_1)_{i=16}^{t'+\Delta t'}} \right]^2 \right\}$$

$$= 0.4 \left[\frac{1.716}{0.634} - 0.7 \left(\frac{0.594}{0.634} \right)^2 \right] = \boxed{0.837}$$

The above numbers for the predicted ρ' and T' will be needed on the corrector step. Before we move on to the corrector step, we need to find the predicted values of F_1, F_2, and F_3 at grid points $i = 15$ and 16; these values for $i = 16$ are based on the predicted values of U_1, U_2, and U_3 found above, and for $i = 15$ are based on the predicted values of U_1, U_2, and U_3 for $i = 15$ (not recorded above in order to not let the length of this section get out of hand). The predicted fluxes, obtained from Eqs. (7.108), (7.110), and (7.112) using $\bar{U}_1$, $\bar{U}_2$, and $\bar{U}_3$, are

$(\bar{F}_1)_{i=16} = 0.594 \qquad (\bar{F}_2)_{i=16} = 0.936 \qquad (\bar{F}_3)_{i=16} = 2.105$
$(\bar{F}_1)_{i=15} = 0.585 \qquad (\bar{F}_2)_{i=15} = 0.915 \qquad (\bar{F}_3)_{i=15} = 2.037$

Corrector step. The predicted time derivatives of U_1, U_2, and U_3 are obtained from Eqs. (7.101a) to (7.101c), respectively, using rearward differences for the x derivatives. From Eq. (7.101a)

$$\overline{\left(\frac{\partial U_1}{\partial t'}\right)}_{i=16}^{t'+\Delta t'} = -\frac{(\bar{F}_1)_{i=16} - (\bar{F}_1)_{i=15}}{\Delta x'}$$

$$= -\frac{0.594 - 0.585}{0.1} = \boxed{-0.0918}$$

From Eq. (7.101b)

$$\overline{\left(\frac{\partial U_2}{\partial t'}\right)}_{i=16}^{t'+\Delta t'} = -\frac{(\bar{F}_2)_{i=16} - (\bar{F}_2)_{i=15}}{\Delta x'} + \frac{1}{\gamma} \bar{\rho}' \bar{T}' \frac{\partial A'}{\partial x'}$$

$$= -\frac{0.936 - 0.915}{0.1} + \frac{1}{1.4}(0.634)(0.837)\left(\frac{1.0 - 1.022}{0.1}\right)$$

$$= \boxed{-0.290}$$

350 NUMERICAL SOLUTIONS OF QUASI-ONE-DIMENSIONAL NOZZLE FLOWS

From Eq. (7.101c)

$$\left(\overline{\frac{\partial U_3}{\partial t'}}\right)^{t'+\Delta t'}_{i=16} = -\frac{(\bar{F}_3)_{i=16} - (\bar{F}_3)_{i=15}}{\Delta x'}$$

$$= -\frac{2.105 - 2.037}{0.1} = \boxed{-0.679}$$

The average time derivatives are formed as follows.

$$\left(\frac{\partial U_1}{\partial t}\right)_{av} = \frac{1}{2}\left[\left(\frac{\partial U_1}{\partial t'}\right)^{t'}_{i=16} + \left(\overline{\frac{\partial U_1}{\partial t'}}\right)^{t'+\Delta t'}_{i=16}\right]$$

$$= 0.5(0 - 0.0918) = \boxed{-0.0459}$$

$$\left(\frac{\partial U_2}{\partial t}\right)_{av} = \frac{1}{2}\left[\left(\frac{\partial U_2}{\partial t'}\right)^{t'}_{i=16} + \left(\overline{\frac{\partial U_2}{\partial t'}}\right)^{t'+\Delta t'}_{i=16}\right]$$

$$= 0.5(0.156 - 0.290) = \boxed{-0.0668}$$

$$\left(\frac{\partial U_3}{\partial t}\right)_{av} = \frac{1}{2}\left[\left(\frac{\partial U_3}{\partial t'}\right)^{t'}_{i=16} + \left(\overline{\frac{\partial U_3}{\partial t'}}\right)^{t'+\Delta t'}_{i=16}\right]$$

$$= 0.5(0.416 - 0.679) = \boxed{-0.131}$$

The final *corrected* values of U_1, U_2, and U_3 at time step $t' + \Delta t'$ (here, since $t' = 0$ to start with, we are calculating the final *corrected* values at time $t' = \Delta t'$) are obtained from

$$(U_1)^{t'+\Delta t'}_{i=16} = (U_1)^{t'}_{i=16} + \left(\frac{\partial U_1}{\partial t'}\right)_{av} \Delta t$$

$$= 0.634 + (-0.0459)(0.0267) = \boxed{0.633}$$

$$(U_2)^{t'+\Delta t'}_{i=16} = (U_2)^{t'}_{i=16} + \left(\frac{\partial U_2}{\partial t'}\right)_{av} \Delta t$$

$$= 0.590 + (-0.0668)(0.0267) = \boxed{0.588}$$

$$(U_3)^{t'+\Delta t'}_{i=16} = (U_3)^{t'}_{i=16} + \left(\frac{\partial U_3}{\partial t'}\right)_{av} \Delta t$$

$$= 1.705 + (-0.131)(0.0267) = \boxed{1.701}$$

Finally, the corrected values of the primitive variables are obtained by decoding U_1, U_2, and U_3, obtained above, via Eqs. (7.102) to (7.105). That is, from Eq. (7.102),

$$(\rho')^{t'+\Delta t'}_{i=16} = (U_1)^{t'+\Delta t'}_{i=16} = \frac{0.633}{1} = \boxed{0.633}$$

From Eq. (7.103)

$$(V')^{t'+\Delta t'}_{i=16} = \left(\frac{U_2}{U_1}\right)^{t'+\Delta t'}_{i=16} = \frac{0.588}{0.633} = \boxed{0.930}$$

From Eq. (7.104)

$$(T')^{t'+\Delta t'}_{i=16} = (\gamma - 1)\left(\frac{U_3}{U_1} - \frac{\gamma}{2}V'^2\right)^{t'+\Delta t'}_{i=16}$$

$$= 0.4\left[\frac{1.701}{0.633} - 0.7(0.930)^2\right] = \boxed{0.833}$$

This brings to an end the calculations of the flow properties at grid point $i = 16$ at time $t' = \Delta t'$. This process is repeated for all the interior grid points distributed along the nozzle. The properties at the inflow and outflow boundaries are calculated as described in the Boundary Conditions subsection of Sec. 7.5.2. By this stage, since you are most likely saturated with numbers, we will spare you the details.

For the sake of reference and so that you can check the numbers from your own computer program, the flow-field variables, including U_1, U_2, and U_3, obtained after the first time step at all the grid points are tabulated in Table 7.10. Comparing the numbers in Table 7.10 with the initial conditions given in Table 7.9, we see that the largest changes over the first time step have taken place near the exit of the nozzle and that the mass flow distribution, originally choosen as constant at $t' = 0$, is no longer constant after the first time step.

7.5.4 Final Numerical Results: The Steady-State Solution

The steady-state results obtained from the time-marching solution of the governing equations in conservation form are essentially the same as those obtained using the nonconservation form (described in Sec. 7.3.3), with a few slight, but notable differences. The present converged solution is tabulated in Table 7.11, which are the results obtained after 1400 time steps. A quick comparison of the numbers given in Tables 7.11 (for conservation form) and 7.3 (for nonconservation form) show little material difference. We conclude that, for all practical purposes, both forms of the governing equations give the same results. This is as it should be; the flow problem studied in both tables is the isentropic, subsonic-supersonic flow through a nozzle, and for such a flow the choice of the form of equations is not important. However, as described in Sec. 2.10, an important numerical distinction between the nonconservation and conservation forms of the equations is related to problems dealing with shock capturing, and we are not capturing any shocks in the present problem.

Let us highlight some of the slight but notable differences mentioned above. The most dramatic difference is in the mass flow distribution. First of all, with the initial conditions assuming a constant mass flow, it is interesting to examine the variation of $\dot{m}$ with x/L at a few different times during the convergence toward the

TABLE 7.10
Flow-field variables at the end of the first time step

$\dfrac{x}{L}$	$\dfrac{A}{A*}$	$\dfrac{\rho}{\rho_0}$	$\dfrac{V}{a_0}$	$\dfrac{T}{T_0}$	$\dfrac{p}{p_0}$	M	$\dot{m}$	U_1	U_2	U_3
0.000	5.950	1.000	0.099	1.000	1.000	0.099	0.588	5.950	0.588	14.916
0.100	5.312	1.000	0.111	1.000	1.000	0.111	0.588	5.312	0.588	13.326
0.200	4.718	1.000	0.125	1.000	1.000	0.125	0.588	4.718	0.588	11.846
0.300	4.168	1.000	0.141	1.000	1.000	0.141	0.587	4.168	0.587	10.478
0.400	3.662	1.000	0.160	1.000	1.000	0.160	0.587	3.662	0.587	9.221
0.500	3.200	0.999	0.187	1.000	0.999	0.187	0.598	3.197	0.598	8.067
0.600	2.782	0.963	0.228	0.983	0.947	0.230	0.611	2.679	0.611	6.682
0.700	2.408	0.927	0.271	0.967	0.897	0.276	0.606	2.233	0.606	5.513
0.800	2.078	0.891	0.325	0.950	0.846	0.333	0.601	1.851	0.601	4.534
0.900	1.792	0.854	0.389	0.934	0.798	0.403	0.596	1.531	0.596	3.735
1.000	1.550	0.818	0.467	0.917	0.750	0.487	0.592	1.268	0.592	3.098
1.100	1.352	0.781	0.557	0.900	0.703	0.587	0.588	1.056	0.588	2.605
1.200	1.198	0.744	0.656	0.883	0.657	0.698	0.585	0.892	0.585	2.238
1.300	1.088	0.707	0.759	0.866	0.613	0.815	0.584	0.770	0.584	1.977
1.400	1.022	0.670	0.854	0.849	0.569	0.927	0.585	0.685	0.585	1.804
1.500	1.000	0.633	0.930	0.833	0.527	1.018	0.588	0.633	0.588	1.701
1.600	1.022	0.594	0.979	0.800	0.475	1.094	0.594	0.607	0.594	1.621
1.700	1.088	0.555	0.992	0.766	0.425	1.134	0.599	0.604	0.599	1.572
1.800	1.198	0.517	0.975	0.731	0.377	1.141	0.604	0.619	0.604	1.542
1.900	1.352	0.478	0.939	0.695	0.333	1.126	0.607	0.647	0.607	1.523
2.000	1.550	0.440	0.893	0.660	0.290	1.099	0.609	0.682	0.609	1.506
2.100	1.792	0.401	0.848	0.625	0.251	1.073	0.610	0.719	0.610	1.485
2.200	2.078	0.362	0.809	0.590	0.214	1.054	0.610	0.753	0.610	1.456
2.300	2.408	0.324	0.781	0.554	0.179	1.049	0.609	0.780	0.609	1.413
2.400	2.782	0.285	0.766	0.519	0.148	1.063	0.607	0.793	0.607	1.354
2.500	3.200	0.246	0.768	0.484	0.119	1.104	0.605	0.788	0.605	1.278
2.600	3.662	0.208	0.791	0.448	0.093	1.182	0.601	0.760	0.601	1.184
2.700	4.168	0.169	0.846	0.412	0.070	1.318	0.595	0.704	0.595	1.078
2.800	4.718	0.131	0.949	0.375	0.049	1.551	0.584	0.616	0.584	0.965
2.900	5.312	0.093	1.133	0.324	0.030	1.990	0.560	0.494	0.560	0.846
3.000	5.950	0.063	1.438	0.200	0.013	3.217	0.536	0.373	0.536	0.726

steady state. This is shown in Fig. 7.19, where the nondimensional mass flow is plotted verus x/L for several different values of time. The dashed line labeled $0\Delta t$ represents the assumed initial conditions. Note that the transient mass flow deviates away from the initial conditions; the result after 100 time steps (labeled $100\Delta t$) shows a somewhat "humped" distribution. After 200 time steps (labeled $200\Delta t$), the mass flow distribution is becoming more constant, and after 700 time steps (labeled $700\Delta t$), it is almost (but not quite) equal to a constant value. Moreover, it is quite close to the exact analytical value of 0.579. Comparing Fig. 7.19 with the corresponding results obtained with the nonconservation form of the equations as plotted in Fig. 7.11, we see that the present variations in mass flow are much less severe. Of course, this is comparing apples and oranges, because Figs. 7.11 and 7.19 correspond to different initial conditions. We can suppose that the milder behavior

TABLE 7.11
Steady-state results, using the conservation form

$\dfrac{x}{L}$	$\dfrac{A}{A*}$	$\dfrac{\rho}{\rho_0}$	$\dfrac{V}{a_0}$	$\dfrac{T}{T_0}$	$\dfrac{p}{p_0}$	M	$\dot{m}$	U_1	U_2	U_3
0.000	5.950	1.000	0.098	1.000	1.000	0.098	0.583	5.950	0.583	14.915
0.100	5.312	0.999	0.110	0.999	0.998	0.110	0.583	5.306	0.583	13.301
0.200	4.718	0.997	0.124	0.999	0.996	0.124	0.583	4.704	0.583	11.798
0.300	4.168	0.995	0.141	0.998	0.993	0.141	0.583	4.147	0.583	10.404
0.400	3.662	0.992	0.161	0.997	0.989	0.161	0.583	3.633	0.583	9.118
0.500	3.200	0.988	0.184	0.995	0.983	0.185	0.583	3.161	0.583	7.941
0.600	2.782	0.982	0.213	0.993	0.975	0.214	0.583	2.732	0.583	6.869
0.700	2.408	0.974	0.249	0.989	0.964	0.250	0.584	2.345	0.584	5.903
0.800	2.078	0.962	0.292	0.985	0.948	0.294	0.584	2.000	0.584	5.043
0.900	1.792	0.946	0.344	0.978	0.926	0.348	0.584	1.696	0.584	4.287
1.000	1.550	0.923	0.408	0.969	0.894	0.415	0.584	1.431	0.584	3.632
1.100	1.352	0.891	0.485	0.955	0.851	0.496	0.585	1.205	0.585	3.075
1.200	1.198	0.847	0.577	0.935	0.792	0.596	0.585	1.015	0.585	2.609
1.300	1.088	0.789	0.682	0.909	0.718	0.715	0.585	0.859	0.585	2.231
1.400	1.022	0.718	0.798	0.874	0.628	0.854	0.586	0.734	0.586	1.932
1.500	1.000	0.648	0.904	0.839	0.544	0.987	0.586	0.648	0.586	1.730
1.600	1.022	0.548	1.046	0.783	0.429	1.182	0.586	0.560	0.586	1.525
1.700	1.088	0.462	1.164	0.731	0.338	1.361	0.585	0.503	0.585	1.396
1.800	1.198	0.384	1.272	0.679	0.261	1.544	0.585	0.460	0.585	1.301
1.900	1.352	0.316	1.368	0.628	0.198	1.726	0.585	0.427	0.585	1.231
2.000	1.550	0.260	1.452	0.581	0.151	1.905	0.584	0.402	0.584	1.178
2.100	1.792	0.214	1.524	0.538	0.115	2.077	0.584	0.383	0.584	1.138
2.200	2.078	0.177	1.586	0.500	0.088	2.243	0.583	0.368	0.583	1.107
2.300	2.408	0.148	1.639	0.466	0.069	2.402	0.583	0.356	0.583	1.083
2.400	2.782	0.124	1.685	0.436	0.054	2.554	0.583	0.346	0.583	1.064
2.500	3.200	0.106	1.725	0.409	0.043	2.698	0.583	0.338	0.583	1.048
2.600	3.662	0.090	1.760	0.384	0.035	2.838	0.582	0.331	0.582	1.035
2.700	4.168	0.078	1.790	0.363	0.028	2.969	0.582	0.325	0.582	1.025
2.800	4.718	0.068	1.817	0.344	0.023	3.100	0.582	0.320	0.582	1.015
2.900	5.312	0.060	1.840	0.327	0.019	3.216	0.582	0.316	0.582	1.008
3.000	5.950	0.052	1.863	0.310	0.016	3.345	0.582	0.312	0.582	1.001

illustrated in Fig. 7.19 is due predominantly to our assumed initial condition of constant mass flow.

Let us compare the *steady-state* variations of mass flow obtained with the nonconservation and conservation forms of the governing equations (both after 1400 time steps—well beyond the time required to converge to the steady state). This comparison is shown in Fig. 7.20, in which the scale of the ordinate for mass flow is greatly magnified. Here we see that the steady-state mass flow distribution predicted by the *conservation* form of the equations is much more satisfactory than that obtained from the nonconservation form, on two accounts:

1. The conservation form gives a distribution that is much closer to being a constant. In contrast, the nonconservation results have (on the magnified scale) a

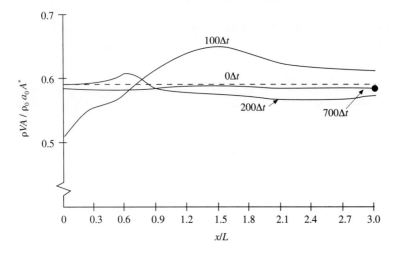

FIG. 7.19
Variation of mass flow distribution through the nozzle at different times during the time-marching process; solution of the conservation form of the governing equations. Circle indicates exact analytical value.

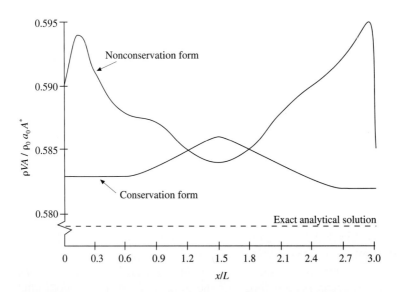

FIG. 7.20
A detailed comparison of the steady-state mass flow variations (on a magnified scale) obtained with the nonconservation and conservation forms of the governing flow equations.

sizeable variation, with some spurious oscillations at both the inflow and outflow boundaries. Of course, on a practical basis, when plotted on the scale shown in Fig. 7.11, these variations are not apparent, and the mass flow essentially appears to be a constant.

2. The steady-state mass flow results obtained with the conservation form are, on the whole, much closer to the exact analytical solution of $\rho'A'V' = 0.579$, shown in Fig. 7.20 by the dashed line.

The comparison shown in Fig. 7.20 illustrates a general advantage of the conservation form of the equations. The conservation form does a better job of preserving mass throughout the flow field, mainly because the mass flow itself is one of the dependent variables in the equations—the mass flow is a primary result from these equations. In contrast, the dependent variables in the nonconservation form of the equations are the primitive variables, and the mass flow is obtained only as a secondary result. Because the conservation form of the equations does a better job of conserving mass throughout the flow field, we can begin to understand why they are labeled the *conservation* form.

Caution: The above discussion does not necessarily establish a definite superiority of the conservation form results over the nonconservation form results. Quite the contrary, let us take a look at the *primitive* variables; in particular, temperature, pressure, and Mach number at the nozzle throat, as tabulated in Table 7.12. The first row gives the exact, analytical results. The second and third rows give the numerical results for the nonconservation and the conservation forms, respectively. Note that the *nonconservation form* results are distinctly closer to the exact values. The last row in Table 7.12 gives conservation form results for a grid with twice as many grid points (61 in comparison to 31 points). A comparison of the last two rows are an indication of *grid independence* for the conservation form results. Note that, by doubling the number of grid points, the steady-state numerical results are slightly closer to the exact, analytical values (but still not as close as the nonconservation form results with half as many grid points). For all practical purposes, we have grid independence with 31 grid points.

TABLE 7.12
Comparison of steady-state results; conservation versus nonconservation form

	$\dfrac{\rho^*}{\rho_0}$	$\dfrac{T^*}{T_0}$	$\dfrac{p^*}{p_0}$	M
Exact analytical solution	0.634	0.833	0.528	1.000
Nonconservation form, numerical results (31 points)	0.639	0.836	0.534	0.999
Conservation form, numerical results (31 points)	0.648	0.839	0.544	0.987
Conservation form, numerical results (61 points)	0.644	0.838	0.540	0.989

The behavior of the residuals for the conservation form is not as good as that for the nonconservation form of the equations. For the nonconservation form, recall from Fig. 7.10 that at early times the residuals are on the order of 10^{-1}, but they decay to about 10^{-6} after 1400 time steps. In contrast, for the conservation form of the equations, the residuals at early times are on the order of 10^{-1} but decay only to about 10^{-3} after 1400 time steps. However, this is sufficient to produce the steady-state results, for all practical purposes.

In summary, for the given flow problem, we cannot establish a clear superiority of the conservation form of the governing equations over the nonconservation form. In essence, from all our previous discussions, we can only make the following observations:

1. The conservation form yields a better mass flow distribution. The conservation form simply does a better job of conserving mass.
2. The nonconservation form leads to smaller residuals. The amount by which the residuals decay is often used as an index of "quality" of the numerical algorithm. In this sense, the nonconservation form does a better job.
3. There is no clear superiority of either form in terms of accuracy of the results. The nonconservation form seems to produce slightly more accurate results for the primitive variables, and the conservation form seems to produce slightly more accurate results for the flux variables. The results in either case are certainly satisfactory.
4. Comparing the amount of calculational effort to achieve a solution, as reflected in our extended discussions in Sec. 7.3 (nonconservation form) and Sec. 7.5 (conservation form), we note that the solution of the conservation form requires marginally more work. Most of this is due to the need to decode the primitive variables from the flux variables; such decoding is not necessary when you are solving the nonconservation form.

7.6 A CASE WITH SHOCK CAPTURING

In Sec. 7.2 we discussed the physical aspects of subsonic-supersonic isentropic flow. We emphasized that for a given nozzle shape there exists only one unique solution; the qualitative aspects of that solution are sketched in Fig. 7.2. Return to Fig. 7.2, and in particular focus on the pressure distribution shown in Fig. 7.2c. The pressure ratio across the nozzle, p_e/p_0, comes out as part of the solution; i.e., we do not have to *specify* it to obtain the solution. (On the other hand, in the laboratory we would have to make certain that this particular pressure ratio somehow is maintained across the nozzle, or else the subsonic-supersonic isentropic solution may not occur.) In contrast, in Sec. 7.4 we discussed the physical aspects of purely subsonic flow through the nozzle and emphasized the fact that there are an infinite number of possible isentropic flow solutions to this problem, each one corresponding to a specific value of the pressure ratio p_e/p_0. In this case, we *have to specify* p_e/p_0 to obtain a unique solution. The qualitative behavior of such subsonic flow solutions is sketched in Fig. 7.13.

Return to Fig. 7.13, and let us ask the question: What happens when the exit pressure is reduced slightly *below* the value $(p_e)_c$? The answer is that the nozzle becomes "choked;" i.e., the flow remains sonic at the throat, and the mass flow becomes a fixed value, no matter how much p_e is reduced below the value $(p_e)_c$. The flow downstream of the nozzle throat goes supersonic, following for a certain length the isentropic flow solution described in Fig. 7.2. Assume that the exit pressure is denoted by $(p_e)_d$, where $(p_e)_d$ is less than $(p_e)_c$ by a relatively small amount. In this case, a normal shock wave must form somewhere in the divergent portion of the nozzle, as sketched in Fig. 7.21. Upstream of the normal shock wave, the flow is given by the subsonic-supersonic isentropic solution. The flow, which is supersonic immediately in front of the shock, becomes subsonic immediately behind the shock. Further downstream, this subsonic flow slows within the divergent duct, with a corresponding increase in pressure. These variations are sketched in Fig. 7.21. The pressure at the exit of the nozzle is equal to $(p_e)_d$, which is the *imposed* pressure at the exit. The *location* of the normal shock wave within the nozzle is just right such that the static pressure increase across the shock wave plus the further static pressure

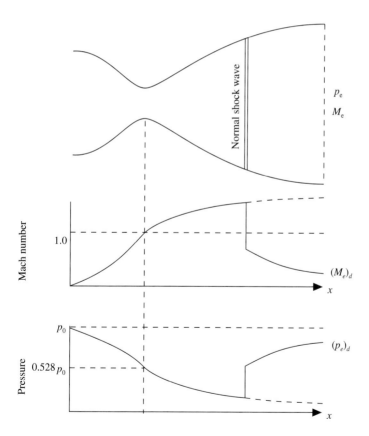

FIG. 7.21
Schematic of a nozzle flow with an internal normal shock wave.

increase downstream of the shock results in precisely $(p_e)_d$ at the exit. (In contrast, the full subsonic-supersonic isentropic solution is shown by the dashed line in Fig. 7.21.) As in the case of the purely subsonic flow case, the present solution depends on the value of $(p_e)_d$. To have a unique solution, $(p_e)_d$ must be *specified*. For more details on the physical nature of this type of flow, see the extensive discussion in Refs. 8 and 21.

In the present section, we will numerically solve a nozzle flow with p_e specified such that a normal shock wave will form within the nozzle. In terms of our overall development of the basics of CFD in this book, this case is important because it will illustrate the aspect of *shock capturing* within a numerical solution of the flow. The nature of shock capturing was described in Sec. 2.10. Make certain to review that section before progressing further; it is important for you to have clearly in mind the idea of shock capturing and why it is necessary to use the conservation form of the governing equations to numerically capture shock waves within a flow field. Also, reexamine Fig. 1.32c, which itemizes those various ideas that feed into this application.

7.6.1 The Setup

Consider the nomenclature shown in Fig. 7.22. The normal shock wave is located at area A_1. Conditions immediately upstream of the shock are denoted with a subscript 1, and those immediately downstream of the shock are denoted with a subscript 2. The flow from the reservoir, where the pressure is p_0, to station 1 is isentropic (with constant entropy s_1). Hence, the total pressure is constant in this flow; that is, $p_{0_1} = p_0$. The total pressure decreases across the shock (due to the entropy increase across the shock). The flow from station 2 downstream of the shock to the nozzle exit is also isentropic (with constant entropy s_2, where $s_2 > s_1$). Hence the total pressure is constant in this portion of the flow, with $(p_0)_e = p_{0_2}$. Keep in mind that $p_{0_2} < p_{0_1}$. For the flow in front of the shock, A_1^* is a constant value, equal to the area of the sonic throat, $A_1^* = A_t$. However, due to the entropy increase across the shock, the value of A^* in the subsonic flow downstream of the shock, denoted by A_2^*, takes on the role of a reference value (just as in the purely subsonic case discussed in Sec. 7.4). Indeed, $A_2^* > A_1^*$.

In this section, we will numerically calculate the flow through a convergent-divergent nozzle under the condition where a normal shock wave exists in the divergent portion. The nozzle shape will be the same as used in Sec. 7.3, namely, that given by Eq. (7.73). We will use the governing equations in conservation form and will employ the philosophy of shock capturing. However, before jumping into the numerical solution, let us examine the exact analytical results.

EXACT ANALYTICAL RESULTS. For the nozzle shape specified by Eq. (7.73), the area of the exit is $A_e/A_t = 5.95$. Let us calculate the flow where p_e is *specified* as follows:

$$\frac{p_e}{p_{0_1}} = 0.6784 \quad \text{(specified)} \tag{7.123}$$

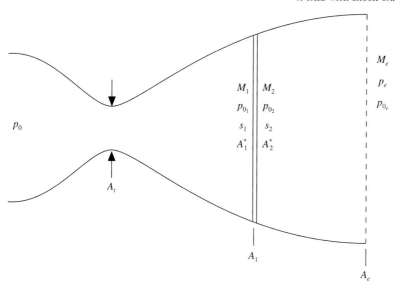

FIG. 7.22
Nomenclature for the normal shock case.

Note that this value is considerably below the values specified in Sec. 7.4 for the purely subsonic case; there, we specified, for example, that $p_e/p_0 = 0.93$ and calculated the corresponding subsonic flow through the nozzle. Also note that the value $p_e/p_{0_1} = 0.6784$ is considerably higher than that which came out of the subsonic-supersonic isentropic solution in Sec. 7.2, where we found that $p_e/p_0 = 0.016$. Hence, the value of $p_e/p_{0_1} = 0.6784$ specified in the current section should be about right to force a normal shock wave to stand somewhere inside the divergent portion of the nozzle. Let us first calculate the precise location, i.e., the precise area ratio inside the nozzle, where the normal shock wave will be located, compatible with the specified exit pressure given in Eq. (7.123). This calculation can be done in a direct fashion as follows.

The mass flow through the nozzle can be expressed as

$$\dot{m} = \frac{p_{0_1} A_1^*}{\sqrt{T_0}} \sqrt{\frac{\gamma}{R} \left(\frac{2}{\gamma+1}\right)^{(\gamma+1)/(\gamma-1)}} \qquad (7.124)$$

See, for example, Refs. 8 and 21 for further discussion. That is, for a given T_0,

$$\dot{m} \propto p_0 A^*$$

Since the mass flow is constant across the normal shock wave in Fig. 7.22, we have

$$p_{0_1} A_1^* = p_{0_2} A_2^* \qquad (7.125)$$

(Keep in mind from our previous conversations that A^* is always defined as the *sonic* throat area; in the supersonic flow ahead of the shock, A_1^* is equal to the actual throat area A_t, because the flow is actually sonic at A_t, whereas behind the shock A_2^* is the area the flow behind the shock would have to be reduced to in order to make it locally sonic. Since the flow behind the shock is always subsonic, then A_2^* never equals the actual physical throat area in the nozzle itself, because the entropy in region 2 is higher than in region 1.) Forming the ratio $p_e A_e / p_{0_e} A_2^*$, where $A_e^* = A_2^*$, and invoking Eq. (7.125), we have

$$\frac{p_e A_e}{p_{0_e} A_e^*} = \frac{p_e A_e}{p_{0_2} A_2^*} = \frac{p_e A_e}{p_{0_1} A_1^*} = \frac{p_e}{p_{0_1}} \frac{A_e}{A_t} \tag{7.126a}$$

The right-hand side of Eq. (7.126a) is known, because p_e/p_{0_1} is specified as 0.6784 and $A_e/A_t = 5.95$. Thus, from Eq. (7.126a)

$$\frac{p_e A_e}{p_{0_e} A_e^*} = 0.6784(5.95) = 4.03648 \tag{7.126b}$$

From the isentropic relations given by Eqs. (7.6) and (7.7), we have, respectively,

$$\frac{A_e}{A_e^*} = \frac{1}{M_e} \left[\frac{2}{\gamma+1} \left(1 + \frac{\gamma-1}{2} M_e^2 \right) \right]^{(\gamma+1)/2(\gamma-1)} \tag{7.127}$$

and

$$\frac{p_e}{p_{0_e}} = \left(1 + \frac{\gamma-1}{2} M_e^2 \right)^{-\gamma/(\gamma-1)} \tag{7.128}$$

Substituting Eqs. (7.127) and (7.128) in (7.126b), we have

$$\frac{1}{M_e} \left(\frac{2}{\gamma+1} \right)^{(\gamma+1)/2(\gamma-1)} \left[1 + \frac{\gamma-1}{2} M_e^2 \right]^{-1/2} = 4.03648 \tag{7.129}$$

Solving Eq. (7.129) for M_e, we have

$$M_e = 0.1431 \tag{7.130}$$

From Eq. (7.128), we have

$$\frac{p_e}{p_{0_e}} = \left[1 + \frac{\gamma-1}{2}(0.1431)^2 \right]^{-3.5} = 0.9858 \tag{7.131}$$

The total pressure ratio across the normal shock can be written as

$$\frac{p_{0_2}}{p_{0_1}} = \frac{p_{0_e}}{p_{0_1}} = \frac{p_{0_e}}{p_e} \frac{p_e}{p_{0_1}} \tag{7.132}$$

Substituting the numbers from Eqs. (7.123) and (7.131) into Eq. (7.132), we have

$$\frac{p_{0_2}}{p_{0_1}} = \frac{0.6784}{0.9858} = 0.6882 \tag{7.133}$$

The total pressure ratio across a normal shock is a function of M_1 in front of the shock, given by (see Ref. 45)

$$\frac{p_{0_2}}{p_{0_1}} = \left[\frac{(\gamma+1)M_1^2}{(\gamma-1)M_1^2+2}\right]^{\gamma/(\gamma-1)} \left[\frac{\gamma+1}{2\gamma M_1^2-(\gamma-1)}\right]^{1/(\gamma-1)} \quad (7.134)$$

Combining Eqs. (7.133) and (7.134) and solving for M_1, we have

$$M_1 = 2.07 \quad (7.135)$$

Substituting Eq. (7.135) into (7.6), we have

$$\frac{A_1}{A_1^*} = \frac{A_1}{A_t} = 1.790 \quad (7.136)$$

The exact, analytical location of the normal shock wave is now known—it stands at a location in the nozzle where the area ratio is 1.79. From Eq. (7.73) for our nozzle shape, this corresponds to a station of $x/L = 2.1$. All other properties across the shock wave now fall out from the result that $M_1 = 2.07$. For example, from Ref. 21, the static pressure ratio across the shock and the Mach number immediately behind the shock are obtained from

$$\frac{p_2}{p_1} = 1 + \frac{2\gamma}{\gamma+1}(M_1^2-1) = 1 + 1.167[(2.07)^2 - 1] = 4.83 \quad (7.137)$$

and

$$M_2 = \left\{\frac{1+[(\gamma-1)/2]M_1^2}{\gamma M_1^2 - (\gamma-1)/2}\right\}^{1/2} = \left[\frac{1+0.2(2.07)^2}{1.4(2.07)^2 - 0.2}\right]^{1/2} = 0.566 \quad (7.138)$$

The exact, analytical solution obtained above will be compared with the numerical solution in subsequent sections.

BOUNDARY CONDITIONS. The subsonic inflow boundary conditions are treated exactly as described in Sec. 7.5.2 and given by Eqs. (7.116) and (7.118); hence, no elaboration will be given here.

The outflow boundary condition for the present problem is also subsonic. A generic discussion of a subsonic outflow boundary was given in Sec. 7.4.1, where we emphasized that the exit pressure p_e must be specified, but all other properties are allowed to float. The same applies to the present calculation. However, in Sec. 7.4.1 we proceeded to couch the details of the numerical implementation of the subsonic outflow boundary condition in terms of the solution of the nonconservation form of the governing equations. In contrast, in the present calculation we are using the conservation form of the equations; hence the numerical implementation is slightly different, as follows. Keep in mind that U_1, U_2, and U_3 are the primary dependent variables in the governing equations. Hence, we obtain U_1 and U_2 at the downstream boundary by linear extrapolation from the adjacent two interior points.

$$(U_1)_N = 2(U_1)_{N-1} - (U_1)_{N-2} \quad (7.139a)$$
$$(U_2)_N = 2(U_2)_{N-1} - (U_2)_{N-2} \quad (7.139b)$$

Next, we decode V'_N from $(U_1)_N$ and $(U_2)_N$ using Eq. (7.103).

$$V'_N = \frac{(U_2)_N}{(U_1)_N} \quad (7.140)$$

The value of U_3 at grid point $i = N$ is determined from the *specified* value of $p'_N = 0.6784$ as follows. From the definition of U_3,

$$U_3 = \rho' \left(\frac{e'}{\gamma - 1} + \frac{\gamma}{2} V'^2 \right) A' \quad (7.141)$$

However, $e' = T'$, and from the equation of state, $p' = \rho' T'$. Hence, Eq. (7.141) becomes

$$U_3 = \frac{p'A'}{\gamma - 1} + \frac{\gamma}{2} \rho' A' V'^2 \quad (7.142)$$

Since $U_2 = \rho' A' V'$, Eq. (7.142) becomes

$$U_3 = \frac{\rho' A'}{\gamma - 1} + \frac{\gamma}{2} U_2 V' \quad (7.143)$$

Evaluating Eq. (7.143) at the downstream boundary, we have

$$(U_3)_N = \frac{p'_N A'}{\gamma - 1} + \frac{\gamma}{2} (U_2)_N V'_N \quad (7.144)$$

Since p'_N is specified as 0.6784, Eq. (7.144) becomes

$$(U_3)_N = \frac{0.6784 A'}{\gamma - 1} + \frac{\gamma}{2} (U_2)_N V'_N \quad (7.145)$$

Equation (7.145) is the manner in which the specified exit pressure is folded into the numerical solution.

INITIAL CONDITIONS. For the present calculations, we choose the following initial conditions, which are qualitatively similar to the final solution. From $x' = 0$ to 1.5, we use the same initial conditions as given by Eqs. (7.120a) to (7.120d). However, for $x' > 1.5$, we use

$$\left. \begin{array}{l} \rho' = 0.634 - 0.702(x' - 1.5) \\ T' = 0.833 - 0.4908(x' - 1.5) \end{array} \right\} \text{ for } 1.5 \leq x' \leq 2.1 \quad \begin{array}{l} (7.146a) \\ (7.146b) \end{array}$$

$$\left. \begin{array}{l} \rho' = 0.5892 + 0.10228(x' - 2.1) \\ T' = 0.93968 + 0.0622(x' - 2.1) \end{array} \right\} \text{ for } 2.1 \leq x' \leq 3.0 \quad \begin{array}{l} (7.146c) \\ (7.146d) \end{array}$$

As before, the initial condition for V' is determined by assuming a constant mass flow; it is calculated from Eq. (7.121).

7.6.2 The Intermediate Time-Marching Procedure: The Need For Artificial Viscosity

Perhaps the most dramatic distinction between the present shock-capturing case and our previous calculations in this chapter is the matter of artificial viscosity. Think back about our calculations so far; they have been carried out with no artificial viscosity explicitly added to the numerical calculations. The solutions of the subsonic-supersonic isentropic flow (Sec. 7.3) and the purely subsonic flow (Sec. 7.4) did not require additional numerical dissipation—there was enough dissipation inherent in the algorithm itself to yield stable and smooth solutions. Furthermore, it made no difference whether the governing equations were used in nonconservation form (Sec. 7.3 and 7.4) or in conservation form (Sec. 7.5). The requirement for artificial viscosity is essentially disconnected with which form of the equations is used. However, as we will see in the next section, when we practice the art of shock capturing, the smoothing and stabilization of the solution by the addition of some type of numerical dissipation is absolutely necessary. At this stage, return to Sec. 6.6 where the matter of artificial viscosity is introduced. Read this section again before proceeding further so that you can more fully understand what we have to do to obtain a reasonable solution for the nozzle flow with a normal shock wave standing inside the nozzle.

To proceed with this solution, we will add artificial viscosity in the manner described in Sec. 6.6. Specifically following Eq. (6.58), we form an expression

$$S^{t'}_i = \frac{C_x |(p')^{t'}_{i+1} - 2(p')^{t'}_i + (p')^{t'}_{i-1}|}{(p')^{t'}_{i+1} + 2(p')^{t'}_i + (p')^{t'}_{i-1}} (U^{t'}_{i+1} - 2U^{t'}_i + U^{t'}_{i-1}) \qquad (7.147)$$

Whereas beforehand we would calculate a predicted value (using MacCormack's technique) from

$$\bar{U}^{t'+\Delta t'}_i = (U)^{t'}_i + \left(\frac{\partial U}{\partial t'}\right)^{t'}_i \Delta t'$$

we now replace this with

$$(\bar{U}_1)^{t'+\Delta t'}_i = (U_1)^{t'}_i + \left(\frac{\partial U_1}{\partial t'}\right)^{t'}_i \Delta t' + (S_1)^{t'}_i \qquad (7.148)$$

$$(\bar{U}_2)^{t'+\Delta t'}_i = (U_2)^{t'}_i + \left(\frac{\partial U_2}{\partial t'}\right)^{t'}_i \Delta t' + (S_2)^{t'}_i \qquad (7.149)$$

$$(\bar{U}_3)^{t'+\Delta t'}_i = (U_3)^{t'}_i + \left(\frac{\partial U_3}{\partial t'}\right)^{t'}_i \Delta t' + (S_3)^{t'}_i \qquad (7.150)$$

where U_1, U_2, and U_3 are our dependent variables in Eqs. (7.101a) to (7.101c) and S_1, S_2, and S_3 in Eqs. (7.148) to (7.150) are obtained from Eq. (7.147) by using, respectively, U_1, U_2, and U_3 on the right-hand side. Similarly, on the corrector step, whereas beforehand we would calculate the corrected values from

$$U_i^{t'+\Delta t'} = U_i' + \left(\frac{\partial U}{\partial t'}\right)_{av} \Delta t'$$

we now replace this with

$$(U_1)_i^{t'+\Delta t'} = (U_1)_i^{t'} + \left(\frac{\partial U_1}{\partial t}\right)_{av} \Delta t' + (\bar{S}_1)_i^{t} \qquad (7.151)$$

$$(U_2)_i^{t'+\Delta t'} = (U_2)_i^{t'} + \left(\frac{\partial U_2}{\partial t}\right)_{av} \Delta t' + (\bar{S}_2)_i^{t} \qquad (7.152)$$

$$(U_3)_i^{t'+\Delta t'} = (U_3)_i^{t'} + \left(\frac{\partial U}{\partial t}\right)_{av} \Delta t' + (\bar{S}_3)_i^{t} \qquad (7.153)$$

where $\bar{S}_1$, $\bar{S}_2$, and $\bar{S}_3$ are obtained from an equation patterned after Eq. (6.59), namely,

$$\bar{S}_i^{t'+\Delta t'} = \frac{C_x |(\bar{p}')_{i+1}^{t'+\Delta t'} - 2(\bar{p}')_i^{t'+\Delta t'} + (\bar{p}')_{i-1}^{t'+\Delta t'}|}{(\bar{p}')_{i+1}^{t'} + 2(\bar{p}')_i^{t'} + (\bar{p}')_{i-1}^{t'}} \\ \times [(\bar{U})_{i+1}^{t'+\Delta t'} - 2(\bar{U})_i^{t'+\Delta t'} + (\bar{U})_{i-1}^{t'+\Delta t'}] \qquad (7.154)$$

The values of $\bar{S}_1$, $\bar{S}_2$, and $\bar{S}_3$ are obtained from Eq. (7.154) by using, respectively, the values of $\bar{U}_1$, $\bar{U}_2$, and $\bar{U}_3$ on the right-hand side.

The rest of the shock-capturing solution proceeds in exactly the same manner as our previously described case in Sec. 7.5; hence no further elaboration will be given here. We will proceed directly to a discussion of the steady-state results.

7.6.3 Numerical Results

The following numerical results were obtained with 61 grid points distributed evenly through the nozzle rather than the 31-point grid used for most of our previous results. Since in the shock-capturing approach using MacCormack's finite-difference technique the captured shock wave is spread over several grid points, it is desirable to have a finer grid so as to more precisely define the location of the shock. Also, for the following results, a Courant number of 0.5 was employed. The conservation form of the governing flow equations was used in exactly the same manner as in Sec. 7.5 (except for the numerical implementation of the downstream boundary conditions, which has already been described in Sec. 7.6.1 and for the addition of artificial viscosity as described in Sec. 7.6.2). The pressure ratio at the nozzle exit is specified as $p'_e = p_e/p_0 = 0.6784$ and is held fixed, invariant of time.

To begin with, it is instructional to examine what happens when *no* artificial viscosity is added to the calculations. Figure 7.23 shows the numerical solution for the pressure distribution through the nozzle (the solid line) compared with the exact,

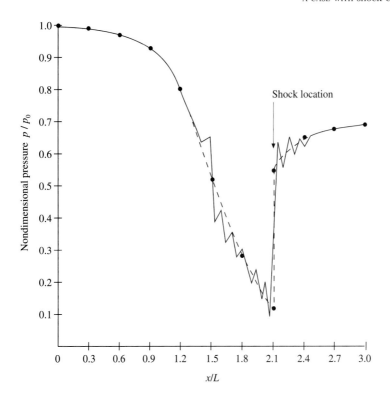

FIG. 7.23
Shock-capturing numerical results (solid line) for the pressure distribution through the nozzle. No artificial viscosity. Results shown are for 1600 time steps. Comparison with the exact analytical results (solid circles connected by dashed curve).

analytical results (the solid circles connected by the dashed curve). The numerical solution is that obtained after 1600 time steps, corresponding to a nondimensional time of 17.2. No artificial viscosity has been added to this calculation. At 1600 time steps, the numerical results are not steady-state results. Although the numerical results are trying to capture the shock wave in about the right location, the residuals are still fairly large—on the order of 10^{-1}. Moreover, as time progresses beyond 1600 time steps, the residuals start to grow instead of decreasing as they should. By 2800 time steps, the attempted solution has not blown up, but the oscillations have grown much more pronounced, and some of the residuals have grown to 10^1. This is a totally unsatisfactory solution, and we will not discuss it further. It needs to be fixed by the addition of artificial viscosity, as discussed below.

When artificial viscosity is added to the calculation via Eqs. (7.147) to (7.154) and the adjustable constant C_x is set equal to 0.2, the following results are obtained. The steady-state pressure distribution through the nozzle is shown in Fig. 7.24. The numerical results (the solid curve) are shown after 1400 time steps–the converged, steady state. The exact, analytical results are given by the solid circles connected by the dashed curve. From Fig. 7.24, we make the following observations:

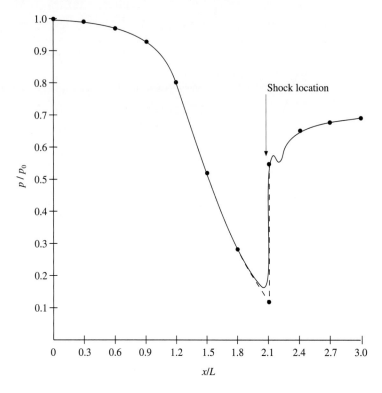

FIG. 7.24
Shock-capturing numerical results (solid curve) for the pressure distribution through the nozzle. With artificial viscosity; $C_x = 0.2$. Results shown are for 1400 time steps. Comparison with exact, analytical results (solid circles connected by dashed curve).

1. The addition of artificial viscosity has just about eliminated the oscillations that were encountered in the case with no artificial viscosity. The contrast between the numerical results in Fig. 7.24 (with $C_x = 0.2$) and those in Fig. 7.23 (with $C_x = 0.0$) is dramatic. This is what artificial viscosity does—smooth the results and decrease (if not virtually eliminate) the oscillations.
2. Close examination of Fig. 7.24 shows that the oscillations are not *completely* eliminated. There is a small oscillation in the pressure distribution just downstream of the shock; however, it is not that bothersome. Results obtained with more artificial viscosity ($C_x = 0.3$) show that even this small oscillation virtually disappears. However, *too much* artificial viscosity can compromise other aspects of the solution, as noted below.
3. The numerical results in Fig. 7.24 show that artificial viscosity tends to smear the captured shock wave over more grid points. The more extreme changes across the shock that are predicted by the exact, analytical results are slightly diminished by the inclusion of artificial viscosity in the numerical results. This increased smearing of the shock wave due to increased artificial viscosity is one of the undesirable aspects of adding extra numerical dissipation to the

solution. Some modern CFD methods (beyond the scope of this book) have successfully improved this situation; by using innovative ideas from applied mathematics, current researchers are able to reap the benefits of adding numerical dissipation when and where in the flow field it is really needed and still preserve the sharpness of the captured shock wave. Such matters are left to your future advanced studies of CFD.

The steady-state Mach number distribution is shown in Fig. 7.25; these results simply reinforce the comments made above.

The detailed steady-state numerical results, obtained after 1400 time steps, are tabulated in Table 7.13 for comparison with numbers obtained with your own computer program. Just as a reminder, these results are obtained by using the conservation form of the governing equations, artificial viscosity where $C_x = 0.2$, a Courant number of 0.5, and 61 points evenly distributed along the nozzle. The solution corresponds to a specified exit pressure ratio $p_e/p_0 = 0.6784$. Scanning down the various columns for ρ', p', etc., in the vicinity of the shock wave (which is theoretically located at grid point $i = 43$, i.e., at $x' = 2.1$) we see just how small is the slight oscillation downstream of the shock. However, focus for a moment on the column for mass flow; here we see that $\dot{m} = \rho'A'V'$ is essentially constant upstream of the shock wave at a value of $\rho'A'V' = 0.582$. (Recall that the exact, analytical

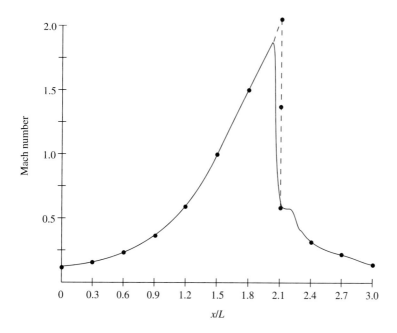

FIG. 7.25
Shock-capturing numerical results (solid curve) for the Mach number distribution through the nozzle. With artificial viscosity; $C_x = 0.2$. Results shown here are for 1400 time steps. Comparison with the exact, analytical results (solid circles connected by dashed curve).

368 NUMERICAL SOLUTIONS OF QUASI-ONE-DIMENSIONAL NOZZLE FLOWS

TABLE 7.13
Shock capturing, steady-state numerical results

I	$\dfrac{x}{L}$	$\dfrac{A}{A^*}$	$\dfrac{\rho}{\rho_0}$	$\dfrac{V}{a_0}$	$\dfrac{T}{T_0}$	$\dfrac{p}{p_0}$	M	$\dot{m}$
1	0.000	5.950	1.000	0.098	1.000	1.000	0.098	0.582
2	0.050	5.626	0.999	0.103	1.000	0.999	0.103	0.582
3	0.100	5.312	0.999	0.110	1.000	0.998	0.110	0.582
4	0.150	5.010	0.998	0.116	0.999	0.997	0.116	0.582
5	0.200	4.718	0.997	0.124	0.999	0.996	0.124	0.582
6	0.250	4.438	0.996	0.132	0.998	0.995	0.132	0.582
7	0.300	4.168	0.995	0.140	0.998	0.993	0.140	0.582
8	0.350	3.910	0.994	0.150	0.997	0.991	0.150	0.582
9	0.400	3.662	0.992	0.160	0.997	0.989	0.160	0.582
10	0.450	3.425	0.990	0.172	0.996	0.986	0.172	0.582
11	0.500	3.200	0.988	0.184	0.995	0.983	0.184	0.582
12	0.550	2.985	0.985	0.198	0.994	0.979	0.198	0.582
13	0.600	2.782	0.982	0.213	0.993	0.975	0.214	0.582
14	0.650	2.589	0.979	0.230	0.991	0.970	0.231	0.582
15	0.700	2.408	0.974	0.248	0.990	0.964	0.249	0.582
16	0.750	2.237	0.969	0.268	0.987	0.957	0.270	0.582
17	0.800	2.078	0.963	0.291	0.985	0.948	0.293	0.582
18	0.850	1.929	0.956	0.316	0.982	0.938	0.319	0.582
19	0.900	1.792	0.947	0.343	0.978	0.926	0.347	0.582
20	0.950	1.665	0.936	0.373	0.974	0.912	0.378	0.582
21	1.000	1.550	0.924	0.407	0.969	0.895	0.413	0.582
22	1.050	1.445	0.909	0.443	0.963	0.875	0.452	0.582
23	1.100	1.352	0.892	0.483	0.955	0.852	0.494	0.583
24	1.150	1.270	0.872	0.526	0.946	0.825	0.541	0.583
25	1.200	1.198	0.848	0.573	0.936	0.794	0.593	0.583
26	1.250	1.138	0.821	0.624	0.924	0.759	0.649	0.583
27	1.300	1.088	0.791	0.678	0.910	0.720	0.710	0.583
28	1.350	1.050	0.757	0.734	0.894	0.677	0.776	0.583
29	1.400	1.022	0.719	0.793	0.876	0.630	0.847	0.583
30	1.450	1.006	0.679	0.854	0.856	0.581	0.923	0.583

value is 0.579—the numerical result is very close.) But in the vicinity of the shock wave, $\dot{m}$ takes a substantial jump and seems to settle in to a value of about 0.632 further downstream of the shock wave.

A further look at this spurious mass flow behavior is provided in Fig. 7.26. Here, the nondimensional mass flow $\rho' A' V'$ is plotted versus distance through the nozzle. The scale of the graph is the same as that used in Fig. 7.19 for the subsonic-supersonic isentropic flow case. The solid line corresponds to the numerical results (obtained after 1600 time steps) for the case with no artificial viscosity ($C_x = 0$); the dashed line gives the numerical results for the case with artificial viscosity ($C_x = 0.2$). Note that with no artificial viscosity the mass flow exhibits a massive, vibratory behavior in the general vicinity of the shock wave—totally unacceptable, as stated before. In contrast, the case with artificial viscosity exhibits excellent mass flow behavior *upstream* of the shock wave, with a quality and accuracy every bit as good as that reflected in the steady-state results for the

I	$\dfrac{x}{L}$	$\dfrac{A}{A^*}$	$\dfrac{\rho}{\rho_0}$	$\dfrac{V}{a_0}$	$\dfrac{T}{T_0}$	$\dfrac{p}{p_0}$	M	$\dot{m}$
31	1.500	1.000	0.633	0.921	0.832	0.527	1.009	0.583
32	1.550	1.005	0.596	0.973	0.812	0.484	1.080	0.583
33	1.600	1.022	0.549	1.040	0.786	0.431	1.173	0.583
34	1.650	1.049	0.507	1.096	0.761	0.386	1.256	0.584
35	1.700	1.088	0.462	1.159	0.734	0.339	1.353	0.583
36	1.750	1.137	0.424	1.210	0.709	0.301	1.437	0.584
37	1.800	1.198	0.383	1.268	0.680	0.261	1.538	0.582
38	1.850	1.269	0.351	1.311	0.658	0.231	1.617	0.584
39	1.900	1.352	0.315	1.368	0.628	0.198	1.725	0.582
40	1.950	1.445	0.289	1.398	0.611	0.177	1.788	0.584
41	2.000	1.550	0.256	1.462	0.574	0.147	1.930	0.581
42	2.050	1.665	0.318	1.207	0.677	0.215	1.467	0.639
43	2.100	1.792	0.524	0.697	0.872	0.457	0.747	0.655
44	2.150	1.929	0.619	0.521	0.925	0.573	0.542	0.622
45	2.200	2.078	0.613	0.501	0.926	0.567	0.521	0.638
46	2.250	2.237	0.643	0.436	0.939	0.604	0.450	0.627
47	2.300	2.408	0.643	0.410	0.943	0.607	0.422	0.635
48	2.350	2.589	0.660	0.368	0.950	0.627	0.378	0.629
49	2.400	2.782	0.662	0.344	0.953	0.631	0.353	0.635
50	2.450	2.985	0.671	0.314	0.959	0.643	0.321	0.629
51	2.500	3.200	0.675	0.294	0.958	0.647	0.300	0.634
52	2.550	3.425	0.680	0.271	0.964	0.655	0.276	0.630
53	2.600	3.662	0.682	0.253	0.965	0.658	0.258	0.633
54	2.650	3.909	0.687	0.235	0.965	0.663	0.239	0.632
55	2.700	4.168	0.687	0.221	0.969	0.666	0.224	0.632
56	2.750	4.437	0.690	0.206	0.970	0.669	0.209	0.631
57	2.800	4.718	0.692	0.194	0.970	0.671	0.197	0.633
58	2.850	5.009	0.694	0.182	0.971	0.674	0.184	0.631
59	2.900	5.312	0.694	0.171	0.973	0.675	0.174	0.631
60	2.950	5.625	0.697	0.161	0.972	0.677	0.164	0.632
61	3.000	5.950	0.698	0.152	0.972	0.678	0.154	0.632

subsonic-supersonic isentropic flow solution shown in Fig. 7.19 and tabulated in Table 7.11. However, in the vicinity of the shock wave, the case with artificial viscosity takes an almost quantum jump in mass flow, leveling out at the nozzle exit to a value about 8.6 percent higher than at the nozzle inlet. Clearly, the artificial viscosity terms added to the numerical solution of the governing equations are acting like a *source* of mass flow in the vicinity of the shock wave. This is not hard to imagine when you again examine Eqs. (7.147) to (7.154). Note that the values of S_i'' obtained from Eq. (7.147) and those for $\bar{S}_i^{t'+\Delta t'}$ obtained from Eq. (7.154) are large in those regions of the flow where the change in pressure gradients are large; this is the role of the leading factor involving pressure in these expressions—it plays the role of a "sensor" which increases the amount of artificial viscosity in those regions where the pressure gradients are changing rapidly (the second derivative of the pressure), such as when the oscillatory behavior discussed earlier tries to occur. Furthermore, these values of $S_i^{t'}$ and $\bar{S}_i^{t'+\Delta t'}$ are directly added to the calculated

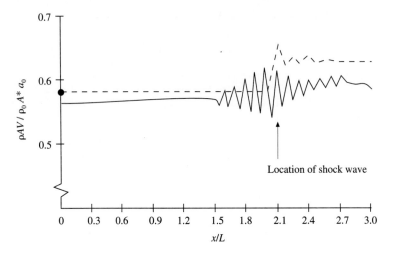

FIG. 7.26
Comparison of mass flow distributions with artificial viscosity (dashed line; $C_x = 0.2$) and without artificial viscosity (solid line; $C_x = 0$). Solid circle indicates exact analytical result.

values of the U_i's via Eqs. (7.148) to (7.150) and (7.151) to (7.154). In particular, recall that $U_2 = \rho'A'V'$ is the mass flow. Therefore, it is no surprise that the artificial viscosity governed by the scheme originally described in Sec. 6.6 would lead to source terms for mass flow.

Is this mass flow behavior for cases with artificial viscosity acceptable? The answer is essentially yes when you consider the alternative. Clearly, shock capturing leads to unacceptable oscillations (and sometimes unstable behavior) when no artificial viscosity is used in the calculations. So we *have* to use artificial viscosity, at least for the explicit MacCormack technique we have been illustrating in this chapter. In general, the results for the *primitive variables* that are obtained via a shock-capturing solution with artificial viscosity are acceptable. This is shown in Tables 7.14 and 7.15. In Table 7.14, we tabulate the steady-state flow-field values at the nozzle throat for the cases with C_x ranging from 0 to 0.3 and compare them with the exact analytical values. The case with no artificial viscosity, $C_x = 0$, has already

TABLE 7.14
Shock-capturing solution; values at the nozzle throat

	$\dfrac{\rho}{\rho_0}$	$\dfrac{V}{a_0}$	$\dfrac{T}{T_0}$	$\dfrac{p}{p_0}$	M	$\dot{m}$
Exact analytical values	0.634	0.913	0.833	0.528	1.0	0.579
Numerical values:						
$C_x = 0$	0.735	0.784	0.879	0.646	0.836	0.576
$C_x = 0.1$	0.629	0.926	0.831	0.523	1.016	0.583
$C_x = 0.2$	0.633	0.921	0.832	0.527	1.009	0.583
$C_x = 0.3$	0.640	0.911	0.836	0.535	0.997	0.583

TABLE 7.15
Shock-capturing solution; values at the nozzle exit

	$\dfrac{\rho}{\rho_0}$	$\dfrac{V}{a_0}$	$\dfrac{T}{T_0}$	$\dfrac{p}{p_0}$	M	$\dot{m}$
Exact analytical values	0.681	0.143	0.996	0.678	0.143	0.579
Numerical values:						
$C_x = 0$	0.672	0.148	1.009	0.678	0.147	0.591
$C_x = 0.1$	0.694	0.151	0.978	0.678	0.153	0.624
$C_x = 0.2$	0.698	0.152	0.972	0.678	0.154	0.632
$C_x = 0.3$	0.698	0.153	0.972	0.678	0.155	0.634

been affected at the throat by the oscillations working their way upstream from the shock wave; the comparison with the exact analytical values shows that the numerical results with no artificial viscosity are totally unacceptable. In contrast, the results with artificial viscosity are quite good. Indeed, the results at the nozzle throat obtained for the case with $C_x = 0.2$ are the most accurate of *any* case we have examined in this chapter! In Table 7.15, we tabulate the steady-state flow-field values at the nozzle exit, downstream of the captured normal shock wave. It is interesting to note (but not too surprising) that as the artificial viscosity is increased, the numerical results for the exit flow-field variables progressively move further away from the exact analytical values. Indeed, the results from the case with no artificial viscosity give the best comparison with the exact analytical results. On the other hand, the case for $C_x = 0$ is tabulated after 1600 time steps; as we have mentioned earlier, this case further deviates away from the steady state as time progresses and may very well blow up after enough time steps are taken. Therefore, the comparison associated with $C_x = 0$ in both Tables 7.14 and 7.15 is really moot.

With this, we end our discussion of shock capturing in a convergent-divergent nozzle. This has been a particularly relevant section, because:

1. It is an illustration of the shock-capturing philosophy as first discussed in Sec. 2.10. This is one of the two basic approaches for handling shock waves in CFD, the other being shock fitting. The shock-capturing philosophy is, by far, the most prevalent in CFD today.
2. It was our first application of artificial viscosity, which allowed us to examine some of the pros and cons of explicitly increasing the amount of numerical dissipation in the solution.
3. It allowed us the opportunity to calculate yet another flow using the conservation form of the governing equations. This form of the equations is, by far, the most prevalent in CFD today.

Also, let us wax philosophical for a moment. In this section, we have calculated a flow which contains a shock wave *without doing anything special to account for the shock*; that is, we have employed a form of the governing Euler equations for an inviscid flow and have imposed boundary conditions across the nozzle that, in nature, calls for a shock wave to be present in the nozzle. The numerical solution of the Euler equations senses this need for a shock wave and establishes it within the flow. Of course, this is the essence of shock capturing. But isn't it rather awesome that a set of equations for an *inviscid* flow, namely, the Euler equations, will allow the solution of such a flow with shock waves without us adding some additional *theoretical* baggage to the equations to alert them to the existence of the shock? Of course, some of the awesomeness is diminished when we realize that the *numerical solution* is really not solving the exact Euler equations, but rather a set of modified differential equations in the spirit of our discussion in Sec. 6.6, and that these modified equations have viscouslike terms on the right-hand side. Moreover, during the numerical solution, we are adding even more numerical dissipation via the artificial viscosity terms. Therefore, what we think is the numerical solution of the Euler equations is really a solution of some "mildly viscouslike" equations, which in turn have the mechanism (through these viscouslike terms) to create a shock wave. In any event, it is still somewhat a marvel to this author that not only will shock waves form in such a numerical solution, but they will be the *correct* shock waves with (more or less) the correct jump conditions across the wave as well as standing at the correct location in the flow.

7.7 SUMMARY

This brings to a conclusion our application of CFD to the time-marching solution of quasi-one-dimensional nozzle flows. Such flows are particularly useful in this regard because, within the framework of a relatively familiar flow problem, many of the important facets of CFD as discussed in Chaps. 1 to 6 can be illustrated. The flow of ideas in the present chapter can be diagrammed on the road map shown in Fig. 7.27. Once again we put the road map at the end of the chapter because it has the most significance *after* we have labored through our various cases. Examining Fig. 7.27, we make the following observations about the content of Chap. 7:

1. It has provided a nonstop illustration of the philosophy of time marching to obtain steady-state solutions in the limit of large times. The use of time-marching solutions in CFD is extensive.
2. Reading across the top row of blocks in Fig. 7.27, we have four of the most important aspects of CFD, namely, the choice between the nonconservation form and the conservation form of the governing equations, the use of the conservation form in conjunction with the shock-capturing philosophy, and the corresponding need for artificial viscosity.
3. We applied both the nonconservation and the conservation forms for solutions of the subsonic-supersonic isentropic nozzle flow and compared the results. For all practical purposes, the results are the same, except that the conservation form yielded a slightly better mass flow distribution. Artificial viscosity is not needed to obtain solutions for this flow, and none was used.
4. The solution of the purely subsonic flow provided an opportunity to explore the effect of the numerical implementation of *boundary conditions*—a vital aspect of

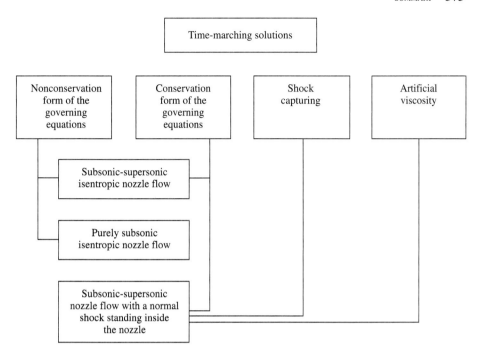

FIG. 7.27
Road map for Chap. 7.

CFD. Here, the subsonic flow case is driven by a *fixed* pressure ratio between the exit and inlet, invariant with time. This case provided a further opportunity to discuss the various aspects of subsonic and supersonic inflow and outflow boundary conditions. We chose to use the nonconservation form of the equations for this solution—we could have just as well used the conservation form.

5. The case with a normal shock wave standing inside the nozzle was an opportunity for the confluence of four important streams in CFD, namely, (*a*) the *necessary* use of the conservation form of the governing equations, (*b*) the application of the shock-capturing philosophy, (*c*) the *necessary* use of artificial viscosity to obtain a quality solution, and (*d*) once again, the way that a subsonic outflow boundary condition can be implemented.

CHAPTER 8

NUMERICAL SOLUTION OF A TWO-DIMENSIONAL SUPERSONIC FLOW: PRANDTL-MEYER EXPANSION WAVE

The error therefore lyeth neither in the abstract nor in geometry, nor in physicks, but in the calculator, that knoweth not how to adjust his accompts.

Galileo Galilei, 1632

8.1 INTRODUCTION

In the above quote, Galileo was expressing a concern with the role of mathematics in the analysis of real physical problems. Prior to the seventeenth century, driven by the concepts of Aristotelian physics, the prevailing method was to accept geometric purity as the explanation for much physical phenomena. The concepts of the physical world were bent and adjusted so as to be in harmony with perfect geometry. For example, a perfect sphere touches a plane at only one point, whereas a real ball (such as a basketball) touches the floor over a finite surface area—the basketball has a small finite flat region in contact with the floor; hence it is not a perfect sphere. Early mathematicians would have assumed the basketball to be a perfect sphere

INTRODUCTION **375**

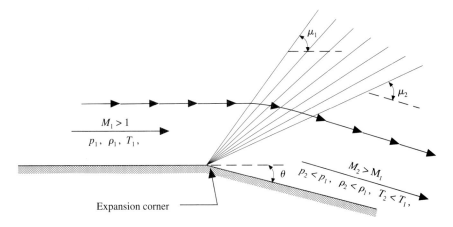

FIG. 8.1
Centered Prandtl-Meyer expansion wave.

touching the floor at only one point; they would not have considered an analysis of a ball with a flat spot to be of any value in mathematics, or in nature. In the early seventeenth century, Galileo was reacting to this altitude. In his *Dialogue Concerning the Two Chief Systems of the World, the Ptolemaic and the Copernican*, from which the above quote is taken, Galileo argued that the role of mathematics is to adjust to the real physical world, and not vice versa. The mathematics for studying the basketball should be adjusted to *account* for the flat spot, not rule it away. The person making the calculation (the "calculator" in the above quote) must know how to adjust his or her mathematical analysis (the "accompts" in the above quote) to match the physics. Little did Galileo realize that he was establishing a basic tenet of CFD, namely, the effort to adjust numerical mathematics to the real physical problem. We will see a graphic example of such a philosophy in the present chapter.

The type of flow highlighted in the present chapter is a two-dimensional, inviscid, supersonic flow moving over a surface. In this type of problem, it is particularly vital to couple the surface boundary condition into the flow-field calculation—to make certain that the inviscid flow readily sees the shape of the surface over which it is flowing. Here we will be seriously concerned with how to "adjust" the numerical mathematics to properly "see" the shape of the boundary.

The discussion in the present chapter is an illustration of the *downstream marching (or space marching)* philosophy described in Sec. 6.4.3. This is in contrast to the time-marching technique illustrated in Chap. 7. Downstream marching is used in many standard CFD codes today, so this chapter has much relevance. Make certain to review Sec. 6.4.3 before proceeding further. MacCormack's space marching technique as described in Sec. 6.4.3 will be applied for the solution of the two-dimensional supersonic flow problem highlighted in the present chapter.

Specifically, we choose to numerically solve the inviscid flow over an expansion corner, as sketched in Fig. 8.1. This problem is in keeping with our philosophy of choosing flow problems for which an exact analytical solution exists in order to obtain a reasonable feeling for the accuracy of the numerical technique.

Finally, the road map for this chapter is given as Fig. 8.9, near the end of the chapter. Make certain to examine this road map as you progress through the various sections. In addition, reexamine Fig. 1.32e, which illustrates the flow of various ideas that impact this application.

8.2 INTRODUCTION TO THE PHYSICAL PROBLEM: PRANDTL–MEYER EXPANSION WAVE—EXACT ANALYTIC SOLUTION

A centered, Prandtl–Meyer expansion wave is illustrated in Fig. 8.1. Here, a supersonic flow is expanded around a sharp expansion corner. An expansion wave, made up of an infinite number of infinitely weak Mach waves, fans out from the corner, as shown in Fig. 8.1. The leading edge of the expansion fan makes an angle μ_1 with respect to the upstream flow direction, and the trailing edge of the wave makes an angle μ_2 with respect to the downstream flow direction. The angles μ_1 and μ_2 are Mach angles, defined by

$$\mu_1 = \sin^{-1}\frac{1}{M_1} \quad \text{and} \quad \mu_2 = \sin^{-1}\frac{1}{M_2}$$

where M_1 and M_2 are the upstream and downstream Mach numbers, respectively. The flow through an expansion wave is isentropic. As the flow passes through the expansion wave, the Mach number increases and the pressure, temperature, and density decrease; these trends are noted in Fig. 8.1. The flow in front of the centered expansion wave is uniform at a Mach number M_1 and is parallel to the wall in front of the wave. The flow behind the expansion wave is also uniform at a Mach number of M_2 and is parallel to the wall behind the wave. Inside the wave itself, the flow properties change smoothly, and the streamlines are curved, as sketched in Fig. 8.1. Inside the wave, the flow is two-dimensional. The only exception to the above discussion is right at the corner itself; this is a singular point at which the streamline at the wall experiences a discontinuous change in direction and where the flow properties are discontinuous. This singularity has some impact on the numerical solution of the flow field, as you might suspect. Such matters will be addressed in a subsequent section. For given supersonic upstream conditions and a given flow deflection angle θ at the corner, the downstream conditions (denoted by a subscript 2) are uniquely defined. For a calorically perfect gas, there is an exact, analytical solution for the conditions behind the expansion wave, as outlined below. Many more details associated with a Prandtl–Meyer expansion can be found in Refs. 8 and 21.

The analytical solution of the flow across a centered expansion wave hinges on the simple relation

$$f_2 = f_1 + \theta \tag{8.1}$$

where f is the Prandtl–Meyer function and θ is the flow deflection angle shown in Fig. 8.1. For a calorically perfect gas, the Prandtl–Meyer function depends on M and γ and is given by

$$f = \sqrt{\frac{\gamma+1}{\gamma-1}} \tan^{-1} \sqrt{\frac{\gamma-1}{\gamma+1}(M^2-1)} - \tan^{-1}\sqrt{M^2-1} \qquad (8.2)$$

The analytical solution proceeds as follows. For the given M_1, calculate f_1 from Eq. (8.2). Then, for the given θ, calculate f_2 from Eq. (8.1). The Mach number in region 2 is then obtained by solving (implicitly, by trial and error) Eq. (8.2) for M_2, using the value of f_2 obtained above. Once M_2 is obtained, then the pressure, temperature, and density behind the wave are calculated from the isentropic flow relations

$$p_2 = p_1 \left\{ \frac{1 + [(\gamma-1)/2]M_1^2}{1 + [(\gamma-1)/2]M_2^2} \right\}^{\gamma/(\gamma-1)} \qquad (8.3)$$

$$T_2 = T_1 \frac{1 + [(\gamma-1)/2]M_1^2}{1 + [(\gamma-1)/2]M_2^2} \qquad (8.4)$$

and the equation of state

$$\rho_2 = \frac{p_2}{RT_2} \qquad (8.5)$$

With Eqs. (8.1) to (8.5), the flow behind the centered expansion wave is completely determined.

8.3 THE NUMERICAL SOLUTION OF A PRANDTL–MEYER EXPANSION WAVE FLOW FIELD

In this chapter we will carry out a downstream marching solution for the supersonic flow over an expansion corner. The solution technique will be MacCormack's predictor-corrector explicit finite-difference method. The details of this downstream (or space) marching approach are given in Sec. 6.4.3. Make certain that you feel comfortable with the contents of Sec. 6.4.3 before progressing further.

8.3.1 The Governing Equations

The governing Euler equations for a steady, two-dimensional flow in strong conservation form can be expressed in the generic form given by Eq. (6.24), repeated below:

$$\frac{\partial F}{\partial x} = J - \frac{\partial G}{\partial y} \qquad (6.24)$$

where F and G are column vectors with elements defined by Eqs. (2.106) and (2.107), respectively, repeated below.

$$F = \begin{Bmatrix} \rho u \\ \rho u^2 + p \\ \rho u v \\ \rho u \left(e + \dfrac{V^2}{2} \right) + pu \end{Bmatrix} \qquad (2.106)$$

$$G = \begin{Bmatrix} \rho v \\ \rho u v \\ \rho v^2 + p \\ \rho v \left(e + \dfrac{V^2}{2} \right) + pv \end{Bmatrix} \qquad (2.107)$$

We are considering isentropic (hence adiabatic) flow with no body forces; therefore, the source term denoted by J in Eq. (6.24) is, from Eq. (2.109), equal to zero. For clarity in our subsequent calculations, we will denote each of the elements in the column vector expressed by Eq. (2.106) as follows:

$$F_1 = \rho u \qquad (8.6a)$$
$$F_2 = \rho u^2 + p \qquad (8.6b)$$
$$F_3 = \rho u v \qquad (8.6c)$$
$$F_4 = \rho u \left(e + \dfrac{u^2 + v^2}{2} \right) + pu \qquad (8.6d)$$

For a calorically perfect gas, it is convenient to eliminate e in Eq. (8.6d) in favor of p and ρ as follows.

$$e = c_v T = \frac{RT}{\gamma - 1} = \frac{1}{\gamma - 1} \frac{p}{\rho}$$

Hence, Eq. (8.6d) can be written as

$$F_4 = \rho u \left(\frac{1}{\gamma - 1} \frac{p}{\rho} + \frac{u^2 + v^2}{2} \right) + pu$$
$$= \frac{1}{\gamma - 1} pu + \rho u \frac{u^2 + v^2}{2} + pu$$

Combining the terms involving pu, we have

$$F_4 = \frac{\gamma}{\gamma - 1} pu + \rho u \frac{u^2 + v^2}{2} \qquad (8.6e)$$

Also, the elements of the column vector expressed by Eq. (2.107) are denoted by

$$G_1 = \rho v \tag{8.7a}$$

$$G_2 = \rho u v \tag{8.7b}$$

$$G_3 = \rho v^2 + p \tag{8.7c}$$

$$G_4 = \rho v \left(e + \frac{u^2 + v^2}{2} \right) + pv \tag{8.7d}$$

In a fashion analogous to that carried out for Eq. (8.6e), we can express Eq. (8.7d) as

$$G_4 = \frac{\gamma}{\gamma - 1} pv + \rho v \frac{u^2 + v^2}{2} \tag{8.7e}$$

The essence of the downstream marching solution, to be discussed shortly, can be presaged by examining the above equations. Note that Eq. (6.24) is written with the x derivative on the left-hand side and the y derivative on the right-hand side. Examining Fig. 8.2, if the flow-field variables are given at location x_0 as a function of y along an *initial data line* (the dashed line in Fig. 8.2), then the y derivative of G in Eq. (6.24) is known along this line. This allows the x derivative of F to be calculated. With this known x derivative, we can advance the flow-field variables to the next vertical line located at $x_0 + \Delta x$. In this fashion, the solution can be carried out by marching in steps of Δx along the x direction in Fig. 8.2, starting with the specified flow field along the initial data line.

We recall from our discussion in Sec. 6.4.3 that, for a downstream marching solution, we *have* to employ the governing equations in the strong conservation form given by Eq. (6.24); this is the only form in which a *single* x derivative can be couched on the left-hand side of the equation. Therefore, as you might suspect from our experience with the strong conservation form of the equations in Chap. 7, there

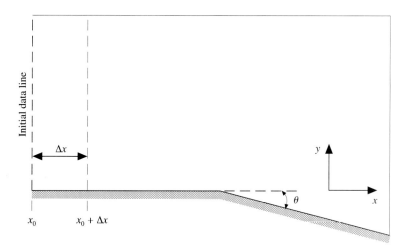

FIG. 8.2
Model for the downstream marching solution.

is some extra baggage that goes along with the numerical solution of this form of the equations, namely, (1) the need to *decode* the primitive variables from the flux variables F_1, F_2, F_3, and F_4, and (2) the corresponding desirability of expressing the elements of the G vector, G_1, G_2, G_3, and G_4, in a "pure" form involving F_1, F_2, F_3, and F_4 rather than the primitive variables as originally defined in Eqs. (8.7a) to (8.7e). Let us proceed with a discussion of these two items.

For decoding the primitive variables from the flux variables, we will simply write down the results because the derivation is assigned as Prob. 2.1; i.e., the answer to Prob. 2.1 is as follows:

$$\rho = \frac{-B + \sqrt{B^2 - 4AC}}{2A} \tag{8.8}$$

where

$$A = \frac{F_3^2}{2F_1} - F_4$$

$$B = \frac{\gamma}{\gamma - 1} F_1 F_2$$

$$C = -\frac{\gamma + 1}{2(\gamma - 1)} F_1^3$$

$$u = \frac{F_1}{\rho} \tag{8.9}$$

$$v = \frac{F_3}{F_1} \tag{8.10}$$

$$p = F_2 - F_1 u \tag{8.11}$$

and from the equation of state

$$T = \frac{p}{\rho R} \tag{8.12}$$

Note that the solution for ρ involves a quadratic equation. Because of this, the decoding for the primitive variables for the present case of the *steady* flow equations in the form of Eq. (6.24) requires a more rigorous derivation than the rather straightforward decoding when the *unsteady* flow equations are used in the form of Eq. (2.99), where the decoding is given by Eqs. (2.100) to (2.104). We took advantage of this more straightforward decoding in Chap. 7 where in part we dealt with the unsteady equations in the form of Eqs. (7.101a) to (7.101c), with the decoding given by Eqs. (7.102) to (7.105).

As we have noted before, when the strong conservation form of the governing equations is used for a numerical solution, specifically when Eq. (6.24) is used, numbers are directly obtained for the fluxes F_1, F_2, F_3, and F_4—not the primitive variables. The corresponding values of ρ, u, v, p, and T have to be obtained after the fact from Eqs. (8.8) to (8.12).

We now address a related matter, namely, the way in which the values of G in Eq. (6.24) are calculated. Since values of F_1, F_2, F_3, and F_4 are directly calculated at a given grid point from our numerical solution of Eq. (6.24), it makes sense to return these numbers back to the equation in the form of $G_1, G_2, G_3,$ and G_4 for the calculation at the next downstream-located grid point. That is, it makes sense to calculate numbers for $G_1, G_2, G_3,$ and G_4 *directly* from the numbers obtained for $F_1, F_2, F_3,$ and F_4, rather than going through the intermediate step of extracting the primitive variables by using Eqs. (8.8) to (8.12) and then synthesizing $G_1, G_2, G_3,$ and G_4 from these primitive variables as given in the definitions of the G's in Eqs. (8.7a) to (8.7e). Indeed, the G's are clearly functions of the F's; let us obtain these functions.

To begin with, from Eqs. (8.7a) and (8.10), we have

$$G_1 = \rho v = \rho \frac{F_3}{F_1} \tag{8.13}$$

In Eq. (8.13), ρ can be expressed in terms of $F_1, F_2, F_3,$ and F_4 via Eq. (8.8); since this is a quadratic relationship, we will not bother to substitute the complicated expression into Eq. (8.13). From Eqs. (8.6c) and (8.7b), we can write directly for G_2,

$$G_2 = F_3 \tag{8.14}$$

From Eqs. (8.7c) and (8.10), we can write

$$G_3 = \rho v^2 + p = \rho \left(\frac{F_3}{F_1}\right)^2 + p \tag{8.15}$$

We can eliminate p from Eq. (8.15) by combining Eqs. (8.6b) and (8.9) as follows:

$$p = F_2 - \rho u^2 = F_2 - \frac{F_1^2}{\rho} \tag{8.16}$$

Substituting Eq. (8.16) into (8.15), we have

$$G_3 = \rho \left(\frac{F_3}{F_1}\right)^2 + F_2 - \frac{F_1^2}{\rho} \tag{8.17}$$

Finally, an expression for G_4 can be constructed as follows. From Eqs. (8.7e), (8.10), and (8.16), we have

$$G_4 = \frac{\gamma}{\gamma - 1} pv + \rho v \frac{u^2 + v^2}{2}$$

$$= \frac{\gamma}{\gamma - 1}\left(F_2 - \frac{F_1^2}{\rho}\right)\frac{F_3}{F_1} + \frac{\rho}{2}\frac{F_3}{F_1}\left[\left(\frac{F_1}{\rho}\right)^2 + \left(\frac{F_3}{F_1}\right)^2\right] \tag{8.18}$$

In summary, Eqs. (8.13), (8.14), (8.17), and (8.18) give expressions for $G_1, G_2, G_3,$ and G_4 as functions of $F_1, F_2, F_3,$ and F_4 [keeping in mind that ρ in these equations is itself a function of $F_1, F_2, F_3,$ and F_4 via Eq. (8.8)]. When the values of $G_1, G_2, G_3,$ and G_4 are calculated from these equations [rather than from the primitive variables by using Eqs. (8.7a) to (8.7e)], then we are using a "purer" formulation of the strong conservation form of the governing equations, in the same spirit as discussed in Sec. 7.5.2.

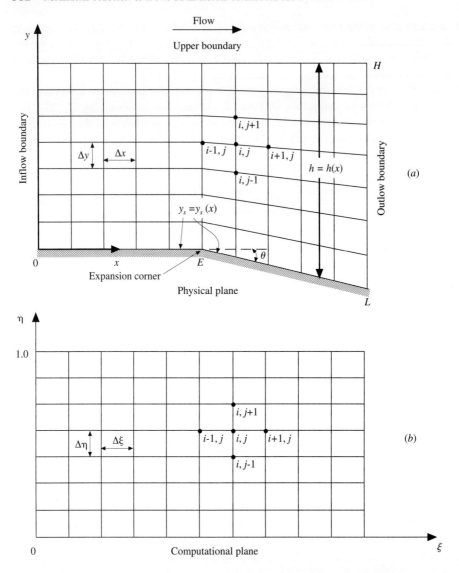

FIG. 8.3
The (a) physical and (b) computational planes for the numerical solution of the centered expansion wave problem.

THE TRANSFORMATION. The present problem affords us an opportunity to exercise some of the aspects of grid generation and equation transformation that were discussed in Chap. 5. In particular, to set up a finite-difference solution for the flow over an expansion corner, we must use a boundary-fitted coordinate system, as sketched in Fig. 8.3. The physical plane using an xy cartesian coordinate system is shown in Fig. 8.3a. The surface including the expansion corner forms the lower boundary in this physical space. The inflow boundary occurs at $x = 0$, and the outflow boundary is at $x = L$. The upper boundary is chosen as a horizontal line at a rather arbitrary value of $y = H$. Clearly, the physical space, due to the downward-sloping wall downstream of the expansion corner, does not lend itself to a completely rectangular grid. Therefore, we must transform the physical plane to a computational plane where the finite-difference grid *is* rectangular, as shown in Fig. 8.3*b*. The computational plane is couched in terms of ξ and η as the independent variables. The bottom surface in the physical plane should correspond to a constant η coordinate curve; i.e., we need to establish a *boundary-fitted coordinate system*. Boundary-fitted coordinate systems are discussed in some detail in Sec. 5.7. In the present application, we need only a *simple* boundary-fitted coordinate system, much along the lines given by Eqs. (5.65) and (5.66). Before continuing further, return to Sec. 5.7 and review the first part having to do with a simple boundary-fitted, algebraically generated grid.

Examining Fig. 8.3*a*, we can readily construct a proper transformation as follows. Let h denote the local height from the lower to the upper boundary in the physical plane; clearly, $h = h(x)$. Denote the y location of the solid surface (the lower boundary in the physical plane) by y_s, where $y_s = y_s(x)$. With this, we define the transformation as

$$\xi = x \tag{8.19}$$

$$\eta = \frac{y - y_s(x)}{h(x)} \tag{8.20}$$

With this transformation, in the computational plane ξ varies from 0 to L and η varies from 0 to 1.0; $\eta = 0$ corresponds to the surface in the physical plane, and $\eta = 1.0$ corresponds to the upper boundary. The lines of constant ξ and η form a regular rectangular grid in the computational plane (Fig. 8.3*b*). The lines of constant ξ and η are also sketched in the physical plane (Fig. 8.3*a*); they form a rectangular grid upstream of the corner and a network of divergent lines downstream of the corner.

As discussed in Chap. 5, we carry out the finite-difference calculations on the rectangular grid in the $\xi\eta$ plane. The partial differential equations for the flow are numerically solved in the transformed space and therefore must be appropriately transformed for use in the transformed, computational plane. That is, Eq. (6.24) must be transformed into terms dealing with ξ and η. The derivative transformation is given by Eqs. (5.2) and (5.3), repeated below.

$$\frac{\partial}{\partial x} = \frac{\partial}{\partial \xi}\left(\frac{\partial \xi}{\partial x}\right) + \frac{\partial}{\partial \eta}\left(\frac{\partial \eta}{\partial x}\right) \tag{5.2}$$

$$\frac{\partial}{\partial y} = \frac{\partial}{\partial \xi}\left(\frac{\partial \xi}{\partial y}\right) + \frac{\partial}{\partial \eta}\left(\frac{\partial \eta}{\partial y}\right) \tag{5.3}$$

The metrics in Eqs. (5.2) and (5.3) are obtained from the transformation given by Eqs. (8.19) and (8.20), that is,

$$\frac{\partial \xi}{\partial x} = 1 \tag{8.21}$$

$$\frac{\partial \xi}{\partial y} = 0 \tag{8.22}$$

$$\frac{\partial \eta}{\partial x} = -\frac{1}{h}\frac{dy_s}{dx} - \frac{\eta}{h}\frac{dh}{dx} \tag{8.23}$$

$$\frac{\partial \eta}{\partial y} = \frac{1}{h} \tag{8.24}$$

The metric $\partial \eta / \partial x$ in Eq. (8.23) can be expressed in a simpler way, as follows. Examining Fig. 8.3a, and denoting the x location of the expansion corner by $x = E$, we have

For $x \leq E$:
$$y_s = 0$$
$$h = \text{constant}$$

For $x \geq E$:
$$y_s = -(x - E)\tan\theta$$
$$h = H + (x - E)\tan\theta$$

Differentiating these expressions, we have

For $x \leq E$:
$$\frac{dy_s}{dx} = 0$$
$$\frac{dh}{dx} = 0$$

For $x \geq E$:
$$\frac{dy_s}{dx} = -\tan\theta$$
$$\frac{dh}{dx} = \tan\theta$$

Hence, the metric $\partial \eta / \partial x$ can be written as

$$\frac{\partial \eta}{\partial x} = \begin{cases} 0 & \text{for } x \leq E \tag{8.25a} \\ (1-\eta)\dfrac{\tan\theta}{h} & \text{for } x \geq E \tag{8.25b} \end{cases}$$

The complete derivative transformation is obtained by substituting Eqs. (8.21), (8.22), (8.24), and (8.25) into (5.2) and (5.3), obtaining

$$\boxed{\frac{\partial}{\partial x} = \frac{\partial}{\partial \xi} + \left(\frac{\partial \eta}{\partial x}\right)\frac{\partial}{\partial \eta}} \tag{8.26}$$

and
$$\boxed{\frac{\partial}{\partial y} = \frac{1}{h}\frac{\partial}{\partial \eta}} \qquad (8.27)$$

where in Eq. (8.26), $\partial \eta/\partial x$ is given by either one of Eqs. (8.25a) or (8.25b), as appropriate.

Return to the governing flow equations in conservation form, given in the physical plane by Eq. (6.24). With $J = 0$, this equation becomes

$$\frac{\partial F}{\partial x} = -\frac{\partial G}{\partial y} \qquad (8.28)$$

Transforming Eq. (8.28) via Eqs. (8.26) and (8.27), we have

$$\frac{\partial F}{\partial \xi} + \left(\frac{\partial \eta}{\partial x}\right)\frac{\partial F}{\partial \eta} = -\frac{1}{h}\frac{\partial G}{\partial \eta}$$

or
$$\boxed{\frac{\partial F}{\partial \xi} = -\left[\left(\frac{\partial \eta}{\partial x}\right)\frac{\partial F}{\partial \eta} + \frac{1}{h}\frac{\partial G}{\partial \eta}\right]} \qquad (8.29)$$

where the metric term $\partial \eta/\partial x$ is given by Eq. (8.25a) or (8.25b), as appropriate. Written in terms of the elements of the column vectors F and G, Eq. (8.29) represents the following system of equations, where the labels are added to remind you of the physical origin of each equation.

$$\text{Continuity}: \quad \frac{\partial F_1}{\partial \xi} = -\left[\left(\frac{\partial \eta}{\partial x}\right)\frac{\partial F_1}{\partial \eta} + \frac{1}{h}\frac{\partial G_1}{\partial \eta}\right] \qquad (8.30)$$

$$x \text{ momentum}: \quad \frac{\partial F_2}{\partial \xi} = -\left[\left(\frac{\partial \eta}{\partial x}\right)\frac{\partial F_2}{\partial \eta} + \frac{1}{h}\frac{\partial G_2}{\partial \eta}\right] \qquad (8.31)$$

$$y \text{ momentum}: \quad \frac{\partial F_3}{\partial \xi} = -\left[\left(\frac{\partial \eta}{\partial x}\right)\frac{\partial F_3}{\partial \eta} + \frac{1}{h}\frac{\partial G_3}{\partial \eta}\right] \qquad (8.32)$$

$$\text{Energy}: \quad \frac{\partial F_4}{\partial \xi} = -\left[\left(\frac{\partial \eta}{\partial x}\right)\frac{\partial F_4}{\partial \eta} + \frac{1}{h}\frac{\partial G_4}{\partial \eta}\right] \qquad (8.33)$$

Equations (8.30) to (8.33) are the governing flow equations which are to be solved numerically in the computational plane sketched in Fig. 8.3b.

Note: Equations (8.30) to (8.33) are in *dimensional* form; we have not bothered to nondimensionalize the variables in the equations, in contrast to the approach taken in Chap. 7. Indeed, in the present solution, we will continue to treat all variables in their dimensional form—just to illustrate that a CFD solution can just as well be carried out using numbers with units attached to them and that the use of dimensionless variables is in no way necessary for the integrity of a CFD solution. In fact, an advantage of using dimensional variables in a numerical solution is that you quickly obtain an engineering feeling for the magnitudes of the physical properties in a given flow problem. The choice of using nondimensional variables is simply up to you; for some problems it makes more sense than others—for example, the convenience of dealing with nondimensional variables for the quasi-one-dimensional flow problem was amply demonstrated in Chap. 7. However, when you use *dimensional* variables, it is vitally important that you *keep the units straight*. For this reason, it is strongly recommended that you use a *consistent set of units* throughout your calculation. The equations discussed in this section hold in their precise form as long as consistent units are employed; i.e., there is no need to insert any "conversion factor" in the equations as would be the case when inconsistent units are employed. See, for example, Chap. 2 of Ref. 1 for a discussion of what is meant by consistent and inconsistent units. Two common sets of consistent units are the English engineering system (pound, slug, foot, second, degree Rankine) and the international, SI system (newton, kilogram, meter, second, kelvin). In this present solution, we will use SI units. Again, keep in mind that when you choose to use dimensional properties in your CFD calculation, you incur the necessity to handle the units correctly.

This finishes our development of the general equations germane to the given problem. Let us now proceed with the solution.

8.3.2 The Setup

We need to establish some details of the particular problem to be solved. We consider the detailed physical plane drawn to scale in Fig. 8.4. The flow at the

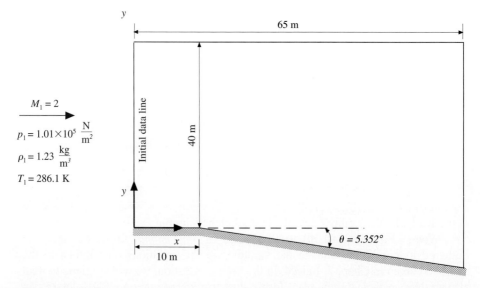

FIG. 8.4
Physical plane, drawn to scale.

upstream boundary is at Mach 2 with a pressure, density, and temperature equal to 1.01×10^5 N/m^2, 1.23 kg/m^3, and 286 K, respectively. The supersonic flow is expanded through an angle of 5.352°, as shown in Fig. 8.4. This is a rather mild expansion angle; the reasons for this choice will be discussed later. The calculations will be made in the domain from $x = 0$ to $x = 65$ m and from the wall to $y = 40$ m, as shown in Fig. 8.4. The location of the expansion corner is at $x = 10$ m. For this geometry, the variation of $h = h(x)$ is given by

$$h = \begin{cases} 40 \text{ m} & 0 \leq x \leq 10 \text{ m} \quad (8.34) \\ 40 + (x - 10) \tan \theta & 10 \leq x \leq 65 \text{ m} \quad (8.35) \end{cases}$$

Equation (8.34) or (8.35), as appropriate, is needed to define the value of the metric $\partial \eta / \partial x$ expressed by Eq. (8.25b).

INITIAL DATA LINE. The initial data line is given by $x = 0$; along this vertical line at each grid point, the initial data are fed in, equal to the uniform upstream flow conditions. The calculation starts at this initial data line and marches downstream in steps of Δx. For our application, we will divide the $x = 0$ initial data line into 40 increments by evenly spacing 41 grid points ($j = 1$ to 41) along this line. To reinforce this picture, the initial data at $x = 0$ is tabulated in Table 8.1 versus j.

FINITE-DIFFERENCE EQUATIONS. We are following the technique set forth in Sec. 6.4.3, which outlines MacCormack's predictor-corrector technique applied to space marching (note that space marching and downstream marching are synonymous terms). Hence, the finite-difference forms of Eq. (8.30) to (8.33) are as follows.

Predictor step. Analogous to Eq. (6.26), Eqs. (8.30) to (8.33) are written in terms of forward differences.

$$\left(\frac{\partial F_1}{\partial \xi}\right)_{i,j} = \left(\frac{\partial \eta}{\partial x}\right) \frac{(F_1)_{i,j} - (F_1)_{i,j+1}}{\Delta \eta} + \frac{1}{h} \frac{(G_1)_{i,j} - (G_1)_{i,j+1}}{\Delta \eta} \quad (8.36a)$$

$$\left(\frac{\partial F_2}{\partial \xi}\right)_{i,j} = \left(\frac{\partial \eta}{\partial x}\right) \frac{(F_2)_{i,j} - (F_2)_{i,j+1}}{\Delta \eta} + \frac{1}{h} \frac{(G_2)_{i,j} - (G_2)_{i,j+1}}{\Delta \eta} \quad (8.36b)$$

$$\left(\frac{\partial F_3}{\partial \xi}\right)_{i,j} = \left(\frac{\partial \eta}{\partial x}\right) \frac{(F_3)_{i,j} - (F_3)_{i,j+1}}{\Delta \eta} + \frac{1}{h} \frac{(G_3)_{i,j} - (G_3)_{i,j+1}}{\Delta \eta} \quad (8.36c)$$

$$\left(\frac{\partial F_4}{\partial \xi}\right)_{i,j} = \left(\frac{\partial \eta}{\partial x}\right) \frac{(F_4)_{i,j} - (F_4)_{i,j+1}}{\Delta \eta} + \frac{1}{h} \frac{(G_4)_{i,j} - (G_4)_{i,j+1}}{\Delta \eta} \quad (8.36d)$$

TABLE 8.1
Initial conditions at $x = 0$

j	u, m/s	v, m/s	ρ, kg/m^3	p, N/m^2	T, K	M
1	.678E+03	.000E+00	.123E+01	.101E+06	.286E+03	.200E+01
2	.678E+03	.000E+00	.123E+01	.101E+06	.286E+03	.200E+01
3	.678E+03	.000E+00	.123E+01	.101E+06	.286E+03	.200E+01
4	.678E+03	.000E+00	.123E+01	.101E+06	.286E+03	.200E+01
5	.678E+03	.000E+00	.123E+01	.101E+06	.286E+03	.200E+01
6	.678E+03	.000E+00	.123E+01	.101E+06	.286E+03	.200E+01
7	.678E+03	.000E+00	.123E+01	.101E+06	.286E+03	.200E+01
8	.678E+03	.000E+00	.123E+01	.101E+06	.286E+03	.200E+01
9	.678E+03	.000E+00	.123E+01	.101E+06	.286E+03	.200E+01
10	.678E+03	.000E+00	.123E+01	.101E+06	.286E+03	.200E+01
11	.678E+03	.000E+00	.123E+01	.101E+06	.286E+03	.200E+01
12	.678E+03	.000E+00	.123E+01	.101E+06	.286E+03	.200E+01
13	.678E+03	.000E+00	.123E+01	.101E+06	.286E+03	.200E+01
14	.678E+03	.000E+00	.123E+01	.101E+06	.286E+03	.200E+01
15	.678E+03	.000E+00	.123E+01	.101E+06	.286E+03	.200E+01
16	.678E+03	.000E+00	.123E+01	.101E+06	.286E+03	.200E+01
17	.678E+03	.000E+00	.123E+01	.101E+06	.286E+03	.200E+01
18	.678E+03	.000E+00	.123E+01	.101E+06	.286E+03	.200E+01
19	.678E+03	.000E+00	.123E+01	.101E+06	.286E+03	.200E+01
20	.678E+03	.000E+00	.123E+01	.101E+06	.286E+03	.200E+01
21	.678E+03	.000E+00	.123E+01	.101E+06	.286E+03	.200E+01
22	.678E+03	.000E+00	.123E+01	.101E+06	.286E+03	.200E+01
23	.678E+03	.000E+00	.123E+01	.101E+06	.286E+03	.200E+01
24	.678E+03	.000E+00	.123E+01	.101E+06	.286E+03	.200E+01
25	.678E+03	.000E+00	.123E+01	.101E+06	.286E+03	.200E+01
26	.678E+03	.000E+00	.123E+01	.101E+06	.286E+03	.200E+01
27	.678E+03	.000E+00	.123E+01	.101E+06	.286E+03	.200E+01
28	.678E+03	.000E+00	.123E+01	.101E+06	.286E+03	.200E+01
29	.678E+03	.000E+00	.123E+01	.101E+06	.286E+03	.200E+01
30	.678E+03	.000E+00	.123E+01	.101E+06	.286E+03	.200E+01
31	.678E+03	.000E+00	.123E+01	.101E+06	.286E+03	.200E+01
32	.678E+03	.000E+00	.123E+01	.101E+06	.286E+03	.200E+01
33	.678E+03	.000E+00	.123E+01	.101E+06	.286E+03	.200E+01
34	.678E+03	.000E+00	.123E+01	.101E+06	.286E+03	.200E+01
35	.678E+03	.000E+00	.123E+01	.101E+06	.286E+03	.200E+01
36	.678E+03	.000E+00	.123E+01	.101E+06	.286E+03	.200E+01
37	.678E+03	.000E+00	.123E+01	.101E+06	.286E+03	.200E+01
38	.678E+03	.000E+00	.123E+01	.101E+06	.286E+03	.200E+01
39	.678E+03	.000E+00	.123E+01	.101E+06	.286E+03	.200E+01
40	.678E+03	.000E+00	.123E+01	.101E+06	.286E+03	.200E+01
41	.678E+03	.000E+00	.123E+01	.101E+06	.286E+03	.200E+01

The predicted values of F are obtained as follows, analogous to Eq. (6.27).

$$(\bar{F}_1)_{i+1,j} = (F_1)_{i,j} + \left(\frac{\partial F_1}{\partial \xi}\right)_{i,j} \Delta\xi \qquad (8.37a)$$

$$(\bar{F}_2)_{i+1,j} = (F_2)_{i,j} + \left(\frac{\partial F_2}{\partial \xi}\right)_{i,j} \Delta\xi \qquad (8.37b)$$

$$(\bar{F}_3)_{i+1,j} = (F_3)_{i,j} + \left(\frac{\partial F_3}{\partial \xi}\right)_{i,j} \Delta\xi \qquad (8.37c)$$

$$(\bar{F}_4)_{i+1,j} = (F_4)_{i,j} + \left(\frac{\partial F_4}{\partial \xi}\right)_{i,j} \Delta\xi \qquad (8.37d)$$

Before proceeding to the corrector step, we need to decode the values of $\bar{F}_{i+1,j}$. This is carried out using Eq. (8.8).

$$(\bar{\rho})_{i+1,j} = \frac{-B + \sqrt{B^2 - 4AC}}{2A} \qquad (8.38)$$

where

$$A = \frac{(\bar{F}_3)^2_{i+1,j}}{2(\bar{F}_1)_{i+1,j}} - (\bar{F}_4)_{i+1,j}$$

$$B = \frac{\gamma}{\gamma - 1}(\bar{F}_1)_{i+1,j}(\bar{F}_2)_{i+1,j}$$

$$C = -\frac{\gamma+1}{2(\gamma-1)}(\bar{F}_1)^3_{i+1,j}$$

With the predicted values of ρ obtained above, we can form the predicted values of G, which are needed for the corrector step. From Eqs. (8.13), (8.14), (8.17), and (8.18), we have, respectively,

$$(\bar{G}_1)_{i+1,j} = \bar{\rho}_{i+1,j}\frac{(\bar{F}_3)_{i+1,j}}{(\bar{F}_1)_{i+1,j}} \qquad (8.39)$$

$$(\bar{G}_2)_{i+1,j} = (\bar{F}_3)_{i+1,j} \qquad (8.40)$$

$$(\bar{G}_3)_{i+1,j} = \bar{\rho}_{i+1,j}\left(\frac{\bar{F}_3}{\bar{F}_1}\right)^2_{i+1,j} + (\bar{F}_2)_{i+1,j} - \frac{(\bar{F}_1)^2_{i+1,j}}{\bar{\rho}_{i+1,j}} \qquad (8.41)$$

$$(\bar{G}_4)_{i+1,j} = \frac{\gamma}{\gamma-1}\left[(\bar{F}_2)_{i+1,j} - \frac{(\bar{F}_1)^2_{i+1,j}}{\bar{\rho}_{i+1,j}}\right]\left(\frac{\bar{F}_3}{\bar{F}_1}\right)_{i+1,j}$$

$$+ \frac{\bar{\rho}_{i+1,j}}{2}\left(\frac{\bar{F}_3}{\bar{F}_1}\right)_{i+1,j}\left[\left(\frac{\bar{F}_1}{\bar{\rho}}\right)^2_{i+1,j} + \left(\frac{\bar{F}_3}{\bar{F}_1}\right)^2_{i+1,j}\right] \qquad (8.42)$$

Corrector step. On the corrector step, we return to Eqs. (8.30) to (8.33), with rearward differences used for the η derivatives. Analogous to Eq. (6.29), we have

$$\left(\frac{\overline{\partial F_1}}{\partial \xi}\right)_{i+1,j} = \left(\frac{\partial \eta}{\partial x}\right) \frac{(\bar{F}_1)_{i+1,j-1} - (\bar{F}_1)_{i+1,j}}{\Delta \eta}$$

$$+ \frac{1}{h} \frac{(\bar{G}_1)_{i+1,j-1} - (\bar{G}_1)_{i+1,j}}{\Delta \eta} \quad (8.43a)$$

$$\left(\frac{\overline{\partial F_2}}{\partial \xi}\right)_{i+1,j} = \left(\frac{\partial \eta}{\partial x}\right) \frac{(\bar{F}_2)_{i+1,j-1} - (\bar{F}_2)_{i+1,j}}{\Delta \eta}$$

$$+ \frac{1}{h} \frac{(\bar{G}_2)_{i+1,j-1} - (\bar{G}_2)_{i+1,j}}{\Delta \eta} \quad (8.43b)$$

$$\left(\frac{\overline{\partial F_3}}{\partial \xi}\right)_{i+1,j} = \left(\frac{\partial \eta}{\partial x}\right) \frac{(\bar{F}_3)_{i+1,j-1} - (\bar{F}_3)_{i+1,j}}{\Delta \eta}$$

$$+ \frac{1}{h} \frac{(\bar{G}_3)_{i+1,j-1} - (\bar{G}_3)_{i+1,j}}{\Delta \eta} \quad (8.43c)$$

$$\left(\frac{\overline{\partial F_4}}{\partial \xi}\right)_{i+1,j} = \left(\frac{\partial \eta}{\partial x}\right) \frac{(\bar{F}_4)_{i+1,j-1} - (\bar{F}_4)_{i+1,j}}{\Delta \eta}$$

$$+ \frac{1}{h} \frac{(\bar{G}_4)_{i+1,j-1} - (\bar{G}_4)_{i+1,j}}{\Delta \eta} \quad (8.43d)$$

Forming the average derivatives analogous to Eq. (6.30), we have

$$\left(\frac{\partial F_1}{\partial \xi}\right)_{av} = \frac{1}{2}\left[\left(\frac{\partial F_1}{\partial \xi}\right)_{i,j} + \left(\frac{\overline{\partial F_1}}{\partial \xi}\right)_{i+1,j}\right] \quad (8.44a)$$

$$\left(\frac{\partial F_2}{\partial \xi}\right)_{av} = \frac{1}{2}\left[\left(\frac{\partial F_2}{\partial \xi}\right)_{i,j} + \left(\frac{\overline{\partial F_2}}{\partial \xi}\right)_{i+1,j}\right] \quad (8.44b)$$

$$\left(\frac{\partial F_3}{\partial \xi}\right)_{av} = \frac{1}{2}\left[\left(\frac{\partial F_3}{\partial \xi}\right)_{i,j} + \left(\frac{\overline{\partial F_3}}{\partial \xi}\right)_{i+1,j}\right] \quad (8.44c)$$

$$\left(\frac{\partial F_4}{\partial \xi}\right)_{av} = \frac{1}{2}\left[\left(\frac{\partial F_4}{\partial \xi}\right)_{i,j} + \left(\frac{\overline{\partial F_4}}{\partial \xi}\right)_{i+1,j}\right] \quad (8.44d)$$

where the derivatives on the right-hand side of Eqs. (8.44a) to (8.44d) are known numbers, known from Eqs. (8.36a) to (8.36d) and Eqs. (8.43a) to (8.43d). Finally, analogous to Eq. (6.25), we have

$$(F_1)_{i+1,j} = (F_1)_{i,j} + \left(\frac{\partial F_1}{\partial \xi}\right)_{av} \Delta\xi \qquad (8.45a)$$

$$(F_2)_{i+1,j} = (F_2)_{i,j} + \left(\frac{\partial F_2}{\partial \xi}\right)_{av} \Delta\xi \qquad (8.45b)$$

$$(F_3)_{i+1,j} = (F_3)_{i,j} + \left(\frac{\partial F_3}{\partial \xi}\right)_{av} \Delta\xi \qquad (8.45c)$$

$$(F_4)_{i+1,j} = (F_4)_{i,j} + \left(\frac{\partial F_4}{\partial \xi}\right)_{av} \Delta\xi \qquad (8.45d)$$

Our calculation of the flow field (via the flux variables F_1 to F_4) at the next downstream location $i + 1$ is now complete, except for one remaining aspect—artificial viscosity. In the present problem, the sharp expansion corner located at $x = 10$ m (see Fig. 8.4) is a singular point; it introduces a discontinuous change in the surface flow properties at that point. The system of finite-difference equations developed above sees this discontinuity through a discontinuous change in the metric term $\partial\eta/\partial x$, which from Eqs. (8.25a) and (8.25b) is zero just ahead of the corner and $(1 - \eta)(\tan\theta)/h$ just behind the corner. Such a discontinuous change always has the potential to introduce oscillations in the numerical solution. Indeed, this author's experience in solving the present problem has shown that such oscillations do indeed develop in the flow field—oscillations which are virtually eliminated by including some artificial viscosity in the solution. The formulation of the artificial viscosity term for the present case follows the discussion given in Sec. 6.6 and is patterned after Eqs. (6.58) to (6.61). For the present case, we formulate the artificial viscosity term as follows. On the predictor step,

$$(SF_1)_{i,j} = \frac{C_y|p_{i,j+1} - 2p_{i,j} + p_{i,j-1}|}{p_{i,j+1} + 2p_{i,j} + p_{i,j-1}}$$
$$\times [(F_1)_{i,j+1} - 2(F_1)_{i,j} + (F_1)_{i,j-1}] \qquad (8.46a)$$

$$(SF_2)_{i,j} = \frac{C_y|p_{i,j+1} - 2p_{i,j} + p_{i,j-1}|}{p_{i,j+1} + 2p_{i,j} + p_{i,j-1}}$$
$$\times [(F_2)_{i,j+1} - 2(F_2)_{i,j} + (F_2)_{i,j-1}] \qquad (8.46b)$$

Similar expressions are obtained for $(SF_3)_{i,j}$ and $(SF_4)_{i,j}$; we do not need to take the space to write the corresponding equations. The values of $(SF_1)_{i,j}$, $(SF_2)_{i,j}$, etc., are added to Eqs. (8.37a) to (8.37d) as follows:

$$(\bar{F}_1)_{i+1,j} = (F_1)_{i,j} + \left(\frac{\partial F_1}{\partial \xi}\right)_{i,j} \Delta\xi + (SF_1)_{i,j} \qquad (8.47a)$$

$$(\bar{F}_2)_{i+1,j} = (F_2)_{i,j} + \left(\frac{\partial F_2}{\partial \xi}\right)_{i,j} \Delta\xi + (SF_2)_{i,j} \qquad (8.47b)$$

and similarly for $[\bar{F}_3]_{i+1,j}$ and $[\bar{F}_4]_{i+1,j}$. Artificial viscosity is also added on the corrector step as follows:

$$(\bar{S}\bar{F}_1)_{i+1,j} = \frac{C_y|\bar{p}_{i+1,j+1} - 2\bar{p}_{i+1,j} + \bar{p}_{i+1,j-1}|}{\bar{p}_{i+1,j+1} + 2\bar{p}_{i+1,j} + \bar{p}_{i+1,j-1}}$$
$$\times [(\bar{F}_1)_{i+1,j+1} - 2(\bar{F}_1)_{i+1,j} + (\bar{F}_1)_{i+1,j-1}] \quad (8.48a)$$

$$(\bar{S}\bar{F}_2)_{i+1,j} = \frac{C_y|\bar{p}_{i+1,j+1} - 2\bar{p}_{i+1,j} + \bar{p}_{i+1,j-1}|}{\bar{p}_{i+1,j+1} + 2\bar{p}_{i+1,j} + \bar{p}_{i+1,j-1}}$$
$$\times [(\bar{F}_2)_{i+1,j+1} - 2(\bar{F}_2)_{i+1,j} + (\bar{F}_2)_{i+1,j-1}] \quad (8.48b)$$

Similar expressions are obtained for $(\bar{S}\bar{F}_3)_{i+1,j}$ and $(\bar{S}\bar{F}_4)_{i+1,j}$. Finally, these values of artificial viscosity are added to Eqs. (8.45a) to (8.45d) as follows:

$$(F_1)_{i+1,j} = (F_1)_{i,j} + \left(\frac{\partial F_1}{\partial \xi}\right)_{av} \Delta\xi + (\bar{S}\bar{F}_1)_{i+1,j} \quad (8.49a)$$

$$(F_2)_{i+1,j} = (F_2)_{i,j} + \left(\frac{\partial F_2}{\partial \xi}\right)_{av} \Delta\xi + (\bar{S}\bar{F}_2)_{i+1,j} \quad (8.49b)$$

and similarly for $(F_3)_{i+1,j}$ and $(F_4)_{i+1,j}$. This completes the addition of artificial viscosity to the above algorithm.

Finally, the primitive variables at grid point $(i + 1, j)$ can be decoded from the values of $(F_1)_{i+1,j}$, $(F_2)_{i+1,j}$, $(F_3)_{i+1,j}$, and $(F_4)_{i+1,j}$ using Eqs. (8.8) to (8.12). This totally completes the calculation of the flow field at the next downstream location $i + 1$ at all the vertically arranged grid points in the internal part of the flow, from grid point $j = 2$ to $j = 40$. We have one remaining item to discuss, namely, the flow solution for the grid points at the boundaries, i.e., at $j = 1$ and 41.

BOUNDARY CONDITIONS. At the wall, the physically proper boundary condition for an inviscid flow is that the flow be tangent to the wall. This is the *only* boundary condition at the wall; all other flow properties at the wall must be obtained as part of the solution. As innocent as this may sound, in terms of the CFD calculation the proper numerical treatment of this wall boundary condition is not always straightforward; indeed, it has been the subject of much research in CFD. In the present case, we will employ a treatment of the wall boundary condition patterned after that suggested by Abbett (Ref. 46). For a steady flow, the steps in Abbett's boundary condition treatment are as follows.

1. Consider point 1 on the wall, as sketched in Fig. 8.5. Calculate trial values of u_1 and v_1 at point 1 using *one-sided differences* in the internal flow algorithm, i.e., using Eqs. (8.36a) to (8.49b), except modifying the corrector step to use forward differences just as on the predictor step. At the wall, this is the only choice, because we have no grid points below the wall and hence no way of forming the rearward differences called for on the corrector step. This use of forward-forward differences on the predictor-corrector sequence at the wall compromises slightly the second-order accuracy of the algorithm at the wall.

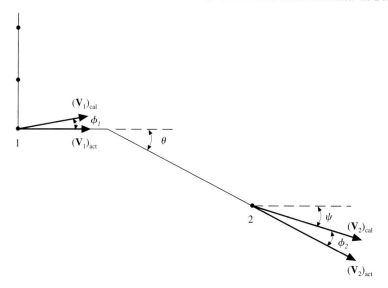

FIG. 8.5
Abbett's boundary condition for a steady flow.

2. The *direction* of the resultant velocity at the wall calculated in step 1 will not necessarily be tangent to the wall due to numerical inaccuracy. Usually, the calculated value of the velocity vector at the wall $(V_1)_{cal}$ in Fig. 8.5, will make an angle ϕ_1 at the wall, where

$$\phi_1 = \tan^{-1}\frac{v_1}{u_1} \tag{8.50}$$

Also, the calculated Mach number at the wall will be

$$(M_1)_{cal} = \frac{\sqrt{(u_1)_{cal}^2 + (v_1)_{cal}^2}}{(a_1)_{cal}} \tag{8.51}$$

Along with this value of $(M_1)_{cal}$ is a corresponding value of the Prandtl–Meyer function f_{cal}, obtained by substituting $(M_1)_{cal}$ into the right-hand side of Eq. (8.2).

3. Assume that the supersonic flow calculated at point 1 in step 2 is *rotated through a local centered Prandtl–Meyer expansion wave* so that the velocity vector is tangent to the wall. That is, $(V_1)_{cal}$ in Fig. 8.5 is rotated through a Prandtl–Meyer expansion wave where the deflection angle through the wave is ϕ_1. This yields a new velocity vector $(V_1)_{act}$, which is assumed to be the actual velocity tangent to the wall. The Mach number associated with $(V_1)_{act}$ is $(M_1)_{act}$, obtained as follows. First, calculate f_{act} which corresponds to $(M_1)_{act}$ from the Prandtl–Meyer relationship given by Eq. (8.1), namely

$$f_{act} = f_{cal} + \phi_1 \tag{8.52}$$

In Eq. (8.52), f_{cal} is known from step 2, and ϕ_1 is the deflection angle shown in Fig. 8.5 and known from Eq. (8.50). The value of f_{act} calculated from Eq. (8.52) is that value corresponding to $(M_1)_{act}$, namely, the Mach number which exists *after* the flow is rotated to be parallel to the wall. The value of $(M_1)_{act}$ must be backed out of Eq. (8.2) by substituting f_{act} into the left-hand side and solving Eq. (8.2) by trial and error for $(M_1)_{cal}$.

4. Let the values of pressure, temperature, and density as originally calculated in step 1 (using one-sided differences) be denoted by p_{cal}, T_{cal}, and ρ_{cal}, respectively. These values must be changed to correspond to the new, "actual" conditions after the calculated velocity vector is rotated through the expansion wave to be parallel to the wall. These new values, denoted by p_{act}, T_{act}, and ρ_{act}, are obtained from Eqs. (8.3) to (8.5), respectively, using M_{cal} and M_{act} as follows:

$$p_{act} = p_{cal} \left\{ \frac{1 + [(\gamma - 1)/2]M_{cal}^2}{1 + [(\gamma - 1)/2]M_{act}^2} \right\}^{\gamma/(\gamma-1)} \tag{8.53}$$

$$T_{act} = T_{cal} \frac{1 + [(\gamma - 1)/2]M_{cal}^2}{1 + [(\gamma - 1)/2]M_{act}^2} \tag{8.54}$$

$$\rho_{act} = \frac{p_{act}}{RT_{act}} \tag{8.55}$$

The values of p_{act}, T_{act}, and ρ_{act} calculated from Eqs. (8.53) to (8.55), respectively, are interpreted to be the final values of p, T, and ρ at grid point 1 at the wall.

Interpretation: What are we really doing by imposing the above boundary condition calculation? Returning to Fig. 8.5, we recall that the velocity at the wall as calculated from the internal flow algorithm using one-sided, forward differences on both the predictor and corrector steps will, in general, *not* be tangent to the wall. That is, there will be a *finite* normal component of velocity at the wall, v_1. The function of Abbett's boundary condition as described above is to simply *cancel* this calculated finite vertical velocity component by means of an imaginary, local, Prandtl–Meyer expansion wave at the wall. This local expansion wave is just an *artifice* which we use in the numerical calculation; it does not say that nature is actually doing this in the real flow. (Indeed, nature always does the right thing and never requires such an artifice.) However, consistent with this artifice of a local expansion wave, we must slightly modify the values of p, T, and ρ originally calculated at point 1 to be somewhat compatible with the cancellation of the finite v_1 at the wall by the expansion wave. Hence, at the flow boundary, namely, at point 1, when the calculation is finished, not only is the velocity now tangent to the wall but the pressure, temperature, and density at the wall are taken to be p_{act}, T_{act}, and ρ_{act} as calculated from Eqs. (8.53) to (8.55).

Note that, in the above procedure, if $(\mathbf{V}_1)_{cal}$ turns out to point *into* the wall, rather than out of the wall as sketched in Fig. 8.5, then a local Prandtl–Meyer isentropic *compression* wave is assumed. This implies only that ϕ_1 is now considered to be a *negative value* in Eq. (8.52); all other steps in the calculation remain the same.

For that portion of the wall behind the expansion corner, the above technique is still the same. Focusing on point 2 in Fig. 8.5, $(\mathbf{V}_2)_{\text{cal}}$ is rotated through the angle ϕ_2 to be tangent to the wall. Equation (8.52) now becomes

$$f_{\text{act}} = f_{\text{cal}} + \phi_2 \tag{8.56}$$

All aspects of the calculation at point 2 are carried out exactly as described above for point 1, with the exception that ϕ_2 is not given by Eq. (8.50). Instead, from the geometry shown at point 2 in Fig. 8.5, we have

$$\psi = \tan^{-1} \frac{|v_2|}{u_2} \tag{8.57}$$

and

$$\phi_2 = \theta - \psi \tag{8.58}$$

CALCULATION OF DOWNSTREAM MARCHING STEP SIZE. As discussed in Sec. 3.4.1, the governing flow equations for steady, inviscid, supersonic flow are hyperbolic; this is why a downstream-marching solution is well-posed. Moreover, in Sec. 4.5 we indicated that the proper stability criterion for linear hyperbolic equations is the CFL (Courant–Friedrichs–Lewy) criterion. An equation for the maximum allowable marching step according to the CFL criterion was developed in Sec. 4.5 for the case of time marching. We have stated that, on a physical basis, the maximum allowable time step for an explicit time-marching solution (based on the CFL criterion) should be less than, or at best equal to, the time required for a sound wave to move from one grid point to the next, adjacent grid point.

With this interpretation involving the propagation of sound waves, we can intuitively develop the CFL criterion for steady flow. Consider the sketch shown in Fig. 8.6, which shows a vertical array of grid points at a given x station. A small disturbance (e.g., a sound wave) introduced at point 1 will propagate along the two characteristic lines through point 1 (recall our discussion in the Steady, Inviscid Supersonic Flow subsection of Sec. 3.4.1); the characteristic lines are Mach lines in the flow, which are at the Mach angle μ relative to the streamline direction. If the angle made by the streamline at point 1 is θ relative to the x axis, then the angles made by the left- and right-running Mach waves relative to the x axis are $\theta + \mu$ and $\theta - \mu$, respectively. In Fig. 8.6, only the left-running Mach line is shown at point 1. Consider a horizontal line through point 2; the left-running characteristic from point 1 intersects this horizontal line at point a. Point a is therefore located a distance $(\Delta x)_1$ from point 2, where

$$\Delta x_1 = \frac{\Delta y}{\tan(\theta + \mu)_1} \tag{8.59}$$

Based on the CFL criterion applied locally at point 2, the downstream value chosen for Δx should be no more than $(\Delta x)_1$ for stability; in this fashion the distance between points 2 and a is less than, or at most equal to, the distance required for a sound wave from point 1 to reach the level defined by the y location of point 2. A

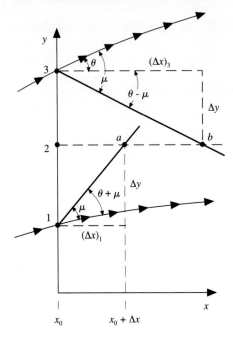

FIG. 8.6
Sketch in the physical plane for the calculation of the marching step size.

similar argument regarding the right-running Mach wave through point 3 shows that it intersects the horizontal line through point 2 at point b. Point b is therefore located a distance $(\Delta x)_3$ from point 2, where

$$(\Delta x)_3 = \frac{\Delta y}{\tan(\theta - \mu)_3} \tag{8.60}$$

For stability of the downstream marching calculations locally at point 2, the chosen value of the step size Δx should be no more than the minimum of $(\Delta x)_1$ and $(\Delta x)_3$. Expanding this argument to *all* the grid points arrayed along the vertical line at x_0, we can express the value of Δx to be chosen for the next downstream marching step at x_0 to be given by

$$\Delta x = \frac{\Delta y}{|\tan(\theta \pm \mu)|_{\max}} \tag{8.61}$$

where $|\tan(\theta + \mu)|_{\max}$ is the *maximum* of the absolute values of $\tan(\theta \pm \mu)$ evaluated for *all* the grid points arrayed along the vertical line at $x = x_0$. Since the transformation defined by Eqs. (8.19) and (8.20) states that $\xi = x$, then the proper step size for downstream marching in the computational plane shown in Fig. 8.3b is

$$\Delta \xi \leq \Delta x \tag{8.62}$$

where Δx is given by Eq. (8.61). Combining Eq. (8.62) with Eq. (8.61) and introducing the Courant number C, we have as our stability criterion for the value of $\Delta \xi$,

$$\Delta \xi = C \frac{\Delta y}{|\tan(\theta \pm \mu)|_{\max}} \tag{8.63}$$

where the CFL criterion states that $C \leq 1$. The value of $\Delta \xi$ obtained from Eq. (8.63) is the value that goes into Eqs. (8.37a) to (8.37d) and Eqs. (8.45a) to (8.45d).

8.3.3 Intermediate Results

Once again, we will follow our philosophy of giving some intermediate results during the course of a calculation so that you can check some intermediate numbers from your own computer program; if you are not writing your own program for this application, the present section still provides educational value for you—it is essentially a glorified flow diagram for the numerical solution.

Starting from the initial conditions given in Table 8.1 at $x = 0$ and using a Courant number $C = 0.5$ in Eq. (8.63), we find that after taking 16 marching steps downstream, we are located at $x = 12.928$ m. Examining Fig. 8.4, this station is located 2.928 m downstream of the expansion corner. Let us focus on the calculations associated with the second grid point at this station, i.e., the grid point labeled $j = 2$ in Fig. 8.7, which shows the local grid in the vicinity of the wall in the region around $\xi = 12.928$ m. In the finite-difference procedure, the station $\xi = 12.928$ m represents the location at which the flow is to be calculated from the known values at the previous station. Hence, $\xi = 12.928$ m corresponds to location $i + 1$ in the finite-difference equations given in Sec. 8.3.2, and the previous station corresponds to location i.

Using the stability criterion given by Eq. (8.63), with $C = 0.5$, the value of $\Delta \xi$ between stations i and $i + 1$ in Fig. 8.7 is $\Delta \xi = 0.818$ m. Hence, at station i, $\xi = 12.928 - 0.818 = 12.11$ m. At this station i, we have from Eq. (8.35),

$$h = 40 + (12.11 - 10) \tan 5.352° = 40.20 \text{ m}$$

Also, the metric $\partial \eta / \partial x$ evaluated at grid point $j = 2$ at station i is, from Eq. (8.25b),

$$\frac{\partial \eta}{\partial x} = (1 - \eta) \frac{\tan \theta}{h} = (1 - 0.025) \frac{\tan 5.352}{40.20} = 2.272 \times 10^{-3} \text{ m}^{-1}$$

At station i, the values of F_1 at points $j = 1$, 2, and 3 are known from the calculations at the previous step. These values are $(F_1)_{i,1} = 0.696 \times 10^3 \frac{\text{kg}}{\text{m}^2 \cdot \text{s}}$,

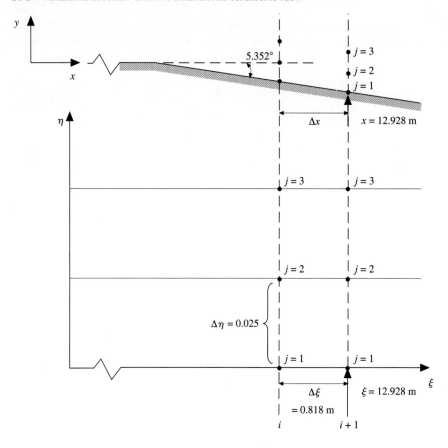

FIG. 8.7
Local grid adjacent to the wall at station $x = \xi = 12.928$ m.

$(F_1)_{i,2} = 0.744 \times 10^3 \frac{\text{kg}}{(\text{m}^2 \cdot \text{s})}$, and $(F_1)_{i,3} = 0.798 \times 10^3 \frac{\text{kg}}{(\text{m}^2 \cdot \text{s})}$. From eq. (8.36a), we have

$$\left(\frac{\partial F_1}{\partial \xi}\right)_{i,2} = \left(\frac{\partial \eta}{\partial x}\right)_{i,2} \frac{(F_1)_{i,2} - (F_1)_{i,3}}{\Delta \eta} + \frac{1}{h_i} \frac{(G_1)_{i,2} - (G_1)_{i,3}}{\Delta \eta}$$

$$= (2.272 \times 10^{-3}) \left[\frac{(0.744 - 0.798) \times 10^3}{0.025}\right]$$

$$+ \frac{1}{40.20} \left[\frac{(-0.435 + 0.193) \times 10^3}{0.025}\right]$$

$$= -4.908 - 24.080 = \boxed{-28.99 \text{ kg}/(\text{m}^3 \cdot \text{s})}$$

From Eq. (8.37a), we have

$$(\bar{F}_1)_{i+1,\,2} = (F_1)_{i,\,2} + \left(\frac{\partial F_1}{\partial \xi}\right)_{i,\,2} \Delta\xi$$
$$= 0.744 \times 10^3 + (-28.99)(0.818)$$
$$= 0.720 \times 10^3 \text{ kg}/(\text{m}^2 \cdot \text{s})$$

At this stage, we add some artificial viscosity. From Eq. (8.46a) we have, where $C_y = 0.6$,

$$(SF_1)_{i+1,\,2} = \frac{C_y|p_{i+1,\,3} - 2p_{i+1,\,2} + p_{i+1,\,1}|}{p_{i+1,\,3} + 2p_{i+1,\,2} + p_{i+1,\,1}}$$
$$\times [(F_1)_{i+1,\,3} - 2(F_1)_{i+1,\,2} + (F_1)_{i+1,\,1}]$$
$$= 0.001 \times 10^3$$

From Eq. (8.47a), we have

$$(\bar{F}_1)_{i+1,\,2} = (F_1)_{i,\,2} + \left(\frac{\partial F}{\partial \xi}\right)_{i,\,2} \Delta\xi + (SF_1)_{i+1,\,2}$$
$$= 0.720 \times 10^3 + 0.001 \times 10^3$$
$$= \boxed{0.721 \times 10^3 \text{ kg}/(\text{m}^2 \cdot \text{s})}$$

Note how small is the value of the artificial viscosity compared to the magnitude of the variable to which it is being added. This is as it should be, since the calculations are being made in a region where only small gradients of the flow-field variables exist, and hence artificial viscosity is not a strong player here. From the same calculations applied at grid points $(i, 1)$ and $(i, 3)$, we have

$$(\bar{F}_1)_{i+1,\,1} = 0.703 \times 10^3 \text{ kg}/(\text{m}^2 \cdot \text{s})$$
$$(\bar{F}_1)_{i+1,\,3} = 0.783 \times 10^3 \text{ kg}/(\text{m}^2 \cdot \text{s})$$

Also, from the sequential application of Eqs. (8.36b) to (8.36d) and Eqs. (8.37b) to (8.37d) we find that

$$(\bar{F}_2)_{i+1,\,2} = 0.585 \times 10^6 \text{ N}/\text{m}^2$$
$$(\bar{F}_3)_{i+1,\,2} = -0.388 \times 10^5 \text{ kg}/(\text{m} \cdot \text{s}^2)$$
$$(\bar{F}_4)_{i+1,\,2} = 0.372 \times 10^9 \text{ N}/(\text{m} \cdot \text{s})$$

The predicted density at point $(i + 1, 2)$ is obtained from Eq. (8.38), where

$$A = \frac{(\bar{F}_3)^2_{i+1,2}}{2(\bar{F}_1)_{i+1,2}} - (\bar{F}_4)_{i+1,2}$$

$$= \frac{(-0.388 \times 10^5)^2}{2(0.721 \times 10^3)} - 0.372 \times 10^9$$

$$= -0.37095 \times 10^9 \frac{\text{N}}{\text{m} \cdot \text{s}}$$

$$B = \frac{\gamma}{\gamma - 1}(\bar{F}_1)_{i+1,2}(\bar{F}_2)_{i+1,2}$$

$$= \frac{1.4}{0.4}(0.721 \times 10^3)(0.585 \times 10^6)$$

$$= 1.476 \times 10^9 \frac{\text{N}^2 \cdot \text{s}}{\text{m}^5}$$

$$C = -\frac{\gamma + 1(\bar{F}_1)^3_{i+1,j}}{2(\gamma - 1)}$$

$$= -\frac{2.4}{2(0.4)}(0.721 \times 10^3)^3 = 1.124 \times 10^9 \frac{\text{kg}}{\text{m}^2 \cdot \text{s}}$$

and

$$\bar{\rho}_{i+1,2} = \frac{-B + \sqrt{B^2 - 4AC}}{2A}$$

$$= \frac{-1.476 \times 10^9 + \sqrt{(1.476 \times 10^9)^2 - 4(0.372 \times 10^9)(1.124 \times 10^9)}}{2(-0.37095 \times 10^9)}$$

$$= \boxed{1.02 \text{ kg/m}^3}$$

With this, we can form the predicted values for G, for example, from Eq. (8.39):

$$(\bar{G}_1)_{i+1,2} = \bar{\rho}_{i+1,2}\frac{(\bar{F}_3)_{i+1,2}}{(\bar{F}_1)_{i+1,2}} = 1.02\left(\frac{-0.388 \times 10^5}{0.721 \times 10^3}\right)$$

$$= \boxed{-0.552 \times 10^2 \text{ kg/(m}^2 \cdot \text{s})}$$

In a similar manner, we find that $[\bar{G}_1]_{i+1,1} = -0.658 \times 10^2$ kg/(m$^2 \cdot$ s).

With the above information, we move to the corrector step. From Eq. (8.43a),

$$\left(\frac{\partial \bar{F}_1}{\partial \xi}\right)_{i+1,2} = \left(\frac{\partial \eta}{\partial x}\right)\frac{(\bar{F}_1)_{i+1,1} - (\bar{F})_{i+1,2}}{\Delta \eta} + \frac{1}{h}\frac{(\bar{G}_1)_{i+1,1} - (\bar{G}_1)_{i+1,2}}{\Delta \eta}$$

At this stage we have a choice to make; in the above equation, do we evaluate $\partial \eta/\partial x$ and h at station i or $i + 1$? It is not immediately obvious which choice to make. It appears that we are evaluating the left-hand side of the equation at station $i + 1$. On the other hand, this equation is simply one element of a calculation that seeks to represent an *average* value of the flow-field derivatives between locations i and $i + 1$, and hence it might be appropriate to treat $\partial \eta/\partial x$ and h consistently as those values at station i on *both* the predictor and corrector step. Faced with this choice, we make the latter. Hence, in the above equation we will use the value of the metric and the value of h as those existing at station i. With this, we obtain

$$\left(\frac{\partial \bar{F}_1}{\partial \xi}\right)_{i+1,2} = (2.272 \times 10^{-3})\left[\frac{(0.703 - 0.721) \times 10^3}{0.025}\right]$$

$$+ \frac{1}{40.2}\left[\frac{(-0.658 + 0.552) \times 10^2}{0.025}\right]$$

$$= \boxed{-0.122 \times 10^2 \text{ kg}/(\text{m}^3 \cdot \text{s})}$$

From Eq. (8.44a), we have

$$\left(\frac{\partial F_1}{\partial \xi}\right)_{av} = \frac{1}{2}\left[\left(\frac{\partial F_1}{\partial \xi}\right)_{i,2} + \left(\frac{\partial \bar{F}}{\partial \xi}\right)_{i+1,2}\right]$$

$$= \tfrac{1}{2}(-28.99 - 12.2) = \boxed{-20.5 \text{ kg}/(\text{m}^3 \cdot \text{s})}$$

From Eq. (8.45a)

$$(F_1)_{i+1,2} = (F_1)_{i,2} + \left(\frac{\partial F}{\partial \xi}\right)_{av} \Delta \xi$$

$$= 0.744 \times 10^3 + (-20.5)(0.818)$$

$$= 0.727 \times 10^3 \text{ kg}/(\text{m}^2 \cdot \text{s})$$

At this stage, we add some artificial viscosity. From Eq. (8.48a), we find that

$$(\bar{S}\bar{F}_1)_{i+1,2} = -0.8$$

Hence, from Eq. (8.49a)

$$(F_1)_{i+1,2} = (F_1)_{i,2} + \left(\frac{\partial F_1}{\partial \xi}\right)_{av} \Delta \xi + (\bar{S}\bar{F}_1)_{i+1,2}$$

$$= \boxed{0.728 \times 10^3 \text{ kg}/(\text{m}^2 \cdot \text{s})}$$

In a similar manner, we obtain

$$(F_2)_{i+1,2} = 0.590 \times 10^6 \frac{\text{N}}{\text{m}^2}$$

$$(F_3)_{i+1,2} = -0.36 \times 10^5 \frac{\text{kg}}{\text{m} \cdot \text{s}^2}$$

$$(F_4)_{i+1,2} = 0.375 \times 10^9 \frac{\text{N}}{\text{m} \cdot \text{s}}$$

Decoding the primitive variables as described earlier, we have from Eq. (8.8)

$$A = \frac{(F_3)^2_{i+1,2}}{2(F_1)^2_{i+1,2}} - (F_4)_{i+1,2}$$

$$= \frac{(-0.36 \times 10^5)^2}{2(0.728 \times 10^3)} - 0.375 \times 10^9$$

$$= -0.374 \times 10^9 \frac{N}{m \cdot s}$$

$$B = \frac{\gamma}{\gamma - 1}(F_1)_{i+1,2}(F_2)_{i+1,2}$$

$$= \left(\frac{1.4}{0.4}\right)(0.728 \times 10^3)(0.590 \times 10^6) = 1.503 \times 10^9 \frac{N \cdot s}{m^5}$$

$$C = \frac{-(\gamma + 1)}{2(\gamma - 1)}(F_1)^3_{i+1,2}$$

$$= \frac{-(2.4)}{0.8}(0.728 \times 10^3)^3 = 1.152 \times 10^9 \left(\frac{kg}{m^2 \cdot s}\right)^3$$

Thus,

$$\rho_{i+1,2} = \frac{-B + \sqrt{B^2 - 4AC}}{2A}$$

$$= \frac{-1.503 \times 10^9 + \sqrt{(1.50 \times 10^9)^2 - 4(-0.374 \times 10^9)(1.152 \times 10^9)}}{2(-0.374 \times 10^9)}$$

$$= \boxed{1.04 \text{ kg/m}^3}$$

From Eq. (8.9)

$$u_{i+1,2} = \frac{(F_1)_{i+1,2}}{\rho_{i+1,2}} = \frac{0.728 \times 10^3}{1.04} = \boxed{701 \text{ m/s}}$$

From Eq. (8.10)

$$v_{i+1,2} = \frac{(F_3)_{i+1,2}}{(F_1)_{i+1,2}} = \frac{-0.36 \times 10^5}{0.728 \times 10^3} = \boxed{-49.4 \text{ m/s}}$$

From Eq. (8.11)

$$p_{i+1,2} = (F_2)_{i+1,2} - (F_1)_{i+1,2} u_{i+1,2}$$

$$= 0.590 \times 10^6 - (0.728 \times 10^3)(701)$$

$$= \boxed{0.795 \times 10^5 \text{ N/m}^2}$$

Finally, from Eq. (8.11) we have

$$T_{i+1,2} = \frac{p_{i+1,2}}{R\rho_{i+1,2}} = \frac{0.795 \times 10^5}{287(1.04)} = \boxed{267 \text{ K}}$$

Return to Fig. 8.7. In the above calculations, we have illustrated how the flow-field values at grid point $j = 2$ located at station $i + 1$ are computed from the known flow field at station i. Let us now concentrate on the calculation of the flow field at the boundary, i.e., at grid point $j = 1$ at station $i + 1$ in Fig. 8.7. To avoid repetitiveness, we will examine the calculation on the corrector step; the treatment of the boundary condition on the predictor step follows the same approach.

We first need to calculate the values of F_1, F_2, etc., at the boundary using one-sided, forward differences on both the predictor and corrector steps. We will pick up the calculation on the corrector step. From Eq. (8.43a), but with forward differences, we have

$$\left(\overline{\frac{\partial F_1}{\partial \xi}}\right)_{i+1,1} = \left(\frac{\partial \eta}{\partial x}\right) \frac{(\bar{F}_1)_{i+1,1} - (\bar{F}_1)_{i+1,2}}{\Delta \eta} + \frac{1}{h} \frac{(\bar{G}_1)_{i+1,1} - (\bar{G}_1)_{i+1,2}}{\Delta \eta}$$

From the predictor step, we have values for the quantities on the right-hand side of the above equations; they are

$$(\bar{F}_1)_{i+1,1} = 0.703 \times 10^3 \frac{\text{kg}}{\text{m}^2 \cdot \text{s}} \qquad (\bar{F}_1)_{i+1,2} = 0.721 \times 10^3 \frac{\text{kg}}{\text{m}^2 \cdot \text{s}}$$

$$(\bar{G}_1)_{i+1,1} = -0.658 \times 10^2 \frac{\text{kg}}{\text{m}^2 \cdot \text{s}} \qquad (\bar{G}_1)_{i+1,2} = -0.552 \times 10^2 \frac{\text{kg}}{\text{m}^2 \cdot \text{s}}$$

Thus,

$$\left(\overline{\frac{\partial F_1}{\partial \xi}}\right)_{i+1,1} = (2.272 \times 10^{-3}) \left[\frac{(0.703 - 0.721) \times 10^3}{0.025}\right]$$

$$+ \frac{1}{40.20} \left[\frac{(-0.658 + 0.552) \times 10^2}{0.025}\right]$$

$$= -1.64 - 10.55 = -12.18 \text{ kg/(m}^3 \cdot \text{s)}$$

Also from the predictor step, we have

$$\left(\frac{\partial F}{\partial \xi}\right)_{i,1} = -26.1 \text{ kg/(m}^3 \cdot \text{s)}$$

From Eq. (8.44a),

$$\left(\frac{\partial F}{\partial \xi}\right)_{av} = \frac{1}{2}\left[\left(\frac{\partial F_1}{\partial \xi}\right)_{i,1} + \left(\overline{\frac{\partial F}{\partial \xi}}\right)_{i+1,1}\right]$$

$$= \tfrac{1}{2}[(-26.1) + (-12.18)] = -19.14 \text{ kg/(m}^3 \cdot \text{s)}$$

From Eq. (8.45a)

$$(F_1)_{i+1,1} = (F_1)_{i,1} + \left(\frac{\partial F}{\partial \xi}\right)_{av} \Delta\xi$$
$$= 0.696 \times 10^3 + (-19.14)(0.818)$$
$$= 0.680 \times 10^3 \text{ kg/(m}^2 \cdot \text{s)}$$

This is the value of F_1 at the boundary as obtained from the algorithm designed for the *internal* flow-field points, modified for one-sided differences at the wall. Analogous results are obtained for F_2, F_3, and F_4 at the boundary. These values are then decoded to obtain the primitive variables at the wall. The results are

$$M_{cal} = 2.22$$
$$p_{cal} = 0.705 \times 10^5 \text{ N/m}^2$$
$$T_{cal} = 255 \text{ K}$$
$$\rho_{cal} = 0.963 \text{ kg/m}^3$$
$$v_{cal} = -74.6 \text{ m/s}$$
$$u_{cal} = 707 \text{ m/s}$$

Note that the calculated values of v_{cal} and u_{cal} yield a velocity vector in the direction defined by the angle ψ in Fig. 8.5. From Eq. (8.57),

$$\psi = \tan^{-1}\frac{|v_{cal}|}{u_{cal}} = \tan^{-1}\frac{74.6}{707} = 6.02°$$

However, the wall downstream of the expansion corner is at an angle $\theta = 5.353°$ (see Fig. 8.5). Thus, after the use of the one-sided differences at the wall as described above, we see that the calculated velocity vector is pointing *into* the wall, since $\psi > \theta$ (again, see Fig. 8.5). From Eq. (8.58), we have

$$\phi_2 = \theta - \psi = 5.352 - 6.02 = -0.668°$$

Hence, we need to *imagine* that the *calculated* supersonic flow at the wall must be rotated through an angle $\phi_2 = -0.668°$ (an upward rotation) in order to be tangent to the wall; this rotation is carried out by means of a local Prandtl–Meyer *compression* wave, since the calculated flow is *into* the wall. From Eq. (8.56)

$$f_{act} = f_{cal} + \phi_2$$

Since $f_{cal} = 32.24°$ for $M_{cal} = 2.22$, we have

$$f_{act} = 32.24 - 0.668 = 31.57°$$

From Eq. (8.2), this yields

$$M_{act} = 2.19$$

The actual values of pressure, temperature, and density at the wall are obtained from Eqs. (8.53) to (8.55), respectively.

$$p_{i+1,1} = p_{act} = p_{cal}\left\{\frac{1+[(\gamma-1)/2]M_{cal}^2}{1+[(\gamma-1)/2]M_{act}^2}\right\}^{\gamma/(\gamma-1)}$$

$$= (0.705 \times 10^5)\left[\frac{1+0.4(2.22)^2}{1+0.4(2.19)^2}\right]^{3.5}$$

$$= \boxed{0.734 \times 10^5 \text{ N/m}^2}$$

$$T_{i+1,1} = T_{act} = T_{cal}\frac{1+[(\gamma-1)/2]M_{cal}^2}{1+[(\gamma-1)/2]M_{act}^2}$$

$$= 255\left[\frac{1+0.4(2.22)^2}{1+0.4(2.19)^2}\right] = \boxed{258 \text{ K}}$$

$$\rho_{i+1,1} = \rho_{act} = \frac{p_{act}}{RT_{act}} = \frac{0.734 \times 10^5}{287(258)} = \boxed{0.992 \text{ kg/m}^3}$$

The use of the local Prandtl–Meyer wave at the wall to rotate the calculated velocity vector so that it becomes tangent to the wall is purely a conceptual matter; it is simply a way to *imagine* that the component of the calculated velocity *perpendicular* to the wall, which is usually a finite value when the one-sided differences are used, is *canceled* by means of the local Prandtl–Meyer wave. (Keep in mind that the proper flow tangency boundary condition can be expressed by stating that the component of velocity normal to the wall must be *zero*.) The actual values of p, T, and ρ obtained above represent a small *adjustment* to the originally calculated values to be consistent with this cancellation of the normal velocity component.

Finally, since the local Prandtl–Meyer wave is functioning to simply cancel the normal component of velocity, and this cancellation involves only a small velocity change, we choose to leave the x-component velocity, as calculated from one-sided differences, alone. That is, we will stipulate that

$$u_{i+1,1} = u_{cal} = 707 \text{ m/s}$$

From this, the corresponding y component of velocity at the wall must be the following value, after the normal component is effectively canceled and the flow is therefore tangent to the wall.

$$v_{i+1,1} = -u_{i+1,1}\tan\theta = -707\tan 5.352° = \boxed{-66.2 \text{ m/s}}$$

Note that this value of v is slightly smaller than $v_{cal} = -74.6$ m/s as listed above, calculated from one-sided differences. The value $v_{i+1,1} = -66.2$ m/s is compatible with the flow tangency boundary condition.

This ends our sample, intermediate calculations. For completeness, the results obtained at all the grid points from $j = 1$ to $j = 41$ at the station located at

$x = \xi = 12.928$ m are tabulated in Tables 8.2 and 8.3. The numbers tabulated here are the final flow-field values obtained from the downstream-marching technique at the station $x = 12.928$ m; this corresponds to the sixteenth marching step starting from the given initial conditions at $x = 0$. We will return to Tables 8.2 and 8.3 during our final analysis of the results as discussed in the next section.

TABLE 8.2
Results at $x = 12.928$ m

j	y, m	η	u, m/s	v, m/s	ρ, kg/m³	p, N/m³
1	−0.274	0.000	.707E+03	−.662E+02	.992E+00	.734E+05
2	0.733	0.025	.701E+03	−.494E+02	.104E+01	.795E+05
3	1.739	0.050	.691E+03	−.266E+02	.112E+01	.891E+05
4	2.746	0.075	.683E+03	−.869E+01	.119E+01	.969E+05
5	3.753	0.100	.679E+03	−.131E+01	.122E+01	.100E+06
6	4.760	0.125	.678E+03	−.148E−01	.123E+01	.101E+06
7	5.767	0.150	.678E+03	.326E−05	.123E+01	.101E+06
8	6.774	0.175	.678E+03	−.167E−03	.123E+01	.101E+06
9	7.781	0.200	.678E+03	.472E−04	.123E+01	.101E+06
10	8.787	0.225	.678E+03	−.702E−04	.123E+01	.101E+06
11	9.794	0.250	.678E+03	−.195E−04	.123E+01	.101E+06
12	10.801	0.275	.678E+03	.180E−04	.123E+01	.101E+06
13	11.808	0.300	.678E+03	−.598E−04	.123E+01	.101E+06
14	12.815	0.325	.678E+03	−.642E−04	.123E+01	.101E+06
15	13.822	0.350	.678E+03	−.325E−13	.123E+01	.101E+06
16	14.829	0.375	.678E+03	.000E+00	.123E+01	.101E+06
17	15.835	0.400	.678E+03	.000E+00	.123E+01	.101E+06
18	16.842	0.425	.678E+03	.000E+00	.123E+01	.101E+06
19	17.849	0.450	.678E+03	.000E+00	.123E+01	.101E+06
20	18.856	0.475	.678E+03	.000E+00	.123E+01	.101E+06
21	19.863	0.500	.678E+03	.000E+00	.123E+01	.101E+06
22	20.870	0.525	.678E+03	.000E+00	.123E+01	.101E+06
23	21.877	0.550	.678E+03	.000E+00	.123E+01	.101E+06
24	22.883	0.575	.678E+03	.000E+00	.123E+01	.101E+06
25	23.890	0.600	.678E+03	.217E−10	.123E+01	.101E+06
26	24.897	0.625	.678E+03	.118E−03	.123E+01	.101E+06
27	25.904	0.650	.678E+03	.120E−03	.123E+01	.101E+06
28	26.911	0.675	.678E+03	.354E−05	.123E+01	.101E+06
29	27.918	0.700	.678E+03	.125E−03	.123E+01	.101E+06
30	28.925	0.725	.678E+03	−.193E−04	.123E+01	.101E+06
31	29.931	0.750	.678E+03	−.607E−04	.123E+01	.101E+06
32	30.938	0.775	.678E+03	.242E−03	.123E+01	.101E+06
33	31.945	0.800	.678E+03	.160E−03	.123E+01	.101E+06
34	32.952	0.825	.678E+03	.161E−03	.123E+01	.101E+06
35	33.959	0.850	.678E+03	.401E−04	.123E+01	.101E+06
36	34.966	0.875	.678E+03	−.848E−04	.123E+01	.101E+06
37	35.973	0.900	.678E+03	−.128E−03	.123E+01	.101E+06
38	36.979	0.925	.678E+03	−.342E−04	.123E+01	.101E+06
39	37.986	0.950	.678E+03	−.107E−03	.123E+01	.101E+06
40	38.993	0.975	.678E+03	−.636E−04	.123E+01	.101E+06
41	40.000	1.000	.678E+03	.000E+00	.123E+01	.101E+06

TABLE 8.3
Flux values at $x = 12.928$ m

j	T, K	M	F_1, kg/(m$^2 \cdot$ s)	F_2, N/m^2	F_3, kg/(m $\cdot$ s^2)	F_4, N/(m $\cdot$ s)
1	.258E+03	.220E+01	.701E+03	.569E+06	−.464E+05	.358E+09
2	.267E+03	.215E+01	.728E+03	.590E+06	−.360E+05	.375E+09
3	.277E+03	.208E+01	.776E+03	.626E+06	−.207E+05	.402E+09
4	.283E+03	.203E+01	.815E+03	.654E+06	−.708E+04	.422E+09
5	.286E+03	.200E+01	.831E+03	.665E+06	−.109E+04	.430E+09
6	.286E+03	.200E+01	.834E+03	.667E+06	−.123E+02	.431E+09
7	.286E+03	.200E+01	.834E+03	.667E+06	.272E−02	.431E+09
8	.286E+03	.200E+01	.834E+03	.667E+06	−.140E+00	.431E+09
9	.286E+03	.200E+01	.834E+03	.667E+06	.394E−01	.431E+09
10	.286E+03	.200E+01	.834E+03	.667E+06	−.586E−01	.431E+09
11	.286E+03	.200E+01	.834E+03	.667E+06	−.162E−01	.431E+09
12	.286E+03	.200E+01	.834E+03	.667E+06	.150E−01	.431E+09
13	.286E+03	.200E+01	.834E+03	.667E+06	−.499E−01	.431E+09
14	.286E+03	.200E+01	.834E+03	.667E+06	−.535E−01	.431E+09
15	.286E+03	.200E+01	.834E+03	.667E+06	−.271E−10	.431E+09
16	.286E+03	.200E+01	.834E+03	.667E+06	.000E+00	.431E+09
17	.286E+03	.200E+01	.834E+03	.667E+06	.000E+00	.431E+09
18	.286E+03	.200E+01	.834E+03	.667E+06	.000E+00	.431E+09
19	.286E+03	.200E+01	.834E+03	.667E+06	.000E+00	.431E+09
20	.286E+03	.200E+01	.834E+03	.667E+06	.000E+00	.431E+09
21	.286E+03	.200E+01	.834E+03	.667E+06	.000E+00	.431E+09
22	.286E+03	.200E+01	.834E+03	.667E+06	.000E+00	.431E+09
23	.286E+03	.200E+01	.834E+03	.667E+06	.000E+00	.431E+09
24	.286E+03	.200E+01	.834E+03	.667E+06	.000E+00	.431E+09
25	.286E+03	.200E+01	.834E+03	.667E+06	.181E−07	.431E+09
26	.286E+03	.200E+01	.834E+03	.667E+06	.988E−01	.431E+09
27	.286E+03	.200E+01	.834E+03	.667E+06	.100E+00	.431E+09
28	.286E+03	.200E+01	.834E+03	.667E+06	.295E−02	.431E+09
29	.286E+03	.200E+01	.834E+03	.667E+06	.104E+00	.431E+09
30	.286E+03	.200E+01	.834E+03	.667E+06	−.161E−01	.431E+09
31	.286E+03	.200E+01	.834E+03	.667E+06	−.506E−01	.431E+09
32	.286E+03	.200E+01	.834E+03	.667E+06	.201E+00	.431E+09
33	.286E+03	.200E+01	.834E+03	.667E+06	.133E+00	.431E+09
34	.286E+03	.200E+01	.834E+03	.667E+06	.134E+00	.431E+09
35	.286E+03	.200E+01	.834E+03	.667E+06	.335E−01	.431E+09
36	.286E+03	.200E+01	.834E+03	.667E+06	−.707E−01	.431E+09
37	.286E+03	.200E+01	.834E+03	.667E+06	−.106E+00	.431E+09
38	.286E+03	.200E+01	.834E+03	.667E+06	−.285E−01	.431E+09
39	.286E+03	.200E+01	.834E+03	.667E+06	−.891E−01	.431E+09
40	.286E+03	.200E+01	.834E+03	.667E+06	−.530E−01	.431E+09
41	.286E+03	.200E+01	.834E+03	.667E+06	.000E+00	.431E+09

8.3.4 Final Results

Let us examine the results of the present downstream marching calculations in a more global fashion. Such a global picture is shown in Fig. 8.8. Here the x component of the velocity, u, is plotted versus the vertical distance y at five different stations in the x direction, namely, $x = 0$, 16.17, 32.31, 48.99, and 66.23 m. The

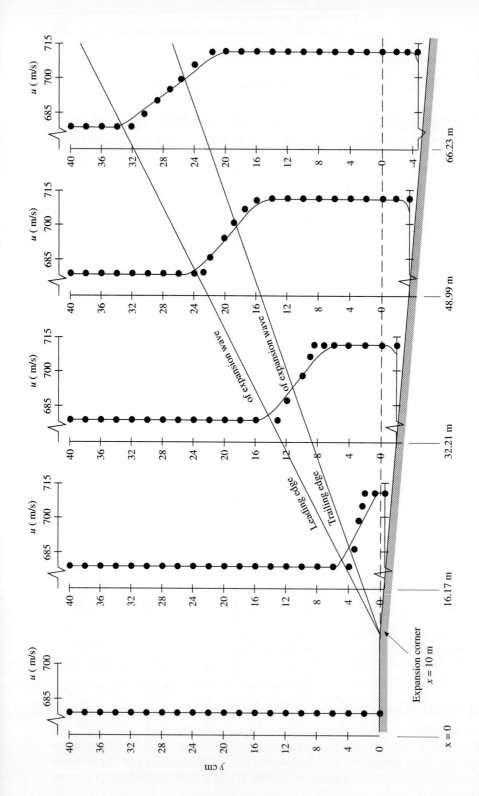

FIG. 8.8
Comparison between the CFD results and the exact, analytical solution (dark circles) for supersonic flow through a centered expansion wave. Solid lines are numerical results.

geometry of the wall, with the expansion corner at $x = 10$ m, is also plotted to scale. Moreover, the leading and trailing edges of the centered expansion wave (exact solution) are drawn to scale and superimposed on the figure. For each of the five velocity profiles shown, a comparison between the exact analytical results (the dark circles) and the numerical downstream marching, finite-difference results (the solid line) is given. Make certain to orient yourself with regard to what is shown in Fig. 8.8 before progressing further.

In Fig. 8.8, the velocity profile at $x = 0$ shows the uniform inflow conditions where $u = 678$ m/s. The velocity profile at $x = 16.17$ m is just slightly downstream of the expansion corner, which is located at $x = 10$ m. The station at $x = 16.17$ m corresponds to 20 marching steps downstream of the inlet. It is intuitively obvious that the most severe demands on the numerical solution are in the vicinity of the expansion corner, which analytically is a singular point. Reflecting back to the governing flow equations expressed by Eqs. (8.30) to (8.33), note that the metric $\partial \eta / \partial x$ experiences a discontinuous change at the expansion corner, as is clearly seen in Eqs. (8.25a) and (8.25b). Also, just behind the expansion corner, there are only a few vertically arrayed grid points between the wall and the trailing edge of the wave, and fewer points yet inside the wave itself. As a result, not only is the numerical solution in this region effectively hit over the head with a sledgehammer (i.e., in the form of the singular point and the discontinuity in the metric), there are also very few grid points available in that region to absorb the blow. The net result is the partial lack of definition of the wave at the $x = 16.17$ m station; the agreement between the numerical calculation and the exact, analytical results inside and behind the wave is not very good. This behavior is one of the reasons why a relatively mild expansion corner with a deflection angle of 5.352° is chosen for the present test case. For a larger deflection angle, the problems discussed above are exacerbated. Consider the case for a corner deflection angle of 23.38° with the same Mach 2 upstream flow in the present example. This flow deflection will expand the flow to Mach 3 downstream of the corner. However, when $\theta = 23.38°$ is fed into the current setup, the calculations develop some strong oscillations and eventually (about 6 or 8 m downstream of the corner) blow up. Presumably there is some combination of number of grid points with an appropriately heavy numerical damping (via large artificial viscosity) that might result in a successful solution downstream of the expansion corner for this large deflection case. This matter is left for you to examine.

At this stage, we recall that Tables 8.2 and 8.3 give the flow-field variables even closer to the expansion corner, namely, at $x = 12.928$ m. The numbers tabulated in those tables simply reinforce the above discussion.

As the numerical solution progressively marches further downstream, where the expansion wave is wider and the distance between the wall and the trailing edge of the wave is larger, the agreement between the exact, analytical results and the numerical computations improves considerably. Concentrating on the velocity profile at the $x = 66.23$ m station, we see good agreement; the numerical solution is capturing the wave in the right location, the variation of the computed velocity inside the wave nicely follows the exact, analytical solution, and the velocity downstream of the wave is uniform at the correct value.

There is a slightly disturbing phenomenon occurring right at the wall as the numerical calculations progress downstream. The exact, analytical value for u_2 downstream of the wave is 710.2 m/s. At $x = 16.17$ m, the numerical value of u_2 right at the wall is 711 m/s—very close to the exact result. However, as we march downstream, the value of u_2 at the wall begins to deviate; at the subsequent three downstream stations ($x = 32.3$, 48.99, and 66.23 m), the respective values of u_2 at the wall are 708, 707, and 705 m/s. This can be seen in Fig. 8.8 as a small "velocity layer" right at the wall, which departs from the exact solution. The thickness of this layer is only one grid increment, i.e., the thickness is only Δy. At the first grid point above the wall, the flow velocity comes into very good agreement with exact results. This small velocity layer, which is certainly a numerically induced phenomenon and *not* a physical result, may be due to the history of the expansion corner singularity being propagated downstream by the numerical computations. It also may be due to a progressive accumulation of numerical error owing to the numerical implementation of the boundary condition. Since the solution uses a downstream marching philosophy, numerical errors that occur upstream are simply carried along as the solution marches downstream. If a certain type of numerical error, no matter how slight, is repeatedly generated at the wall, it will have a tendency to accumulate as we march downstream. Perhaps a slight improvement in our numerical implementation of the flow tangency boundary condition is in order. This would be an interesting matter for you to examine.

Note from Fig. 8.8 that the leading edge of the exact expansion wave exits the computational domain through the *downstream* boundary, i.e., at about a height $y = 32.5$ m. The geometry of our computational space is chosen intentionally to allow this to happen. Recall that the boundary condition we are imposing at the *upper* boundary (at $\eta = 1.0$) is simply the specification of uniform conditions equal to those in the uniform flow upstream of the expansion wave. This is appropriate as long as the expansion wave completely exits the computational space along the vertical, downstream boundary at $x = 66.23$ m. Imagine what would happen if we were to continue marching downstream, say to a station $x = 100$ m. By examining Fig. 8.8, we can easily see that the leading edge and part of the internal portion of the expansion wave will exit through the upper boundary. If and when such a situation exits, we must use a different boundary condition along the upper boundary—different from that which we have used so far. In this case, what changes would you make along the upper boundary? A simple thought is the calculation of the flow properties along the upper boundary from the flow equations using one-sided differences (in this case requiring one-sided rearward differences on both the predictor and corrector steps). Another possibility is to extrapolate the values to the upper boundary from information at the internal grid points; however, in this case, rather than using a linear extrapolation in the vertical direction, it would be more appropriate to extrapolate along *characteristic lines* from the internal grid points. This is something else with which you might want to experiment.

Return to the results shown in Fig. 8.8. For the sake of completeness, all the computed flow-field variables at the $x = 66.23$ station (which corresponds to 80 marching steps from the inlet) are tabulated in Tables 8.4 and 8.5. When examining these tabulated results, it is useful to note that the exact, analytical values

TABLE 8.4
Results at $x = 66.278$ m

j	y, m	η	u, m/s	v, m/s	ρ, kg/m³	p, N/m³
1	−5.272	0.000	.705E+03	−.661E+02	.109E+01	.731E+05
2	−4.140	0.025	.710E+03	−.682E+02	.107E+01	.730E+05
3	−3.009	0.050	.711E+03	−.690E+02	.969E+00	.732E+05
4	−1.877	0.075	.711E+03	−.688E+02	.977E+00	.731E+05
5	−0.745	0.100	.711E+03	−.689E+02	.976E+00	.731E+05
6	0.387	0.125	.711E+03	−.688E+02	.976E+00	.731E+05
7	1.519	0.150	.711E+03	−.689E+02	.976E+00	.731E+05
8	2.650	0.175	.711E+03	−.690E+02	.976E+00	.731E+05
9	3.782	0.200	.711E+03	−.690E+02	.976E+00	.731E+05
10	4.914	0.225	.711E+03	−.688E+02	.977E+00	.731E+05
11	6.046	0.250	.711E+03	−.686E+02	.977E+00	.732E+05
12	7.178	0.275	.711E+03	−.688E+02	.977E+00	.731E+05
13	8.309	0.300	.711E+03	−.694E+02	.975E+00	.729E+05
14	9.441	0.325	.711E+03	−.696E+02	.974E+00	.729E+05
15	10.573	0.350	.711E+03	−.690E+02	.976E+00	.731E+05
16	11.705	0.375	.711E+03	−.678E+02	.980E+00	.735E+05
17	12.837	0.400	.711E+03	−.672E+02	.982E+00	.737E+05
18	13.968	0.425	.711E+03	−.683E+02	.978E+00	.733E+05
19	15.100	0.450	.712E+03	−.708E+02	.970E+00	.725E+05
20	16.232	0.475	.713E+03	−.732E+02	.963E+00	.717E+05
21	17.364	0.500	.713E+03	−.740E+02	.960E+00	.714E+05
22	18.496	0.525	.713E+03	−.726E+02	.964E+00	.719E+05
23	19.627	0.550	.711E+03	−.693E+02	.975E+00	.730E+05
24	20.759	0.575	.709E+03	−.647E+02	.990E+00	.746E+05
25	21.891	0.600	.707E+03	−.591E+02	.101E+01	.765E+05
26	23.023	0.625	.705E+03	−.531E+02	.103E+01	.787E+05
27	24.155	0.650	.702E+03	−.468E+02	.105E+01	.810E+05
28	25.287	0.675	.699E+03	−.405E+02	.107E+01	.834E+05
29	26.418	0.700	.696E+03	−.343E+02	.110E+01	.859E+05
30	27.550	0.725	.693E+03	−.283E+02	.112E+01	.883E+05
31	28.682	0.750	.690E+03	−.227E+02	.114E+01	.907E+05
32	29.814	0.775	.688E+03	−.175E+02	.116E+01	.930E+05
33	30.946	0.800	.685E+03	−.129E+02	.118E+01	.950E+05
34	32.077	0.825	.683E+03	−.901E+01	.119E+01	.968E+05
35	33.209	0.850	.681E+03	−.591E+01	.121E+01	.982E+05
36	34.341	0.875	.680E+03	−.361E+01	.121E+01	.993E+05
37	35.473	0.900	.679E+03	−.203E+01	.122E+01	.100E+06
38	36.605	0.925	.679E+03	−.105E+01	.123E+01	.100E+06
39	37.736	0.950	.678E+03	−.499E+00	.123E+01	.101E+06
40	38.868	0.975	.678E+03	−.229E+00	.123E+01	.101E+06
41	40.000	1.000	.678E+03	.000E+00	.123E+01	.101E+06

TABLE 8.5
Flux values at $x = 66.278$ m

j	T, K	M	F_1, kg/(m$^2 \cdot$ s)	F_2, N/m^2	F_3, kg/(m $\cdot$ s^2)	F_4, N/(m $\cdot$ s)
1	.233E+03	.231E+01	.769E+03	.616E+06	−.508E+05	.374E+09
2	.237E+03	.231E+01	.760E+03	.612E+06	−.519E+05	.374E+09
3	.263E+03	.220E+01	.689E+03	.563E+06	−.475E+05	.358E+09
4	.261E+03	.221E+01	.694E+03	.567E+06	−.478E+05	.359E+09
5	.261E+03	.221E+01	.694E+03	.567E+06	−.478E+05	.359E+09
6	.261E+03	.221E+01	.694E+03	.567E+06	−.478E+05	.359E+09
7	.261E+03	.221E+01	.694E+03	.567E+06	−.478E+05	.359E+09
8	.261E+03	.221E+01	.694E+03	.567E+06	−.479E+05	.359E+09
9	.261E+03	.221E+01	.694E+03	.567E+06	−.479E+05	.359E+09
10	.261E+03	.221E+01	.694E+03	.567E+06	−.478E+05	.359E+09
11	.261E+03	.221E+01	.695E+03	.567E+06	−.477E+05	.359E+09
12	.261E+03	.221E+01	.695E+03	.567E+06	−.478E+05	.359E+09
13	.261E+03	.221E+01	.693E+03	.566E+06	−.481E+05	.359E+09
14	.261E+03	.221E+01	.693E+03	.566E+06	−.483E+05	.358E+09
15	.261E+03	.221E+01	.694E+03	.567E+06	−.479E+05	.359E+09
16	.261E+03	.220E+01	.697E+03	.569E+06	−.472E+05	.360E+09
17	.261E+03	.220E+01	.698E+03	.569E+06	−.469E+05	.361E+09
18	.261E+03	.221E+01	.696E+03	.568E+06	−.475E+05	.360E+09
19	.260E+03	.221E+01	.691E+03	.564E+06	−.489E+05	.357E+09
20	.259E+03	.222E+01	.686E+03	.561E+06	−.502E+05	.355E+09
21	.259E+03	.222E+01	.685E+03	.560E+06	−.506E+05	.354E+09
22	.260E+03	.222E+01	.687E+03	.562E+06	−.499E+05	.356E+09
23	.261E+03	.221E+01	.694E+03	.566E+06	−.481E+05	.359E+09
24	.262E+03	.219E+01	.703E+03	.573E+06	−.454E+05	.363E+09
25	.264E+03	.218E+01	.713E+03	.581E+06	−.422E+05	.369E+09
26	.266E+03	.216E+01	.725E+03	.590E+06	−.385E+05	.375E+09
27	.269E+03	.214E+01	.737E+03	.599E+06	−.345E+05	.382E+09
28	.271E+03	.212E+01	.750E+03	.608E+06	−.304E+05	.388E+09
29	.273E+03	.210E+01	.763E+03	.617E+06	−.262E+05	.394E+09
30	.275E+03	.209E+01	.775E+03	.625E+06	−.220E+05	.401E+09
31	.277E+03	.207E+01	.786E+03	.634E+06	−.178E+05	.407E+09
32	.279E+03	.205E+01	.797E+03	.641E+06	−.139E+05	.412E+09
33	.281E+03	.204E+01	.807E+03	.648E+06	−.104E+05	.417E+09
34	.283E+03	.203E+01	.815E+03	.654E+06	−.734E+04	.422E+09
35	.284E+03	.202E+01	.822E+03	.658E+06	−.485E+04	.425E+09
36	.285E+03	.201E+01	.826E+03	.661E+06	−.298E+04	.428E+09
37	.285E+03	.201E+01	.830E+03	.664E+06	−.169E+04	.429E+09
38	.286E+03	.200E+01	.832E+03	.665E+06	−.877E+03	.430E+09
39	.286E+03	.200E+01	.833E+03	.666E+06	−.416E+03	.431E+09
40	.286E+03	.200E+01	.834E+03	.666E+06	−.191E+03	.431E+09
41	.286E+03	.200E+01	.834E+03	.667E+06	.000E+00	.431E+09

downstream of the expansion wave are

$$M_2 = 2.20$$
$$p_2 = 0.739 \times 10^5 \text{ N/m}^2$$
$$\rho_2 = 0.984 \text{ kg/m}^3$$
$$T_2 = 262 \text{ K}$$
$$u_2 = 710 \text{ m/s}$$
$$v_2 = -66.5 \text{ m/s}$$

In Tables 8.4 and 8.5, if we compare the numerical results in the uniform region downstream of the wave (say between $j = 2$ and 23) with the exact analytical results listed above, we find the following percentage errors.

Quantity	% error
M_2	0.45
p_2	1.08
ρ_2	0.813
T_2	0.038
u_2	0.141
v_2	3.76

This agreement between the exact analytical solution for the flow behind an expansion wave and the corresponding numerical results is reasonable; indeed, the percentage errors listed above are on par with those obtained in Chap. 7 for our time-marching solutions of nozzle flows. The only disturbing feature shown in Tables 8.4 and 8.5 are the values at the wall ($j = 0$). Here we see the presence of some type of "error layer" at the wall, as discussed earlier in regard to the velocity profiles in Fig. 8.8. The velocity is not the only variable affected by this phenomena; the other flow variables exhibit a slight change at the wall compared to the values immediately above it (except for the pressure, which is virtually constant in the region of the wall, including the wall point.) Our earlier discussion on this matter is sufficient; we will not repeat it here. Suffice it to say that such behavior is an example that *CFD is not perfect*—a fact which is important for you to appreciate.

In regard to the matter of grid independence, a solution was carried out where the number of grid points in the y direction was doubled; i.e., the value of Δy (hence $\Delta \eta$) was halved. This led to 81 points being distributed in the y direction. Also, since the marching step $\Delta \xi$ is related to $\Delta \eta$ through the stability criterion (see Eq. 8.63), this also doubled the number of grid points in the ξ direction. The net result was an increase in the number of grid points by a factor of 4. The calculated results for the flow field in this case were not materially different from those discussed earlier. Therefore, the earlier results basically reflect grid independence.

As a final comment in this section, note that the geometric units chosen to describe the *size* of the computational space for the above calculations, namely, a height of about 40 m and a length of about 65 m, is irrelevant to the answer. Instead of using meters, we could just as well as chosen millimeters, with a computational space of 40 mm by 65 mm, or any other length units for that matter. The flow problem of the supersonic flow through an expansion wave does not depend on any particular length scale. Since in the present calculations we chose to solve the governing equations in *dimensional* variables, we had to stipulate some geometric length; to maintain *consistent* units, we chose the unit of meters. So if a length of 65 m sounds very large to you, do not worry; it is totally irrelevant to the solution of the problem.

8.4 SUMMARY

The major items discussed and illustrated in this chapter are diagrammed in the road map shown in Fig. 8.9. The main thrust of this chapter is to highlight the philosophy of space marching in contrast to that of time marching discussed in Chap. 7. Such space marching required us to use the conservation form of the steady flow equations. The geometry of the problem requires a boundary-fitted coordinate system, so this gave us a chance to work with some aspects of grid generation and to use the governing equations in the transformed space. Moreover, we used the technique of wave capturing, albeit here we captured an expansion wave rather than a shock wave as done in Chap. 7; with wave capturing, we already appreciate the need to use the conservation form of the governing equations. We also applied some artificial viscosity to smooth the results; this is mainly needed in the vicinity of the expansion corner, which itself is a mathematical singularity. For the rest of the expansion wave, we most likely could do without the artificial viscosity. Finally, for the inviscid flow at the boundary, we utilized Abbett's numerical treatment of the boundary condition, which involved the use of a local, imaginary, Prandtl–Meyer wave at the wall to rotate the calculated velocity vector to be parallel to the wall. All

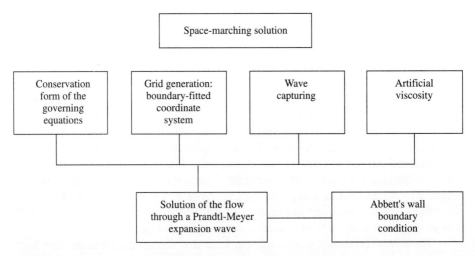

FIG. 8.9
Road map for Chap. 8.

these elements went into the calculation of the supersonic flow through a centered, Prandtl–Meyer expansion wave, which was the featured flow problem in this chapter.

At this stage, we are reminded that the CFD techniques used in both Chaps. 7 and 8 are *explicit* finite-difference techniques. To expand our horizons, it is time for us to explore an *implicit* solution applied to an appropriate flow problem. This leads us directly to the next chapter.

CHAPTER 9

INCOMPRESSIBLE COUETTE FLOW: NUMERICAL SOLUTIONS BY MEANS OF AN IMPLICIT METHOD AND THE PRESSURE CORRECTION METHOD

The most dangerous of our calculations are those we call illusions.

George Bernanos, from *Dialogue des Carmelites*, 1949

9.1 INTRODUCTION

The numerical techniques illustrated in Chaps. 7 and 8 are *explicit* finite-difference methods. Moreover, the mathematical nature of the governing equations for the problems treated in both chapters is *hyperbolic*. For an explicit solution of hyperbolic partial differential equations, we have seen that the CFL stability criterion essentially limits the size of the marching step (Δt in Chap. 7 and Δx in Chap. 8). Furthermore, the flows studied in Chaps. 7 and 8 have been *inviscid* flows.

The present chapter provides a contrast to the previous two chapters in the following respects:

1. It deals with an *implicit* finite-difference solution to the governing equations.
2. The governing equations for the present problem are *parabolic* partial differential equations.
3. The present problem is a *viscous* flow.

In particular, we will deal with incompressible Couette flow, which represents an exact analytical solution of the Navier–Stokes equations. Couette flow is perhaps the simplest of all viscous flows, while at the same time retaining much of the same physical characteristics of a more complicated boundary-layer flow. The numerical technique that we will employ for the solution of the Couette flow is the Crank–Nicolson implicit method discussed in Sec. 4.4. As discussed in Chap. 3, parabolic partial differential equations lend themselves to a marching solution; in addition, the use of an implicit technique allows a much larger marching step size than would be the case for an explicit solution. Hence, in the present chapter we will have the opportunity to explore some aspects of CFD different from those discussed in the previous two chapters.

Near the end of this chapter, we will carry out a second solution of Couette flow, this time using the pressure correction technique described in Sec. 6.8. We will deal with the two-dimensional Navier–Stokes equations for incompressible flow and set up a solution of these equations for the incompressible flow between two parallel plates in relative motion to each other using the pressure correction method. This method is an iterative approach, and we will set up the initial conditions to be a two-dimensional flow field. This will give us the opportunity to examine the behavior of the pressure correction method to an incompressible flow problem which is treated as two-dimensional during the iterative solution but which converges to an answer that is a function of only the vertical coordinate across the flow—it converges to the Couette flow solution.

9.2 THE PHYSICAL PROBLEM AND ITS EXACT ANALYTICAL SOLUTION

Couette flow is defined as follows. Consider the viscous flow between two parallel plates separated by the vertical distance D, as sketched in Fig. 9.1. The upper plate is moving at the velocity u_e, and the lower plate is stationary; i.e., its velocity is $u = 0$. The flow in the xy plane is sketched in Fig. 9.1. The flow field between the two plates is driven exclusively by the shear stress exerted on the fluid by the moving upper plate, resulting in a velocity profile across the flow, $u = u(y)$, as sketched in Fig. 9.1.

The governing equation for this flow is the x-momentum equation, given by Eq. (2.50a), repeated below.

$$\rho \frac{Du}{Dt} = -\frac{\partial p}{\partial x} + \frac{\partial \tau_{xx}}{\partial x} + \frac{\partial \tau_{yx}}{\partial y} + \frac{\partial \tau_{zx}}{\partial z} + \rho f_x \qquad (2.50a)$$

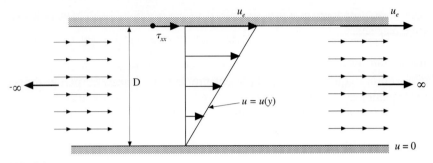

FIG. 9.1
Schematic of Couette flow.

When applied to Couette flow, this equation is greatly simplified, as follows. Examining Fig. 9.1, we note that the model for Couette flow stretches to plus and minus infinity in the x direction. Since there is no beginning or end of this flow, the flow-field variables must be independent of x; that is, $\partial/\partial x = 0$ for all quantities. Moreover, from the continuity equation, Eq. (2.25), written for steady flow,

$$\frac{\partial(\rho u)}{\partial x} + \frac{\partial(\rho v)}{\partial y} = 0 \tag{9.1}$$

Since $\partial(\rho u)/\partial x = 0$ for Couette flow, Eq. (9.1) becomes

$$\frac{\partial(\rho v)}{\partial y} = \rho \frac{\partial v}{\partial y} + v \frac{\partial \rho}{\partial y} = 0 \tag{9.2}$$

Evaluated at the lower wall, where $v = 0$ at $y = 0$, Eq. (9.2) yields

$$\left(\rho \frac{\partial v}{\partial y}\right)_{y=0} = 0$$

or

$$\left(\frac{\partial v}{\partial y}\right)_{y=0} = 0 \tag{9.3}$$

If we expand v in a Taylor series about the point $y = 0$, we have

$$v(y) = v(0) + \left(\frac{\partial v}{\partial y}\right)_{y=0} y + \left(\frac{\partial^2 v}{\partial y^2}\right)_{y=0} \frac{y^2}{2} + \cdots \tag{9.4}$$

Evaluated at the upper wall, Eq. (9.4) becomes

$$v(D) = v(0) + \left(\frac{\partial v}{\partial y}\right)_{y=0} D + \left(\frac{\partial^2 v}{\partial y^2}\right)_{y=0} \frac{D^2}{2} + \cdots \tag{9.5}$$

Since both $v(D) = 0$ and $v(0) = 0$, as well as $[\partial v/\partial y]_{y=0} = 0$ from Eq. (9.3), the only result that makes sense in Eq. (9.5) is that $[\partial^n v/\partial y^n]_{y=0} = 0$ for all n, and hence

$$v = 0 \tag{9.6}$$

everywhere. This is a physical characteristic of Couette flow, namely, that there is no vertical component of velocity anywhere. This states that the streamlines for Couette flow are straight, parallel streamlines—a result which is almost intuitively obvious simply by inspecting Fig. 9.1. Finally, from the y-momentum equation, Eq. (2.50b), repeated below,

$$\rho \frac{Dv}{Dt} = -\frac{\partial p}{\partial y} + \frac{\partial \tau_{xy}}{\partial x} + \frac{\partial \tau_{yy}}{\partial y} + \frac{\partial \tau_{zy}}{\partial z} + \rho f_y \tag{2.50b}$$

we have, for Couette flow with no body forces,

$$0 = -\frac{\partial p}{\partial y} + \frac{\partial \tau_{yy}}{\partial y} \tag{9.7}$$

where, from Eq. (2.57b),

$$\tau_{yy} = \lambda \left(\frac{\partial u}{\partial x} + \frac{\partial v}{\partial y} \right) + 2\mu \frac{\partial v}{\partial y} = 0 \tag{9.8}$$

Hence, with $\tau_{yy} = 0$, Eq. (9.7) yields

$$\frac{\partial p}{\partial y} = 0 \tag{9.9}$$

Conclusion: For Couette flow, there are no pressure gradients in either the x or y direction. With all the above information, we return to the x-momentum equation, displayed earlier as Eq. (2.50a). From this equation, for steady, two-dimensional flow with no body forces, we have

$$\rho u \frac{\partial u}{\partial x} + \rho v \frac{\partial u}{\partial y} = -\frac{\partial p}{\partial x} + \frac{\partial \tau_{xx}}{\partial x} + \frac{\partial \tau_{yx}}{\partial y} \tag{9.10}$$

From Eqs. (2.57a) and (2.57d), applied for the case of Couette flow, we have

$$\tau_{xx} = \lambda \left(\frac{\partial u}{\partial x} + \frac{\partial v}{\partial y} \right) + 2\mu \frac{\partial u}{\partial x} = 0 \tag{9.11}$$

$$\tau_{yx} = \mu \left(\frac{\partial v}{\partial x} + \frac{\partial u}{\partial y} \right) = \mu \frac{\partial u}{\partial y} \tag{9.12}$$

Substituting Eqs. (9.11) and (9.12) into (9.10), we have, for Couette flow,

$$0 = \frac{\partial}{\partial y} \left(\mu \frac{\partial u}{\partial y} \right) \tag{9.13}$$

At this stage, we will now assume an incompressible, constant-temperature flow for which μ = constant. With this, Eq. (9.13) becomes

$$\frac{\partial^2 u}{\partial y^2} = 0 \qquad (9.14)$$

Equation (9.14) is the governing equation for incompressible, constant-temperature, Couette flow.

The exact analytic solution of Eq. (9.14) is straightforward. Integrating twice with respect to y, we have

$$u = c_1 y + c_2 \qquad (9.15)$$

where c_1 and c_2 are constants of integration; their values are found by applying the boundary conditions. Specifically, at the lower plate (see Fig. 9.1), we know that $u = 0$ for $y = 0$. From Eq. (9.15), this yields $c_2 = 0$. At the upper plate, we know that $u = u_e$ for $y = D$. From Eq. (9.15), this yields $c_1 = u_e/D$. With these values for c_1 and c_2, Eq. (9.15) becomes

$$\frac{u}{u_e} = \frac{y}{D} \qquad (9.16)$$

Equation (9.16) is the exact, analytical solution for the velocity profile for incompressible Couette flow. Note from Eq. (9.16) that the exact result is a *linear profile*; u varies directly as y. Such a linear profile is sketched in Fig. 9.1.

We now proceed to set up a numerical solution for this flow; the exact analytical result given by Eq. (9.16) will be used as a standard of comparison for the numerical results.

9.3 THE NUMERICAL APPROACH: IMPLICIT, CRANK–NICOLSON TECHNIQUE

We will pose the numerical solution as follows. Imagine that we assume a velocity profile which is not linear, i.e., a different velocity profile than the exact solution given by Eq. (9.16). Specifically, let us assume a velocity profile defined as

$$u = \begin{cases} 0 & \text{for } 0 \leq y < D \\ u_e & \text{for } y = D \end{cases} \qquad \begin{array}{c} (9.17a) \\ (9.17b) \end{array}$$

This will be identified as our *initial* profile; it is shown by the solid line in Fig. 9.2a. We will consider this to be the *initial condition* at time $t = 0$. We will set up a time-marching solution for the flow field, starting from this initial condition. We would expect to see the velocity profile change in steps of time, as reflected in Fig. 9.2b and c. Finally, after enough time steps are taken, the velocity profile will approach its steady-flow value, as sketched in Fig. 9.2d.

THE NUMERICAL APPROACH: IMPLICIT, CRANK–NICOLSON TECHNIQUE   **421**

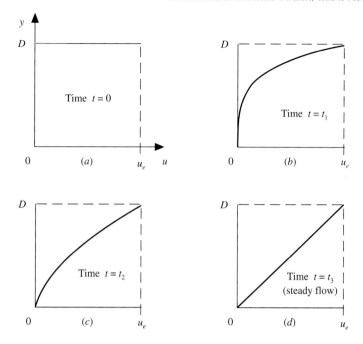

FIG. 9.2
Schematic of the velocity profiles at various times in an unsteady Couette flow.

The flow illustrated by the timewise-changing velocity profiles in Fig. 9.2 is *unsteady* Couette flow. The governing equation for this flow is obtained from Eq. (2.50a), making the Couette flow assumptions of $\partial/\partial x = 0$ and $v = 0$ but carrying along the time derivative. The resulting governing equation, the x-momentum equation for unsteady, incompressible, Couette flow, is

$$\rho \frac{\partial u}{\partial t} = \mu \frac{\partial^2 u}{\partial y^2} \qquad (9.18)$$

Equation (9.18) is a parabolic partial differential equation; hence a time-marching solution represents a well-posed problem.

9.3.1 The Numerical Formulation

It will be convenient to deal with a nondimensional form of Eq. (9.18). Defining the following nondimensional variables

$$u' = \frac{u}{u_e} \qquad y' = \frac{y}{D} \qquad t' = \frac{t}{D/u_e}$$

Eq. (9.18) is nondimensionalized as follows.

$$\rho \frac{\partial (u/u_e)}{\partial [t/(D/u_e)]} \left(\frac{u_e^2}{D}\right) = \mu \frac{\partial^2 (u/u_e)}{\partial (y/D)^2} \left(\frac{u_e}{D^2}\right)$$

or
$$\frac{\partial u'}{\partial t'} = \frac{\mu}{\rho u_e D} \frac{\partial^2 u'}{\partial y'^2} \tag{9.19}$$

However, in Eq. (9.19), we recognize the quantity

$$\frac{\mu}{\rho u_e D} \equiv \frac{1}{\text{Re}_D}$$

where Re_D is the Reynolds number based on the height D between the two plates. Thus, Eq. (9.19) becomes

$$\boxed{\frac{\partial u'}{\partial t'} = \frac{1}{\text{Re}_D} \frac{\partial^2 u'}{\partial y'^2}} \tag{9.20}$$

Equation (9.20) is the equation for which we will obtain a numerical solution.

We choose to use an *implicit* finite-difference technique for this numerical solution; specifically, we will employ the Crank–Nicolson method introduced in Sec. 4.4 (in conjunction with Eq. (4.40)). In the present calculation, we will find that the incompressible Couette flow solution illustrates all the pertinent features of an implicit solution using the Crank–Nicolson technique. Make certain to review Sec. 4.4 before progressing further; it is important that you understand the basic ideas behind the Crank–Nicolson technique.

As we have done several times in previous sections of this book, for simplicity of notation we will *drop the primes* in Eq. (9.20) and treat all subsequent variables in the remainder of this section as the *nondimensional* variables. That is, we will write Eq. (9.20) as

$$\frac{\partial u}{\partial t} = \frac{1}{\text{Re}_D} \frac{\partial^2 u}{\partial y^2} \tag{9.21}$$

where u, y, and t are *identically* the *nondimensional* variables u', y', and t' that appear in Eq. (9.20).

Following the Crank–Nicolson technique, the finite-difference representation of Eq. (9.21) is

$$\frac{u_j^{n+1} - u_j^n}{\Delta t} = \frac{1}{\text{Re}_D} \frac{\frac{1}{2}(u_{j+1}^{n+1} + u_{j+1}^n) + \frac{1}{2}(-2u_j^{n+1} - 2u_j^n) + \frac{1}{2}(u_{j-1}^{n+1} + u_{j-1}^n)}{(\Delta y)^2}$$

or

$$u_j^{n+1} = u_j^n + \frac{\Delta t}{2(\Delta y)^2 \text{Re}_D} (u_{j+1}^{n+1} + u_{j+1}^n - 2u_j^{n+1} - 2u_j^n + u_{j-1}^{n+1} + u_{j-1}^n)$$

$$\tag{9.22}$$

Grouping all terms at time level $n + 1$ in Eq. (9.22) on the left-hand side and factoring both sides appropriately, Eq. (9.22) becomes

$$\left[-\frac{\Delta t}{2(\Delta y)^2 \text{Re}_D}\right] u_{j-1}^{n+1} + \left[1 + \frac{\Delta t}{(\Delta y)^2 \text{Re}_D}\right] u_j^{n+1} + \left[-\frac{\Delta t}{2(\Delta y)^2 \text{Re}_D}\right] u_{j+1}^{n+1}$$

$$= \left[1 - \frac{\Delta t}{(\Delta y)^2 \text{Re}_D}\right] u_j^n + \frac{\Delta t}{2(\Delta y)^2 \text{Re}_D} (u_{j+1}^n + u_{j-1}^n) \quad (9.23)$$

Equation (9.23) is of the form

$$A u_{j-1}^{n+1} + B u_j^{n+1} + A u_{j+1}^{n+1} = K_j \quad (9.24)$$

where

$$A = -\frac{\Delta t}{2(\Delta y)^2 \text{Re}_D} \quad (9.25a)$$

$$B = 1 + \frac{\Delta t}{(\Delta y)^2 \text{Re}_D} \quad (9.25b)$$

$$K_j = \left[1 - \frac{\Delta t}{(\Delta y)^2 \text{Re}_D}\right] u_j^n + \frac{\Delta t}{2(\Delta y)^2 \text{Re}_D} (u_{j+1}^n + u_{j-1}^n) \quad (9.25c)$$

Equation (9.24) is solved on a grid such as that sketched in Fig. 9.3. The vertical distance (the y direction) across the duct is divided into N equal increments of length Δy by distributing $N + 1$ grid points over the height D, that is,

$$\Delta y = \frac{D}{N} \quad (9.26)$$

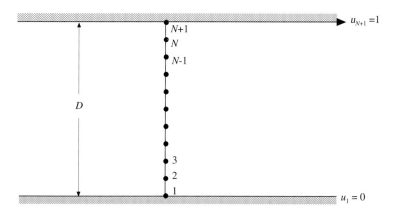

FIG. 9.3
Labeling of points for the grid.

From the boundary conditions, u_1 and u_{N+1} are known:

$$u_1 = 0 \tag{9.27a}$$

$$u_{N+1} = 1 \tag{9.27b}$$

(Keep in mind that in Eqs. (9.21) to (9.27), u denotes the *nondimensional* velocity.) Hence, the system of equations represented by Eq. (9.24) represents $N - 1$ equations with $N - 1$ unknowns, namely, $u_2, u_3, \ldots, u_N$. We can write this system in more detail as follows. The first equation is

$$Au_1^{n+1} + Bu_2^{n+1} + Au_3^{n+1} = K_2 \tag{9.28}$$

However, $u_1 = 0$. Thus, Eq. (9.28) becomes

$$Bu_2^{n+1} + Au_3^{n+1} = K_2 \tag{9.29}$$

The last equation in the system represented by Eq. (9.24) is

$$Au_{N-1}^{n+1} + Bu_N^{n+1} + Au_{N+1}^{n+1} = K_N \tag{9.30}$$

However, $u_{N+1} = 1$. Thus, Eq. (9.30) becomes

$$Au_{N+1}^{n+1} + Bu_N^{n+1} = K_N - Au_e \tag{9.31}$$

With this, the system of equations represented by Eq. (9.24) can be written, in matrix form, as

$$\begin{bmatrix} B & A & 0 & 0 & 0 & 0 & 0 & 0 \\ A & B & A & 0 & 0 & 0 & 0 & 0 \\ 0 & A & B & A & 0 & 0 & 0 & 0 \\ 0 & 0 & A & B & A & 0 & 0 & 0 \\ & & & \cdots & & & & \\ & & & & \cdots & & & \\ & & & & & \cdots & & \\ 0 & 0 & 0 & 0 & 0 & A & B & A \\ 0 & 0 & 0 & 0 & 0 & 0 & A & B \end{bmatrix} \begin{bmatrix} u_2^{n+1} \\ u_3^{n+1} \\ u_4^{n+1} \\ u_5^{n+1} \\ \cdot \\ \cdot \\ \cdot \\ u_{N-1}^{n+1} \\ u_N^{n+1} \end{bmatrix} = \begin{bmatrix} K_2 \\ K_3 \\ K_4 \\ K_5 \\ \cdot \\ \cdot \\ \cdot \\ K_{N-1} \\ K_N - Au_e \end{bmatrix} \tag{9.32}$$

Clearly, the system represented by Eq. (9.32) is in tridiagonal form. It can be solved using Thomas' algorithm, derived in App. A. Thomas' algorithm has been mentioned in Chap. 4 in connection with the Crank–Nicolson method, but now is the first chance we have had to actually solve a specific problem using this algorithm. Therefore, stop where you are, turn to App. A, and study the derivation of Thomas' algorithm before proceeding further; it will make all the difference in the world in your mental comfort for the ensuing sections dealing with the numerical solution of Couette flow.

After applying Thomas' algorithm to the system of equations represented by Eq. (9.32), we have the solution for $u_2^{n+1}, u_3^{n+1}, \ldots, u_n^{n+1}$. These are the values of the velocities at the time step $n + 1$. The whole process is then repeated for a number of time steps until the velocity profile converges to a steady state, as illustrated in Fig. 9.2.

9.3.2 The Setup

For our specific solution, we choose to use 21 grid points across the flow; i.e., in Fig. 9.3, $N + 1 = 21$. Since y is nondimensional, it varies from 0 to 1; hence

$$\Delta y = \tfrac{1}{20}$$

For initial conditions, we will use Eqs. (9.17a) and (9.17b), which yield

$$u_1, u_2, u_3, \cdots, u_{20} = 0$$
$$u_{21} = 1 \quad \text{at } t = 0$$

The calculation of the "proper" time step Δt for the present solution is not as stringent as that in our previous applications in Chap. 7 or for the spatially marching step in Chap. 8. Unlike the explicit methods used in Chaps. 7 and 8, our current application involves an *implicit* method. A stability analysis similar to our discussion in Sec. 4.5 shows that the Crank–Nicolson technique is *unconditionally stable*; i.e., it is stable for *all* values of Δt. This is the main advantage of an implicit method, as described in Sec. 4.4. That is, stability considerations tell us that we can use as large a value of Δt as we wish. On the other hand, if we would want to simulate with any accuracy the actual transient variation of the flow field starting from the given initial conditions, we should keep Δt small in order to minimize the truncation error with respect to time. Of course, when we are interested in the steady state only, timewise accuracy is not a major concern. So what shall we choose for Δt? For an answer, we will be guided by the stability criterion based on an explicit basis. For our unsteady Couette flow, the governing equation, Eq. (9.20), is a parabolic partial differential equation in exactly the same form as one of the model equations used in Chap. 3, namely, Eq. (3.28). The corresponding finite-difference equation for an explicit method is given by Eq. (4.36). In turn, the stability criterion for Eq. (4.36) was found to be, from Eq. (4.77), repeated below,

$$\frac{\alpha \Delta t}{(\Delta x)^2} \leq \frac{1}{2} \qquad (4.77)$$

By analogy with Eq. (9.20) and continuing with our modified notation that all variables in the following equation are nondimensional, we can write, for an explicit method,

$$\frac{1}{\text{Re}_D} \frac{\Delta t}{(\Delta y)^2} \leq \frac{1}{2} \qquad (9.33)$$

or
$$\Delta t \leq \tfrac{1}{2} \text{Re}_D (\Delta y)^2 \qquad (9.34)$$

Taking a cue from Eq. (9.34), for our present *implicit* method, we will calculate Δt as

$$\Delta t = E \, \text{Re}_D (\Delta y)^2 \qquad (9.35)$$

where E is a parameter. Since the Crank–Nicolson technique is unconditionally stable, we could choose E to be *any* value. Indeed, in the next section, we will examine the results obtained from a numerical experiment where E is varied from 1 to 4000.

The definition of E given in Eq. (9.35), namely,

$$E = \frac{\Delta t}{\text{Re}_D(\Delta y)^2} \tag{9.36}$$

and its subsequent use as a parameter simplify the coefficients that appear in Eq. (9.24). In particular, inserting Eq. (9.36) into Eqs. (9.25a) to (9.25c), we have

$$A = -\frac{E}{2} \tag{9.37a}$$

$$B = 1 + E \tag{9.37b}$$

$$K_j = (1-E)u_j^n + \frac{E}{2}(u_{j+1}^n + u_{j-1}^n) \tag{9.37c}$$

Another aspect of the definition of E from Eq. (9.36) is that it includes the Reynolds number Re_D. The final steady-state velocity profile for Couette flow is independent of Re_D; notice that the exact, analytical result given by Eq. (9.16) does not contain Re_D. On the other hand, the *transient approach* to the steady state *does* depend on Re_D, and it is interesting to note that this Reynolds number effect is buried exclusively in the definition of E.

9.3.3 Intermediate Results

Let us examine the calculation of the velocity profile for the first time step $n = 1$. We choose $E = 1$ and $\text{Re}_D = 5000$. Also, since we are using 21 grid points across the flow, $\Delta y = \frac{1}{20} = 0.05$. With these values, we have for Δt from Eq. (9.35),

$$\Delta t = E \, \text{Re}_D (\Delta y)^2 = 1(5000)(0.05)^2 = 12.5$$

From Eqs. (9.37a) and (9.37b),

$$A = -\frac{E}{2} = -0.5$$

$$B = 2$$

These are the values of A and B that appear in the system of equations represented by Eq. (9.32).

We now invoke Thomas' algorithm as given in App. A. Using the notation of App. A and keeping Eq. (9.32) always in eyesight, the first line in Eq. (9.32) is unchanged, i.e.,

$$2u_2^{n+1} - 0.5u_3^{n+1} = K_2 \tag{9.38}$$

From Eq. (9.37c) and the fact that $u_1^n = u_2^n = u_3^n = 0$, we have

$$K_2 = (1-E)u_2^n + \frac{E}{2}(u_3^n + u_1^n) = 0$$

So Eq. (9.38) becomes
$$2u_2^{n+1} - 0.5u_3^{n+1} = 0 \qquad (9.39)$$

The second line in Eq. (9.32) is
$$-0.5u_2^{n+1} + 2u_3^{n+1} - 0.5u_4^{n+1} = K_3 \qquad (9.40)$$

where, since $u_2^n = u_3^n = u_4^n = 0$,
$$K_3 = (1 - E)u_3^n + \frac{E}{2}(u_4^n + u_2^n) = 0$$

and Eq. (9.40) becomes
$$-0.5u_2^{n+1} + 2u_3^{n+1} - 0.5u_4^{n+1} = 0 \qquad (9.41)$$

However, using the nomenclature of App. A, Eq. (A.21), repeated below,
$$d_i' = d_i - \frac{b_i a_{i-1}}{d_{i-1}'} \qquad (A.21)$$

when applied to Eq. (9.41) becomes
$$d_3' = d_3 - \frac{b_3 a_2}{d_2'} \qquad (9.42)$$

From the coefficients in Eqs. (9.39) and (9.41), we have $d_3 = 2$, $b_3 = -0.5$, $a_2 = -0.5$, and $d_2' = 2$. Thus, from Eq. (9.42), we have
$$d_3' = 2 - \frac{-0.5(-0.5)}{2} = 1.875$$

Also, Eq. (A.22) in App. A, repeated below,
$$c_i' = c_i - \frac{c_{i-1}' b_i}{d_{i-1}'} \qquad (A.22)$$

when applied to Eq. (9.41) becomes
$$c_3' = c_3 - \frac{c_2' b_2}{d_2'} \qquad (9.43)$$

From the coefficients in Eqs. (9.39) and (9.41), we have $c_3 = 0$, $c_2' = 0$, $b_2 = 0$, and $d_2' = 2$. Hence, from Eq. (9.43),
$$c_3' = 0$$

From the values of d_3' and c_3' obtained above, the new, bidiagonal form of Eq. (9.41) becomes
$$1.875u_3^{n+1} - 0.5u_4^{n+1} = 0 \qquad (9.44)$$

Let us proceed to the third line of Eq. (9.32), which is
$$-0.5u_3^{n+1} + 2u_4^{n+1} - 0.5u_5^{n+1} = K_4 \qquad (9.45)$$

Since $u_3^n = u_4^n = u_5^n = 0$, then $K_4 = 0$, and Eq. (9.45) becomes

$$-0.5u_3^{n+1} + 2u_4^{n+1} - 0.5u_5^{n+1} = 0 \tag{9.46}$$

Equation (9.46) is put in bidiagonal form as follows. From Eq. (A.21) applied to Eq. (9.46),

$$d_4' = d_4 - \frac{b_4 a_3}{d_3'} \tag{9.47}$$

The values in Eq. (9.47) are obtained from the coefficients in Eqs. (9.46) and (9.44) as follows: $d_4 = 2$, $b_4 = -0.5$, $a_3 = -0.5$, and $d_3' = 1.875$. Hence

$$d_4' = 2 - \frac{-0.5(-0.5)}{1.875} = 1.867$$

Also, from Eq. (A.22) applied to Eq. (9.46),

$$c_4' = c_4 - \frac{c_3' b_4}{d_3'} \tag{9.48}$$

where from Eqs. (9.46) and (9.44), we have $c_4 = 0$, $c_3' = 0$, $b_4 = -0.5$, and $d_3' = 1.875$. Thus, from Eq. (9.48), we have

$$c_4' = 0$$

From the values of d_4' and c_4' obtained above, the new, bidiagonal form of Eq. (9.46) becomes

$$1.867 u_4^{n+1} - 0.5\, u_5^{n+1} = 0 \tag{9.49}$$

For the remaining bidiagonalization of the system given by Eq. (9.32), the coefficients turn out to be exactly the same (to three significant figures) as calculated above, except for the last line in Eq. (9.32). This last line corresponds to the equation

$$-0.5 u_{19}^{n+1} + 2 u_{20}^{n+1} = K_{20} - (-0.5)u_e \tag{9.50}$$

where, from Eq. (9.37c),

$$K_{20} = (1 - E)u_{20}^n + \frac{E}{2}(u_{21}^n + u_{19}^n) \tag{9.51}$$

From the initial conditions, we have $u_{19}^n = 0$, $u_{20}^n = 0$, and $u_{21}^n = 1$. Hence, from Eq. (9.51),

$$K_{20} = 0.5$$

and Eq. (9.50) becomes

$$-0.5 u_{19}^{n+1} + 2 u_{20}^{n+1} = 1.0 \tag{9.52}$$

Equation (A.21) applied to Eq. (9.52) is written as

$$d_{20}' = d_{20} - \frac{b_{20} a_{19}}{d_{19}'} \tag{9.53}$$

where $d_{20} = 2$, $b_{20} = -0.5$, $a_{19} = -0.5$, and $d'_{19} = 1.866$. Thus

$$d'_{20} = 2 - \frac{-0.5(-0.5)}{1.866} = 1.866 \tag{9.54}$$

From Eq. (A.22) applied to Eq. (9.52), we have

$$c'_{20} = c_{20} - \frac{c'_{19} b_{20}}{d'_{19}} \tag{9.55}$$

where in Eq. (9.55) we have $c_{20} = 1.0$, $c'_{19} = 0$, $b_{20} = -0.5$, and $d'_{19} = 1.866$. Thus, from Eq. (9.55), we have

$$c'_{20} = 1.0$$

With this, Eq. (9.52) becomes, in upper bidiagonal form,

$$1.866 u_{20}^{n+1} = 1.0 \tag{9.56}$$

We are now ready to solve for the velocities u_j^{n+1}, $j = 2$ to 20. Obviously, u_{20}^{n+1} can be obtained directly from Eq. (9.56) as

$$u_{20}^{n+1} = \frac{1.0}{1.866} = \boxed{0.536}$$

Note that this result is *exactly* the same that is obtained by using Eq. (A.25), repeated below.

$$u_m = \frac{c'_m}{d'_m} \tag{A.25}$$

This should be no surprise; the above calculation has essentially followed the same path as the derivation of Eq. (A.25) in App. A. The next (and final) step of Thomas' algorithm is to calculate the other unknown velocities using the recursion formula given by Eq. (A.27), repeated below.

$$u_i = \frac{c'_i - a_i u_{i+1}}{d'_i} \tag{A.27}$$

For example, from Eq. (A.27),

$$u_{19}^{n+1} = \frac{c'_{19} - a_{19} u_{20}}{d'_{19}} \tag{9.57}$$

In Eq. (9.57), we have $c'_{19} = 0$, $a_{19} = -0.5$, $u_{20} = 0.536$, and $d'_{19} = 1.866$. Thus, from Eq. (9.57),

$$u_{19}^{n+1} = \frac{0 - (-0.5)(0.536)}{1.866} = \boxed{0.144}$$

The remaining velocities, $u_{18}, u_{17}, \ldots, u_2$, are calculated in the same fashion.

The numbers for b_j, d'_j, a_j, and c'_j are tabulated versus grid point j in Table 9.1, along with the corresponding calculated velocities u_j. (Note that the subscript i was used in App. A, whereas the subscript j is used in our present calculation. This is done intentionally to reinforce the fact that i and j are simply running indices; what

TABLE 9.1
Velocity profile after the first time step

j	y/D	u/u_e	b_j	d'_j	a_j	c'_j
1	.000E+00	.000E+00				
2	.500E−01	.252E−01	.000E+00	.200E+01	.500E+00	.000E+00
3	.100E+00	.101E−09	−.500E+00	.188E+01	−.500E+00	.000E+00
4	.150E+00	.378E−09	−.500E+00	.187E+01	−.500E+00	.000E+00
5	.200E+00	.141E+00	−.500E+00	.187E+01	−.500E+00	.000E+00
6	.250E+00	.527E+08	−.500E+00	.187E+01	−.500E+00	.000E+00
7	.300E+00	.197E−07	−.500E+00	.187E+01	−.500E+00	.000E+00
8	.350E+00	.734E−07	−.500E+00	.187E+01	−.500E+00	.000E+00
9	.400E+00	.274E−06	−.500E+00	.187E+01	−.500E+00	.000E+00
10	.450E+00	.102E−05	−.500E+00	.187E+01	−.500E+00	.000E+00
11	.500E+00	.382E−05	−.500E+00	.187E+01	−.500E+00	.000E+00
12	.550E+00	.142E−04	−.500E+00	.187E+01	−.500E+00	.000E+00
13	.600E+00	.531E−00	−.500E+00	.187E+01	−.500E+00	.000E+00
14	.650E+00	.198E−03	−.500E+00	.187E+01	−.500E+00	.000E+00
15	.700E+00	.740E−03	−.500E+00	.187E+01	−.500E+00	.000E+00
16	.750E+00	.276E−02	−.500E+00	.187E+01	−.500E+00	.000E+00
17	.800E+00	.103E−01	−.500E+00	.187E+01	−.500E+00	.000E+00
18	.850E+00	.385E−01	−.500E+00	.187E+01	−.500E+00	.000E+00
19	.900E+00	.144E+00	−.500E+00	.187E+01	−.500E+00	.000E+00
20	.950E+00	.536E+00	−.500E+00	.187E+01	.000E+00	.100E+01
21	.100E+01	.100E+01				

symbol is used for the running index is irrelevant.) Note that the answers obtained in the above calculation appear in the table; for example, in Table 9.1, reading across the line for $j = 20$, we see entered that $u_{20} = 0.536$, $b_{20} = -0.5$, $d'_{20} = 1.866 \approx 1.87$ (to three significant figures), $a_{20} = 0$, and $c'_{20} = 1.0$—all as calculated above. Reading across the line for $j = 19$ in Table 9.1, we see that $u_{19} = 0.144$, $b_{19} = -0.5$, $d'_{19} = 1.87$, $a_{19} = -0.5$, and $c'_{19} = 0$. And so forth, for the remainder of the grid points.

The velocities tabulated in Table 9.1, calculated as shown above for $j = 1, 2, \ldots, 21$ (including the known boundary values at $j = 1$ and 21), represent the velocity profile in the unsteady Couette flow after time $t = \Delta t$, starting from the specified initial conditions. The above calculations are subsequently carried out for a number of time steps until the velocity profile reaches a steady state.

9.3.4 Final Results

Starting from the assumed initial conditions given by Eqs. (9.27a) and (9.27b), the velocity is calculated in steps of time, using the approach described in Secs. 9.3.1 and 9.3.2. Some results for the velocity profiles are various stages in the time-marching process are shown in Fig. 9.4. The initial conditions at time $t = 0$ are given in Fig. 9.4 and labeled as $0\Delta t$. The velocity profile after two time steps is labeled $2\Delta t$; note that the velocity is changing most rapidly near the upper plate, as

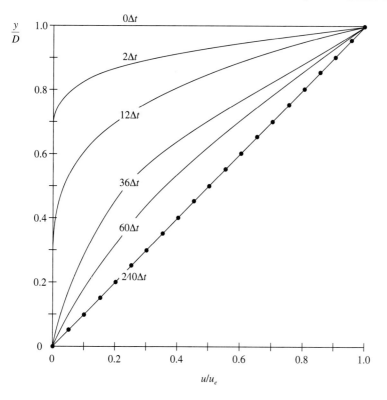

FIG. 9.4
Velocity profiles for unsteady Couette flow at various stages in the time-stepping process. Solid circles are exact analytical solution (steady state); solid lines are numerical solutions.

to be expected. Other profiles are shown in Fig. 9.4 after 12, 36, 60, and 240 time steps, labeled $12\Delta t$, $36\Delta t$, $60\Delta t$, and $240\Delta t$, respectively. The driving influence of the shear stress exerted by the upper plate is gradually communicated to the rest of the fluid, resulting in a final, steady-state profile after 240 time steps. This steady-state profile is linear, as to be expected; it agrees perfectly with the exact, analytical solution. To provide a more direct comparison of your computations with the present calculations, Table 9.2 gives a tabulation of the velocity profiles at a number of different times throughout the time-marching solution.

All the above calculations were carried out with $E = 1$. *Question*: What is the effect of using a larger time step; i.e., reflecting on Eq. (9.35), what is the effect of using a larger value of E? In terms of stability, there should be no difference in the behavior of the solutions—the Crank–Nicolson technique is unconditionally stable. However, when E is increased, the *accuracy* of the *transients* may be compromised, and the number of marching steps required to obtain a steady state may change, for better or for worse. To address these matters, a numerical experiment is carried out wherein a number of different cases are calculated, each with a different value of E, with E ranging as high as 4000. From Eq. (9.35), we can interpret the effect of

TABLE 9.2
Velocity profiles at later time steps

		Velocity profiles u/u_e					
j	y/D	12Δt	36Δt	60Δt	120Δt	240Δt	360Δt
1	.000E+00	.000E+00	.000E+00	.000E+00	.000E+00	.000E+00	.000E+00
2	.500E−01	.124E−03	.119E−01	.276E−01	.448E−01	.497E−01	.500E−01
3	.100E+00	.313E−03	.245E−01	.557E−01	.898E−01	.995E−01	.100E+00
4	.150E+00	.661E−03	.386E−01	.849E−01	.135E+00	.149E+00	.150E+00
5	.200E+00	.132E−02	.549E−01	.116E+00	.181E+00	.199E+00	.200E+00
6	.250E+00	.254E−02	.741E−01	.148E+00	.227E+00	.249E+00	.250E+00
7	.300E+00	.474E−02	.970E−01	.184E+00	.273E+00	.299E+00	.300E+00
8	.350E+00	.859E−02	.124E+00	.222E+00	.321E+00	.348E+00	.350E+00
9	.400E+00	.151E−01	.157E+00	.263E+00	.369E+00	.398E+00	.400E+00
10	.450E+00	.256E−01	.194E+00	.307E+00	.417E+00	.448E+00	.450E+00
11	.500E+00	.422E−01	.238E+00	.355E+00	.467E+00	.498E+00	.500E+00
12	.550E+00	.672E−01	.289E+00	.407E+00	.517E+00	.548E+00	.550E+00
13	.600E+00	.103E+00	.346E+00	.462E+00	.569E+00	.598E+00	.600E+00
14	.650E+00	.154E+00	.409E+00	.520E+00	.621E+00	.648E+00	.650E+00
15	.700E+00	.221E+00	.479E+00	.582E+00	.673E+00	.699E+00	.700E+00
16	.750E+00	.308E+00	.556E+00	.647E+00	.727E+00	.749E+00	.750E+00
17	.800E+00	.414E+00	.637E+00	.714E+00	.781E+00	.799E+00	.800E+00
18	.850E+00	.540E+00	.724E+00	.783E+00	.835E+00	.849E+00	.850E+00
19	.900E+00	.683E+00	.814E+00	.855E+00	.890E+00	.899E+00	.900E+00
20	.950E+00	.838E+00	.906E+00	.927E+00	.945E+00	.950E+00	.950E+00
21	.100E+01	.100E+01	.100E+01	.100E+01	.100E+01	.100E+01	.100E+01

increasing E the same as increasing Δt for fixed Δy and Re_D. We will use this interpretation; whenever we refer to our increase in E, it will be synonymous with taking a larger time step, i.e., a larger Δt.

With this in mind, consider the velocity profiles tabulated in Table 9.3. Three profiles are given, one each for $E = 1$, 5, and 10. These are *transient* profiles, all corresponding to the same nondimensional time $t = 1.5 \times 10^3$; this is an intermediate time—the steady-state profile corresponds to a nondimensional time on the order of $t = 4.5 \times 10^3$. Of course, since different values of E correspond to different values of Δt, then the three velocity profiles given in Table 9.3, which correspond to the same value of t, consequently correspond to a *different* number of time steps. Specifically, in Table 9.3 the column labeled $E = 1$ corresponds to the results obtained after 120 time-marching steps, $E = 5$ corresponds to 24 time-marching steps, and $E = 10$ corresponds to 12 steps. Examine these three columns carefully. The columns labeled $E = 1$ and $E = 5$ are exactly the same. Since $E = 1$ corresponds to a relatively small time step—one that is only twice the value allowed for an explicit solution [see Eq. (9.34)]—we can readily construe the results for $E = 1$ as being relatively time-accurate. This is reinforced by the comparison in Table 9.3 for $E = 1$ and $E = 5$, which give identical transient results at $t = 1.5 \times 10^3$. We can feel comfortable that a value as high as $E = 5$ provides timewise accuracy for the present implicit calculations. However, examine the last

TABLE 9.3
Comparison of transient velocity profiles

		u/u_e		
j	y/D	E = 1	E = 5	E = 10
1	.000E+00	.000E+00	.000E+00	.000E+00
2	.500E−01	.448E−01	.448E−01	.449E−01
3	.100E+00	.898E−01	.898E−01	.899E−01
4	.150E+00	.135E+00	.135E+00	.135E+00
5	.200E+00	.181E+00	.181E+00	.181E+00
6	.250E+00	.227E+00	.227E+00	.227E+00
7	.300E+00	.273E+00	.273E+00	.274E+00
8	.350E+00	.321E+00	.321E+00	.321E+00
9	.400E+00	.369E+00	.369E+00	.369E+00
10	.450E+00	.417E+00	.417E+00	.418E+00
11	.500E+00	.467E+00	.467E+00	.467E+00
12	.550E+00	.517E+00	.517E+00	.518E+00
13	.600E+00	.569E+00	.569E+00	.569E+00
14	.650E+00	.621E+00	.621E+00	.622E+00
15	.700E+00	.673E+00	.673E+00	.674E+00
16	.750E+00	.727E+00	.727E+00	.725E+00
17	.800E+00	.781E+00	.781E+00	.777E+00
18	.850E+00	.835E+00	.835E+00	.838E+00
19	.900E+00	.890E+00	.890E+00	.915E+00
20	.950E+00	.945E+00	.945E+00	.905E+00
21	.100E+01	.100E+01	.100E+01	.100E+01

column in Table 9.3 for $E = 10$. There are some differences between these results and those for $E = 1$ and 5, especially near the upper wall (e.g., for $j = 19$ and 20). Apparently $E = 10$ corresponds to a large-enough value of Δt to cause some noticeable inaccuracy for the transient results. This inaccuracy continues to grow as E is further increased.

Let us examine an extreme case, namely, one for $E = 4000$. Here, the value of Δt is so large that no timewise accuracy can be expected, and none is obtained. Some results are plotted in Fig. 9.5. Two intermediate, transient velocity profiles are shown, one after 40 time steps and the other after 200 time steps; both profiles exhibit totally nonphysical behavior, especially near the top plate. Compare these results in Fig. 9.5, obtained for $E = 4000$, with the more realistic transient results shown in Fig. 9.4, obtained for $E = 1$—there is no real comparison. The transient results in Fig. 9.5 are clearly nonphysical. However, after a very large number of time steps—on the order of 1000—the implicit solution will finally converge to the proper steady-state velocity profile, given by the solid circles in Fig. 9.5.

This last statement highlights another aspect of the implicit solution as E is increased to large values, an aspect involving the number of marching time steps required to obtain the steady-state solution. For $E = 1$, over 240 time steps are required to obtain the steady state; this is reflected in the results shown in Table 9.2. For $E = 5$, only about 50 steps are necessary to obtain the steady state, a

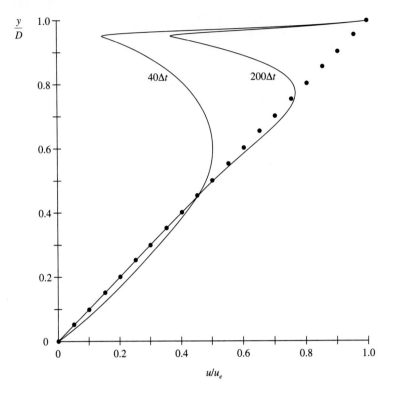

FIG. 9.5
Totally nonphysical transient velocity profiles obtained when $E = 4000$; comparison with exact steady-state results (dark circles).

tremendous savings in calculation time. For $E = 10$, the steady state is obtained after 36 time steps, even better yet. However, for large values of E, the story reverses itself. For $E = 20$, about 60 steps are required; for $E = 40$, more than 120 steps are required. And it gets worse as E is further increased.

From the above numerical experiment associated with increasing the value of Δt (via increasing E), we can make the following two conclusions regarding the behavior of the Crank–Nicolson implicit method as applied to the present problem:

1. Time accuracy is lost when Δt is made too large; for the present results, time accuracy is lost when E is about 10 or greater. This is no surprise, because the truncation error with respect to time is greatly increased when Δt is increased. We can conclude from this result that *implicit methods* with large values of Δt are not the methods to use for problems wherein the transients are of interest. Of course, time accuracy *can* be obtained for smaller values of Δt, but at the cost of requiring more steps to reach the steady state. Recall that the practical value of implicit methods is that a large step size can be used and still maintain stability, thus resulting in fewer marching steps to achieve the steady state. Therefore,

when using small values of Δt in an implicit solution, we are totally compromising the very advantage that makes implicit methods attractive. *Note*: At the time of writing, there are major efforts being made to develop new implicit methods that are time-accurate and to modify existing implicit methods to obtain better time accuracy—all for the application of implicit methods to the study of transient fluid dynamic problems. This is currently a state-of-the-art research problem.

2. By simply increasing the value of Δt (that is, increasing E), we first see a *reduction* in the number of time steps required to obtain a steady-state; this is consistent with the practical advantage of using an implicit method. However, for a large-enough value of Δt (in the present results, for $E > 20$), the trend reverses itself, and as E increases further, *more* (not less) time steps are required to obtain the steady state. When we reach this condition, the practical value of using an implicit method is lost. In other words, there is some *optimum* value of E which leads to the most efficient implementation of the Crank–Nicolson method. For the present results, that optimum value of E is about 10.

9.4 ANOTHER NUMERICAL APPROACH: THE PRESSURE CORRECTION METHOD

The pressure correction technique is described in Sec. 6.8. It is recommended that you review Sec. 6.8 before progressing further. In the present section, we will apply this method to the solution of the incompressible, viscous flow between two parallel plates as sketched in Fig. 9.6. The upper plate is located a distance D above the lower plate and is in motion with velocity u_e relative to the lower plate. Although the plates are theoretically infinite in extent, the computational domain is finite, with length L and height D, as shown by the shaded region in Fig. 9.6. We will treat the boundary conditions around this finite computational domain in the same fashion as described in Sec. 6.8.6, with p and v fixed and u allowed to float at the inflow boundary, and with only p fixed at the outflow boundary.

The pressure correction method is an iterative method, starting from arbitrarily assumed initial conditions. We will induce a two-dimensional flow within the

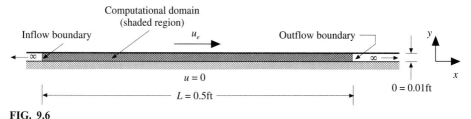

FIG. 9.6
The finite computational domain for the application of the pressure correction method for the solution of the incompressible flow between two plates in relative motion.

436 INCOMPRESSIBLE COUETTE FLOW

computational domain by setting the initial conditions to be an arbitrary two-dimensional flow field. Then, during the iterative procedure, we will watch this originally two-dimensional flow field converge to the exact solution for Couette flow.

9.4.1 The Setup

The physical problem is sketched in Fig. 9.6. We will carry out the present solution in dimensional terms, rather than nondimensionalizing the governing equations and dealing with a nondimensional space. We offer this calculation in part as an example that CFD solutions are frequently carried out using dimensional terms throughout the calculation. Hence, as shown in Fig. 9.6, we will treat a computational domain which is 0.5 ft long in the x direction and 0.01 ft high in the y direction. The upper plate is moving with velocity u_e, and the lower plate is stationary. The fluid is air at standard sea level conditions, with a density $\rho = 0.002377$ slug/ft^3. Since we will employ a very coarse grid for the example, we treat the case of a low velocity; e.g., we set $u_e = 1$ ft/s for the present calculation. At this low velocity, there is absolutely no doubt about the assumption that the flow is incompressible. Also, nothing is to be gained in terms of the objective of this example by considering higher values of u_e. Based on the height D of 0.01 ft, the Reynolds number for this case is 63.6.

The computational grid is shown in Fig. 9.7. Based on the reasons discussed in Sec. 6.8.2, we choose a staggered grid. There are three systems of grid points shown in Fig. 9.7; the solid points are where p is calculated, the open points are where u is calculated, and the points denoted by $\times$ are where v is calculated. The use

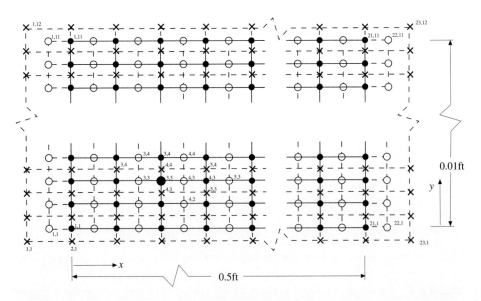

FIG. 9.7
Staggered computational grid. p points, solid circles; u points, open circles; v points, $\times$.

of a staggered grid requires careful attention to the indexing system that identifies each set of points, and somewhat complicates the coding of the computer program. There are various ways of setting up the logic dealing with the proper bookkeeping for a staggered grid. In Fig. 9.7, each set of points has its own independent indexing. For example, the "p points" run from 1 to 21 in the x direction and from 1 to 11 in the y direction, the "u points" run from 1 to 22 in the x direction and from 1 to 11 in the y direction, and the "v points" run from 1 to 23 in the x direction and from 1 to 12 in the y direction.

The pressure correction method is an iterative solution for the flow field. Hence, we need to set the initial conditions for the flow variables in order to start the iterative process. The choice is arbitrary. For the present calculation, we set the following *initial conditions* on all *interior* points, except for point $(i, j) = (15, 5)$, which will be addressed later:

$$u = v = 0$$
$$p^* = p' = 0$$

The specification of the initial conditions for the pressure correction p' equal to zero seems reasonable. But why is the pressure itself, p^*, set equal to zero? The answer is—simply for convenience. An examination of the x- and y-momentum equations, Eqs. (6.94) and (6.95), respectively, shows that only the pressure *difference* between adjacent grid points appears. Therefore, the individual values of p^* are not so important—it is the pressure difference that counts. Therefore, it is totally appropriate to set $p^* = 0$ for the initial conditions, because the pressure *difference* will be dictated by the values of the pressure correction calculated for subsequent iterations.

The *boundary values* are as follows:

$$\left. \begin{array}{l} u = u_e \\ v = 0 \end{array} \right\} \text{ at the upper wall}$$

$$u = v = 0 \quad \text{at the lower wall}$$

$$\left. \begin{array}{l} p' = 0 \\ v = 0 \end{array} \right\} \text{ at the inflow boundary}$$

$$p' = 0 \quad \text{at the outflow boundary}$$

These boundary values are constants, held fixed during the iteration process. In the present example, we make a slight modification to the pressure boundary condition at the upper and lower walls. Instead of employing the zero-pressure gradient condition as expressed by Eq. (6.108), we simply assume $p' = 0$ at the wall. This is done for mathematical convenience; in this fashion, p' is specified over the complete boundary of the domain, as opposed to a *mixed* boundary condition of pressure specified on the outflow and inflow boundaries and the pressure gradient specified at the walls. A constant-pressure boundary condition is allowable in the present example because the final steady-state flow is one where the pressure is uniform. It is not appropriate, however, for a more general flow, where the pressure at the wall varies with distance along the wall and is one of the unknowns to be numerically obtained.

438   INCOMPRESSIBLE COUETTE FLOW

To complete our discussion of the initial conditions for the present calculation, we note that v is set equal to zero at all interior grid points *except* at the point $(i, j) = (15, 5)$. Here, we initially set $v = 0.5$ ft/s, one-half the magnitude of the upper plate velocity. This "velocity spike" is inserted at point (15, 5) to produce a two-dimensional flow during the iterative process. The location and magnitude of the vertical velocity spike are arbitrarily chosen. We are interested in examining the behavior of the pressure correction method for a two-dimensional flow; hence the insertion of the velocity spike in the initial conditions guarantees the existence of such a two-dimensional flow. Moreover, the subsequent dampening and eventual total decay of this velocity spike is an excellent demonstration that the pressure correction philosophy is working as intended. We also want to apply the pressure correction method to a problem for which we have an exact analytical solution—hence the choice of Couette flow. This is in keeping with the philosophy throughout all the applications chapters in this book.

Combining the above initial and boundary conditions, we see that the iterative process begins from a state of zero velocity everywhere, except at the upper boundary where the velocity is $u_e = 1$ ft/s and except for the v velocity spike at $(i, j) = (15, 5)$. Also, the pressure field is uniform throughout the domain and is set equal to zero. Hence, the iterative process starts with the picture of an impulsively started upper plate at velocity u_e, with no flow everywhere else except for the v velocity spike at point (15, 5). Keep in mind that, although we identify the starting values at the beginning of the first iteration as "initial conditions," the pressure correction method is not a time-accurate method. The calculation of the flow field at each subsequent iteration is *analogous* to a time-marching procedure, but the calculated flow values are not accurate representations of the actual flow transients. You are reminded that the pressure correction method is simply an iterative approach to obtain the *steady* flow field.

We now follow the steps outlined in Sec. 6.8.5.

Step 1. Guess at values of p^* at all interior grid points. Also, arbitrarily set values of $(\rho u^*)^n$ and $(\rho v^*)^n$ at all the appropriate grid points. As stated above, p^*, ρu^*, and ρv^* are all set to zero for the beginning of the iterative process, except for $u_e = 1$ ft/s at the upper wall and for $v^*_{15,5} = 0.5$ ft/s at the velocity spike.

Step 2. Solve for $(\rho u^*)^{n+1}$ from Eq. (6.94) and $(\rho v^*)^{n+1}$ from Eq. (6.95) at all interior grid points. Let us set up this calculation by first repeating Eqs. (6.94) and (6.95) below:

$$(\rho u^*)^{n+1}_{i+1/2, j} = (\rho u^*)^{n}_{i+1/2, j} + A^* \, \Delta t - \frac{\Delta t}{\Delta x}(p^*_{i+1,j} - p^*_{i,j}) \quad (6.94)$$

and

$$(\rho v^*)^{n+1}_{i, j+1/2} = (\rho v^*)^{n}_{i, j+1/2} + B^* \, \Delta t - \frac{\Delta t}{\Delta y}(p^*_{i,j+1} - p^*_{i,j}) \quad (6.95)$$

where

$$A^* = -\left[\frac{(\rho u^2)^n_{i+3/2,j} - (\rho u^2)^n_{i-1/2,j}}{2\Delta x} + \frac{(\rho u \bar{v})^n_{i+1/2,j+1} - (\rho \bar{\bar{v}})^n_{i+1/2,j-1}}{2\Delta y}\right]$$
$$+ \mu \left[\frac{u^n_{i+3/2,j} - 2u^n_{i+1/2,j} + u^n_{i-1/2,j}}{(\Delta x)^2} + \frac{u^n_{i+1/2,j+1} - 2u^n_{i+1/2,j} + u^n_{i+1/2,j-1}}{(\Delta y)^2}\right]$$

$$\bar{v} = \tfrac{1}{2}(v^n_{i,j+1/2} + v^n_{i+1,j+1/2})$$
$$\bar{\bar{v}} = \tfrac{1}{2}(v^n_{i,j-1/2} + v^n_{i+1,j-1/2})$$

and

$$B^* = -\left[\frac{(\rho v \bar{u})^n_{i+1,j+1/2} - (\rho v \bar{\bar{u}})^n_{i-1,j+1/2}}{2\Delta x} + \frac{(\rho v^2)^n_{i,j+3/2} - (\rho v^2)^n_{i,j-1/2}}{2\Delta y}\right]$$
$$+ \mu \left[\frac{v^n_{i+1,j+1/2} - 2v^n_{i,j+1/2} - u^n_{i-1,j+1/2}}{(\Delta x)^2} + \frac{v^n_{i,j+3/2} - 2v^n_{i,j+1/2} + v^n_{i,j-1/2}}{(\Delta y)^2}\right]$$

$$\bar{u} = \tfrac{1}{2}(u_{i+1/2,j} + u_{i+1/2,j+1})$$
$$\bar{\bar{u}} = \tfrac{1}{2}(u_{i-1/2,j} + u^n_{i-1/2,j+1})$$

Return to Fig. 9.7, and let us write the above equations using the pressure grid point (3, 3) as a focus. This grid point is drawn oversize in Fig. 9.7. Using this grid point to represent the pressure point (i, j), Eq. (6.94) is written in the following form, keeping in mind the three different indexing systems for the staggered grid shown in Fig. 9.7.

$$(\rho u^*)^{n+1}_{4,3} = (\rho u^*)^n_{4,3} + A^* \Delta t - \frac{\Delta t}{\Delta x}(p^*_{4,3} - p^*_{3,3}) \qquad (9.58)$$

$$A^* = -\left[\frac{(\rho u^2)^n_{5,3} - (\rho u^2)^n_{3,3}}{2\Delta x} + \frac{(\rho u \bar{v})^n_{4,4} - (\rho u \bar{\bar{v}})_{4,2}}{2\Delta y}\right]$$
$$+ \mu \left[\frac{u^n_{5,3} - 2u^n_{4,3} + u^n_{3,3}}{(\Delta x)^2} + \frac{u^n_{4,4} - 2u^n_{4,3} + u^n_{4,2}}{(\Delta y)^2}\right]$$

$$\bar{v} = \tfrac{1}{2}(v^n_{4,4} + v^n_{5,4})$$
$$\bar{\bar{v}} = \tfrac{1}{2}(v_{4,3} + v_{5,3})$$

Using the same pressure point $(i, j) = (3, 3)$ as the focus, Eq. (6.95) is written in the following form.

$$(\rho v^*)_{4,4}^{n+1} = (\rho v^*)_{4,4}^n + B^* \Delta t - \frac{\Delta t}{\Delta y}(p_{3,4}^* - p_{3,3}^*) \tag{9.59}$$

$$B^* = -\left[\frac{(\rho v \bar{u})_{5,4}^n - (\rho v \bar{u})_{3,4}^n}{2\Delta x} + \frac{(\rho v^2)_{4,5}^n - (\rho v^2)_{4,3}^n}{2\Delta y}\right]$$

$$+ \left[\frac{v_{5,4}^n - 2v_{4,4}^n + v_{3,4}^n}{(\Delta x)^2} + \frac{v_{4,5}^n - 2v_{4,4}^n + v_{4,3}^n}{(\Delta y)^2}\right]$$

$$\bar{u} = \tfrac{1}{2}(u_{4,3}^n + u_{4,4}^n)$$
$$\bar{\bar{u}} = \tfrac{1}{2}(u_{33} + u_{34})$$

Keep in mind when examining the above equations in light of Fig. 9.7 that the indexing on p^* corresponds to the solid points, the indexing on u corresponds to the open points, and the indexing on v corresponds to the × points. The points that appear in the above equations are explicitly numbered in Fig. 9.7. (Although straightforward, you can already sense the extra bookkeeping necessary to deal with a staggered grid in comparison to a conventional single grid.)

After ρu^* and ρv^* are obtained at all interior grid points, values for u^* and v^* are obtained by dividing these values by ρ. Then, the values of u^* at the inflow boundary (which are being allowed to float) are obtained by zeroth-order extrapolation; i.e.,

$$u_{1,j}^* = u_{2,j}^* \qquad \text{for all } j$$

Similarly, the values of u^* and v^* at the outflow boundary (which are being allowed to float) are obtained by zeroth-order extrapolation; i.e.,

$$u_{22,j}^* = u_{21,j}^*$$
$$v_{23,j}^* = v_{22,j}^* \qquad \text{for all } j$$

In the above equations, the values of Δx, Δy, and Δt for the present calculations are

$$\Delta x = \frac{0.5}{20} = 0.025 \text{ ft}$$
$$\Delta y = \frac{0.01}{10} = 0.001 \text{ ft}$$
$$\Delta t = 0.001 \text{ s}$$

The value of Δt was chosen somewhat arbitrarily. However, if Δt is chosen to be too large, experience with the present calculation shows that the calculation becomes unstable. Examining Eqs. (6.94) and (6.95), we see that Δt plays the role of a "relaxation factor"; the larger Δt, the larger is the change in ρu^* and ρv^* from one iteration to the next. It seems reasonable that if this change becomes too large, instabilities could arise. The value $\Delta t = 0.001$ s was found to be acceptable for the present calculation; no effort was made to optimize this value.

Step 3. Using the values for ρu^* and ρv^* obtained from step 2, solve for p' from the pressure correction formula, Eq. (6.104), repeated below,

$$ap'_{i,j} + bp'_{i+1,j} + bp'_{i-1,j} + cp'_{i,j+1} + cp'_{i,j-1} + d = 0 \qquad (6.104)$$

where

$$a = 2\left[\frac{\Delta t}{(\Delta x)^2} + \frac{\Delta t}{(\Delta y)^2}\right]$$

$$b = -\frac{\Delta t}{(\Delta x)^2}$$

$$c = -\frac{\Delta t}{(\Delta y)^2}$$

$$d = \frac{1}{\Delta x}[(\rho u^*)_{i+1/2,j} - (\rho u^*)_{i-1/2,j}] + \frac{1}{\Delta y}[(\rho v^*)_{i,j+1/2} - (\rho v^*)_{i,j-1/2}]$$

Again, we illustrate the above equation by focusing on the pressure grid point (3, 3) shown in Fig. 9.7. At this point, Eq. (6.104) becomes, after solving for $p'_{i,j}$,

$$p'_{3,3} = -\frac{1}{a}(bp'_{4,3} + bp'_{2,3} + cp'_{3,4} + cp'_{3,2} + d) \qquad (9.60)$$

where

$$d = \frac{1}{\Delta x}[(\rho u^*)_{4,3} - (\rho u^*)_{3,3}] + \frac{1}{\Delta y}[(\rho v^*)_{4,4} - (\rho v^*)_{4,3}] \qquad (9.61)$$

Equations like (9.60) are solved for $p'_{i,j}$ at every interior grid point by means of a relaxation approach, as described in Sec. 6.5. This, too, is an iterative process, one that takes place nested within the main iterative sequence being described here. Experience with the present problem shows that after approximately 200 relaxation steps, the values of $p_{i,j}$ have converged.

Step 4. Calculate p^{n+1} at all internal grid points from Eq. (6.106),

$$p_{i,j}^{n+1} = (p^*)_{i,j}^n + \alpha_p p' \qquad (6.106)$$

where α_p is an underrelaxation factor. In the present calculations, the value of α_p was set as 0.1, which is conservatively lower than that suggested in Sec. 6.8.5. For the present calculations, no effort was made to optimize α_p.

Step 5. Designate the values of $p_{i,j}^{n+1}$ obtained from step 4 as the *new* values of $(p^*)^n$ to be inserted into the equivalents of Eq. (9.58) and (9.59) written at all interior grid points. Return to step 2, and repeat steps 2 to 5 until convergence is achieved. For the present calculation, convergence of this primary iteration loop was achieved after approximately 300 iterations. Again, no effort was made to optimize the calculation so as to obtain the smallest number of iterations necessary for convergence.

9.4.2 Results

Because of the insertion of the v velocity spike at point $(i, j) = (15, 5)$ in the initial conditions, the flow field is two-dimensional during the ensuing iterations. This is illustrated in Fig. 9.8, which shows v profiles as a function of distance y across the duct at the axial station where $i = 15$. Hence, these profiles include the grid point (15, 5) where the initial value of the velocity spike, $v_{15,5} = 0.5$ ft/s, was inserted. Indeed, this velocity spike is shown in Fig. 9.8 by the dashed line at $y = 0.004$ ft extending to a value of $v = 0.5$ ft/s. In Fig. 9.8, K denotes the iteration number; hence the velocity spike at the zeroth iteration (the initial conditions) is denoted by $K = 0$. Three other velocity profiles are shown in Fig. 9.8, each one corresponding to the results obtained after K iterations. Note that the peak value of v has already been reduced to 0.343 ft/s after only one iteration, as seen in the profile labeled $K = 1$. The profile labeled $K = 4$ shows that the peak value of v continues to be

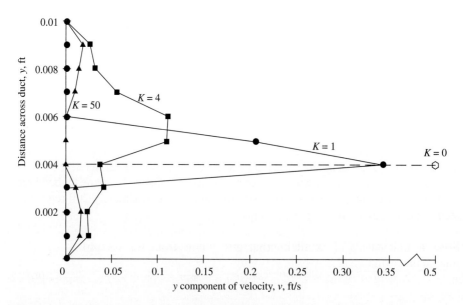

FIG. 9.8
Profiles of the y component of velocity v across the duct at the axial station denoted by $i = 15$. Profiles are shown at various stages during the iterative process. The iteration number is denoted by K.

reduced and that finite values of v are spreading both upward and downward away from grid point (15, 5); indeed, the region of two-dimensional flow introduced by the velocity spike is spreading throughout the flow field in both the x and y directions, although the magnitude of v progressively gets smaller as the iterations progress. Note in Fig. 9.8 that the profile labeled $K = 50$ shows a major reduction in v by the end of the fiftieth iteration. Finally, after 300 iterations, when convergence is obtained throughout the entire flow field, v has essentially gone to zero at all grid points. Reflecting on the results shown in Fig. 9.8, the pressure correction formula, Eq. (9.60) and its equivalent at each grid point, is certainly doing its intended job—it is setting up a pressure field that pushes the velocity field in the correct direction, in this case in the direction of v going toward zero.

In Sec. 6.8.5, the mass source term d was identified as a valuable diagnostic to determine when the pressure correction method has converged to the correct velocity field. As introduced in Eq. (6.104), and as expressed in Eq. (9.61) for grid point (3, 3), d acts as a mass source term in the continuity equation when a velocity field is present that does not satisfy the continuity equation. The object of the pressure correction method is to modify the velocity field through a series of iterations which, when converged, *will* satisfy the continuity equation. When this occurs, the mass source term goes to zero; that is $d = 0$. Hence, examining the variation of d at each grid point throughout the iterative process is a reasonable way of ascertaining when convergence is achieved. An example from the present calculation is shown in Fig. 9.9. Here, the mass source term at grid point (15, 5)—the point at which the initial velocity spike was introduced—is shown as a function of iteration number. Three different sets of iteration numbers are shown in Fig. 9.9. The first set pertains to the early part of the iterative process and gives the values of $d_{15,5}$ for the first five iterations. Note that d is relatively large for these early iterations, as would be expected, and that it exhibits a rather wide variation from one iteration to the next. The second set of iteration numbers covers the range from $K = 8$ to 20. Here we see a general reduction of d compared to the earlier iterations, but the values of $d_{15,5}$ in this set are still relatively large. The third set of iteration numbers covers the range from $K = 50$ to 300 and shows $d_{15,5}$ converging to zero at $K = 300$. (In reality, $d_{15,5} = -0.172 \times 10^{-5}$ at $K = 300$, close enough to zero for our purposes.) Reflecting on Fig. 9.9, we again see that the pressure correction method is doing its job—driving the velocity field to a distribution which satisfies the continuity equation and hence resulting in the mass source term going to zero.

Finally, let us examine some profiles of the x component of velocity, u, across the duct. Figure 9.10 illustrates such profiles at the axial location corresponding to $i = 15$. Note that as the iterations progress, the velocity profiles monotonically approach a linear variation across the duct; i.e., they approach the exact Couette flow solution. Indeed, the numerical iterative process has solidly converged to the Couette flow solution at $K = 300$. It is interesting to note that the numerical solution has also converged to the same Couette flow result at *all* axial stations along the duct, from $i = 1$ to 22, including at the inflow and outflow boundaries. Reflecting on the results shown in Fig. 9.10, we feel quite comfortable that the pressure correction method is working as intended—all aspects of the numerical solution have converged to the exact analytical solution for incompressible Couette flow.

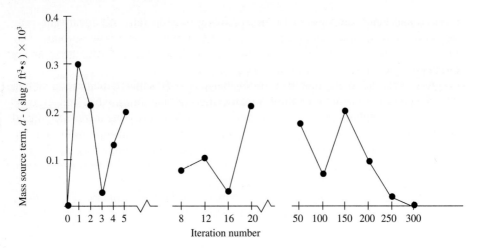

FIG. 9.9
Variation of the mass source term at grid point $(i, j) = (15, 5)$ as a function of iteration number.

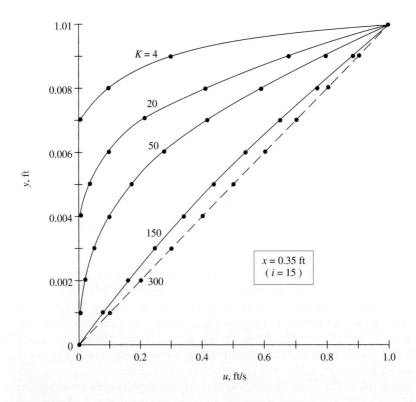

FIG. 9.10
Velocity profiles for the x component of velocity as a function of vertical distance across the duct. Profiles are shown for various iteration numbers, ranging from 4 to 300. At $K = 300$, the velocity profile has converged to the Couette flow solution.

In summary, in the present section we have illustrated the use and behavior of the pressure correction method for the solution of an incompressible viscous flow. From our results, it is interesting to observe the relative roles played by pressure and viscosity in the formulation of the velocity field. In Fig. 9.8 we see the vertical velocity spike decaying fairly rapidly; the values of v throughout the whole flow field became small after about 50 iterations. The rapid decay of v is due to pressure gradients being set up in the flow and propagating via pressure waves that move rapidly throughout the flow field. Again, the calculated pressure corrections are acting to rapidly reduce v. In contrast, in Fig. 9.10 we see the horizontal velocity profiles more slowly converging to the proper solution. Here, the values of u are dominated by viscosity (shear stress), the effects of which are propagated more slowly than those due to pressure waves. Indeed, the values of u do not converge to the proper solution until about 300 iterations, well after the values of v have become very small. This numerical behavior is directly analogous to actual physical behavior in real flows. Flow fields are driven under the impetus of pressure gradients and shear stress, and generally the influence of pressure propagates more rapidly throughout the flow field than that of viscosity.

9.5 SUMMARY

The primary purpose of this chapter was to illustrate the use of an *implicit* finite-difference method for the solution of a fluid flow problem; this is in contrast to the explicit methods demonstrated in Chaps. 7 and 9. In addition, the flow problem chosen in the present chapter was a *viscous* flow; this is in contrast to the inviscid flows calculated in Chaps. 7 and 8. The major results of the application of the Crank–Nicolson implicit method to the solution of incompressible Couette flow in the present chapter underscored the following trends:

1. Theoretically, the method is unconditionally stable; this situation is clearly supported by the present calculations, wherein stable results were obtained even when Δt was abnormally large (equivalent to $E = 4000$).
2. A general advantage of an implicit over an explicit method is that much larger marching steps can be used, hopefully resulting in fewer steps needed to reach the steady state. For the present calculations, an optimum value of Δt which led to the shortest convergence time was found to be about equal to 20 times the maximum value of Δt allowed by an explicit approach. The implicit method used here becomes less efficient (i.e., requiring more marching steps, hence more computer time, to reach a steady state) when Δt is either made too small or too large.
3. Timewise accuracy is a problem with implicit methods. This problem disappears when Δt is made small enough. On the other hand, the calculations wherein Δt was large produced some nonphysical transient results. If all that you are interested in is the final steady-state values, such nonphysical transients are not a problem.

This has been our first detailed calculation using an implicit method, with Thomas' algorithm employed for the solution of the governing equations. For simplicity, we intentionally chose to solve a simple flow problem: incompressible Couette flow. However, this simple problem illustrated the main aspects of implicit finite-difference calculations. You are reminded that many modern CFD calculations employ implicit methods, and therefore it is well worth your while to feel at home with the basic concept.

Another primary purpose of this chapter was to illustrate the use of the pressure correction technique for the solution of the two-dimensional, incompressible, Navier–Stokes equations. We set up a solution of these equations for the incompressible flow between two parallel plates in relative motion to each other. The pressure correction method is an iterative solution. We set up the initial conditions to be a two-dimensional flow field, and hence the pressure correction method was carried out in this chapter for a two-dimensional flow during the iterative process. However, the physical problem was that of Couette flow, and the pressure correction method converged to the proper Couette flow solution. With this example, we have illustrated that the pressure correction method is a viable technique for the solution of incompressible, viscous flows.

PROBLEM

9.1 Solve the Couette flow problem using an *explicit* finite-difference approach. Compare the computer time required for both the implicit and explicit solutions.

CHAPTER 10

SUPERSONIC FLOW OVER A FLAT PLATE: NUMERICAL SOLUTION BY SOLVING THE COMPLETE NAVIER–STOKES EQUATIONS*

Capstone: The crowning or final stroke; culmination.

From *The American Heritage Dictionary of the English Language*, 1969

10.1 INTRODUCTION

If you waded into the CFD solutions in the previous chapters, you are indeed "dirty"—congratulations! You are now in a position to take the next step, that is, to apply your experience to solve the complete Navier–Stokes equations.

---

* This chapter was written by Lt. Col. Wayne Hallgren, a professor in the department of aeronautics at the U.S. Air Force Academy. Colonel Hallgren field-tested the preliminary manuscript of this book at the Academy and kindly agreed to write this chapter in the spirit of the rest of the book. This chapter is a capstone chapter for the applications in Part III; also, it contains some helpful programming hints which are not mentioned in any of the previous chapters. The author is indebted to Colonel Hallgren for contributing this chapter to the present book.

447

448   SUPERSONIC FLOW OVER A FLAT PLATE

In the present chapter a two-dimensional, laminar, viscous, supersonic flow over a flat plate, at zero incidence, is examined. In effect, this problem serves as a capstone to your understanding (at the level of this text), in the following ways:

1. You just solved the classic incompressible Couette flow problem. That problem introduced the effect of viscosity. This problem also includes viscous effects but now accounted for in both the x and y directions (two-dimensional). Additionally, thermal conduction is included in the flow equations.
2. By solving the conservative form of the governing equations, your solution will capture the leading-edge shock wave. This is analogous to your capturing the expansion fan in the Prandtl–Meyer expansion wave problem presented in Chap. 8.
3. Based on your experience from Chaps. 7 and 8, you should feel relatively comfortable with MacCormack's explicit finite-difference technique. Because his technique is "student-friendly," it is used as well in this chapter. However, the complexity of considerably more terms is added. If not already, you will soon appreciate this point. You have an understanding of numerical stability; once again, you will be exposed to this key aspect of explicit numerical approaches.
4. Recall that despite the mixed mathematical nature of the complete Navier–Stokes equations, a time-marching solution is well-posed; hence, this approach is again taken.

The supersonic flow over a flat plate is a classic fluid dynamic problem. However, *no* exact analytical solution exists! A flat plate at zero incidence is a simple geometry. Isn't it surprising that no one has solved this problem without making limiting assumptions? Herein lies the real benefit of CFD. Traditionally, a boundary-layer-solution technique has been used to "solve" this problem (for example, see Ref. 8). Although results obtained from boundary-layer techniques are reasonably good for certain applications, their approximate nature is extremely limiting in terms of flight condition and geometry. Navier–Stokes solutions overcome this inherent shortfall.

As you can see by now, we have slightly diverged from our philosophy of solving problems that have exact solutions. This serves two purposes:

1. This is a capstone problem to tie in your previous learning. At the same time, it pushes you toward greater understanding.
2. This chapter provides a logical and clear link between Chaps. 7 to 9 (relatively simple numerical schemes and physical problems) and the following chapter, which serves to summarize some of the latest, more sophisticated computational techniques and challenges.

As a final note, due to the relative complexity of this problem, this chapter is organized differently from the preceding. A section of "intermediate results" is simply not practical considering the number of equations and steps involved in solving the complete Navier–Stokes equations. That's not to say you are on your own ... indeed, detail is provided when discussing the more challenging aspects of the solution process. Additionally, flowcharts, containing a considerable amount of detail, are provided to help you structure your code. Recognizing that you are experienced in applying MacCormack's technique, the emphasis here is to highlight the more difficult facets of the problem and give you sufficient direction to ensure your success. You have come to a crossroad—simply read the chapter to get a sense of understanding of what a complete Navier–Stokes solution entails (remember, this is arguably the easiest application), or *start wading*!

10.2 THE PHYSICAL PROBLEM

Consider the supersonic flow over a thin sharp flat plate at zero incidence and of length L, as sketched in Fig. 10.1. A laminar boundary layer develops at the leading edge of the flat plate and remains laminar for the case of relatively low Reynolds number. The oncoming freestream no longer "sees" a sharp flat plate. Rather, due to the presence of the viscous boundary layer, the plate possesses a fictitious curvature. Consequently, a curved induced shock wave, as shown in Fig. 10.1, is generated at the leading edge (Ref. 2).

The region between the surface and the shock is called the *shock layer*. Depending on (for example) Mach number, Reynolds number, and surface temperature, the shock layer can be characterized by a region of viscous flow and inviscid flow (refer to Fig. 10.2a), or the entire layer can be fully viscous, a so-called merged shock layer (Fig. 10.2b). Furthermore, dissipation of kinetic energy within the boundary layer (*viscous dissipation*) can cause high flow-field temperatures and thus high heat-transfer rates. *Bottom line*: Although we are dealing with a simple geometry, capturing and understanding the physics of this problem is indeed a significant challenge. So, let us get on with it!

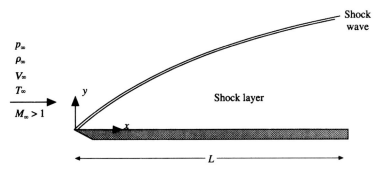

FIG. 10.1
Illustration of supersonic flow over a sharp leading-edged flat plate at zero incidence.

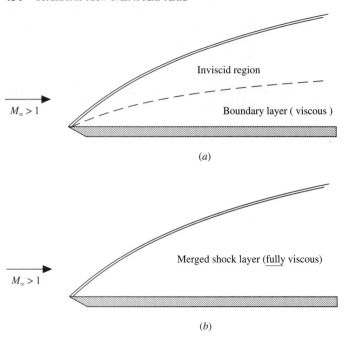

FIG. 10.2
(a) Supersonic flow over a flat plate with a distinct boundary layer and region of inviscid flow. (b) Supersonic flow over a flat plate with a merged shock layer.

10.3 THE NUMERICAL APPROACH: EXPLICIT FINITE-DIFFERENCE SOLUTION OF THE TWO-DIMENSIONAL COMPLETE NAVIER–STOKES EQUATIONS

This problem is packed with interesting fluid phenomena. The advantage of using a time-dependent Navier–Stokes approach is its inherent ability to evolve to the correct steady-state solution. Of course, in the process the shock location is determined, as well as the physical characteristics of the shock layer.

10.3.1 The Governing Flow Equations

Neglecting body forces and volumetric heating, the two-dimensional forms of the Navier–Stokes equations are repeated below. (*Note*: It may be useful at this point to review Sec. 2.8.)

Continuity: $\quad \dfrac{\partial \rho}{\partial t} + \dfrac{\partial}{\partial x}(\rho u) + \dfrac{\partial}{\partial y}(\rho v) = 0 \quad$ (2.33)

x Momentum: $\quad \dfrac{\partial}{\partial t}(\rho u) + \dfrac{\partial}{\partial x}(\rho u^2 + p - \tau_{xx}) + \dfrac{\partial}{\partial y}(\rho u v - \tau_{yx}) = 0 \quad$ (2.56a) *y Momentum* : $\quad \dfrac{\partial}{\partial t}(\rho v) + \dfrac{\partial}{\partial x}(\rho u v - \tau_{xy}) + \dfrac{\partial}{\partial y}(\rho v^2 + p - \tau_{yy}) = 0 \qquad (2.56b)$

Energy:

$$\dfrac{\partial}{\partial t}(E_t) + \dfrac{\partial}{\partial x}[(E_t + p)u + q_x - u\tau_{xx} - v\tau_{xy}] + \dfrac{\partial}{\partial y}[(E_t + p)v + q_y - u\tau_{yx} - v\tau_{yy}] = 0$$
$$(2.81)$$

In the above equation, E_t is physically the sum of the kinetic energy and internal energy e per unit volume; it is defined as

$$E_t = \rho\left(e + \dfrac{V^2}{2}\right) \qquad (10.1)$$

The shear and normal stress, expressed in terms of velocity gradients, are repeated below for convenience:

$$\tau_{xy} = \tau_{xy} = \mu\left(\dfrac{\partial u}{\partial y} + \dfrac{\partial v}{\partial x}\right) \qquad (2.57d)$$

$$\tau_{xx} = \lambda(\nabla \cdot \mathbf{V}) + 2\mu\dfrac{\partial u}{\partial x} \qquad (2.57a)$$

$$\tau_{yy} = \lambda(\nabla \cdot \mathbf{V}) + 2\mu\dfrac{\partial v}{\partial y} \qquad (2.57b)$$

Likewise, the components of the heat flux vector (from Fourier's law of heat conduction) are repeated below:

$$\dot{q}_x = -k\dfrac{\partial T}{\partial x}$$

$$\dot{q}_y = -k\dfrac{\partial T}{\partial y}$$

Let us pause for a minute. At this point, the system consists of four equations: continuity, x and y momentum, and energy. Nine unknowns are embedded in the equations: ρ, u, v, $|\mathbf{V}|$, p, T, e, μ, and k. To close the system, five additional equations are needed, as described below:

1. A perfect gas is assumed. From Chap. 2, the equation of state is

$$p = \rho R T$$

2. Furthermore, if we assume the air is *calorically* perfect, then the following relation holds (also from Chap. 2):

$$e = c_v T$$

3. The x and y components of the velocity vector are u and v, respectively, such that

$$|\mathbf{V}| = \sqrt{u^2 + v^2} \qquad (10.2)$$

4. To evaluate the viscosity, assuming a calorically perfect gas, Sutherland's law is typically used. μ_0 and T_0 are reference values at standard sea level conditions (Ref. 8).

$$\mu = \mu_0 \left(\frac{T}{T_0}\right)^{3/2} \frac{T_0 + 110}{T + 110} \tag{10.3}$$

5. One additional equation is required. If the Prandtl number, defined below, is assumed constant (approximately equal to 0.71 for calorically perfect air), thermal conductivity can be calculated from the equation (Ref. 8)

$$\text{Pr} = 0.71 = \frac{\mu c_p}{k}$$

where c_p is the specific heat at constant pressure (like c_v, a constant as long as the air is assumed calorically perfect).

The system of equations is now closed: nine equations with nine unknowns. As in Sec. 2.10, the governing equations, expressed in vector notation, are especially suited for numerical application. In a slightly different form, the equations are repeated below:

$$\frac{\partial U}{\partial t} + \frac{\partial E}{\partial x} + \frac{\partial F}{\partial y} = 0 \tag{10.4a}$$

where U, E, and F are column vectors given by

$$U = \begin{Bmatrix} \rho \\ \rho u \\ \rho v \\ E_t \end{Bmatrix} \tag{10.4b}$$

$$E = \begin{Bmatrix} \rho u \\ \rho u^2 + p - \tau_{xx} \\ \rho u v - \tau_{xy} \\ (E_t + p)u - u\tau_{xx} - v\tau_{xy} + q_x \end{Bmatrix} \tag{10.4c}$$

$$F = \begin{Bmatrix} \rho v \\ \rho u v - \tau_{xy} \\ \rho v^2 + p - \tau_{xy} \\ (E_t + p)v - u\tau_{xy} - v\tau_{yy} + q_y \end{Bmatrix} \tag{10.4d}$$

10.3.2 The Setup

Now for the problem at hand. Consider the computational domain, in this case a rectangular structured grid, as shown in Fig. 10.3. The flow at the upstream boundary ($x = 0.0$ or $i = 1 =$ IMIN) is at Mach 4 with pressure, temperature, and speed of sound equal to their respective sea level values.

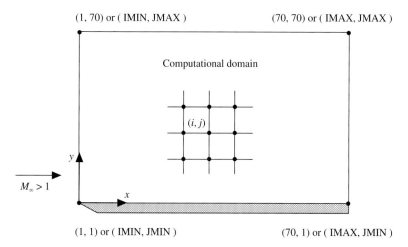

FIG. 10.3
Computational domain.

The length of the plate is 0.00001 m. This is extremely small, but large compared to the mean free path of the oncoming air molecules, and sufficient to capture the desired physics. The Reynolds number is about 1000. We want a low Reynolds number in order to keep computational running times *relatively* short compared to high Reynolds number applications requiring considerably finer grids (Ref. 13).

10.3.3 The Finite-Difference Equations

In Chap. 7, you applied MacCormack's time-marching technique in one spatial direction: down the length of a convergent-divergent nozzle. In Chap. 8, you used his technique to march downstream, effectively solving for flow properties in two spatial dimensions without any time dependency. This problem takes you one step further. As in the convergent-divergent nozzle problem, you will march in time to a steady-state solution but in the process solve the flow properties at every (i, j) spatial location. So, a third dimension is introduced.

Following the presentation in Chap. 6, the key steps in applying MacCormack's technique are shown below. The governing equation [Eq. (10.4a)] is rewritten, in vector notation as

$$\frac{\partial U}{\partial t} = -\frac{\partial E}{\partial x} - \frac{\partial F}{\partial y} \tag{10.4a}$$

By means of a Taylor series expansion, the flow-field variables are advanced at each grid point (i, j) in steps of time, as shown below:

$$U_{i,j}^{t+\Delta t} = U_{i,j}^{t} + \left(\frac{\partial U}{\partial t}\right)_{\text{av}} \Delta t \tag{10.5}$$

where, once again, U is a flow-field variable (from the governing equations) assumed known at time t, either from initial conditions or as a result from the previous iteration in time. $(\partial U/\partial t)_{av}$ is defined as

$$\left(\frac{\partial U}{\partial t}\right)_{av} = \frac{1}{2}\left[\left(\frac{\partial U}{\partial t}\right)^{t}_{i,j} + \left(\frac{\partial U}{\partial t}\right)^{t+\Delta t}_{i,j}\right] \quad (10.6)$$

To obtain a value of $(\partial U/\partial t)_{av}$ (above) so that the solution can be advanced, the following steps are taken:

1. $(\partial U/\partial t)^{t}_{i,j}$ is calculated using forward spatial differences on the right-hand side of the governing equations (refer to the vector form above) from the known flow field at time t.
2. From step 1, PREDICTED values of the flow-field variables (denoted by a bar) can be obtained at time $t + \Delta t$, as follows:

$$\bar{U}^{t+\Delta t}_{i,j} = U^{t}_{i,j} + \left(\frac{\partial U}{\partial t}\right)^{t}_{i,j}\Delta t \quad (10.7)$$

Combining steps 1 and 2, predicted values are determined as follows:

$$\bar{U}^{t+\Delta t}_{i,j} = U^{t}_{i,j} - \frac{\Delta t}{\Delta x}(E^{t}_{i+1,j} - E^{t}_{i,j}) - \frac{\Delta t}{\Delta y}(F^{t}_{i,j+1} - F^{t}_{i,j}) \quad (10.8)$$

3. Using rearward spatial differences, the predicted values (from step 2) are inserted into the governing equations such that a predicted time derivative $\overline{(\partial U/\partial t)}^{t+\Delta t}_{i,j}$ can be obtained.
4. Finally, substitute $\overline{(\partial U/\partial t)}^{t+\Delta t}_{i,j}$ (from step 3) into Eq. (10.6) to obtain CORRECTED second-order-accurate values of U at time $t + \Delta t$. As in Eq. (10.8), steps 3 and 4 are combined as follows:

$$U^{t+\Delta t}_{i,j} = \frac{1}{2}\left[U^{t}_{i,j} + \bar{U}^{t+\Delta t}_{i,j} - \frac{\Delta t}{\Delta x}(\bar{E}^{t+\Delta t}_{i,j} - \bar{E}^{t+\Delta t}_{i-1,j}) - \frac{\Delta t}{\Delta y}(\bar{F}^{t+\Delta t}_{i,j} - \bar{F}^{t+\Delta t}_{i,j-1})\right] \quad (10.9)$$

Steps 1 to 4 are repeated until the flow-field variables approach a steady-state value; this is the desired steady-state solution.

To maintain second-order accuracy, the x-derivative terms appearing in E are differenced in the opposite direction to that used for $\partial E/\partial x$, while the y-derivative terms are approximated with central differences. Likewise, the y-derivative terms appearing in F are differenced in the opposite direction to that used for $\partial F/\partial y$, while the x-derivative terms in F are approximated with central differences (Ref. 13). For example, in the predictor step (see step 1 above), $\partial E/\partial x$ is forward-differenced. However, E has terms like τ_{xy}, which includes derivatives of velocity in both the x and y directions [refer to Eq. (2.57d)]. Therefore, in the predictor step, $\partial v/\partial x$ is rearward-differenced and $\partial u/\partial y$ is central-differenced. (*Note*: use first-order differencing in both cases.)

After each predictor or corrector step, the primitive variables are obtained by decoding the U vector, as shown below; U_4 is reserved for a three-dimensional application (another challenge for you) and is therefore omitted.

$$\rho = U_1 \tag{10.10a}$$

$$u = \frac{\rho u}{\rho} = \frac{U_2}{U_1} \tag{10.10b}$$

$$v = \frac{\rho v}{\rho} = \frac{U_3}{U_1} \tag{10.10c}$$

$$E_t = \rho\left(e + \frac{V^2}{2}\right) = U_5$$

or

$$e = \frac{U_5}{U_1} - \frac{u^2 + v^2}{2} \tag{10.10d}$$

With ρ, u, v, and e determined, the remaining flow-field properties can be obtained by using the equations in Sec. 10.3.1 as follows:

$$T = \frac{e}{c_v}$$

$$p = \rho R T$$

μ and k are functions of temperature T. μ can be determined by applying Sutherland's law. Once μ is known, a constant Prandtl number assumption leads directly to k, as shown below.

$$k = \frac{\mu \, c_p}{\text{Pr}}$$

10.3.4 Calculation of Step Sizes in Space and Time*

As shown in Fig. 10.3 the domain is 70 × 70. The following notation describes the grid size in the streamwise direction:

$$\text{IMIN} = 1 \quad \text{(inflow } x \text{ location)}$$
$$\text{IMAX} = 70 \quad \text{(outflow } x \text{ location)}$$

With the length of the plate known (LHORI), the step size in the x direction (Δx) is determined as follows:

$$\Delta x = \frac{\text{LHORI}}{\text{IMAX} - 1} \tag{10.11}$$

---

* Several of the suggestions in this section were provided by James Weber, a graduate fellow at the University of Maryland.

Likewise, JMIN = 1 and JMAX = 70 describe the grid normal to the plate's surface (JMIN = 1 is the surface and JMAX the upper boundary of the domain). Refer again to Fig. 10.3. To obtain an accurate solution, the shock wave must lie within the computational domain. It is reasonable to assume that a domain of at least five times the height of a boundary layer, as predicted by a Blasius calculation at the trailing edge, will satisfy this computational constraint (refer to Fig. 10.4). Therefore, the vertical height of the domain (LVERT) is determined as follows:

$$\text{LVERT} = 5 \times \delta \tag{10.12a}$$

where δ is given by

$$\delta = \frac{5(\text{LHORI})}{\sqrt{\text{Re}_L}} \tag{10.12b}$$

Therefore, the step size in the y direction is

$$\Delta y = \frac{\text{LVERT}}{\text{JMAX} - 1} \tag{10.13}$$

Using the grid size above (plate length of 0.00001 m) results in a step size in the x and y directions of 0.145×10^{-6} and 0.119×10^{-6}, respectively. How do you know if your grid is sized correctly? Cell Reynolds numbers (defined below) in the x and y directions are calculated at each point, for each time step.

$$\text{Re}_{\Delta x} \equiv \frac{\rho_{i,j} u_{i,j} \Delta x}{\mu_{i,j}} \tag{10.14a}$$

$$\text{Re}_{\Delta y} \equiv \frac{\rho_{i,j} v_{i,j} \Delta y}{\mu_{i,j}} \tag{10.14b}$$

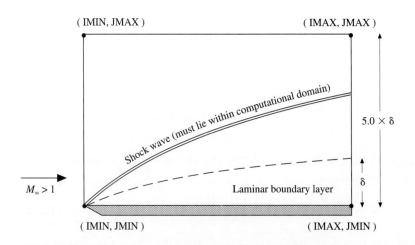

FIG. 10.4
Illustration of how to size the computational domain.

The magnitude of the cell Reynolds numbers provides insight into correctly sizing the computational grid; for this problem cell Reynolds numbers of the following order are used:

$$\mathrm{Re}_{\Delta x} \leq 30\text{–}40 \qquad (10.15a)$$

$$\mathrm{Re}_{\Delta y} \leq 3\text{–}4 \qquad (10.15b)$$

Note that the requirement in the y direction is much smaller; this is a consequence of the stronger gradients in the direction normal to the plate. Thus, to capture the flow field, especially near the surface, typically more grid points are necessary perpendicular to the surface—makes sense!

Because this method is an explicit formulation, the time step is subject to a stability criterion. To determine the size of the time step, the following version of the Courant–Friedrichs–Lewy (CFL) criterion (Ref. 74) is used. $a_{i,j}$ is the local speed of sound in meters per second, and K is the Courant number. K acts as a "fudge factor," if you will, to make sure the solution remains stable.

$$(\Delta t_{\mathrm{CFL}})_{i,j} = \left[\frac{|u_{i,j}|}{\Delta x} + \frac{|v_{i,j}|}{\Delta y} + a_{i,j}\sqrt{\frac{1}{\Delta x^2} + \frac{1}{\Delta y^2}} + 2v'_{i,j}\left(\frac{1}{\Delta x^2} + \frac{1}{\Delta y^2}\right) \right]^{-1}$$

where

$$v'_{i,j} = \max\left[\frac{\frac{4}{3}\mu_{i,j}(\gamma\mu_{i,j}/\mathrm{Pr})}{\rho_{i,j}}\right] \qquad (10.16)$$

$$\Delta t = \min[K(\Delta t_{\mathrm{CFL}})_{i,j}]$$

for $0.5 \leq K \leq 0.8$.

10.3.5 Initial and Boundary Conditions

We are solving a system of partial differential equations. They are first-order in time and second-order in space. Therefore, initial and boundary conditions (on velocity and temperature) are necessary.

Because our solution is marched from a set of initial conditions, we must specify the flow properties at each (i, j) location at time $t = 0.0$. Except as noted below, properties at each grid point are initialized at their respective freestream values. At the surface (JMIN = 1), the no-slip boundary condition is enforced and the wall temperature T_w is a given value.

$$u = v = 0.0 \qquad (2.87)$$

$$T = T_w \qquad (2.88)$$

Having specified the initial conditions ($t = 0.0$), our equations are marched in time to the steady-state solution. In that process, conditions must be enforced at the boundary of our computational domain. Referring to Fig. 10.5 (cases 1–4, as noted below), the following boundary conditions are detailed:

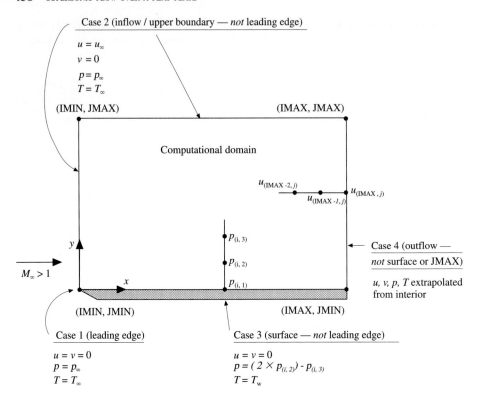

FIG. 10.5
Application of boundary conditions.

Case 1. At the leading edge [(IMIN, JMIN) or (1, 1)], no-slip is enforced ($u_{(1,1)} = v_{(1,1)} = 0.0$) and the temperature ($T_{(1,1)}$) and pressure ($p_{(1,1)}$) are assumed to be their respective freestream values.

Case 2. At the left-hand side (except the leading edge) and upper boundaries of the domain, the x component of velocity, u, temperature, and pressure are assumed to be their respective freestream values; the y component of velocity, v, is assumed to equal zero.

Case 3. At the surface of the plate, no-slip is specified on velocity ($u = v = 0.0$). Temperature (except at the leading edge: see case 1 above) is assumed to equal the wall temperature T_w value. Pressure, (except at the leading edge: see step 1 above) is calculated at the wall by extrapolating from the values at the two points ($j = 2$ and $j = 3$) above the surface, as shown below:

$$p_{(i,1)} = (2p_{(i,2)}) - p_{(i,3)} \tag{10.17}$$

Case 4. Finally, all properties at the right-hand side of the domain (*not including* JMIN = 1 and JMAX = 70) are calculated based on an extrapolation from the two interior points, at the same j location. For example, u is determined as follows:

$$u_{(\text{IMAX},j)} = (2u_{(\text{IMAX}-1,j)}) - u_{(\text{IMAX}-2,j)} \qquad (10.18)$$

From these known values, the balance of the flow properties at the boundaries are calculated from the additional equations presented in Sec. 10.3.1. For example, density is calculated from the equation of state.

In the above, a constant-temperature wall boundary condition is specified. As mentioned in Chap. 2, this is the easiest boundary condition to enforce on temperature. As an aside, one of the most significant advantages of CFD is the ability to make simple changes to, for example, freestream and boundary conditions, and then watch to see what happens. By conducting numerical experiments, you gain a better physical understanding of the implications of changing a flow parameter. Therefore, it makes sense to structure your code in such a fashion that simplifies your natural interest to pursue further numerical experiments. For example, a convenient and well-written subroutine to apply boundary conditions allows you to relatively easily make code changes to see the impact on, for example, an adiabatic wall (refer to Sec. 2.9) boundary condition.

10.4 ORGANIZATION OF YOUR NAVIER–STOKES CODE

10.4.1 Overview

At this point, you have a much better appreciation for what is behind a complete Navier–Stokes numerical solution. With the finite-difference equations in place, as well as step-size constraints and initial and boundary conditions, we are in a position to discuss how you might proceed with organizing your code. As mentioned in Sec. 10.1, flowcharts are used to guide you. By the way, if you typically approach coding without flowcharting, or at least using some form of pseudocode, now is the time to reconsider your process. Codes involving this much detail, with the extensive passing of values between subprograms, require thorough organization *up front*.

In terms of the "big picture," you may want to structure your code as shown in Fig. 10.6. Let us proceed by highlighting key components of the code. Keep in mind that a considerable amount of effort is involved in writing each subprogram.

1. The MAIN program drives the entire code. Its primary functions are:
    a. Establishing flow conditions, sizing the computational domain, and initializing the flow properties at each (i, j) spatial location following the presentation in Sec. 10.3.5.
    b. Marching the code in time and calling the following subroutines:
        i. TSTEP to determine the proper time step (as described in Sec. 10.3.4)

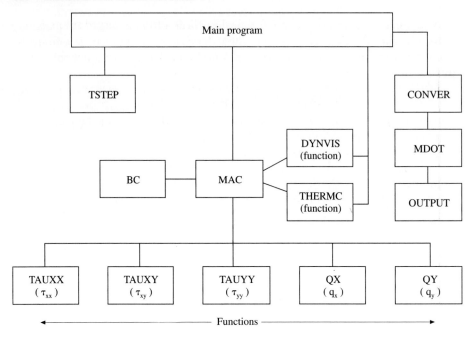

FIG. 10.6
An approach to structuring your code.

    ii. MAC (for MacCormack) to update the flow properties at each (i, j) location by using the predictor-corrector technique, as detailed in Sec. 10.3.3

    iii. CONVER to check for flow-field convergence

2. DYNVIS and THERMC are function subprograms; they are called to solve for dynamic viscosity and thermal conductivity at each (i, j) location following the discussion in Sec. 10.3.3. The MAIN program calls these functions only during the flow-field initialization process. As shown, MAC uses these functions each time it is called.

3. Viscous effects are accounted for in the five functions TAUXX, TAUXY, TAUYY, QX, and QY. Each time one of these stress or heat conduction terms needs to be evaluated, the respective function is called. For example, in determining E_3 [the third component of the E vector: refer to Eq. (10.4c)], TAUXY is called. Remember from our discussion in Sec. 10.3.3 that the derivatives in these equations are differenced either forward, central, or backward depending on your position within MacCormack's scheme (e.g., in the predictor step); more later.

4. Once the internal flow-field properties are determined (either through the predictor or the corrector step), boundary conditions are applied by calling subroutine BC, which implements the boundary conditions as described in Sec. 10.3.5.

5. After each time step, CONVER is called to check for a converged solution. The solution is considered converged when the density, at each (i, j) grid point, changes no more than 1.0×10^{-8} between time steps. Subroutine CONVER also "asks" if the main program has reached a specified number of maximum iterations. If it has, even though the solution has not converged, the subroutine calls MDOT and OUTPUT to assess how the solution is progressing.

6. MDOT provides a check of the validity of the numerical solution. An integration scheme (trapezoidal rule) is used to confirm conservation of mass. The rate of mass inflow across the entrance to the computational domain is compared to the rate of mass outflow across the exit plane. For this case, the deviation between mass flow rate at the entrance and exit is less than 1 percent.

7. Finally, subroutine OUTPUT generates data files for plotting results.

10.4.2 The Main Program

A recommended approach to organizing your main program is provided in Fig. 10.7. IMAX and JMAX size the computational grid. MAXIT is a maximum number of iterations you want the code to execute prior to "kicking out" and stopping; this technique is convenient for testing your code through a couple of iterations prior to extended running times. To begin code execution, freestream conditions, as well as several thermodynamic constants, must be specified (or calculated). For the results presented in the next section (as "set up" in Sec. 10.3.3), the following values were used (SI units):

Mach number = 4.0
Plate length (LHORI) = 0.00001 m
Sea level values for the freestream speed of sound, pressure, and temperature, respectively = 340.28 m/s, 101325.0 N/m², 288.16 K
The ratio of wall temperature to freestream temperature (T_w/T_∞) was set equal to 1.0; this ratio is convenient for investigating the impact of changing wall-temperature boundary conditions.
The ratio of specific heats $(\gamma) = 1.4$
The Prandtl number (Pr) = 0.71
Reference values (sea level) for dynamic viscosity and temperature, respectively = 1.7894×10^{-5} kg/(m · s), 288.16 K
Specific gas constant (R) = 287 J/(kg · K)

Once the above values are specified, the remaining constants are determined by using the equations as shown in Fig. 10.7.

As seen in the flowchart, TSTEP is called prior to executing MacCormack's algorithm. For this code, K [the fudge factor in Eq. (10.16)] is set equal to 0.6. Only internal points are used to determine the appropriate time step in applying MacCormack's technique.

462 SUPERSONIC FLOW OVER A FLAT PLATE

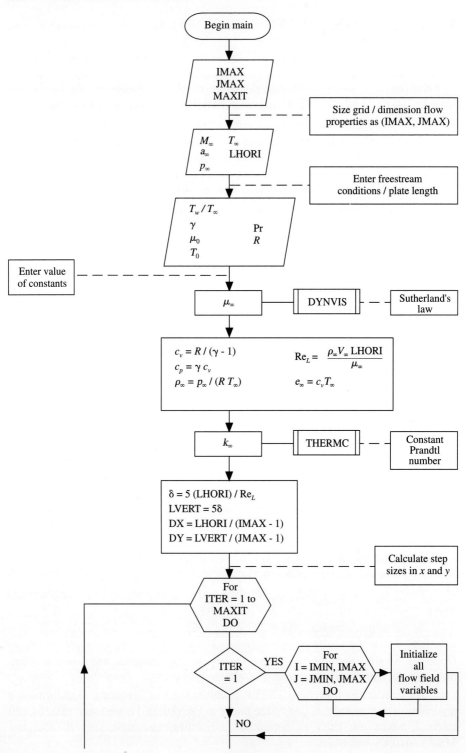

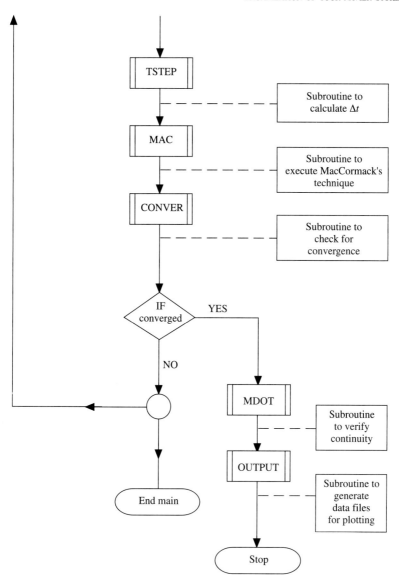

FIG. 10.7
Flowchart for the main program (MAIN).

10.4.3 The MacCormack Subroutine

You are well-versed in applying MacCormack's technique but not to the extent required to perform a full Navier–Stokes solution. So, Fig. 10.8 may help you organize your subroutine. If you follow the structure, as presented in Fig. 10.6, you will find this is, quite naturally, your longest subroutine; it is on the order of 150 lines of executable code, not including the subprograms (e.g., TAUXX, BC, and DYNVIS) that are called in the process of executing the algorithm.

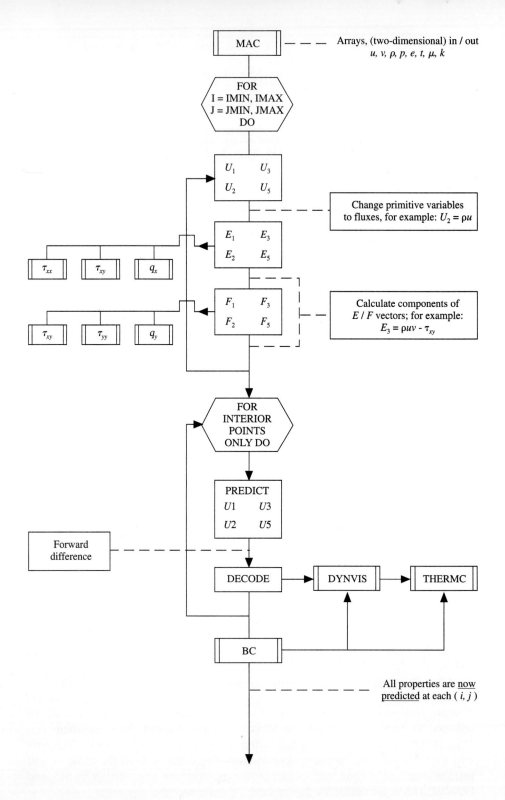

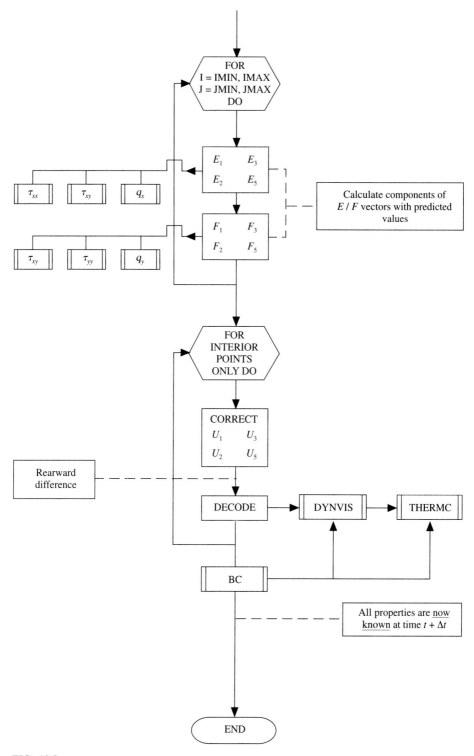

FIG. 10.8
Flowchart for MacCormack's (MAC) subroutine.

The flowchart follows directly from the discussion in Sec. 10.3.3. U_5, E_2, and F_1, for example, correspond to terms found in the U, E, and F vectors [refer to Eqs. (10.4a) to (10.4d)]. Recall that the fourth subscript is reserved for a three-dimensional application, consistent with Eqs. (10.10a) to (10.10d). Therefore, for example, U_5 is equal to E_t. The flowchart is self-explanatory; however, a couple of pointers follow:

1. In addition to the nine flow-field properties, you should dimension $U_{1,2,3,5}$, $E_{1,2,3,5}$, and $F_{1,2,3,5}$ as (IMAX, JMAX). Additionally, predicted values of U (e.g., U_1P) for all four (1, 2, 3, 5) components should be dimensioned as (IMAX, JMAX).
2. Be careful to difference the derivatives found in the shear stress and heat conduction terms in accordance with Sec. 10.3.3. This becomes rather messy! To help give you an idea of how to go about this, Fig. 10.9 is an approach to organizing function TAUXY, the function most frequently called by MAC. Case 1 is executed when τ_{xy} is needed to evaluate E_3 and E_5 in the *predictor* step (refer to the example in Sec. 10.3.3 on how second-order accuracy is maintained—this is precisely the same).
3. When you decode your flux terms to obtain primitive variables, follow the procedure outlined at the end of Sec. 10.3.3.

10.4.4 Final Remarks

Now you are *really* ready! For most, this is the longest code you have ever tackled. A few hints: (1) start (logically) by coding the main program, (2) use *lots* of comment statements, (3) put all the "call" statements in place and write short, dummy subprograms to simply return you to the main program, and (4) one by one begin building your subprograms. *Flowchart* (or pseudocode) each subprogram in detail, *code*, and then *test* to make sure each one is doing precisely what you expect. Once you are confident that each specific piece of code is accurately in place, proceed with the next. For all practical purposes, there is no other way to take on this problem!

10.5 FINAL NUMERICAL RESULTS: THE STEADY-STATE SOLUTION

Before discussing the specific steady-state results, a few general comments are in order:

1. The solution converged after 4339 time steps (6651 for the adiabatic wall case described below). Remember, this case is for a 70 × 70 grid. Reducing the number of grid points to, for example, 40 × 50, speeds up convergence while still adequately capturing the physics of the flow field. As an aside, to ensure your code is grid-independent, running it with different grid sizes is always appropriate.

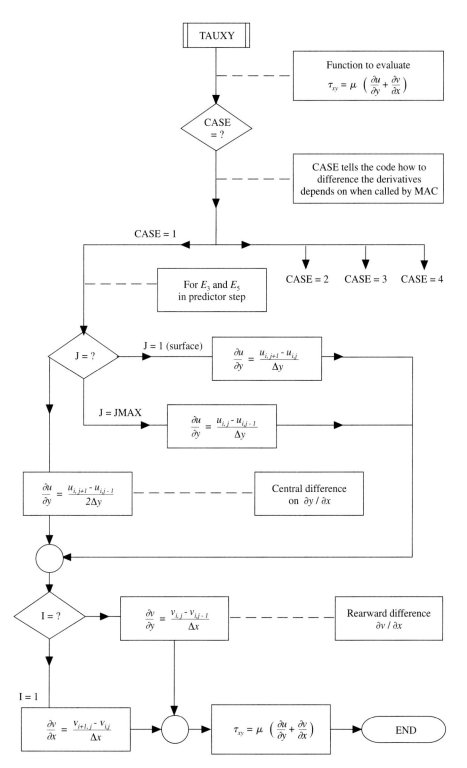

FIG. 10.9
Flowchart for TAUXY function subprogram.

2. In plotting the profiles of various flow-field properties, a normalized y distance, is used. Suggested by Van Driest (Ref. 75), $\bar{y}$ is defined as

$$\bar{y} = \frac{y}{x}\sqrt{\mathrm{Re}_x} \tag{10.19}$$

3. Throughout, profiles of various flow-field properties are presented, frequently nondimensionalized by the freestream value (for example, p/p_∞). This is in contrast to using boundary-layer edge conditions, as frequently done in boundary-layer-type analysis. Here, the entire shock layer is treated as fully viscous. Depending on flow conditions, a well-defined boundary layer is sometimes indistinguishable. Furthermore, even in the case of a distinct boundary layer, the definition of "edge conditions" is somewhat vague.

4. Results are also presented for an adiabatic wall case. This serves two purposes:
   a. An adiabatic wall boundary condition causes significant changes in the flow field. When compared to the constant-temperature wall case, interesting physical insight is gained.
   b. Remember, CFD allows you to "throw switches." Once you have a code in place, numerical experiments are relatively easy to perform. These results are typical of the "next step" in expanding your code for further study. The boundary condition is applied precisely as explained in Sec. 2.9. Mathematically, an adiabatic wall is enforced as follows:

$$\left(\frac{\partial T}{\partial n}\right)_w = 0 \tag{2.91}$$

5. A Mach 25 flow at 200,000 ft (LHORI = 0.005 m) case is presented. Once again, a well-designed code allows you to take a look at other interesting cases.

Representative results are presented in Figs. 10.10a to 10.17. Certainly, you can think of other plots that are of specific interest to you.

1. Figure 10.10a: Nondimensional surface pressure distribution is plotted as a function of distance from the leading edge. A few notes about the figure:
   a. Oscillations are apparent in the leading-edge region; conventional rationalization is that this is a consequence of a continuum assumption in a noncontinuum region. Whether the oscillations are a real phenomenon or a numerical effect is not clear. Although this point is of academic interest, results show the effect of these oscillations, aft of the leading edge, is unimportant.
   b. Note that the adiabatic wall tends to increase the overall pressure above (by about 30 percent) the constant-temperature wall case. Physically, an adiabatic wall increases the boundary-layer temperatures above the constant-temperature wall values. (*Note*: the usual assumption is made that the adiabatic wall temperature is much greater than the constant-temperature value). The result is a relatively lower density and hence thicker boundary layer. Therefore, the oncoming flow "sees" a blunter body, thus creating a

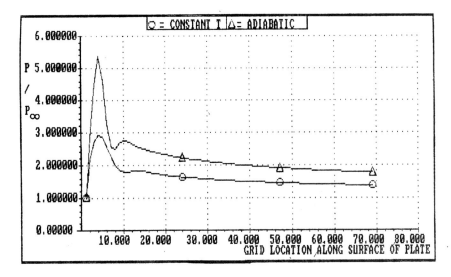

(a)

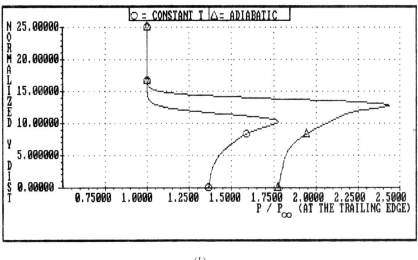

(b)

FIG. 10.10
Mach 4 at sea level. (a) Normalized surface pressure distributions; (b) normalized pressure profiles.

stronger leading-edge shock wave; this, in turn, increases the pressure within the shock layer. Furthermore, pressure is driven upward by higher flow temperatures.

2. Figure 10.10b: This is a nondimensional pressure profile at the plate's trailing edge. Again, an adiabatic wall tends to increase overall pressure within the shock layer. The shock jump for the adiabatic case is about 35 percent, indicating that the oncoming flow has passed through a stronger shock (as mentioned above).

The typical assumption of a zero pressure gradient through the boundary layer is questionable (about a 15 percent change) under these conditions.

3. *Viscous interaction* is a name given to a flow field in which considerable interaction takes place between a growing boundary layer and the outer inviscid flow (for detail, refer to Ref. 2). Figure 10.11 (from Ref. 76) shows good agreement between this solution for both the constant-temperature wall case (solid triangle) and the adiabatic wall case (solid square).

4. Figure 10.12a, b: These are the temperature profiles at the trailing edge. Figure 10.12a has an expanded ordinate; note that the profiles capture the leading-edge shock wave, as well as show classic boundary-layer behavior near the wall (for more detail, refer to Ref. 8). As expected, the temperature gradient is zero at the wall for the adiabatic case; also, temperatures within the thermal boundary layer are about three times higher! Figure 10.12b is presented for comparison to Van Driest's results, as shown in Fig. 10.13a, b. Qualitatively, the agreement is excellent. An interesting point—Van Driest's solution, based on classic supersonic boundary-layer theory, stops short (perpendicular to plate) of the leading-edge shock. This is a direct consequence of the "self-similar" boundary-layer techniques typical of the 1950s and 1960s. In direct contrast to this Navier–Stokes solution, these approximate methods typically required coupling an

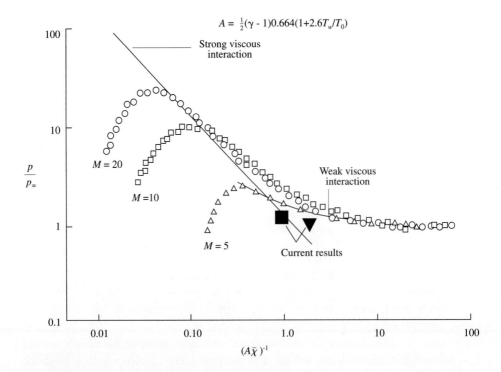

FIG. 10.11
Induced pressures on a flat plate (*Ref.* 76.)

FINAL NUMERICAL RESULTS: THE STEADY-STATE SOLUTION   **471**

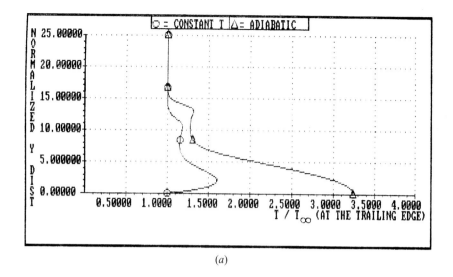

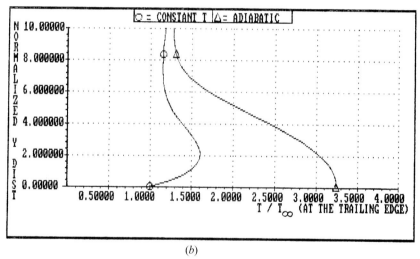

FIG. 10.12
Mach 4 at sea level. (*a*) Normalized temperature profiles through the entire flow field; (*b*) normalized temperature profiles near the surface.

inviscid solution to a boundary layer-solution (note that the temperatures are normalized by the edge conditions in Fig. 10.13*a*, *b*.

5. Figure 10.14*a*, *b*: Similarly, the *u* component of velocity is plotted. The boundary layer is indeed thicker for the adiabatic case.

6. Figure 10.15: Mach number profiles graphically illustrate the relative strength of the two leading-edge shock waves.

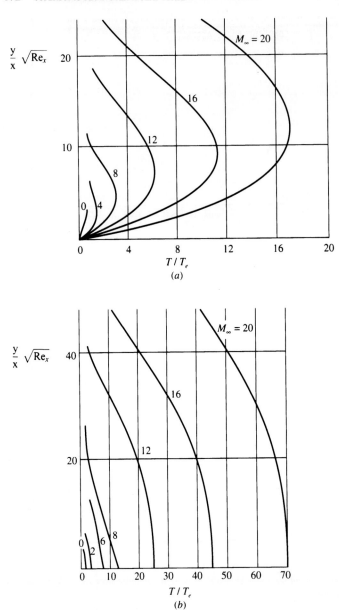

FIG. 10.13
Temperature profiles in a laminar compressible boundary layer. (*a*) Constant-temperature wall; (*b*) adiabatic wall. (*Ref. 75.*)

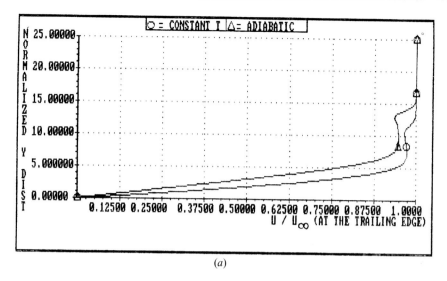

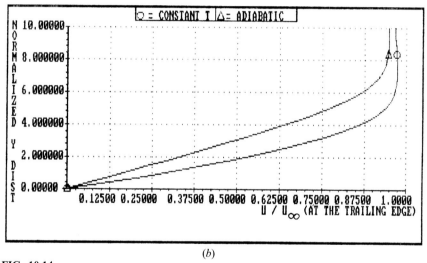

FIG. 10.14
Mach 4 at sea level. (a) Normalized velocity profiles through the entire flow field; (b) normalized velocity profiles near the surface.

7. Figure 10.16a, b: The temperature profiles for the Mach 25 case show a distinct and important difference from those at Mach 4. The shock layer is now fully viscous, a consequence of the combined conditions of high Mach number and low Reynolds number. There is no longer a distinct boundary layer near the wall. Van Driest's results, based on a boundary layer patched to an inviscid flow, is simply inadequate for this problem. The solution evolves naturally when the

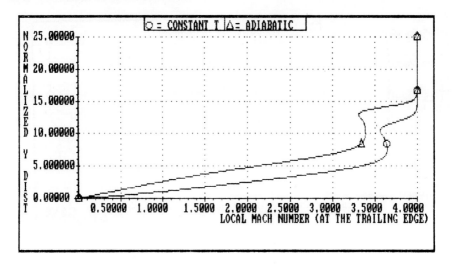

FIG. 10.15
Mach 4 at sea level: local Mach number profiles at the trailing edge.

complete Navier–Stokes equations are used. Mach number profiles are shown (Fig. 10.17) for this case as well. The leading-edge shock is well-defined. Again, the adiabatic wall case results in a considerably stronger shock wave.

10.6 SUMMARY

The primary goal of this chapter was to introduce you to a complete Navier–Stokes solution for the supersonic flow over a flat plate. A flat plate, at zero incidence, is a simple geometry, yet the solution unveils an enormous amount of interesting physics. Building on your previous work, MacCormack's explicit time-marching technique was used to march the flow field to the steady-state solution. All viscous terms were included, and the flow was allowed to vary in both the x and y directions.

Even if you do not go on to code this problem, you have benefited immensely. You certainly have an idea of the magnitude of effort behind planning and implementing a full Navier–Stokes numerical solution. And remember, this is a relatively easy problem!

Sufficient guidance is in place to ensure your success in solving this problem. If you have made the decision to, one more time, go wading: (1) be very deliberate in your approach, and (2) good luck!

SUMMARY **475**

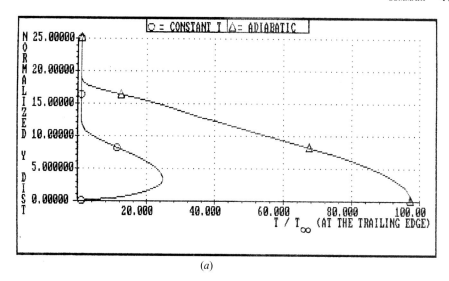

(a)

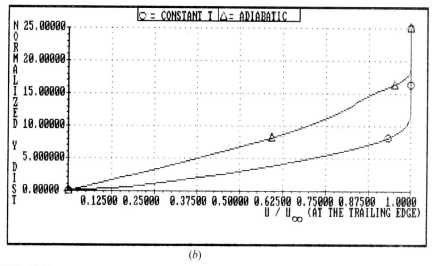

(b)

FIG. 10.16
Mach 25 at 200,000 ft. (a) Normalized temperature profiles; (b) normalized velocity profiles.

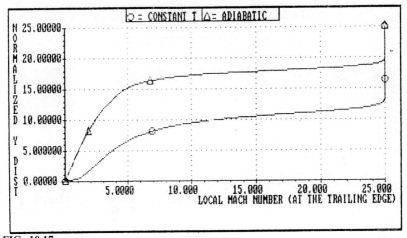

FIG. 10.17
Mach 25 at 200,000 ft: local Mach number profiles at the trailing edge.

PART IV

OTHER TOPICS

The main purpose of this book is to introduce the reader to the basic philosophy and some of the elementary concepts of CFD. In a study of CFD, this book is just a *beginning*. Indeed, it represents a launching platform from which the reader can now progress to more advanced concepts in the form of more sophisticated courses in CFD and/or the computational realities of the workplace. The purpose of Part IV is to enhance this launching process. In particular, Chap. 11 deals with some topics which are more advanced than those discussed previously in this book but which constitute the *essence* of *modern* algorithms in CFD. It is well beyond the scope of this book to present in detail such advanced topics—they await your attention in your future studies. Instead, such aspects are simply *discussed* in Chap. 11 just to give you a preview of coming attractions and to acquaint you with some of the ideas and vocabulary of the most modern CFD techniques being developed today. Finally, Chap. 12 examines the future of CFD and somewhat closes the loop of the book by extending some of the motivational ideas first discussed in Chap. 1. Chapter 12 is intended to focus your thoughts on the expanding future of CFD, with its impact on all aspects of fluid dyanmics. The application of CFD is a *growth industry*, and you are encouraged to grow with it.

CHAPTER 11

SOME ADVANCED TOPICS IN MODERN CFD: A DISCUSSION

> *Over the last twenty to thirty years considerable progress has been achieved, and the field of Computational Fluid Dynamics (CFD) is reaching a mature stage, where most of the basic methodology is, and will remain, well established.*
>
> Charles Hirsch, Professor of Fluid Mechanics,
> Vrije Universiteit Brussel, Belgium, 1990

11.1 INTRODUCTION

In Chaps. 1 through 10, you were introduced to the basic philosophy and some of the elementary concepts of CFD. To repeat some of the thoughts given in Sec. 6.1, where we began to introduce some simple CFD techniques, our purpose has been to develop some CFD tools that are not overly sophisticated—tools which can be appreciated and understood at the introductory level adopted for this book but which are utilitarian enough to allow the solution of a variety of flow problems typified by those presented in detail in Part III.

In comparison to those techniques discussed earlier, we find that the modern world of CFD is awash with relatively new and exciting algorithms which we have not highlighted in this book, until now. For the most part, these modern techniques have been the product of advanced applications of applied mathematics in order to correct some of the deficiencies of the older methods and to greatly increase the speed at which a given problem can be solved on a given computer. Because these modern algorithms are steeped in the principles of applied mathematics—much more so than the methods we have discussed so far—they are properly the subject of considerations which are more advanced than the scope of this book. On the other hand, we would be remiss not to introduce some of the ideas associated with these modern techniques, just to give you some clue as to what you may encounter in your future studies and work in CFD.

Therefore, the purpose of this chapter is to provide a window into your future exposure to CFD. We cannot go into anything like the detail we have presented in the preceding chapters; rather, the present chapter simply *discusses* some of the more modern ideas. We hope to give you some of the essential thoughts and the nomenclature—that is all. The details await you in your future studies.

Finally, we note that the modern CFD techniques of today are all a product of the basic principles which we have covered in the present book. This chapter provides a window for you to look into your future studies in CFD, but all the previous chapters provide the solid footing for you to stand on as you leap through that window.

11.2 THE CONSERVATION FORM OF THE GOVERNING FLOW EQUATIONS REVISITED: THE JACOBIANS OF THE SYSTEM

Most of the CFD state-of-the-art numerical algorithms in use today have their origins deeply embedded in the *mathematical* properties of the governing flow equations. We have touched on these mathematical properties in Chap. 3. In particular, we described in Sec. 3.3 how a system of quasi-linear partial differential equations can be described by examining the eigenvalues of the system. If the eigenvalues are all real and distinct, the equations are hyperbolic; if the eigenvalues are real and equal, the equations are parabolic; if the eigenvalues are all imaginary, the equations are elliptic. If the eigenvalues are a mixed set of the above, then the system of partial differential equations is of a mixed nature. Moreover, we gave an example at the end of Sec. 3.3 which showed that the eigenvalues themselves are the slopes of the characteristic lines; i.e., the eigenvalues themselves give the characteristic directions for the system of partial differential equations. It is important for you to review Sec. 3.3 before proceeding further, because we need to expand this line of thought to the actual governing flow equations as derived in Chap. 2.

We will concentrate on the conservation form of the governing equations; because of the prevalence of CFD applications to high-speed flows with shock waves, and the preference of most investigators to calculate such flows using the shock-capturing philosophy, which virtually mandates the use of the governing equations in conservation form (see the discussion at the end of Sec. 2.10), we find that most CFD applications today use the conservation form of the equations. This seems to be frequently the case even when shock waves are *not* part of the flow picture; it has simply become more or less a habit, and many standard finite-difference codes today are based on the conservation form of either the Euler or the Navier–Stokes equations.

Therefore, let us consider the conservation form of the governing equations represented by the generic form given by Eq. (2.93), repeated below.

$$\frac{\partial U}{\partial t} + \frac{\partial F}{\partial x} + \frac{\partial G}{\partial y} + \frac{\partial H}{\partial z} = J \qquad (2.93)$$

Recall that U, F, G, H, and J are column vectors involving the flux variables, as displayed in Eqs. (2.94) to (2.98) for the Navier–Stokes equations and Eqs. (2.105) to (2.109) for the Euler equations. Also recall that the dependent variables for the system are those contained in the solutions vector U, namely, ρ, ρu, ρv, ρw, and $\rho[e + (u^2 + v^2 + w^2)/2]$. The flux vectors F, G, and H are obviously not equal to U, but the *elements* of F, G, and H can be expressed as functions of the elements U, that is, as functions of ρ, ρu, ρv, ρw, and $\rho[e + (u^2 + v^2 + w^2)/2]$. You can easily see this by inspection of the elements of F, G, and H as displayed in Eqs. (2.106) to (2.108) for the Euler equations. Hence, we can write $F = F(U)$, $G = G(U)$, and $H = H(U)$. These are generally nonlinear functions, and for this reason the form of Eq. (2.93) shown above is *not* that of a quasi-linear equation, in the spirit described in Chap. 3. To examine the mathematical characteristics of Eq. (2.93), we must first cast it in quasi-linear form, as follows.

Since F, G, and H are functions of U, then Eq. (2.93) can be written as

$$\frac{\partial U}{\partial t} + \frac{\partial F}{\partial U}\frac{\partial U}{\partial x} + \frac{\partial G}{\partial U}\frac{\partial U}{\partial y} + \frac{\partial H}{\partial U}\frac{\partial U}{\partial z} = J \qquad (11.1)$$

In Eq. (11.1), the terms $\partial F/\partial U$, $\partial G/\partial U$, and $\partial H/\partial U$ are called the *jacobian matrices* of the flux vectors F, G, and H, respectively. As shorthand notation, we will designate the following:

$$A \equiv \frac{\partial F}{\partial U} \qquad B \equiv \frac{\partial G}{\partial U} \qquad C \equiv \frac{\partial H}{\partial U} \qquad (11.2)$$

where A, B, and C designate the respective jacobian matrices identified in Eq. (11.2). [Note that these jacobians are totally different entities than those defined in Chap. 5 associated with the inverse transformation. For example, the jacobian determinant defined by Eq. (5.22a) is the *jacobian of a given transformation*, whereas the jacobian matrices defined by Eqs. (11.2) are the *jacobians of the flux vectors*—something totally different. When you are dealing with the CFD literature, be on the lookout for these two different uses of the word "jacobian".] With the definitions given in Eqs. (11.2), Eq. (11.1) can be written as

$$\boxed{\frac{\partial U}{\partial t} + A\frac{\partial U}{\partial x} + B\frac{\partial U}{\partial y} + C\frac{\partial U}{\partial z} = J} \qquad (11.3)$$

where A, B, and C are the jacobian matrices. Recall that Eq. (11.3) represents five equations: continuity; x, y, and z components of the momentum equation; and the energy equation. Hence, U is a 1×5 column vector, and A, B, and C are 5×5 matrices. For example, looking at the elements that appear in U and F given by Eqs.

(2.105) and (2.106), respectively, the 25 elements that appear in the 5×5 A matrix are obtained by taking each one of the given elements in F and differentiating them one by one by each of the five elements in U, leading to 25 separate elements making up the A matrix. We will not take the time or space to develop the A, B, and C jacobian matrices. They are displayed in detail for the Euler equations by Hirsch in Ref. 17.

The advantage of Eq. (11.3) is that the derivatives of the dependent variables (the elements of U) appear *linearly*; hence, Eq. (11.3) is in *quasi-linear form*, similar to the form of the model equations treated in Chap. 3. As a result, following the argument given in Sec. 3.3, we can accept the fact that the *mathematical nature of Eq. (11.3) is dictated by the values of the eigenvalues of the jacobian matrices A, B, and C*. In the development of many of the modern techniques in CFD, these eigenvalues play a vital role.

11.2.1 Specialization to One-Dimensional Flow

For a general unsteady, three-dimensional flow as treated above, the development of the jacobian matrices A, B, and C, and especially their eigenvalues, is labor-intensive; we will spare you the effort. Instead, we will illustrate the above thoughts by specializing to an unsteady, one-dimensional, inviscid flow with no body forces, for which the Euler equations in conservation form, from Eqs. (2.93), (2.105), and (2.106), are (with E denoting the total energy per unit mass $= e + V^2/2$)

Continuity : $\quad \dfrac{\partial \rho}{\partial t} + \dfrac{\partial (\rho u)}{\partial x} = 0$ (11.4)

Momentum : $\quad \dfrac{\partial (\rho u)}{\partial t} + \dfrac{\partial (\rho u^2 + p)}{\partial x} = 0$ (11.5)

Energy : $\quad \dfrac{\partial (\rho E)}{\partial t} + \dfrac{\partial (\rho u E + pu)}{\partial x} = 0$ (11.6)

Equations (11.4) to (11.6), written in the form of Eq. (2.93), are

$$\dfrac{\partial U}{\partial t} + \dfrac{\partial F}{\partial x} = 0 \qquad (11.7)$$

where

$$U = \begin{Bmatrix} \rho \\ \rho u \\ \rho E \end{Bmatrix} \qquad (11.8)$$

and

$$F = \begin{Bmatrix} \rho u \\ \rho u^2 + p \\ \rho u E + pu \end{Bmatrix} \qquad (11.9)$$

To help us remember that the elements of U, namely, ρ, ρu, and ρE, are the dependent variables, let us introduce the more compact notation

$$\rho u = m \quad (11.10a)$$
$$\rho E = \varepsilon \quad (10.10b)$$

With this, the column vectors U and F defined by Eqs. (11.8) and (11.9), respectively, become

$$U = \begin{Bmatrix} \rho \\ m \\ \varepsilon \end{Bmatrix} \quad (11.11)$$

and

$$F = \begin{Bmatrix} m \\ \dfrac{m^2}{\rho} + p \\ \dfrac{m(\varepsilon + p)}{\rho} \end{Bmatrix} \quad (11.12)$$

We can eliminate p in the column vector F in favor of ρ, m, and ε, as follows. From the calorically perfect gas relations $c_v = R/(\gamma - 1)$ and $e = c_v T$, the perfect gas equation of state can be written as

$$p = \rho R T = (\gamma - 1) \frac{R}{\gamma - 1} \rho T = (\gamma - 1)\rho c_v T = (\gamma - 1)\rho e \quad (11.13)$$

From the definitions of ε and E, we have

$$\varepsilon = \rho E = \rho \left(e + \frac{u^2}{2} \right) = \rho e + \frac{\rho u^2}{2} \quad (11.14)$$

Solving Eq. (11.14) for ρe, we have

$$\rho e = \varepsilon - \frac{\rho u^2}{2} = \varepsilon - \frac{m^2}{2\rho} \quad (11.15)$$

Substituting Eq. (11.15) in (11.13), we have

$$p = (\gamma - 1)\left(\varepsilon - \frac{m^2}{2\rho} \right) \quad (11.16)$$

This expression for p is substituted into the flux column vector, Eq. (11.12), yielding

$$F = \begin{Bmatrix} m \\ \dfrac{m^2}{\rho} + (\gamma - 1)\left(\varepsilon - \dfrac{m^2}{2\rho} \right) \\ \dfrac{m}{\rho}\left[\varepsilon + (\gamma - 1)\left(\varepsilon - \dfrac{m^2}{2\rho} \right) \right] \end{Bmatrix} \quad (11.17)$$

The governing system of equations for unsteady, one-dimensional flow is now expressed by Eq. (11.7), with U and F given by Eqs. (11.11) and (11.17), respectively. Analogous to the general form of the equations given by Eq. (11.1) written in terms of the jacobian matrices, Eq. (11.7) can be expressed as

$$\frac{\partial U}{\partial t} + A \frac{\partial U}{\partial x} = 0 \tag{11.18}$$

where, for completeness, we write

$$\frac{\partial U}{\partial t} = \left\{ \begin{array}{c} \dfrac{\partial \rho}{\partial t} \\ \dfrac{\partial m}{\partial t} \\ \dfrac{\partial \varepsilon}{\partial t} \end{array} \right\} \tag{11.19}$$

and

$$\frac{\partial U}{\partial x} = \left\{ \begin{array}{c} \dfrac{\partial \rho}{\partial x} \\ \dfrac{\partial m}{\partial x} \\ \dfrac{\partial \varepsilon}{\partial x} \end{array} \right\} \tag{11.20}$$

The jacobian matrix A in Eq. (11.18) is obtained by differentiating each of the flux terms in Eq. (11.17) one by one by each of the independent variables in Eq. (11.11). That is, if we use the following shorthand notation for two of the three elements in Eq. (11.17),

$$M = \frac{m^2}{\rho} + (\gamma - 1)\left(\varepsilon - \frac{m^2}{2\rho}\right) \tag{11.21a}$$

$$N = \frac{m}{\rho}\left[\varepsilon + (\gamma - 1)\left(\varepsilon - \frac{m^2}{2\rho}\right)\right] \tag{11.21b}$$

then the jacobian matrix in Eq. (11.18) is

$$A = \begin{bmatrix} \left(\dfrac{\partial m}{\partial \rho}\right)_{m,\varepsilon} & \left(\dfrac{\partial m}{\partial m}\right)_{\rho,\varepsilon} & \left(\dfrac{\partial m}{\partial \varepsilon}\right)_{\rho,m} \\ \left(\dfrac{\partial M}{\partial \rho}\right)_{m,\varepsilon} & \left(\dfrac{\partial M}{\partial m}\right)_{\rho,\varepsilon} & \left(\dfrac{\partial M}{\partial \varepsilon}\right)_{\rho,m} \\ \left(\dfrac{\partial N}{\partial \rho}\right)_{m,\varepsilon} & \left(\dfrac{\partial N}{\partial m}\right)_{\rho,\varepsilon} & \left(\dfrac{\partial N}{\partial \varepsilon}\right)_{\rho,m} \end{bmatrix} \tag{11.22}$$

where the subscripts on the partial derivatives are added to remind you which independent variables are held constant for a given partial derivative. Each of these partial derivatives is evaluated as follows.

$$\left(\frac{\partial m}{\partial \rho}\right)_{m,\varepsilon} = 0 \tag{11.23a}$$

$$\left(\frac{\partial m}{\partial m}\right)_{\rho,\varepsilon} = 1 \tag{11.23b}$$

$$\left(\frac{\partial m}{\partial \varepsilon}\right)_{\rho,m} = 0 \tag{11.23c}$$

From Eq. (11.21a) we have

$$\left(\frac{\partial M}{\partial \rho}\right)_{m,\varepsilon} = -\frac{m^2}{\rho^2} + (\gamma - 1)\frac{m^2}{2\rho^2} = \left(\frac{\gamma}{2} - \frac{3}{2}\right)\frac{m^2}{\rho^2}$$

$$= (\gamma - 3)\frac{(\rho u)^2}{2\rho^2} = (\gamma - 3)\frac{u^2}{2} \tag{11.23d}$$

$$\left(\frac{\partial M}{\partial m}\right)_{\rho,\varepsilon} = \frac{2m}{\rho} - (\gamma - 1)\frac{m}{\rho} = -(\gamma - 3)\frac{m}{\rho}$$

$$= (3 - \gamma)\frac{\rho u}{\rho} = (3 - \gamma)u \tag{11.23e}$$

$$\left(\frac{\partial M}{\partial \varepsilon}\right)_{\rho,m} = \gamma - 1 \tag{11.23f}$$

From Eq. (11.21b) we have

$$\left(\frac{\partial N}{\partial \rho}\right)_{m,\varepsilon} = \frac{m}{\rho}\left[(\gamma - 1)\frac{m^2}{2\rho^2}\right] + \left[\varepsilon + (\gamma - 1)\left(\varepsilon - \frac{m^2}{2\rho}\right)\right]\left(-\frac{m}{\rho^2}\right)$$

$$= 2(\gamma - 1)\frac{m^3}{2\rho^3} - \gamma \varepsilon \frac{m}{\rho^2} = (\gamma - 1)u^3 - \gamma u E \tag{11.23g}$$

$$\left(\frac{\partial N}{\partial m}\right)_{\rho,\varepsilon} = \frac{m}{\rho}\left[-(\gamma - 1)\frac{m}{\rho}\right] + \left[\varepsilon + (\gamma - 1)\left(\varepsilon - \frac{m^2}{2\rho}\right)\right]\frac{1}{\rho} - (\gamma - 1)\frac{3m^2}{2\rho^2} + \gamma\frac{\varepsilon}{\rho}$$

$$= -(\gamma - 1)\frac{3(\rho u)^2}{2\rho^2} + \gamma\frac{\rho E}{\rho}$$

$$= -\tfrac{3}{2}(\gamma - 1)u^2 + \gamma E \tag{11.23h}$$

$$\left(\frac{\partial N}{\partial \varepsilon}\right)_{\rho,m} = \frac{m}{\rho} + (\gamma - 1)\frac{m}{\rho} = \gamma\frac{m}{\rho} = \gamma\frac{\rho u}{\rho} = \gamma u \tag{11.23i}$$

Equations (11.23a) to (11.23i) give the nine elements of the jacobian matrix; in light of Eq. (11.22), this matrix can now be displayed as

$$A = \begin{bmatrix} 0 & 1 & 0 \\ (\gamma-3)\dfrac{u^2}{2} & (3-\gamma)u & \gamma-1 \\ (\gamma-1)u^3 - \gamma u E & -\tfrac{3}{2}(\gamma-1)u^2 + \gamma E & \gamma u \end{bmatrix} \quad (11.24)$$

To close the loop on the above equations, let us return to the governing flow equations in the form of Eq. (11.18), where U is given by Eq. (11.8) and A by Eq. (11.24). Fully displayed in terms of all its elements, Eq. (11.18) is

$$\frac{\partial}{\partial t}\begin{Bmatrix} \rho \\ \rho u \\ \rho E \end{Bmatrix} + \begin{bmatrix} 0 & 1 & 0 \\ (\gamma-3)\dfrac{u^2}{2} & (3-\gamma)u & \gamma-1 \\ (\gamma-1)u^3 - \gamma u E & -\tfrac{3}{2}(\gamma-1)u^2 + \gamma E & \gamma u \end{bmatrix} \times \frac{\partial}{\partial x}\begin{Bmatrix} \rho \\ \rho u \\ \rho E \end{Bmatrix} = 0 \quad (11.25)$$

Using the rules for matrix multiplication, Eq. (11.25) becomes

$$\begin{Bmatrix} \dfrac{\partial \rho}{\partial t} + \dfrac{\partial(\rho u)}{\partial x} \\ \dfrac{\partial(\rho u)}{\partial t} + (\gamma-3)\dfrac{u^2}{2}\dfrac{\partial \rho}{\partial x} + (3-\gamma)u\dfrac{\partial(\rho u)}{\partial x} + (\gamma-1)\dfrac{\partial(\rho E)}{\partial x} \\ \dfrac{\partial(\rho E)}{\partial t} + [(\gamma-1)u^3 - \gamma u E]\dfrac{\partial \rho}{\partial x} + [\gamma E - \tfrac{3}{2}(\gamma-1)u^2]\dfrac{\partial(\rho u)}{\partial x} + \gamma u\dfrac{\partial(\rho E)}{\partial x} \end{Bmatrix} = 0$$

$$(11.26)$$

The expressions in Eq. (11.26) can be simplified using Eq. (11.13) and the definition $\rho E = \rho(e + u^2/2)$, that is,

$$p = (\gamma-1)\rho e = (\gamma-1)\left(\rho E - \rho\frac{u^2}{2}\right)$$

Hence,
$$\rho E = \frac{p}{\gamma-1} + \frac{\rho u^2}{2} \quad (11.27)$$

Inserting Eq. (11.27) into the expressions in Eq. (11.26) and simplifying, we obtain (the details are left to you as Prob. 11.1):

$$\begin{cases} \dfrac{\partial \rho}{\partial t} + \dfrac{\partial (\rho u)}{\partial x} \\ \dfrac{\partial (\rho u)}{\partial t} + \dfrac{\partial (\rho u^2 + p)}{\partial x} \\ \dfrac{\partial (\rho E)}{\partial t} + \dfrac{\partial (\rho u E + p u)}{\partial x} \end{cases} = 0 \qquad (11.28)$$

This column vector expression represents the following three scalar equations:

$$\frac{\partial \rho}{\partial t} + \frac{\partial (\rho u)}{\partial x} = 0 \qquad (11.29)$$

$$\frac{\partial (\rho u)}{\partial t} + \frac{\partial (\rho u^2 + p)}{\partial x} = 0 \qquad (11.30)$$

$$\frac{\partial (\rho E)}{\partial t} + \frac{\partial (\rho u E + p u)}{\partial x} = 0 \qquad (11.31)$$

Compare Eqs. (11.29) to (11.31) with the original governing equations for unsteady, one-dimensional flow given by Eqs. (11.4) to (11.6); *they are identical*, as they should be. We have just demonstrated that the writing of the governing equations in the quasi-linear form given by (Eq. 11.18) involving the jacobian matrix A, along with our explicit evaluation of the elements of A given by Eq. (11.24), is totally consistent with the original equations. By dealing with the forms of the equations in terms of the jacobian, we have lost nothing—the original equations are still preserved.

Finally, let us examine the eigenvalues of the jacobian matrix. These are found from

$$|A - \lambda I| = 0 \qquad (11.32)$$

where I is the identity matrix and λ is, *by definition*, an eigenvalue of the matrix A. The jacobian A is given by Eq. (11.24). Hence, Eq. (11.32) becomes

$$\begin{vmatrix} -\lambda & 1 & 0 \\ (\gamma - 3)\dfrac{u^2}{2} & (3-\gamma)u - \lambda & \gamma - 1 \\ (\gamma - 1)u^3 - \gamma u E & -\tfrac{3}{2}(\gamma - 1)u^2 + \gamma E & \gamma u - \lambda \end{vmatrix} = 0$$

Expanding the above determinant, we obtain

$$-\lambda \{[(3-\gamma)u - \lambda](\gamma u - \lambda) - (\gamma - 1)[-\tfrac{3}{2}(\gamma - 1)u^2 + \gamma E]\}$$
$$- \left\{ (\gamma - 3)\frac{u^2}{2}(\gamma u - \lambda) - (\gamma - 1)[(\gamma - 1)u^3 - \gamma u E] \right\} = 0 \qquad (11.33)$$

Equation (11.33) is a cubic equation in terms of the unknown λ; hence there are three solutions for λ:

$$\lambda_1 = u \qquad (11.34a)$$
$$\lambda_2 = u + c \qquad (11.34b)$$
$$\lambda_3 = u - c \qquad (11.34c)$$

where c is the speed of sound. That Eqs. (11.34a) to (11.34c) are solutions of Eq. (11.33) can be verified by substituting these answers into Eq. (11.33) and noting that Eq. (11.33) is satisfied.

The eigenvalues of the jacobian play an important role in understanding the mathematical characteristics of the governing equations. As noted in Sec. 3.3, they serve to classify the equations; in the present example, since λ_1, λ_2, and λ_3 are real and distinct, the system of governing equations for unsteady, one-dimensional inviscid flow given by Eqs. (11.4) to (11.6) are *hyperbolic*. This provides the statement originally made on faith in the subsection of Sec. 3.4.1 dealing with unsteady, inviscid flow. Moreover, the eigenvalues give the slopes of the characteristic lines in the *xt* space, as sketched in Fig. 11.1. At a given point in the *xt* plane, there are *three* characteristic lines with slopes $dt/dx = 1/\lambda_1 = 1/u$, $1/\lambda_2 = 1/(u + c)$, and $1/\lambda_3 = 1/(u - c)$, respectively. On a physical basis, the eigenvalues give the directions in which information is propagated in the physical plane. In the present example, $\lambda_1 = u$ tells us that information is carried by a fluid element moving at velocity u; in Fig. 11.1, the curve with local slope equal to $1/u$ is called a *particle path*. Also, $\lambda_2 = u + c$ and $\lambda_3 = u - c$ tell us that information is propagated to the right and left, respectively, along the x axis at the local speed of sound relative to the moving fluid element; in Fig. 11.1, the curves with slopes $1/(u + c)$ and $1/(u - c)$ are right- and left-running Mach waves. *It is important to*

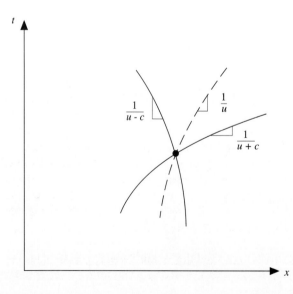

FIG. 11.1
Illustration of the characteristic lines for unsteady, one-dimensional flow.

recognize that the directions in which information travels in a flow field are given to us by the eigenvalues of the jacobian. Because many of the modern CFD techniques have differencing schemes that are associated with the direction of propagation of information in the flow, then the eigenvalues become of primary importance in the development of such schemes. Your appreciation of this fact will grow as you extend your study and application of CFD. This is the primary reason why we have chosen to discuss the jacobian and its eigenvalues at some length in this section.

11.2.2 Interim Summary

Because the jacobians and their eigenvalues play a strong role in the fundamental aspects of modern CFD algorithms, such matters have been discussed at some length in the present section. In particular, we have:

1. Introduced the form of the governing equations in terms of the jacobian matrices. The advantage of this form is that it is quasi-linear, involving the derivatives of the dependent variables appearing linearly. In this form, the mathematical characteristics of the governing flow equations are directly revealed, in the spirit discussed in Chap. 3.
2. Discussed the meaning and the structure of the jacobian matrix and illustrated the specific terms of the jacobian for unsteady, one-dimensional inviscid flow.
3. Showed that the eigenvalues of the jacobian give the direction and velocities of the propagation of information throughout the flow. These eigenvalues play a strong role in the theoretical development of many of the modern CFD techniques in use today.

11.3 ADDITIONAL CONSIDERATIONS FOR IMPLICIT METHODS

The contrast between explicit and implicit methods was first introduced in Sec. 4.4, where the one-dimensional heat conduction equation, Eq. (3.28), was used as a model equation to illustrate both the explicit and implicit approaches. With a constant value of thermal diffusivity α, Eq. (3.28) is a linear equation with one spatial dimension, namely, x; this equation can be solved by means of a marching solution where the marching variable is time t. The implicit differencing applied to this linear equation in Sec. 4.4 was the Crank–Nicolson form, given by Eq. (4.40). Using this technique, we presented a detailed implicit solution to Couette flow in Chap. 9. We emphasize that in this solution we dealt with finite-difference equations which were *linear*, and when centered about a grid point (i, j), only *three* points were needed to support the finite-difference expression. Both of these aspects are necessary for the system of algebraic finite-difference equations to be in tridiagonal form. In turn, it is the ready solution of this tridiagonal form which makes such implicit solutions practical.

What happens when the governing flow equations for a given problem are nonlinear? Also, what happens in a multidimensional problem where there is *more* than just one spatial variable in addition to the marching variable? One or both of the above situations will destroy the convenient, linear, tridiagonal system described earlier, and the computational work required for an implicit solution will increase astronomically—unless something is done about it. Fortunately, several novel ideas have been used to "do something about it" and to preserve the tridiagonal nature of the implicit solution in spite of the nonlinearity and multidimensionality of a given problem. The purpose of this section is to discuss these ideas.

11.3.1 Linearization of the Equations: The Beam and Warming Method

For simplicity, let us consider an inviscid flow, where the governing flow equations are the Euler equations itemized in Sec. 2.8.2. First, consider the Euler equations in nonconservation form, as given by Eqs. (2.82), (2.83a) to (2.83c), and (2.85). Let us choose one of these equations for display, say Eq. (2.83a); any of the others would do. Repeating Eq. (2.83a),

$$\rho \frac{Du}{Dt} = -\frac{\partial p}{\partial x} + \rho f_x \qquad (2.83a)$$

and writing out the substantial derivative, we have

$$\rho \frac{\partial u}{\partial t} + \rho u \frac{\partial u}{\partial x} + \rho v \frac{\partial u}{\partial y} + \rho w \frac{\partial u}{\partial z} = -\frac{\partial p}{\partial x} + \rho f_x \qquad (11.35)$$

Note in Eq. (11.35) that the dependent variables are the *primitive* variables and that they appear in a *linear* form *inside* the derivatives. This will be true for all the equations in nonconservation form; they can all be written such that the primitive variables appear linearly inside the derivatives. The same is true for the nonconservation form of the Navier–Stokes equations as well. In both cases, the governing equations are *nonlinear*, but with the primitive variables appearing linearly inside the derivatives, and with these derivatives multiplied by coefficients which are made up of the primitive variables (or made up of functions of the primitive variables). As a result, when an implicit numerical solution of these equations is set up, the resulting algebraic difference equations can be made linear by evaluating the coefficients using known values at the previous step; this is called the "lagging coefficients" method.

In stark contrast is the situation that results when the conservation form of the governing equations is used, which employs the flux variables as the dependent variables. For example, consider the momentum equation given by Eq. (11.5), which is in conservation form, repeated below.

$$\frac{\partial(\rho u)}{\partial t} + \frac{\partial(\rho u^2 + p)}{\partial x} = 0 \qquad (11.5)$$

Since $\rho u = m$ is one of the dependent variables, we choose to write Eq. (11.5) as

$$\frac{\partial m}{\partial t} + \frac{\partial(m^2/\rho + p)}{\partial x} = 0 \qquad (11.36)$$

An implicit finite-difference expression for Eq. (11.36) would involve terms such as $[m^2/\rho + p]_{i-1}^{n+1}$ and $[m^2/\rho + p]_{i+1}^{n+1}$ at time level $n + 1$. These terms involve the dependent variables m and ρ appearing in a nonlinear fashion, namely, m^2/ρ. This will make the practical solution of the resulting system of algebraic difference equations virtually out of the question. Therefore, something must be done in this situation to "linearize" the finite-difference equations.

A widely used method for achieving this linearization was first suggested by Beam and Warming in 1976 (Ref. 77). For purposes of discussion, let us consider the system of Euler equations given by Eqs. (11.4) to (11.6) for unsteady, one-dimensional flow and represented by Eq. (11.7), repeated below.

$$\frac{\partial U}{\partial t} + \frac{\partial F}{\partial x} = 0 \qquad (11.7)$$

where $F = F(U)$. Using the Crank–Nicolson differencing scheme (see Sec. 4.4), Eq. (11.7) can be written in finite-difference form as

$$U_i^{n+1} = U_i^n - \frac{\Delta t}{2}\left[\left(\frac{\partial F}{\partial x}\right)_i^n + \left(\frac{\partial F}{\partial x}\right)_i^{n+1}\right] \qquad (11.37)$$

(Sometimes the representation of the spatial derivatives as an *average* between time levels n and $n + 1$, as in Eq. (11.37), is called the *trapezoidal rule*.) Equation (11.37), as it stands, is a *nonlinear* difference equation. However, the Beam and Warming approach leads to a local linearization as follows. Expand F in a series expansion around time level n, that is,

$$F_i^{n+1} = F_i^n + \left(\frac{\partial F}{\partial U}\right)_i^n (U_i^{n+1} - U_i^n) + \cdots \qquad (11.38)$$

where the higher-order terms are neglected. The term $\partial F/\partial U$ is recognized as the jacobian as defined in Sec. 11.2.

$$\left(\frac{\partial F}{\partial U}\right)_i^n \equiv A_i^n \equiv \text{jacobian of } F \text{ at time level } n$$

Thus, Eq. (11.38) becomes

$$F_i^{n+1} = F_i^n + A_i^n(U_i^{n+1} - U_i^n) \qquad (11.39)$$

Substituting Eq. (11.39) for F_i^{n+1} which appears in Eq. (11.37), we have

$$U_i^{n+1} = U_i^n - \frac{\Delta t}{2}\left\{\left(\frac{\partial F}{\partial x}\right)_i^n + \frac{\partial}{\partial x}[F_i^n + A_i^n(U_i^{n+1} - U_i^n)]\right\}$$

or

$$U_i^{n+1} = U_i^n - \frac{\Delta t}{2}\left\{2\left(\frac{\partial F}{\partial x}\right)_i^n + \frac{\partial}{\partial x}[A_i^n(U_i^{n+1} - U_i^n)]\right\} \qquad (11.40)$$

Replacing the x derivatives in Eq. (11.40) with central differences, we have

$$U_i^{n+1} = U_i^n - \Delta t \left(\frac{F_{i+1}^n - F_{i-1}^n}{2\Delta x} \right) - \frac{\Delta t}{2} \left(\frac{A_{i+1}^n U_{i+1}^{n+1} - A_{i-1}^n U_{i-1}^{n+1}}{2\Delta x} \right)$$

$$+ \frac{\Delta t}{2} \left(\frac{A_{i+1}^n U_{i+1}^n - A_{i-1}^n U_{i-1}^n}{2\Delta x} \right) \quad (11.41)$$

Putting the unknowns at time level $n+1$ on the left-hand, Eq. (11.41) becomes

$$\boxed{\begin{array}{c} \dfrac{\Delta t}{4\Delta x} A_{i+1}^n U_{i+1}^{n+1} + U_i^{n+1} - \dfrac{\Delta t}{4\Delta x} A_{i-1}^n U_{i-1}^{n+1} \\ = U_i^n - \dfrac{\Delta t}{2\Delta x} (F_{i+1}^n - F_{i-1}^n) + \dfrac{\Delta t}{4\Delta x} (A_{i+1}^n U_{i+1}^n - A_{i-1}^n U_{i-1}^n) \end{array}} \quad (11.41a)$$

Note in Eq. (11.41a) that the right-hand side is a known value at time level n and the left-hand side contains three unknowns at time level $n+1$, namely, U_{i+1}^{n+1}, U_i^{n+1}, and U_{i-1}^{n+1}. Of most importance is that Eq. (11.41a) is *linear*. Moreover, it is in the familiar tridiagonal form which, for example, can be solved by means of Thomas' algorithm.

Therefore, we have achieved what we wanted. We have taken a nonlinear difference equation, namely, Eq. (11.37), and by means of a Taylor series expansion have linearized this equation, obtaining the *linear* difference equation given by Eq. (11.41a). This is one way of achieving the linearization; there are others. However, the purpose of this subsection is to emphasize that implicit finite-difference solutions of the conservation form of the governing flow equations lead to nonlinear difference equations which must in some fashion be linearized before a practical numerical solution can be obtained.

It should be noted that a similar idea for linearization was carried out by Briley and McDonald (Ref. 78). In contrast to Beam and Warming, who treated the function, Briley and McDonald treated the time derivative; the results are effectively the same.

11.3.2 The Multi-Dimensional Problem: Approximate Factorization

The second question addressed in the present section is the following: For an implicit solution of a multidimensional problem involving more than one spatial variable in addition to the marching variable, how do we arrange the finite-difference algorithm to still be of a tridiagonal nature? An illustration of the problem is given by considering unsteady, two-dimensional flow with the governing equations written in the conservation form given by Eq. (2.93), namely,

$$\frac{\partial U}{\partial t} + \frac{\partial F}{\partial x} + \frac{\partial G}{\partial y} = 0 \quad (11.42)$$

We can set up an implicit difference equation by averaging $\partial F/\partial x$ and $\partial G/\partial y$ between time levels n and $n+1$ using the trapezoidal rule; i.e., from Eq. (11.42) we have

$$U^{n+1} = U^n - \frac{\Delta t}{2}\left[\left(\frac{\partial F}{\partial x} + \frac{\partial G}{\partial y}\right)^n + \left(\frac{\partial F}{\partial x} + \frac{\partial G}{\partial y}\right)^{n+1}\right] \quad (11.43)$$

This is a nonlinear difference equation; it can be linearized using the procedure developed in Sec. 11.3.1 as follows:

$$F^{n+1} = F^n + \left(\frac{\partial F}{\partial U}\right)^n (U^{n+1} - U^n) = F^n + A^n(U^{n+1} - U^n) \quad (11.44a)$$

and

$$G^{n+1} = G^n + \left(\frac{\partial G}{\partial U}\right)^n (U^{n+1} - U^n) = G^n + B^n(U^{n+1} - U^n) \quad (11.44b)$$

where A^n and B^n are the corresponding jacobians at time level n. Substituting Eqs. (11.44a) and (11.44b) into (11.43), we have

$$U^{n+1} = U^n - \frac{\Delta t}{2}\left[\left(\frac{\partial F}{\partial x} + \frac{\partial G}{\partial y}\right)^n\right] - \frac{\Delta t}{2}\left[\left(\frac{\partial F}{\partial x}\right)^n + \frac{\partial}{\partial x}(A^n U^{n+1})\right.$$
$$\left. - \frac{\partial}{\partial x}(A^n U^n) + \left(\frac{\partial G}{\partial y}\right)^n + \frac{\partial}{\partial y}(B^n U^{n+1}) - \frac{\partial}{\partial y}(B^n U^n)\right] \quad (11.45)$$

Grouping all terms involving U^{n+1} on the left side, Eq. (11.45) becomes

$$U^{n+1} + \frac{\Delta t}{2}\frac{\partial}{\partial x}(A^n U^{n+1}) + \frac{\Delta t}{2}\frac{\partial}{\partial y}(B^n U^{n+1})$$
$$= U^n - \frac{\Delta t}{2}\left[\left(\frac{\partial F}{\partial x} + \frac{\partial G}{\partial y}\right)^n\right] - \frac{\Delta t}{2}\left(\frac{\partial F}{\partial x}\right)^n + \frac{\Delta t}{2}\frac{\partial}{\partial x}(A^n U^n)$$
$$- \frac{\Delta t}{2}\left(\frac{\partial G}{\partial y}\right)^n + \frac{\Delta t}{2}\frac{\partial}{\partial y}(B^n U^n) \quad (11.46)$$

Introducing the identity matrix I,

$$I = \begin{bmatrix} 1 & 0 & 0 & \cdots & \cdots \\ 0 & 1 & 0 & \cdots & \cdots \\ \cdots & \cdots & \cdots & \cdots & \cdots \\ 0 & 0 & 0 & \cdots & 1 \end{bmatrix}$$

Eq. (11.46) can be written as

$$\left\{I + \frac{\Delta t}{2}\left[\frac{\partial}{\partial x}(A^n) + \frac{\partial}{\partial y}(B^n)\right]\right\}U^{n+1}$$
$$= \left\{I + \frac{\Delta t}{2}\left[\frac{\partial}{\partial x}(A^n) + \frac{\partial}{\partial y}(B^n)\right]\right\}U^n - \Delta t\left[\left(\frac{\partial F}{\partial x} + \frac{\partial G}{\partial y}\right)^n\right] \quad (11.47)$$

Equation (11.47) is written in *operator form*; for example, the expression

$$\left[\frac{\partial}{\partial x}(A^n) + \frac{\partial}{\partial y}(B^n)\right]$$

is an *operator* which, when acting on U^{n+1} as on the left-hand side of Eq. (11.47), represents

$$\left[\frac{\partial}{\partial x}(A^n) + \frac{\partial}{\partial y}(B^n)\right]U^{n+1} \equiv \frac{\partial}{\partial x}(A^n U^{n+1}) + \frac{\partial}{\partial y}(B^n U^{n+1})$$

and similarly on the right-hand side of Eq. (11.47). Examining Eq. (11.47) more closely, we note that the right-hand side is a known number at time level n; all the unknowns are on the left-hand side. *Question*: How many unknowns do we have on the left-hand side? The answer, of course, depends on what type of finite-difference expression we choose to represent the derivatives. For example, if we use the familiar central difference form, then because of *both* the x and y derivatives appearing on the left-hand side, a *five*-point difference module will be needed to support the difference scheme, as sketched in Fig. 11.2. In turn, we will have *five* unknowns on the left-hand side of Eq. (11.47), namely, $U_{i-1,j}^{n+1}$, $U_{i,j}^{n+1}$, $U_{i+1,j}^{n+1}$, $U_{i,j+1}^{n+1}$, and $U_{i,j-1}^{n+1}$. Clearly, we have *lost* the tridiagonal form; in the above expression, in addition to the terms involving the three diagonals, namely, $U_{i-1,j}^{n+1}$, $U_{i,j}^{n+1}$, and $U_{i+1,j}^{n+1}$, we also have terms off the three diagonals, namely, $U_{i,j+1}^{n+1}$ and $U_{i,j-1}^{n+1}$. Indeed, these terms lead to a *pentadiagonal* matrix. The matrix manipulation associated with the solution of such a system is very computational-intensive—we have lost the tremendous computational advantage of the tridiagonal form. The reason for this problem is simply the multidimensional nature of the equations: the simultaneous appearance of both the x and y derivatives in Eq. (11.47).

A solution to this problem involves the idea of *approximate factorization*, described below. This idea has its roots in the classic alternating-direction-implicit

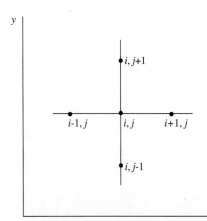

FIG. 11.2 Five-point difference module for Eq. (11.47).

(ADI) procedure developed in the middle 1950s by Peaceman, Rackford, and Douglas (Refs. 79 and 80). The ADI method is discussed in Sec. 6.7; this method essentially splits the unsteady two-dimensional problem described by Eq. (11.42) into two separate one-dimensional problems at each time step: the first stage deals with the unknowns associated with the x derivative evaluated at an intermediate time level $n + \frac{1}{2}$, namely, $U_{i-1,j}^{n+1/2}$, $U_{i,j}^{n+1/2}$, and $U_{i+1,j}^{n+1/2}$, which yields an easily solved tridiagonal form; the second stage deals with the unknowns associated with the y derivatives evaluated at time level $n + 1$, namely, $U_{i,j+1}^{n+1}$, $U_{i,j}^{n+1}$, and $U_{i,j-1}^{n+1}$, which also yields an easily solved tridiagonal form. In this process, the time marching from time level n to level $n + 1$ is achieved by *two* applications of the tridiagonal solution procedure. The concept of the ADI method is nicely described in Ref. 13, which should be consulted for more details.

The ADI philosophy described above has been extended to the solution of the governing flow equations via the Beam and Warming scheme described in Sec. 11.3.1, leading to a procedure called *approximate factorization*. In this procedure, we represent Eq. (11.47) in a somewhat "factored form," namely,

$$\left[I + \frac{\Delta t}{2}\frac{\partial}{\partial x}(A^n)\right]\left[I + \frac{\Delta t}{2}\frac{\partial}{\partial y}(B^n)\right]U^{n+1}$$
$$= \left[I + \frac{\Delta t}{2}\frac{\partial}{\partial x}(A^n)\right]\left[I + \frac{\Delta t}{2}\frac{\partial}{\partial y}(B^n)\right]U^n - \Delta t\left(\frac{\partial F}{\partial x} + \frac{\partial G}{\partial y}\right)^n \quad (11.48)$$

If you mentally carry out the multiplication of the two factors on both the left and right sides of Eq. (11.48), you will see that Eq. (11.48) is not precisely the same as Eq. (11.47); indeed, Eq. (11.48) has some extra terms, namely,

$$\frac{(\Delta t)^2}{4}\left[\frac{\partial}{\partial x}(A^n)\frac{\partial}{\partial y}(B^n)\right]U^{n+1}$$

and

$$\frac{(\Delta t)^2}{4}\left[\frac{\partial}{\partial x}(A^n)\frac{\partial}{\partial y}(B^n)\right]U^n$$

which do not appear in Eq. (11.47). On the other hand, these terms involve $(\Delta t)^2$, and they do not affect the second-order accuracy which is originally embodied by Eq. (11.47). Therefore, we can replace Eq. (11.47) with Eq. (11.48). The factored form appearing in Eq. (11.48) is called *approximate factorization* (approximate because of the aforementioned leftover terms that we simply live with).

Before we underscore the advantage of Eq. (11.48), let us introduce the notation

$$\Delta U^n \equiv U^{n+1} - U^n \quad (11.49)$$

The factors on the left and right sides of Eq. (11.48) are identical; hence, bringing the right-hand-side factors to the left-hand side, we can factor out the expression $U^{n+1} - U^n$, and using the notation of Eq. (11.49), we can write (11.48) as

$$\left[I + \frac{\Delta t}{2}\frac{\partial}{\partial x}(A^n)\right]\left[I + \frac{\Delta t}{2}\frac{\partial}{\partial y}(B^n)\right]\Delta U^n = -\Delta t\left(\frac{\partial E}{\partial x} + \frac{\partial G}{\partial y}\right)^n \quad (11.50)$$

Equation (11.50) is in delta form, so-called because the dependent variable is no longer U but rather the *change in U*, namely, ΔU. A numerical solution of Eq. (11.50) provides numbers for ΔU^n; in turn, the value of U^{n+1} at each time step is obtained from Eq. (11.49), written as

$$U^{n+1} = U^n + \Delta U^n \tag{11.51}$$

The final step in this process is to write Eq. (11.50) as

$$\left[I + \frac{\Delta t}{2}\frac{\partial}{\partial x}(A^n)\right]\overline{\Delta U} = -\Delta t\left(\frac{\partial E}{\partial x} + \frac{\partial G}{\partial y}\right)^n \tag{11.52}$$

where

$$\left[I + \frac{\Delta t}{2}\frac{\partial}{\partial y}(B^n)\right]\Delta U^n = \overline{\Delta U} \tag{11.53}$$

(It is noted that there are other possible factorizations; the above is just one example.) Equations (11.52) and (11.53) represent a two-step process for solving Eq. (11.50), as follows:

1. Solve Eq. (11.52) for $\overline{\Delta U}$. Since the spatial operator in Eq. (11.52) contains only a derivative with respect to x, and if this were replaced by a central difference, then a tridiagonal system in terms of $\overline{\Delta U}$ is obtained, which can be readily solved for $\overline{\Delta U}$.
2. Insert the above results for $\overline{\Delta U}$ into Eq. (11.53). Since the spatial operator in Eq. (11.53) contains only a derivative with respect to y, and if this were replaced by a central difference, then a tridiagonal system in terms of ΔU^n is obtained, which can be readily solved for ΔU^n.

Hence, Eqs. (11.52) and (11.53) represent a two-step process for the solution of ΔU^n, and therefore for U^{n+1} via Eq. (11.51). The underlying advantage of this process is that only a tridiagonal form is encountered at each step, hence allowing a relatively straightforward solution to the multidimensional flow.

11.3.3 Block Tridiagonal Matrices

We need to expand our concept of the tridiagonal form mentioned in the previous sections. If Eq. (11.42) represents a single equation with one unknown, then a purely tridiagonal matrix will result from an implicit scheme. This was the situation for the Couette flow calculations made in Chap. 9. On the other hand, if Eq. (11.42) represents a system of equations, such as the continuity, the three momentum, and the energy equations for fluid flow, then U is a solutions vector with five elements as given by Eq. (2.94). Each one of these equations with their particular solution element will give rise to a tridiagonal matrix, and the whole system of equations therefore leads to a large matrix with three diagonals, where each of the elements of these diagonals is itself a tridiagonal matrix associated with the particular element of U. Such a matrix is called a *block* tridiagonal matrix. This type of matrix is solved in a standard manner, albeit requiring a much lengthier algorithm and calculation than Thomas' algorithm derived in App. A. When you continue your studies and work in CFD, you most likely will have to deal with such block tridiagonal matrices. A FORTRAN subroutine for solving block tridiagonal matrices is given in App. B of Ref. 13.

11.3.4 Interim Summary

This section has discussed two problems that arise from the attempt to set up an implicit finite-difference solution of the governing flow equations. The first problem, that of dealing with nonlinear difference equations, can be handled by a local linearization process such as the Beam and Warming method discussed in Sec. 11.3.1. The second problem, that of dealing with multidimensional flows and thus suffering the apparent loss of the tridiagonal structure of the algorithm, can be handled by the approximate factorization method discussed in Sec. 11.3.2, which recaptures the tridiagonal form via a splitting of the spatial operators and which requires a two-step process at each time level, sweeping first in the x direction and then in the y direction. Also, it is important to note that Eqs. (11.52) and (11.53) are written in a somewhat generic form; the x and y derivatives are not written in any specific finite-difference form—you can choose whatever expression you want: central differences, one-sided differences, upwind differences (to be discussed in Sec. 11.4), etc.

11.4 UPWIND SCHEMES

Recall the discussion in Chap. 3 concerning the definition of characteristic curves and the emphasis in Sec. 11.2 that information concerning a flow field travels along these characteristic curves. Moreover, we have seen that the eigenvalues of the jacobian matrices give the slopes of the characteristic lines; for an unsteady flow, the values of these eigenvalues represent the velocity and direction of propagation of information. It would seem natural that a numerical scheme for solving the flow equations should be consistent with the velocity and direction with which information propagates throughout the flow field. Indeed, this is nothing more than obeying the physics of the flow.

Strictly speaking, the central difference schemes which have been highlighted throughout this book do not always follow the proper flow of information throughout the flow field. In many cases, they draw numerical information from outside the domain of dependence of a given grid point; as discussed at the end of Sec. 4.5, this can compromise the accuracy of the solution. For flow fields which involve smooth, continuous variations of the flow-field variables, this does not appear to cause a major problem. We have seen some examples where a central difference scheme works quite well: the shock-free nozzle flows in Chap. 7, the continuous expansion wave in Chap. 8, and the smoothly varying Couette flow in Chap. 9. In all these cases, a central difference scheme (such as MacCormack's scheme in Chaps. 7 and 8) works reasonably well. In fact, there is a mathematical reason for this, dealing with the analytic continuation properties of smooth functions which validates the Taylor series expansions upon which the central difference schemes are based.

On the other hand, when discontinuities exist in the flow, such as shock waves treated in the context of a shock-capturing approach, central difference schemes do not work quite so well. Witness the undesirable severe oscillations around the shock wave as shown in Fig. 7.23, which result from shock capturing with a central difference scheme with no explicit artificial viscosity. Even with the addition of artificial viscosity, the results given in Figs. 7.24 to 7.26 still show some oscillations, albeit much smaller than in Fig. 7.23.

It is this problem which has paced the development of *upwind* difference schemes in modern CFD. Upwind schemes (or simply *upwinding*) are designed to numerically simulate more properly the direction of the propagation of information in a flow field along the characteristic curves. As a result, if the upwinding is carried out in a proper fashion, the calculation of very sharp discontinuities (spread over only two grid points) with no oscillations is possible.

Perhaps the simplest illustration of the philosophy of upwind differencing can be given in conjunction with the first-order wave equation, Eq. (4.78), repeated below.

$$\frac{\partial u}{\partial t} + c \frac{\partial u}{\partial x} = 0 \qquad (4.78)$$

For a positive value of c, this equation describes the propagation of a wave in the positive direction along the x axis, as sketched in Fig. 11.3. There is a discontinuity in u across the wave, as also sketched in Fig. 11.3. On a physical basis, properties at grid point i in Fig. 11.3 should depend only on the upstream flow field, i.e., on properties at grid point $i - 1$. Grid point $i - 1$ is within the domain of dependence of point i. The properties at grid point $i + 1$ do not physically influence point i, and a proper numerical scheme should reflect this fact. However, if $\partial u/\partial x$ is replaced with a central difference, then the properties at point $i + 1$ *are made* to influence point i by virtue of the numerics. Such central differencing is shown by Eq. (4.80). However, as described in the text surrounding Eq. (4.80), the difference equation given by Eq. (4.80) leads to an unstable solution. Here, the improper propagation of

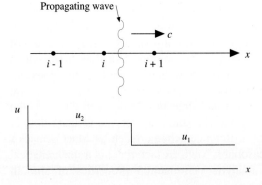

FIG. 11.3
Propagation of a wave in the positive x direction. Sketch shows the flow field at a given instant of time.

information forced by the central differencing causes the solution to blow up. In contrast, if in Eq. (4.78) a one-sided difference is used, i.e., if

$$\frac{\partial u}{\partial x} = \frac{u_i - u_{i-1}}{\Delta x} \quad (11.54)$$

then the resulting difference equation

$$\frac{u_i^{n+1} - u_i^n}{\Delta t} = -c \frac{u_i^n - u_{i-1}^n}{\Delta x} \quad (11.55)$$

is *stable*. The one-sided difference in Eq. (11.54) is an *upwind difference*—it contains points only within the domain of dependence of grid point i. Hence, Eq. (11.55) is a suitable difference equation for the original first-order wave equation, Eq. (4.78).

The use of Eq. (11.55) will result in a numerical calculation with no oscillations in the vicinity of the discontinuity. However, Eq. (11.55) has some disadvantages. It is first-order-accurate and is also highly diffusive. This means that, as a function of time, the original discontinuity at time $t = 0$ will spread out, as sketched in Fig. 11.4. Although the numerical results show a *monotone* variation (no oscillations), the diffusive property is undesirable.

To reduce or eliminate this undesirable property, while at the same time retaining the inherent advantages of an upwind scheme, some rather mathematically elegant algorithms have been developed over the past decade. These modern algorithms have introduced such terminology as total-variation-diminishing (TVD) schemes, flux splitting, flux limiters, Godunov schemes, and approximate Riemann solvers. These ideas are all broadly classified as upwind schemes since they attempt to properly account for the propagation of information throughout the flow. The mathematical rigor behind these schemes is well beyond this book; indeed, the mechanics of the schemes themselves are generally beyond our present scope—these matters are left to your more advanced studies of CFD. Instead, in the

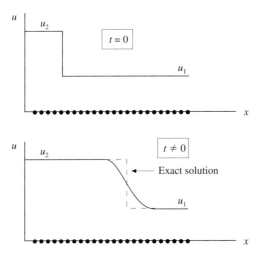

FIG. 11.4
Diffusive properties of the difference equation given by Eq. (11.55).

following subsections, we will only *discuss* the general nature of these ideas so as to familiarize you with just the essence of each. The purpose of these discussions is to ease your transition to more advanced studies.

11.4.1 Flux-Vector Splitting

To introduce the idea of flux-vector splitting, we need to examine some additional matrix properties from linear algebra. The definition of the eigenvalues λ_j of the matrix A is given by Eq. (11.32). We can further *define* the *eigenvector* associated with a specific eigenvalue λ_j as the column vector L^j which is a solution of the equation

$$[L^j]^T [A - \lambda_j I] = 0 \qquad (11.56)$$

where $[L^j]^T$ is the transverse of the column vector L^j; hence $[L^j]^T$ is a *row* vector. Since A and λ_j are known in Eq. (11.56), the elements of L^j are obtained directly by solving Eq. (11.56). For *each* different eigenvalue of the matrix A, there will be a different eigenvector L^j. To be more specific, since $[L^j]^T$ appears on the left of Eq. (11.56), then L^j is called a *left* eigenvector of the matrix A, There are as many eigenvectors as there are eigenvalues, each one defined by Eq. (11.56). Let us now define a matrix T whose inverse T^{-1} has elements that are the elements of *all* the eigenvectors. Specifically, the *j*th row of T^{-1} consists of the elements of the left eigenvector for λ_j. The matrix T has the property of "diagonalizing" the matrix A through the equation

$$T^{-1} A T = [\lambda] \qquad (11.57)$$

where $[\lambda]$ is a diagonal matrix with the eigenvalues of A as the diagonal terms. For example, if there are three eigenvalues associated with A, then

$$[\lambda] = \begin{bmatrix} \lambda_1 & 0 & 0 \\ 0 & \lambda_2 & 0 \\ 0 & 0 & \lambda_3 \end{bmatrix} \qquad (11.58)$$

We will not prove Eq. (11.57); you will have to take it on faith or appeal to a study of linear algebra for its proof. Multiplying the matrix equation given by Eq. (11.57), first by T on the left of both sides and then by T^{-1} on the right of both sides, we have

$$A = T[\lambda] T^{-1} \qquad (11.59)$$

Hence, the matrix A can be recovered by taking the matrix of eigenvalues and multiplying on the left and right by T and T^{-1}, respectively.

Independent of the above formalism, we note an interesting property of the jacobian matrix A for the Euler equations. Consider the Euler equations for unsteady, one-dimensional flow written as Eq. (11.7), repeated below.

$$\frac{\partial U}{\partial t} + \frac{\partial F}{\partial x} = 0 \qquad (11.7)$$

As described by Eq. (11.18), A is the Jacobian of F; $A = \partial F/\partial U$. For an inviscid flow, the flux vector F can be expressed directly in terms of its jacobian as

$$F = AU \qquad (11.60)$$

This relation can be proven by direct substitution of Eq. (11.24) for A and Eq. (11.8) for U into Eq. (11.60), obtaining an expression for F from Eq. (11.60) that is identical to that for F given by Eq. (11.9). (This is left for you as Prob. 11.2.)

The two lines of thought expressed in the above two paragraphs can be combined as follows. Let us define two matrices $[\lambda^+]$ and $[\lambda^-]$ made up of the positive and negative eigenvalues of A, respectively. For example, if we have a *subsonic* flow, then from Eqs. (11.34a) to (11.34c), we have $\lambda_1 = u$ and $\lambda_2 = u + c$, both positive values, and $\lambda_3 = u - c$, a negative value. Therefore, in this case, by definition

$$[\lambda^+] = \begin{bmatrix} u & 0 & 0 \\ 0 & u+c & 0 \\ 0 & 0 & 0 \end{bmatrix}$$

and

$$[\lambda^-] = \begin{bmatrix} 0 & 0 & 0 \\ 0 & 0 & 0 \\ 0 & 0 & u-c \end{bmatrix}$$

From Eq. (11.59), we can define A^+ and A^- as

$$A^+ = T[\lambda^+]T^{-1} \qquad (11.61)$$

and

$$A^- = T[\lambda^-]T^{-1} \qquad (11.62)$$

With this, we can split the flux vector F into two parts, F^+ and F^-:

$$F = F^+ + F^- \qquad (11.63)$$

where F^+ and F^- are defined from Eq. (11.60) as

$$F^+ = A^+ U \qquad (11.64)$$
$$F^- = A^- U \qquad (11.65)$$

Hence, Eq. (11.7) can now be written as

$$\boxed{\frac{\partial U}{\partial t} + \frac{\partial F^+}{\partial x} + \frac{\partial F^-}{\partial x} = 0} \qquad (11.66)$$

where F^+ and F^- are defined by Eqs. (11.64) and (11.65), respectively. Equation (11.66) is an example of *flux-vector splitting*.

In Eq. (11.66), F^+ corresponds to a flux in the positive x direction, with information being propagated from left to right by the positive eigenvalues $\lambda_1 = u$ and $\lambda_2 = u + a$. Hence, when $\partial F^+/\partial x$ is replaced by a difference expression, a backward (rearward) difference should be used since F^+ is associated only with information coming from *upstream* of grid point (i, j). Similarly, F^- corresponds to a flux in the negative x direction, with information being propagated from right to left by the negative eigenvalue $\lambda_3 = u - a$. Hence, when $\partial F^-/\partial x$ is replaced by a difference expression, a forward difference should be used since F^- is associated only with information coming from *downstream* of grid point (i, j). This is why the flux-vector-splitting scheme described by Eq. (11.66) is a type of upwind scheme; flux-vector splitting is a numerical algorithm which attempts to account for the physically proper transfer of information throughout the flow.

There are various improvisations on flux-vector splitting in the modern CFD literature. One such example is Van Leer's flux splitting which imposes certain conditions on F^+ and F^- to improve the performance of the numerical scheme for local Mach numbers near 1. The details, which are beyond the scope of this book, can be found in Ref. 17.

11.4.2 The Godunov Approach

In 1959, S. K. Godunov suggested an approach for the numerical solution of fluid flows (Ref. 81) which is *philosophically* completely different than the finite-difference solutions that we have discussed so far in this book. Instead of solving a general flow field by implementing directly a numerical solution of the Euler equations written in partial differential equation form (discretized by the finite-difference approach), Godunov suggested that *exact solutions* of the Euler equations for a *local* region of the flow be *pieced together* to synthesize the *general* flow field. Imagine that you "step into" a flow field at some *local* point; if you look around at a small region surrounding that point, you will see a localized exact solution for the flow, valid in just that local region. If you then patch together these localized exact solutions for all regions of the flow, a picture of the complete solution of the general flow field can be obtained. The operative concept here is that you are constructing a general flow field from elements that are themselves *solutions* of the Euler equations in a *local* region of the flow. To construct the general flow field, you are piecing together *local solutions* of a smaller problem, rather than visualizing a widely sweeping solution of the governing partial differential equations or integral equations over the *whole space* of the flow as we have considered in all other parts of this book.

Question: What is the exact solution of the *local region* of the flow? As strange as it may seem, the answer is related to what is called the *shock tube problem*. Therefore, before we proceed further, let us examine the shock tube problem.

THE SHOCK TUBE PROBLEM. The flow process in a shock tube is usually a subject of study in advanced courses in compressible flow. In this book, we assume that most readers are not familiar with shock tubes or their flow processes. Therefore, the purpose of this subsection is to provide a brief description of the salient aspects of shock tube flows. An extended discussion of the shock tube and its flow properties, starting from first principles, is given in Chap. 7 of Ref. 21.

A shock tube is a closed tube initially divided into a high-pressure section (the driver section with pressure p_4) and a low-pressure section (the driven section with pressure p_1), as sketched in Fig. 11.5a. A fixed diaphragm divides the high- and low-pressure sections. The pressure distribution in the tube for this case is sketched in Fig. 11.5b. There is no flow velocity anywhere; both the high- and low-pressure sections are initially at velocity $u = 0$. The situation sketched in Fig. 11.5a and b is the initial condition at time $t = 0$.

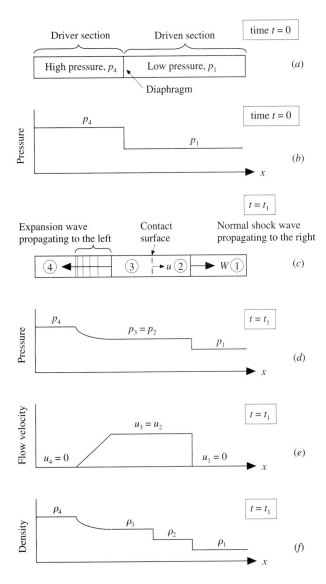

FIG. 11.5
Schematic of flow in a shock tube.

Imagine that the diaphragm is instantaneously removed. The initial pressure discontinuity will propagate to the right in the form of an unsteady normal shock wave traveling at a wave velocity W, as sketched in Fig. 11.5c. Simultaneously, an unsteady, isentropic expansion wave will propagate to the left, as also sketched in Fig. 11.5c. The gas in the tube is now divided into four regions as shown in Fig. 11.5c: region 1, which is the undisturbed portion of the driven section at pressure p_1; region 2, which has been processed by the shock wave propagating through it and which is now at pressure p_2, equal to the pressure behind the normal shock; region 3, which has been processed by the expansion wave propagating through it and which is also now at pressure $p_3 = p_2$ since the gas in regions 2 and 3 cannot support any pressure discontinuities; and region 4, which is the undisturbed portion of the driver section at pressure p_4. Regions 2 and 3 are at the same velocity and pressure; however, since region 2 is processed by a shock, and region 3 by an isentropic expansion wave, the entropy, temperature, and density in regions 2 and 3 are different. Thus, regions 2 and 3 are divided by a contact surface as sketched in Fig. 11.5c. The picture in Fig. 11.5c to f pertains to some value of time $t = t_1$, where $t_1 > 0$. The corresponding pressure distribution is sketched in Fig. 11.5d, and the flow velocity induced by the passage of the waves through the initially stagnant gas is sketched in Fig. 11.5e. Note that both p and u change discontinuously across the shock wave, but their variations are finite and continuous through the expansion wave. As the shock wave propagates to the right, it remains a discontinuity. As the expansion wave propagates to the left, it becomes wider (the expansion wave literally expands with time). The gas between the backs of the shock and expansion waves (regions 2 and 3) is set into motion toward the right by the passage of the waves, with an induced velocity equal to $u_2 = u_3$. The velocity increases discontinuously across the shock, whereas it increases continuously (indeed, linearly) across the expansion wave. Note that the density changes discontinuously across the contact surface; that is, $\rho_3 > \rho_2$, as sketched in Fig. 11.5f.

In Fig. 11.5c, the instantaneous location of the shock wave, contact surface, and expansion wave at time $t = t_1$ is shown. The paths which these waves and contact surface follow as a *function of time* are sketched in Fig. 11.6, which is called a *wave diagram*; it is sometimes called an *xt* diagram. The picture of the shock tube at time $t = 0$ is given in Fig. 11.6a, and the paths of the waves and contact surface for later time $t > 0$ are shown in the wave diagram sketched in Fig. 11.6b.

The *solution* of the flow field in the shock tube as sketched in Figs. 11.5 and 11.6 is frequently called the *Riemann problem*, named after the German mathematician G. F. Bernhard Riemann who first attempted its solution in 1858. The Riemann problem lends itself to a direct analytic solution of the unsteady, one-dimensional Euler equations, as given in detail in many compressible flow texts such as Ref. 21. The precise aspects of this exact solution are left to your future studies.

RELATION OF THE SHOCK TUBE PROBLEM TO THE GODUNOV APPROACH. Think about the nature of the numerically discretized solutions that we have discussed throughout this book. With finite-difference solutions, we have calculated the flow-field properties at discrete *points* in space. The numerical

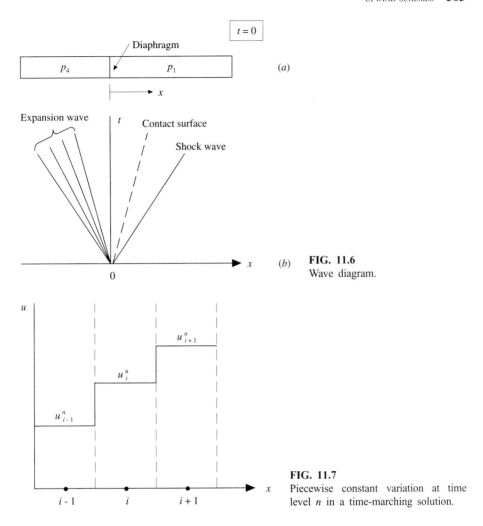

FIG. 11.6
Wave diagram.

FIG. 11.7
Piecewise constant variation at time level n in a time-marching solution.

solution is essentially a *piecewise-constant* distribution in space; i.e., the flow-field variables are treated essentially as step changes from the vicinity of one grid point to another. This variation is sketched in Fig. 11.7. Here we illustrate the piecewise distribution of the velocity u in the x direction through some arbitrary flow field; this is essentially the nature of the finite-difference and finite-volume numerical solutions that have been discussed previously. Figure 11.7 is drawn in the spirit of a certain spatial variation at time level n, within the course of a time-marching solution of the flow field.

Examine Fig. 11.7 closely. If the distribution of u shown here were to actually exist in real life, it would trigger a *series of mini-shock-tube flows*, each one of the nature described in the previous Shock Tube Problem subsection. This is sketched in Fig. 11.8, which shows some miniwave diagrams superimposed on the piecewise variation of u. For example, across the interface a, we have a weak shock wave propagating to the right into the region centered about point i. Across the interface

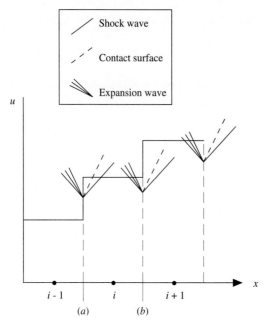

FIG. 11.8
The existence of the Riemann problem at each interface.

b, we have an expansion wave propagating to the left into the same region. Hence, the value of u_i^{n+1} at grid point i at time level $n + 1$ can be calculated as an *average* of the properties created by these waves coming from the left and right. This is illustrated in Fig. 11.9, where the new values of u_{i-1}^{n+1}, u_i^{n+1}, and u_{i+1}^{n+1} at time level $n + 1$ (solid lines) are compared with the old values at time level n (dashed lines).

Look at what is happening here! The numerical solution of the general flow field is being constructed by a *local* application of *exact solutions* of the Riemann problem (the shock tube problem), wherein the Riemann problem itself is an exact solution of the unsteady, one-dimensional Euler equations in a *local* region of the

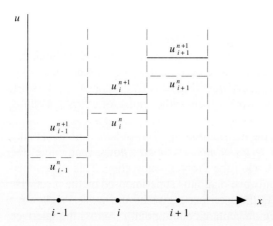

FIG. 11.9
Piecewise constant variation at time level $n + 1$.

flow. This is precisely the philosophy of the Godunov method described at the beginning of this section. And the answer to our question posed just prior to the previous subsection is now clear. What is the exact solution of the local region of the flow? *Answer*: The solution of the localized Riemann problem.

INTERIM SUMMARY. This is as far as we will take this line of reasoning. The actual implementation of the Godunov approach requires attention to details beyond the scope of this book; only the general philosophy is discussed here. See Ref. 17 for a good presentation of such details. However, it is important to note that the Godunov approach is a *type of upwind method*. By applying solutions of the Riemann problem in local regions of the general flow field, the physically proper propagation of information throughout the flow is being accounted for in the numerical solution.

We should mention that the local solution of the Riemann problem involves the solution of the Euler equations, which are *nonlinear*. Such solutions take computer time. In efforts to reduce this computer time, several investigators have suggested that *approximate* solutions to the Riemann problem be applied within the Godunov approach, where these approximate solutions are computationally more efficient. Of particular note are the *approximate Riemann solvers* developed by Stanley Osher in 1980 and by Philip Roe, also in 1980. See Ref. 17 for details.

11.4.3 General Comment

The upwind schemes discussed in this section are all *first-order* methods. The advantage of these first-order methods is that a monotone variation is achieved for the numerical flow-field properties in the vicinity of discontinuities (shock waves and contact surfaces); i.e., *no oscillations* appear in the numerical solutions around these discontinuities. This is good! However, these first-order schemes are diffusive and tend to smear out the flow-field variables, particularly in the vicinity of contact surfaces. This is bad! (Hirsch presents some excellent examples of such diffusive results in Ref. 17.) An approach to mitigate this diffusive effect is to go to the *second-order upwind schemes*. This is discussed in the next section.

11.5 SECOND-ORDER UPWIND SCHEMES

In the upwind schemes discussed in Sec. 11.4, the first-order accuracy is due to the use of *first-order* one-sided differences in the flux-vector-splitting method or to the assumption of a constant variation of flow properties across a grid cell in the Godunov approach. These constraints can be removed as follows.

In the case of one-sided differences, we can simply employ *second-order* one-sided differences. For example, Eq. (4.29) is a second-order-accurate one-sided difference. Based on this result, when written in the x direction, we have

$$\left(\frac{\partial u}{\partial x}\right)_i = \frac{-3u_i + 4u_{i+1} - u_{i+2}^n}{2\Delta x} \qquad (11.67)$$

which is appropriate for information propagated from right to left into point i, and similarly,

$$\left(\frac{\partial u}{\partial x}\right)_i = \frac{3u_i - 4u_{i-1} + u_{i-2}}{2\Delta x} \tag{11.68}$$

which is appropriate for information propagated from left to right into point i. For example, in place of Eq. (11.55) as a finite-difference representation of the first-order wave equation, we could instead write

$$\frac{u_i^{n+1} - u_i^n}{\Delta t} = -c\frac{3u_i^n - 4u_{i-1}^n + u_{i-2}^n}{2\Delta x} \tag{11.69}$$

Equation (11.69) is an example of a *second-order* upwind difference formula.

In the case of the Godunov scheme, second-order accuracy can be obtained by assuming a *linear* variation of flow properties across a given grid cell. For example, the variation originally sketched in Fig. 11.7 can be replaced with the piecewise-linear variation shown in Fig. 11.10. The local Riemann solver is then applied to this variation in an appropriate fashion.

There is a problem in regard to these second-order upwind schemes. Numerical results obtained with such schemes exhibit the oscillatory behavior in the vicinity of discontinuities similar to that encountered with second-order central difference schemes. *Hence, the disappearance of oscillations when the first-order upwind schemes discussed in Sec. 11.4 are employed is more due to the first-order accuracy than to the philosophy of upwinding.* When second-order upwinding is employed as an effort to diminish the diffusive character of the solution, then oscillations reappear. (This just reinforces the old adage that in life, nothing is simple.) However, one should never give up. Indeed, the CFD community, when faced with this situation, did not give up. Instead, the resolution of this problem has led to a class of new algorithms called *high-resolution schemes*. The idea of these schemes is discussed in the next section.

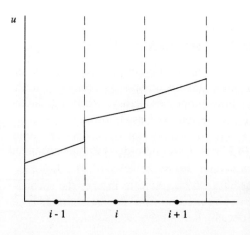

FIG. 11.10
Piecewise linear variation at time level n (second-order Godunov method.)

11.6 HIGH-RESOLUTION SCHEMES: TVD AND FLUX LIMITERS

For simplicity, consider a model equation somewhat like the Euler equations in conservation form, that is,

$$\frac{\partial u}{\partial t} + \frac{\partial f}{\partial x} = 0 \qquad (11.70)$$

where $f = f(u)$. Visualize the variation of u with x at a given time level n. At any given point along the x axis, both u and its derivative, $\partial u/\partial x$, are known at time level n. An important and rather interesting property of physical solutions governed by Eq. (11.70) is that $|\partial u/\partial x|$ integrated over the entire domain on the x axis does not increase with time. This integrated quantity is called the total variation, denoted by TV. That is,

$$\text{TV} = \int \left|\frac{\partial u}{\partial x}\right| dx \qquad (11.71)$$

Hence, for a physically proper solution, TV does not increase with time. In terms of a *numerical* solution of Eq. (11.70), where $\partial u/\partial x$ can be discretized by $(u_{i+1} - u_i)/\Delta x$, then Eq. (11.71) can be written as

$$\boxed{\text{TV}(u) \equiv \sum_i |u_{i+1} - u_i|} \qquad (11.72)$$

Indeed, Eq. (11.72) defines the *total variation in x* of a discrete numerical solution. If $\text{TV}(u^{n+1})$ and $\text{TV}(u^n)$ represent Eq. (11.72) evaluated at time level $n + 1$ and n, respectively, and if

$$\boxed{\text{TV}(u^{n+1}) \leq \text{TV}(u^n)} \qquad (11.73)$$

the numerical algorithm is said to be *total-variation-diminishing* (TVD). From the above discussion, if a numerical solution is to properly follow the physical behavior of a given flow field, then the scheme should be a TVD scheme.

A physical flow field with discontinuities, such as shock waves, does not exhibit oscillations in the vicinity of the discontinuities. On the other hand, many numerical schemes, when used to solve such flow fields, do exhibit such oscillations. These oscillations are of purely numerical origin. In light of the above discussion, any numerical scheme that gives rise to such oscillations does *not* satisfy the TVD condition. For example, the central difference schemes we have emphasized in earlier chapters are not TVD schemes. The second-order upwind schemes referred to in Sec. 11.5 are not TVD schemes. On the other hand, the first-order upwind schemes discussed in Sec. 11.4, which do not result in oscillations in the vicinity of discontinuities, can readily be shown to obey the TVD condition.

To enjoy the advantages of a second-order upwind scheme, while at the same time not generating any nonphysical oscillations, we need to modify the second-order approach such that it obeys the TVD condition. This has been the goal of many researchers in CFD over the past decade; their efforts have resulted in several modern TVD schemes with second-order (and in a few cases, higher-than-second-order) accuracy. These schemes constitute the cutting edge of CFD algorithms today. Although the details of these state-of-the-art schemes are beyond the scope of this book, they remain a very inviting area for your future studies in CFD.

In regard to the philosophy of TVD schemes, we note a distinct difference between their role and that of artificial viscosity (discussed in Sec. 6.6). When a scheme incorporates the TVD feature, numerically induced oscillations are simply *prevented* from happening; this is due to the nature in which the TVD feature is incorporated into the basic differencing procedure. This is in contrast to the role of artificial viscosity, such as discussed in Sec. 6.6. For example, central difference schemes, which do not reflect TVD behavior, result in oscillations no matter what, and the addition of artificial viscosity simply suppresses these oscillations but does not totally eliminate them. In this sense, artificial viscosity acts something like a "filter" *after* the oscillations are produced by the basic numerical scheme.

Finally, we note that one way to take a second-order scheme and make it TVD is to simply multiply selected elements of the difference equation—those elements involving the flux terms—by a nonlinear function and then find appropriate forms for these functions by forcing the difference equation to satisfy the TVD condition described by Eq. (11.73). The purpose of these nonlinear functions is to restrict the amplitude of the gradients appearing in the original second-order difference equations so as to make certain that the TVD condition holds. Since these functions are intended to limit gradients by modifying the flux terms in the difference equations, they are called, quite naturally, *flux limiters*. The use of flux limiters in modern CFD algorithms is quite widespread; you will encounter them frequently in your future studies of CFD.

11.7 SOME RESULTS

Return to the flow process in a shock tube, as illustrated in Fig. 11.5. Assuming a one-dimensional flow, this flow field can be calculated by numerically solving the unsteady, one-dimensional Euler equations using the shock-capturing philosophy. (For a calorically perfect gas, there is also a closed-form analytical solution; see Chap. 7 of Ref. 21 for a development of this exact analytical solution.) Study Fig. 11.5 carefully. Note that the flow includes a shock wave, a contact surface, and an expansion wave; hence, it makes a wonderful model problem on which to study the performance of various numerical schemes for the solution of the Euler equations.

In light of our present discussion on first- and second-order upwind difference schemes, let us examine solutions of the shock tube problem using these schemes in order to assess their various attributes. These solutions are taken from Hirsch (Ref. 17), which should be consulted for more details. Results obtained from the first-order upwind flux-vector-splitting scheme discussed in Sec. 11.4.1 are shown in Fig. 11.11a to d, which compares the numerical results (the discrete data points) with exact analytical results (the solid lines) for the pressure, density, velocity, and Mach number distributions, respectively, as a function of x at a time equal to 6.2 ms after the removal of the diaphram. These numerical results exhibit the following behavior:

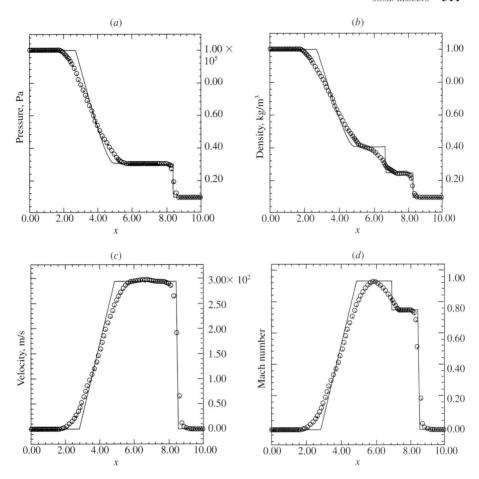

FIG. 11.11
Solution for the shock tube problem using flux-vector splitting. Solution at time = 6.2 ms. (*After Ref. 17.*)

1. There are *no* oscillations in the numerical results; in particular, the variations of the flow properties in the vicinity of the two discontinuities—the shock wave and the contact surface—are monotonic, with no oscillations. (Note that the shock wave is a discontinuity in *all* the flow properties, whereas across the contact surface pressure and velocity are unchanged, but that discontinuities in density and Mach number occur.) The oscillation-free numerical solution is a hallmark of the first-order upwind schemes, as discussed earlier.

2. There is a slight smearing of the numerical results across the shock and a substantial smearing across the contact surface. This smearing is due to the diffusive character of first-order solutions and is not a desirable quality of the scheme.

In contrast, results obtained from a second-order upwind scheme using flux limiters to achieve the TVD property are shown in Fig. 11.12a to d, which compares the numerical results (the discrete data points) with exact analytical results (the solid lines) for the pressure, density, velocity, and Mach number distributions, respectively, as a function of x at a time equal to 6.1 ms (the difference between 6.1 ms here and 6.2 ms for Fig. 11.11 is small and does not compromise the comparison of results between the two cases). These numerical results exhibit the following behavior:

1. There are no oscillations in the numerical results; the oscillations which would ordinarily be produced in the second-order solution are completely prevented by the flux limiters.
2. At the same time, the second-order scheme does not have the massive diffusive behavior exhibited by the first-order results shown in Fig. 11.11. As a result, the

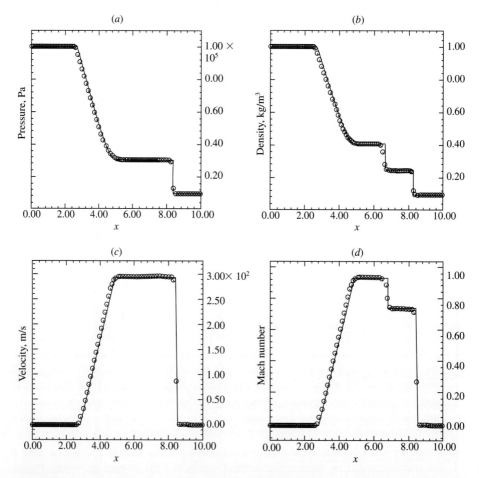

FIG. 11.12
Solution for the shock tube problem using a TVD scheme. Solution at time = 6.1 ms. (*After Ref. 17.*)

agreement between the numerical results and the exact analytical solution is excellent, as clearly seen in Fig. 11.12.

By comparing the numerical results shown in Figs. 11.11 and 11.12, the second-order TVD results shown in Fig. 11.12 are clearly superior, especially in the vicinity of the contact surface. Indeed, the results shown in Fig. 11.12 are typical of the high-resolution TVD schemes that represent the cutting edge of CFD research at the time of this writing. These results are indicative of the best that CFD has to offer today.

11.8 MULTIGRID METHOD

The vast majority of CFD techniques utilize some type of iterative or time-stepping approach that requires multiple sweeps through the flow field. The time-marching methods described in Secs. 6.2 and 6.3 and applied in Chap. 7 are one such example. The relaxation method described in Sec. 6.5 is another example. The convergence of these techniques can be greatly enhanced by the use of a technique called *multigridding*. The multigrid method has been used in many different solutions of a vast array of flow fields; it has become a fixture in some areas of modern CFD, especially for the solution of transonic flow fields.

The philosophy of the multigrid method is to carry out the early iterations on a fine grid and then to progressively transfer these results to a series of coarser grids. Since the coarser grids have fewer grid points, fewer calculations are needed for a given sweep of the flow field, and hence computer time is saved. Then the results on the coarsest grid are transfered back to the fine grid, and the process is repeated a sufficient number of times until satisfactory convergence on the fine grid is obtained.

On a mathematical basis, the advantage of the multigrid method is associated with the exhanced damping of numerical errors through the flow field. Recall the discussion of numerical errors at the beginning of Sec. 4.5 and the fact that a whole spectrum of such errors can propagate through a numerical solution of a flow field; considering a one-dimensional problem in x, the wavelength of an error can vary from the smallest value of $\lambda_{min} = 2\Delta x$ to the largest value of $\lambda_{max} = L$, where Δx is the increment in x between two grid points and L is the length of the whole domain along the x axis. Errors associated with wavelengths near λ_{min} are called *high-frequency errors*, and those associated with wavelengths near λ_{max} are called *low-frequency errors*. For a stable solution, the errors of all frequencies—high, low, and in-between—are damped during the course of the iterative or stepping process. However, in most cases the high-frequency errors are reduced much faster than the low-frequency errors. Therefore, the speed of convergence would be enhanced if something could be done to increase the damping of the low-frequency errors. Now imagine that, after you carry out a few iterations on a fine grid you transfer the intermediate results to a coarser grid. The high-frequency errors are essentially lost, or hidden, in the coarse grid, and the low-frequency errors, because of the larger Δx and hence larger $\lambda_{min} = 2\Delta x$, begin to be damped at a faster rate than would have taken place in the fine grid. Therefore, by going to progressively coarser grids, the low-frequency errors are more readily damped. Then, when the intermediate coarse grid results are transfered back to the fine grid, the low-frequency errors are smaller than they would have been for an equal number of sweeps just on the fine grid itself.

The details of just how you can transfer data properly between fine and coarse grids are beyond the scope of this book. See Ref. 16 for such details.

11.9 SUMMARY

The purpose of this chapter was to introduce you to some of the concepts and nomenclature associated with the modern CFD of today. We have identified and discussed such matters as:

Local linearization
Jacobian matrices and their eigenvalues
Approximate factorization
Upwind schemes
Flux-vector splitting
Wave-based flux methods: The Riemann approach
Total-variation-diminishing (TVD) schemes
Flux limiters
Multigrid method

As you glance over this list, if the basic ideas behind each item do not easily come to mind, then return to the appropriate sections and review our discussion.

The material in the first 10 chapters constitutes a basic foundation of concepts in CFD—an essential foundation on which to build the pillars of modern CFD. The present chapter provided some windows into the modern CFD. Our purpose in this chapter was not to give you every detail; in fact, you are not expected to run right out and implement these modern algorithms based on just the information provided in this chapter. Rather, our purpose was simply to give you some clue as to what to expect when you undertake more advanced studies and readings in CFD. The present chapter has been essentially a discussion chapter, an effort to introduce some of the modern CFD concepts with a minimum of detail. We encourage you with all vigor to pursue these more advanced concepts further.

PROBLEMS

11.1. Starting with the form given by Eq. (11.26), derive the form given by Eq. (11.28).
11.2. Verify Eq. (11.60) using the expressions given by Eqs. (11.8), (11.9), and (11.24).

CHAPTER 12

THE FUTURE OF CFD

> *We should all be concerned about the future because we will have to spend the rest of our lives there.*
>
> Charles F. Kettering, 1949

> *It is expected that the next decade will witness the emergence of CFD as the critical technology for aerodynamic design. There should be a dramatic change and shortening of the design process, which will enhance and enable concurrent engineering and the optimization of air vehicle systems in terms of overall economic performance. This will require a significant advance in CFD algorithm research and code development.*
>
> From *Aeronautical Technologies for the Twenty-First Century*, National Research Council, 1992

12.1 THE IMPORTANCE OF CFD REVISITED

Now that you have reached a certain plateau in your understanding of and appreciation for CFD, let us reiterate some of the philosophy discussed right at the beginning of our studies, namely, Sec. 1.1. (Indeed, you are encouraged to read again *all* of Chap. 1 at this stage—it will mean so much more to you now than when you first read it.) In particular, we emphasize that CFD is without a doubt a new "third dimension" in fluid dynamics, equally sharing the stage with the other dimensions of pure theory and pure experiment. Computational fluid dynamics is with us to stay; it will only grow in importance with time. Your understanding of CFD at the level presented in this book should stand you in good stead, no matter what direction you take in the future. Whether you eventually work as an experimentalist, a theoretician, a manager, or a teacher, or in whatever capacity dealing in any aspect of fluid dynamics, your life will be impacted by CFD. If you choose to proceed further with your studies of CFD, and become a CFD specialist, the material in this book simply becomes a first steppingstone for you. In any event, this author feels strongly that all your efforts to learn CFD from this book will serve you well in your present or future professional career. The importance and massive proliferation of CFD virtually ensures the validity of this feeling.

In the remaining sections of this chapter, we will reflect on the future of CFD. In a sense, this chapter is simply a continuation of Chap. 1, fleshed out by the material contained in the chapters in between.

12.2 COMPUTER GRAPHICS IN CFD

We interject a somewhat parenthetical thought at this stage, but an important one nonetheless. The types of flow fields calculated and discussed in Chaps. 7 to 10 are either one- or two-dimensional flows. Hence, the amount of data collected during the calculations is reasonably moderate, and the graphical and tabular displays of this data are relatively straightforward. However, the story is different for three-dimensional flows; the addition of the third dimension increases the amount of data by orders of magnitude, and the proper graphical displays of this data require much thought and effort (tabular displays of three-dimensional data are totally impractical). This situation has driven much research and development in the discipline of *computer graphics*, the art of displaying quantitative data on a two-dimensional computer screen in a clear and meaningful fashion. Computer graphics is a subject by itself; whole books have been written on it. This subject is discussed in Sec. 6.9. However, we remind you of the importance of good computer graphics to the effective practice of CFD. As you progress further in your studies and work in CFD, you will quickly come to appreciate the value of good graphics packages (software) when you want to examine your CFD data. As you read through the remainder of this chapter, note the various styles in which data are presented in the figures; these are examples of modern computer graphics mated with CFD. Of particular note are *contour plots*, so frequently used in presenting CFD results. Keep in mind that contour plots are simply lines of constant properties drawn in two- or three-dimensional space; pressure contours are lines of constant pressure, density contours are lines of constant density, etc. Regions in which contour lines are bunched together are simply regions where the flow property is rapidly changing; i.e., dark regions in a contour plot are regions of *high gradients* in the flow. Hence, in addition to displaying quantitative data, contour plots are wonderful flow-field visualization pictures. You will see a lot of various contour plots in this chapter.

Finally, we note that modern CFD is a great user of *color graphics*, the displaying of different magnitudes of flow-field properties by different colors. In color contour plots, the contour lines are replaced by a continuous changing in color shades so that the entire flow-field picture becomes a continuous "painting." Some of these color graphic results can be absolutely spectacular—literally works of art.

12.3 THE FUTURE OF CFD: ENHANCING THE DESIGN PROCESS

Computational fluid dynamics has already had a major impact on airplane design, and a recent prediction from the National Research Council calls for CFD to become the *critical technology* for aerodynamic design over the next decade (see the second quotation at the beginning of this chapter). There is no doubt that a major focus of CFD is to enhance the design process for *any* machine that deals with fluid flow. The design role played by CFD was discussed in Sec. 1.3, which should be reviewed before progressing further. The purpose of the present section is to elaborate further on the design matters set forth in Sec. 1.3.

Today, CFD is used to calculate complete three-dimensional flow fields over real airplanes. An excellent example is illustrated in Figs. 1.6 and 1.7, where the flow over a Northrop F-20 is shown as calculated from a solution of the unsteady, three-dimensional Euler equations by means of an explicit finite-volume scheme. Such complete flow-field calculations over entire airplane configurations is a major step in enhancing the overall airplane design process. In this fashion, the amount of experimental wind tunnel testing required for the development of a new airplane is greatly reduced; the burden of "testing" various design options and parameters is instead shouldered by CFD.

The calculations shown in Figs. 1.6 and 1.7, albeit for a complete airplane, are for an inviscid flow (they are from an Euler solution). The next major step forward is the complete solution of an entire airplane flow field using the Navier–Stokes equations, i.e., a fully viscous flow solution. Such solutions have been carried out. Historically, the first complete Navier–Stokes solution for a complete airplane configuration was carried out by Shang and Scherr in 1986 (see Ref. 47). The airplane was the X-24C hypersonic test vehicle shown in Fig. 12.1. As a sample of the results obtained, the calculated surface streamlines are shown in Fig. 12.2; here, only half the airplane is shown since the other half is symmetrical. The calculations were made using the time-marching explicit MacCormack finite-difference technique as described in Sec. 6.3 and as utilized throughout Chap. 7. An elliptically generated grid was used, as described in Sec. 5.7; over 500,000 grid points were employed in the calculation. The pioneering aspect of this calculation cannot be emphasized enough; it achieved a major goal sought by the CFD community—a complete Navier–Stokes solution of an entire airplane flow field. Today, a number of such calculations exist, but the results of Shang and Scherr were the first. (This author takes pride in having had Joe Shang as a valued classmate while we were both graduate students at The Ohio State University.)

A recent example of a complete airplane Navier–Stokes solution is that by Schroder and Mergler from Messerschmett-Bolkow-Blohm in Germany (see Ref. 48). This is indeed a "double plane calculation" in the sense that a multivehicle configuration was used based on the Sanger concept in Germany. Called generically

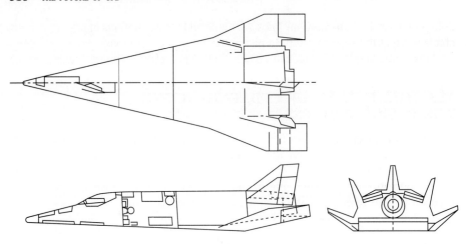

FIG. 12.1
Three-dimensional view of the X-24C hypersonic test vehicle.

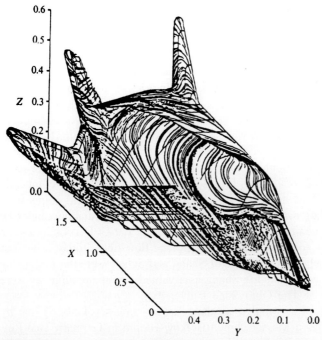

FIG. 12.2
Computed surface streamlines over the X-24C. (*After Ref. 47.*)

a *space-transportation system* (STS), a three-dimensional graphical rendition of this configuration is shown in Fig. 12.3. Here you see a large first-stage carrier vehicle, with a smaller second stage which is intended to go into orbit around the earth; the second stage is mounted above the first stage. This concept for a two-stage aerodynamic lifting vehicle designed to go into orbit was first advanced in 1929 by Eugen Sanger, an Austrian engineer who pursued the idea for more than a decade until World War II broke out. The idea has been revived in Germany in recent years. Recent CFD calculations of the hypersonic flow field over the configuration shown in Fig. 12.3 are illustrated in Fig. 12.4, taken from Ref. 48. Both inviscid (Euler solution) and viscous (Navier–Stokes solution) cases are compared in Fig. 12.4 for a freestream Mach number of 6. These calculations are made with a high-resolution second-order accurate TVD scheme using Roe averaging as discussed in Sec. 11.6. At the left of Fig. 12.4 is a side view of the STS, where the upper-stage vehicle is inclined at three different angles of attack relative to the lower stage vehicle, namely, $\Delta\alpha = 0$, 2, and 4°, respectively, top to bottom. The angle of attack of the lower stage relative to the freestream is zero. Figure 12.4 gives density contours in the flow field. The side view in Fig. 12.4*a* shows the shock wave pattern on the two stages, and illustrates how the bow shock from the lower stage impinges on the nose of the upper stage for $\Delta\alpha = 2$ and 4° but passes above the upper stage where $\Delta\alpha = 0°$. Moreover, the reflecting and interacting shock pattern in the gap between the two stages is clearly evident. The results shown in Fig. 12.4*a* are for inviscid flow; viscous flow results are also presented in Ref. 48 and show very little difference in the shock pattern from that shown in Fig. 12.4*a*. This is because the Reynolds number for these calculations is quite high, namely, Re = 2.98×10^7 based on a total vehicle length of 71.1 m. Now imagine that you take a plane perpendicular to the page of Fig. 12.4*a* and cut the flow field at a location $x = 68.42$ m downstream from the nose of the lower stage. The density contours you see in this perpendicular

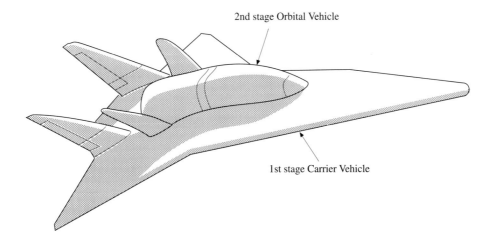

FIG. 12.3
A generic two-stage Space Transportation System.

520 THE FUTURE OF CFD

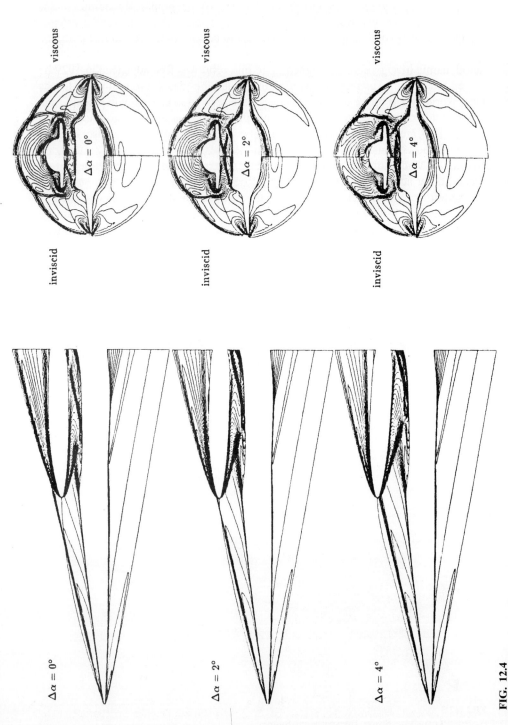

FIG. 12.4
Density contours for the solution of the flow field around the two-stage STS. Freestream Mach number is 6. (*a*) Side view. Flow is from left to right. (*b*) View in a cross-flow plane; comparison between inviscid and viscous results. (*From Ref. 48. Reprinted with permission from Elsevier Science*

plane, called a *cross-flow plane*, are shown in Fig. 12.4b for the same three cases of $\Delta\alpha = 0$, 2, and 4°. Also compared in Fig. 12.4b are the viscous results (right half of each picture) and the inviscid results (left half of each picture). The major difference between the inviscid and viscous results is the appearance of the viscous boundary layer on the body surface, as clearly seen in the right half of these pictures (the dark regions adjacent to the body surface). The type of CFD results illustrated in Fig. 12.4 are vital to the proper design of the mating of the two stages, and they are simply another example of the role of CFD in the overall design process.

Another example of three-dimensional flow-field calculations is that carried out by Turkel et al. in Ref. 49. Using an explicit Runge-Kutta scheme in combination with the multigrid technique (as discussed in Sec. 11.8), the three-dimensional Navier–Stokes equations were solved for the flow field over a blunt-nosed biconic shape at an angle of attack, as shown in Fig. 12.5. The freestream Mach number is 6, and the Reynolds number referred to the base diameter is 2.89×10^5. Figure 12.5 is chosen for display because it represents a good example of computer graphics (see Sec. 12.2) as applied to CFD results. In Fig. 12.5 we see a three-dimensional perspective of the body shape with an overlay of constant-pressure contour lines in the flow field. Moreover, we also see in perspective two planes perpendicular to the body axis on which are drawn pressure contour lines. In this fashion, the pressure variation in three-dimensional space is clearly seen, although the figure is simply a picture in the two-dimensional plane of the page. In particular, the three-dimensional shape of the bow shock wave is clearly evident. Moreover, the calculations presented in Ref. 49, with Fig. 12.5 being just a sample of the results, is another example of modern CFD being applied to Navier–Stokes solutions over three-dimensional bodies.

The airplane design process is aided not only by using CFD to calculate the flow field about a complete airplane configuration but also by concentrating on smaller elements of an airplane. For example, consider the two-dimensional

FIG. 12.5
Three-dimensional view of pressure contours over a blunt-nosed biconic configuration. Mach number = 6. Angle of attack is 5°. (*From Ref. 49. Copyright © 1991 AIAA. Reprinted with permission.*)

compressible flow over an airfoil with a flap, as shown in Fig. 12.6, taken from Ref. 50. The results shown in Fig. 12.6 are from a Navier–Stokes solution by Vilsmeier and Hanel, employing an unstructured grid (such as discussed in Sec. 5.10) and using a finite-volume algorithm with Runge-Kutta time stepping. The freestream Mach number is 0.3. The unstructured grid is shown in Fig. 12.6a, and the Mach number contours are shown in Fig. 12.6b. Note the upward flow through the gap between the main wing and the flap; also note the vortices which are formed at the trailing edge of the flap. Details of the grid and the Mach number contours in the vicinity of the gap are shown in Fig. 12.6c and d, respectively. These calculations are for a low Reynolds number of 10^4, which puts them in the same category as the low Reynolds number airfoil Navier–Stokes calculations by Kothari and Anderson as described in Sec. 1.2. The usefulness of the detailed application of CFD over an element of an airplane, such as the flapped airfoil in Fig. 12.6, is that it can show flow imperfections in a localized region, which can then be sometimes corrected by proper modifications of the design. For example, Fig. 12.6b clearly shows flow separation from the top and bottom surface ahead of the midchord location of the airfoil. Moreover, Vilsmeier and Hanel make reference that "an *almost* steady flow establishes," indicating some degree of flow unsteadiness in the calculation. These phenomena are associated with the physical aspects of low Reynolds number laminar flow over an airfoil—they are very similar to the results obtained by Kothari and Anderson as discussed in Sec. 1.2 and described in detail in Ref. 6.

Another application of CFD to a local element of an airplane is shown in Fig. 12.7, which gives the pressure contours in the vicinity of the engine-pylon-wing region of the McDonnell Douglas Tri-Jet transport aircraft, from the calculations of Vassberg and Dailey (Ref. 51). An unstructured grid was used for the calculations. The mutual aerodynamic interaction between the engine nacelle, the pylon attaching it to the wing, and the wing itself is an important consideration in airplane design. The application of CFD to this geometry, as shown in Fig. 12.7, is an invaluable aid to the design of the nacelle-pylon-wing configuration.

Figure 12.7 shows an application of CFD to the *outside* of a jet engine. In contrast, Fig. 12.8 is an application of CFD to the *inside* of a jet engine. Indeed, the use of CFD in the calculation of internal flows through compressors, burners, and turbine blades is not as mature as that for external flows over airplane components and is just now receiving serious attention by the world's aircraft engine manufacturers. Great strides are being made in this regard, and the three largest engine makers—Pratt and Whitney, General Electric, and Rolls-Royce—have very active CFD groups. The application of CFD to turbomachinery flows is particularly challenging; such flows are inherently unsteady, and viscous effects are particularly important. As an illustration of such turbomachinery flows, Fig. 12.8 shows the Mach number contours around two adjacent turbine blades. These calculations were made by Petot and Fourmaux (Ref. 52) and involved a solution of the Euler equations using a finite-volume scheme with explicit Lax–Wendroff time marching (the Lax–Wendroff method is described in Sec. 6.2). The fishtail shock wave pattern at the trailing edge of each blade in Fig. 12.8 is a standard feature of such flows and is readily captured by the numerical technique.

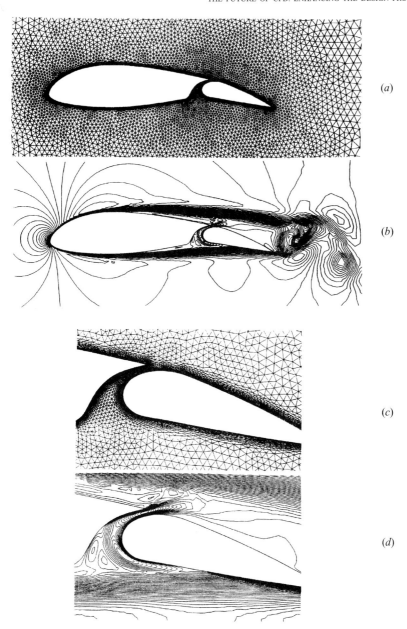

FIG. 12.6
Navier–Stokes solution for the flow over an airfoil with a flap. Re = 10^4; M_∞ = 0.3. (*a*) Unstructured grid. (*b*) Mach number contours. (*c*) Details of grid in the vicinity of the flap-airfoil junction. (*d*) Mach number contours for the region shown in part (*c*). (*From Ref. 50. Reprinted with permission from Elsevier Science Publishers, Amsterdam.*)

FIG. 12.7
Pressure contours in the nacelle-pylon-wing region on the MDC tri-jet.

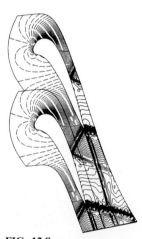

FIG. 12.8
Mach number contours around two adjacent turbine blades in an axial flow jet engine. (*From Ref. 52. Reprinted with permission from Elsevier Science Publishers, Amsterdam.*)

Let us consider another design application of CFD, namely, the *design of experiments and experimental apparatus*. An example is Fig. 12.9, which shows the configuration of a model supersonic engine inlet connected to a supersonic wind tunnel nozzle. In Fig. 12.9, flow is from left to right. The results in Fig. 12.9 were obtained by Enomoto and Arakawa (Ref. 53) using an implicit Beam-Warming scheme with approximate factorization (see Sec. 11.3) for the solution of the three-

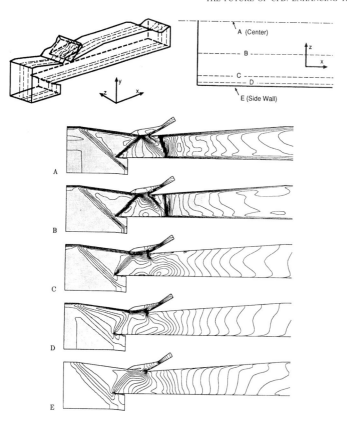

FIG. 12.9
Density contours in a model supersonic engine inlet connected to a supersonic wind tunnel nozzle (the nozzle is out of sight to the left). (*From Ref. 53. Reprinted with permission from Elsevier Science Publishers, Amsterdam.*)

dimensional Navier–Stokes equations. Sections A, B, C, D, and E in Fig. 12.9 correspond to the laterally spaced planes sketched at the top of the figure. The density contours shown in these five planes are different and hence clearly show a three-dimensional effect of the flow in different lateral planes along the test facility. The mainstream Mach number at the entrance to the test section (the extreme left of the sections A–E) is 1.85. The wind tunnel nozzle is not shown in Fig. 12.9; only the test section with the model of the supersonic inlet is shown. The results obtained from this type of CFD application can be used to help design the physical test apparatus itself, to help establish proper running conditions for the test facility, and to help interpret the data when the experiments are run.

We end this section with an example of how CFD is being coupled with other disciplines to enhance the design process on a broader base. When CFD is used to predict the pressure and shear stress distribution over an airplane wing, and thus to predict the aerodynamic loading on the wing, the actual design of the wing does not stop there. The wing is a physical structure which can bend and flap back and forth under the influence of the aerodynamic loading. As the shape of the wing is distorted, the aerodynamic flow is affected, and the aerodynamic load changes. Hence, there is a mutual coupling and feedback mechanism between the structural and aerodynamic behavior of the wing—this is the essence of the discipline of *aeroelasticity*. An aspect of the modern CFD is its application to such problems which are coupled with other disciplines, i.e., multidisciplinary applications. An example of the aeroelastic wing problem described above is shown in Fig. 12.10, obtained from Ref. 54. The double image of a wing in Fig. 12.10 shows the wing's deflection away from its unloaded position when placed under an aerodynamic load. The picture shown in Fig. 12.10 is in reality only an intermediate result during the course of an iterative solution; it involves the application of a CFD calculation, then a structural analysis application, then a repeated CFD calculation, then a repeated structural analysis application, and so forth, until a converged solution is obtained. The whole design process is integrated with a computer-aided design (CAD) software package.

In summary, the purpose of this section has been to illustrate the modern use of CFD in the design process and to use this to indicate the wide-open future of CFD in design. No matter how mature the techniques of CFD may become, the array of future and challenging *applications* of CFD is limitless. *Applied* CFD is clearly a growth industry.

12.4 THE FUTURE OF CFD: ENHANCING UNDERSTANDING

A major role of CFD is that of a research tool, a tool to enhance our understanding of the basic physical nature of fluid dynamics. This perspective is discussed in Sec. 1.2 where the role of CFD in carrying out *numerical experiments* is emphasized. In the present section, we will elaborate on this aspect of CFD—enhancing our understanding.

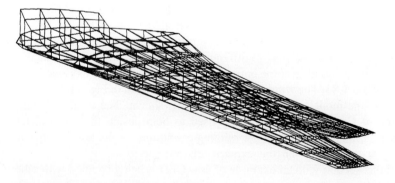

FIG. 12.10
Deflected wing shape due to aerodynamic loads. A multidisciplinary calculation combining CFD with a structural analysis code. (*From Ref. 54. Reprinted with kind permission from Elsevier Science Ltd., The Boulevard, Langford Lane, Kidlington OX5 1GB, UK.*)

For example, consider the flow through a convergent-divergent nozzle, where the pressure ratio across the nozzle is large enough to produce a region of supersonic flow downstream of the throat but small enough that shock waves appear somewhere in the divergent section—the case of the overexpanded nozzle flow as described in Sec. 7.6. Indeed, a calculation was made in Sec. 7.6 wherein a standing normal shock wave exists in the divergent section. A qualitative sketch of this flow field is given in Fig. 7.21, which shows a straight, normal shock wave reaching from top to bottom across the nozzle. However, this picture is consistent only with our assumption of inviscid, quasi-one-dimensional flow, which was the case treated throughout Chap. 7.

In reality, the real flow through a convergent-divergent nozzle is multidimensional, and for the overexpanded case viscous effects can be important. Let us expand our understanding of such nozzle flows by again using CFD, but this time assuming a two-dimensional viscous flow inside the nozzle. An example of such a calculation is shown in Fig. 12.11, taken from Ref. 55. Here we see Mach number contours from a Navier–Stokes solution calculated by means of a finite-volume

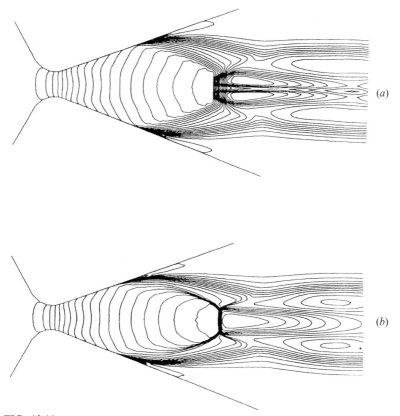

FIG. 12.11
Two-dimensional viscous flow in an overexpanded supersonic nozzle. Mach number contours. Conditions of the flow: $p_0/p_e = 50$, $A_e/A^* = 7$, $\gamma = 1.2$, $T_0 = 1800$ K. (*From Ref. 55. Reprinted with permission from Elsevier Science Publishers, Amsterdam.*)

scheme using second-order upwinding but which is reduced to first-order upwinding in the vicinity of the shock waves. Two solutions are shown in Fig. 12.11; the results in Fig. 12.11a are for a fixed mesh, whereas the results in Fig. 12.11b are for an adaptive mesh (adaptive grids are described in Sec. 5.8). Examining Fig. 12.11, clearly the adaptive mesh results in a sharper definition of the flow structure. From this point of view, Fig. 12.11 simply augments our discussion in Sec. 5.8. However, for the purpose of the present section, we are interested in the physical flow structure shown in Fig. 12.11. It is completely different from that resulting from the simple quasi-one-dimensional flow assumption shown in Fig. 7.21. The flow shown in Fig. 12.11 is the real case that will actually occur in nature for the conditions listed in the figure caption. It is characterized by flow separation from the walls of the nozzle, curved oblique shocks that transit into a normal shock in the middle of the flow (a region called the *Mach disk*), and a pocket of subsonic flow downstream of the disk. The resulting flow that leaves the exit of the nozzle is simply a supersonic jet full of wave structure, where the diameter of the jet is much smaller than the exit of the nozzle. This is quite a complex flow. When you add questions as to the effect of turbulent flow versus laminar flow, the complexity is further increased. (The calculations shown in Fig. 12.11 are for a turbulent flow using a two-equation turbulence model.) Figure 12.11 is simply an example of how numerical experiments run with CFD can be used to enhance our understanding of the basic nature of flow fields.

Another interesting flow is that associated with a vortex which passes through a shock wave, and the question as to whether the shock wave will cause breakdown of the vortex downstream of the shock. Some CFD calculations of this flow field are shown in Fig. 12.12, taken from Ref. 56. Shown in Fig. 12.12 are streamlines in a flow which is moving from left to right through a cylindrical duct, where the flow has a swirling component in the plane perpendicular to the page. Hence, this swirling, cylindrical flow field simulates a vortex. Immediately to the left of the duct is a shock wave which spans the inlet to the duct; this shock is not seen in Fig. 12.12. The shock wave is bending and pulsing with time, and each frame shown in Fig. 12.12 is a "snapshot" of the flow at various times, with time increasing from top to bottom. The calculations were made by Kandil, Kandil, and Liu (Ref. 56) wherein the full compressible Navier–Stokes equations are solved using an implicit, upwind, finite-volume scheme based on the approximate Riemann solver of Roe (see the discussion in the Interim Summary subsection of Sec. 11.4.2). The interaction of the shock wave at the inlet with the swirling flow (the vortex) creates a vortex breakdown bubble which can be seen in the snapshot labeled $t = 3$. This bubble subsequently splits into multiple bubbles as it is convected downstream (snapshot $t = 8$). New bubbles are formed behind the shock, with the same type of behavior as they flow downstream (snapshots $t = 10$ to 36). Finally, the shock wave at the inlet becomes steady, and no new bubbles are formed (snapshot $t = 45$). These results once again show how CFD can be used to enhance our basic understanding of the physical nature of flows. See Ref. 56 for a detailed discussion of this very interesting problem.

One of the greatest unsolved problems in fluid dynamics, indeed in all of classical physics, is the understanding and prediction of turbulence. Here is where

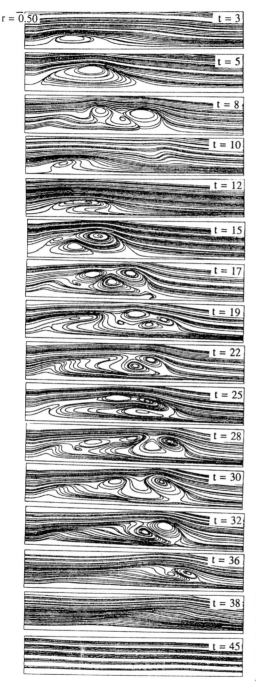

FIG. 12.12
Streamlines for a swirling flow that has passed through a shock wave (which is out of sight to the left). Flow is from left to right. Results show multibubble breakdown (vortex breakdown). (*From Ref. 56. Copyright © 1991 AIAA. Reprinted with permission.*)

CFD can make, perhaps, its greatest contribution in the future to our understanding of fluid dynamics. This hope rests upon the idea that turbulence, with all its complicated large- and small-scale structure, is nothing other than a viscous flow that locally obeys the Navier–Stokes equations, and that if a grid fine enough is used, all the details of this turbulent flow can be calculated *directly* from the Navier–Stokes equations with no artificial "modeling" of the effects of turbulence. This class of CFD calculation is called *direct numerical simulation* (DNS) of turbulent flows. An excellent recent example of DNS calculations can be found in Ref. 57. Here, Rai and Moin solve the three-dimensional Navier–Stokes equations for the flow over a flat plate using an upwind-biased finite-difference scheme. An exceptionally fine grid was used for the calculations in order to resolve the smallest scale of the turbulence structure. Some results are shown in Fig. 12.13. In this figure, we are looking at a side view of a flat plate, where the surface of the plate is the bottom horizontal line in each picture. Flow is from left to right, with a freestream Mach number of 0.1. The axial location along the plate, instead of being given in terms of distance x, is quoted in terms of the local Reynolds number, $Re_x = \rho_\infty V_\infty x / \mu_\infty$. Shown in Fig. 12.13 are the contours of local vorticity. (Recall that local vorticity, by definition, is equal to $\nabla \times \mathbf{V}$, and what is given in Fig. 12.13 are contours of the *component* of vorticity perpendicular to the page.) Each segment of Fig. 12.13 (parts *a–d*) corresponds to a later time. The Reynolds number range shown here corresponds to the region of the flat plate where *transition* from laminar to turbulent flow is taking place. These plots illustrate the rather random, transient nature of the flow process. Another perspective of this same flow is shown in Fig. 12.14. Here we are looking down on the *top* of the plate. Flow is still from left to right. Contours of the same component of vorticity are plotted in the figure, except that we are looking down from the top. In actuality, the pictures shown in Fig. 12.14 are in a plane parallel to the surface of the flat plate but elevated a small distance above the surface. Hence, the contours shown in Fig. 12.14 are in the flow and not right at the surface of the plate. Moreover, the data shown in Fig. 12.14 are all for the same time; they are simply plotted for regions of the flat plate that progressively are located downstream. For example, in Fig. 12.14*a*, the flow is still mainly of a laminar nature, with only isolated patches of vorticity. Further downstream, Fig. 12.14*b* shows the transition process, where the flow at the right (the downstream side) of Fig. 12.14*b* is essentially all turbulent. Finally, Fig. 12.14*c* applies even further downstream, where the flow is clearly fully turbulent. Note a very important physical property of turbulence shown in Fig. 12.14; although the laminar viscous flow over a flat plate is theoretically two-dimensional (properties vary only in the directions along the flow and also perpendicular to the plate surface), turbulence is clearly *three-dimensional*, no matter what the geometry of the body and the external flow. Note particularly the spanwise variations of vorticity across the plate shown in Fig. 12.14*a* and *b*, in spite of the flat surface geometry and a uniform freestream above the plate.

The above results, wherein the transition from laminar to turbulent flow over a flat plate is calculated directly from a solution of the Navier–Stokes equations, look encouraging. But here is the problem: The number of grid points used by Rai and Moin for these calculations was 16,975,196, and the amount of computer execution

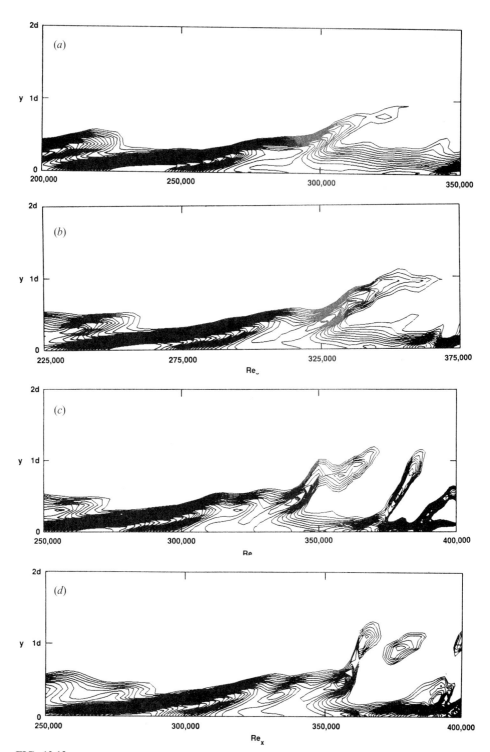

FIG. 12.13
Direct numerical simulation (DNS) of turbulent flow. Vorticity contours at four different times during the calculation of flow over a flat plate: (*a*) $t = 20.5$; (*b*) $t = 41$; (*c*) $t = 51.25$; (*d*) $t = 61.5$. Time is nondimensionalized by δ^*/V_∞, where δ^* is the displacement thickness of the boundary layer. Side view of the flat plate flow. (*From Ref. 57. Copyright © 1991 AIAA. Reprinted with permission.*)

(a)

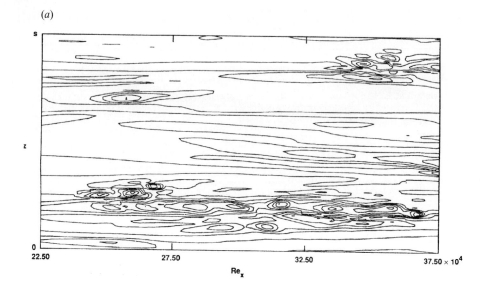

(b)

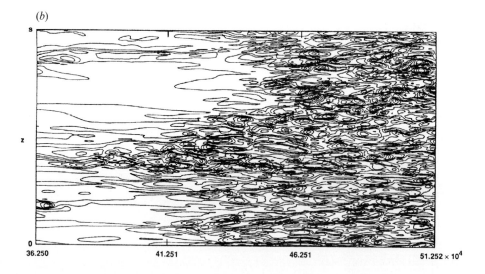

time was 400 hours on a CRAY-YMP! Clearly, these astronomical computer requirements currently prevent DNS calculations from becoming a practical technique for practical configurations. Clearly, the future is ripe for CFD breakthroughs on this problem.

12.5 CONCLUSION

On the above note, we end our discussion of the future of CFD. We repeat that CFD is a *growth industry*, with an unlimited number of new applications and new ideas just waiting in the future.

(c)

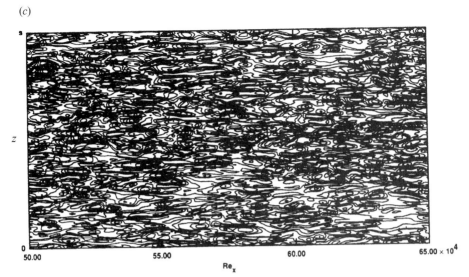

FIG. 12.14
Direct numerical simulation (DNS) of turbulent flow. Vorticity contours looking down at the top of the plate. Parts (*a*) to (*c*) correspond to the same time but cover sequentially downstream portions of the plate. (*a*) Mostly laminar flow; (*b*) transition from laminar to turbulent flow (transition region); (*c*) turbulent flow. (*From Ref. 57. Copyright © 1991 AIAA. Reprinted with permission.*)

With this, we also end this book. We hope that this book has opened the horizons of CFD to you and that you have a deeper appreciation for the basic ideas than you had before we started. This author wishes you the best of success in your future contacts and interaction with CFD. In this modern world, you will most certainly have such interactions no matter what path your career takes.

APPENDIX A

THOMAS' ALGORITHM FOR THE SOLUTION OF A TRIDIAGONAL SYSTEM OF EQUATIONS

Consider a system of M linear, simultaneous algebraic equations with M unknowns, $u_1, u_2, u_3, \ldots, u_M$, given in the form below.

$$d_1 u_1 + a_1 u_2 = c_1 \qquad \text{(A.1)}$$
$$b_2 u_1 + d_2 u_2 + a_2 u_3 = c_2 \qquad \text{(A.2)}$$
$$b_3 u_2 + d_3 u_3 + a_3 u_4 = c_3 \qquad \text{(A.3)}$$
$$\vdots$$
$$b_{M-1} u_{M-2} + d_{M-1} u_{M-1} + a_{M-1} u_M = c_{M-1} \qquad \text{(A.4)}$$
$$b_M u_{M-1} + d_M u_M = c_M \qquad \text{(A.5)}$$

This is a tridiagonal system, i.e., a system of equations with finite coefficients only on the main diagonal (the d_i's), the lower diagonal (the b_i's), and the upper diagonal (the a_i's).

A standard method for solving a system of linear, algebraic equations is gaussian elimination. Thomas' algorithm is essentially the result of applying gaussian elimination to the tridiagonal system of equations. Specifically, we wish to eliminate the lower-diagonal term (the b_i's), as follows. Multiply Eq. (A.1) by b_2.

$$b_2 d_1 u_1 + b_2 a_1 u_2 = c_1 b_2 \tag{A.6}$$

Multiply Eq. (A.2) by d_1.

$$d_1 b_2 u_1 + d_1 d_2 u_2 + d_1 a_2 u_3 = c_2 d_1 \tag{A.7}$$

Subtract Eq. (A.6) from (A.7).

$$(d_1 d_2 - b_2 a_1) u_2 + d_1 a_2 u_3 = c_2 d_1 - c_1 b_2 \tag{A.8}$$

Divide Eq. (A.8) by d_1.

$$\left(d_2 - \frac{b_2 a_1}{d_1} \right) u_2 + a_2 u_3 = c_2 - \frac{c_1 b_2}{d_1} \tag{A.9}$$

Note that Eq. (A.9) no longer has a lower-diagonal term—it has been eliminated by our multiplication and subtraction process above. Let us denote some of the coefficients in Eq. (A.9) as follows:

$$d'_2 = d_2 - \frac{b_2 a_1}{d_1} \tag{A.10}$$

and

$$c'_2 = c_2 - \frac{c_1 b_2}{d_1} \tag{A.11}$$

Then Eq. (A.9) is written in a simpler form as

$$d'_2 u_2 + a_2 u_3 = c'_2 \tag{A.12}$$

Let us continue with our elimination process by multiplying Eq. (A.12) by b_3.

$$b_3 d'_2 u_2 + b_3 a_2 u_3 = b_3 c'_2 \tag{A.13}$$

Multiply Eq. (A.3) by d'_2.

$$d'_2 b_3 u_2 + d'_2 d_3 u_3 + d'_2 a_3 u_4 = d'_2 c_3 \tag{A.14}$$

Subtract Eq. (A.13) from (A.14).

$$(d'_2 d_3 - b_3 a_2) u_3 + d'_2 a_3 u_4 = d'_2 c_3 - b_3 c'_2 \tag{A.15}$$

Divide Eq. (A.15) by d'_2.

$$\left(d_3 - \frac{b_3 a_2}{d'_2} \right) u_3 + a_3 u_4 = c_3 - \frac{b_3 c'_2}{d'_2} \tag{A.16}$$

Note that Eq. (A.16) no longer has a lower-diagonal term; it has been eliminated in the same fashion as was the case for Eq. (A.9).

Before we go any further, notice the pattern that is developing here. Equation (A.9) can be viewed as obtained from Eq. (A.2) by dropping the first term (the term involving u_1), replacing the main-diagonal coefficient with

$$d_2 - \frac{b_2 a_1}{d_1} \tag{A.17}$$

instead of d_2, keeping the third term unchanged ($a_2 u_3$), and replacing the term on the right-hand side of the equation by

$$c_2 - \frac{c_1 b_2}{d_1} \tag{A.18}$$

instead of c_2. Comparing Eqs. (A.16) and (A.3), we see exactly the same pattern, where in Eq. (A.3) the first term is dropped ($b_3 u_2$), the diagonal coefficient is replaced by

$$d_3 - \frac{b_3 a_2}{d'_2} \tag{A.19}$$

the third term remains unchanged ($a_3 u_4$), and the right-hand side is replaced by

$$c_3 - \frac{c'_2 b_3}{d'_2} \tag{A.20}$$

The pattern is clear. Compare the forms given by (A.17) and (A.19); they are the same. Compare the forms given by (A.18) and (A.20); they are the same. Starting at the top of our system of equations represented by Eqs. (A.1) through (A.5), we leave Eq. (A.1) alone, but in all the following equations we drop the first term, replace the coefficient of the main-diagonal term by

$$\boxed{d'_i = d_i - \frac{b_i a_{i-1}}{d'_{i-1}} \quad i = 2, 3, \ldots, M} \tag{A.21}$$

and replace the term on the right-hand side of the equation by

$$\boxed{c'_i = c_i - \frac{c'_{i-1} b_i}{d'_{i-1}} \quad i = 2, \ldots, 3, , M} \tag{A.22}$$

This will result in an upper bidiagonal form of equations given by

$$d_1 u_1 + a_1 u_2 = c_1$$
$$d'_2 u_2 + a_2 u_3 = c'_2$$
$$d'_3 u_3 + a_3 u_4 = c'_3$$
$$\vdots$$
$$d'_{M-1} u_{M-1} + a_{M-1} u_M = c'_{M-1} \tag{A.23}$$
$$d'_M u_M = c'_M \tag{A.24}$$

Examining the above system of equations, we note that the last equation, Eq. (A.24), contains only one unknown, namely, u_M; hence

$$u_M = \frac{c'_M}{d'_M} \qquad (A.25)$$

The solution of the remaining unknowns is obtained by working *upward* in the above system. For example, after u_M is obtained from Eq. (A.25), the value of u_{M-1} can be found from Eq. (A.23) as

$$u_{M-1} = \frac{c'_{M-1} - a_{M-1} u_M}{d'_{M-1}} \qquad (A.26)$$

Indeed, by inspection we can see that Eq. (A.26) can be replaced by the general recursion formula

$$u_i = \frac{c'_i - a_i u_{i+1}}{d'_i} \qquad (A.27)$$

for the calculation of u_i, where u_{i+1} has already been calculated from the previous application of Eq. (A.27).

In summary, Thomas' algorithm is as follows. Given a system of linear, simultaneous, algebraic equations in tridiagonal form represented by Eqs. (A.1) to (A.5), we first *change* this system into an upper bidiagonal form by dropping the first term in each equation (involving the b_i's), replacing the coefficient of the main-diagonal term by Eq. (A.21), and replacing the right-hand side with Eq. (A.22). This will result in the last equation in the system in having only one unknown, namely, u_M. Solve for u_M from Eq. (A.25). Then, all other unknowns are found in sequence from Eq. (A.27), starting with $u_i = u_{M-1}$ and ending with $u_i = u_1$.

For your reference, the computer listing used to solve the Couette flow problem described in Sec. 9.3 is listed below. This computer program is essentially a program for Thomas' algorithm and can be used as a guide to construct your own computer program for Thomas' algorithm.

FORTRAN Computer Listing: solution of Couette flow by means of Thomas' algorithm

```
      REAL U(41),A(41),B(41), D(41), Y(41), C(41)
      N=20
      NN=N+1
      Y(1)=0.0
      DEL=1.0/FLOAT(N)
      RE=5.0E+03
      EE=1.0
      TIME=0.0
      DELTIM=EE*RE*DEL**2
C     BOUNDARY CONDITIONS
```

```
            U(1)=0.0
            U(NN)=1.0
            AA=-0.5*EE
            BB=1.0+EE
            KKEND=2
            KKMOD=1
C           INITIAL CONDITIONS
            DO1 J=2,N
            U(J)=0.0
1           CONTINUE
            A(1)=1.0
            B(1)=1.0
            C(1)=1.0
            D(1)=1.0
            DO5 KK=1,KKEND
C           SET ORIGINAL COEFFICIENTS
            DO2 J=2,N
            Y(J)=Y(J-1)+DEL
            A(J)=AA
            IF(J.EQ.N) A(J)=0.0
            D(J)=BB
            B(J)=AA
            IF(J.EQ.2) B(J)=0.0
            C(J)=(1.0-EE)*U(J)+0.5*EE*(U(J+1)+U(J-1))
            IF(J.EQ.N) C(J)=C(J)-AA*U(NN)
2           CONTINUE
C           UPPER BIDIAGONAL FORM
            DO3 J=3,N
            D(J)=D(J)-B(J)*A(J-1)/D(J-1)
            C(J)=C(J)-C(J-1)*B(J)/D(J-1)
3           CONTINUE
C           CALCULATION OF U(J)
            DO4 K=2,N
            M=N-(K-2)
            U(M)=(C(M)-A(M)*U(M+1))/D(M)
4           CONTINUE
            Y=(1)=0.0
            Y(NN)=Y(N)+DEL
            TIME=TIME+DELTIM
            TEST=MOD(KK,KKMOD)
            IF(TEST.GT.0.01) GO TO 5
            WRITE(6,100) KK,TIME,DELTIM
            WRITE(*,100) KK,TIME,DELTIM
            WRITE(6,101)
            WRITE(*,101)
            WRITE(6,102) (J,Y(J),U(J),B(J),D(J),A(J),C(J),J=1,NN)
            WRITE(*,102) (J,Y(J),U(J),B(J),D(J),A(J),C(J),J=1,NN)
5           CONTINUE
100         FORMAT(5X//5X, 'SOLUTION AT',5X,'KK=',I3,5X, 'TIME=',E10.3,5X
    +       'DELTIM=',E10.3//)
101         FORMAT(3X, 'J',6X,'Y',9X,'U',9X,'B',9X,'D',9X, 'A',9X,'C')
102         FORMAT(2X,I3,6E10.3)
            END``˜Ñ
```

REFERENCES

1. Anderson, John D., Jr.: *Introduction to Flight*, 3d ed., McGraw-Hill, New York, 1989.
2. Anderson, John D., Jr.: *Hypersonic and High Temperature Gas Dynamics*, McGraw-Hill, New York, 1989.
3. Rouse, Hunter, and Simon Ince: *History of Hydraulics*, Iowa Institute of Hydraulic Research, Ames, Iowa 1957.
4. Tokaty, G. A.: *A History and Philosophy of Fluid Mechanics*, G. T. Foulis, Henly-on-Thames, England, 1971.
5. Anderson, John D., Jr.: *The History of Aerodynamics, and Its Impact on Flying Machines*, Cambridge University Press, New York (in preparation).
6. Kothari, A. P., and J. D. Anderson, Jr.: "Flows Over Low Reynolds Number Airfoils—Compressible Navier–Stokes Numerical Solutions," AIAA paper 85-0107, presented at AIAA 23rd Aerospace Sciences Meeting, Reno, Nev., Jan. 14–17, 1985.
7. Pohlen, L. J., and T. J. Mueller: "Boundary Layer Characteristics of the Miley Airfoil at Low Reynolds Numbers," *J. Aircr.*, vol. 21, no. 9, pp. 658–664, September 1984.
8. Anderson, John D., Jr.: *Fundamentals of Aerodynamics*, 2d ed., McGraw-Hill, New York, 1991.
9. Bush, Richard J., Jr., Merle Jager, and Brad Bergman: "The Application of Computational Fluid Dynamics to Aircraft Design," AIAA paper 86-2651, 1986.
10. Jameson, A., W. Schmidt, and E. Turkel: "Numerical Solutions of the Euler Equations by Finite Volume Methods Using Runge-Kutta Time Stepping Schemes," AIAA paper 81-1259, 1981.
11. Chapman, Dean R.: "Computational Aerodynamics Development and Outlook," *AIAA J.*, vol. 17, no. 12, pp. 1293–1313, December 1979.
12. Moretti, G., and M. Abbett: "A Time-Dependent Computational Method for Blunt Body Flows," *AIAA J.*, vol. 4, no. 12, pp. 2136–2141, December 1966.
13. Anderson, Dale A., John C. Tannehill, and Richard H. Pletcher: *Computational Fluid Mechanics and Heat Transfer*, McGraw-Hill, New York, 1984.
14. Fletcher, C. A.: *Computational Techniques for Fluid Dynamics*, vol. I: *Fundamental and General Techniques*, Springer-Verlag, Berlin, 1988.
15. Fletcher, C. A.: *Computational Techniques for Fluid Dynamics*, vol. II: *Specific Techniques for Different Flow Categories*, Springer-Verlag, Berlin, 1988.
16. Hirsch, Charles: *Numerical Computation of Internal and External Flows*, vol. I: *Fundamentals of Numerical Discretization*, Wiley, New York, 1988.
17. Hirsch, Charles: *Numerical Computation of Internal and External Flows*, vol. II: *Computational Methods for Inviscid and Viscous Flows*, Wiley, New York, 1990.
18. Hoffmann, K. A.: *Computational Fluid Dynamics for Engineers*, Engineering Education System, Austin, Tex., 1989.
19. Hildebrand, Francis B.: *Advanced Calculus for Applications*, 2d ed., Prentice-Hall, Englewood Cliffs, N.J., 1976.
20. Schlichting, H.: *Boundary Layer Theory*, 7th ed., McGraw-Hill, New York, 1979.
21. Anderson, John D., Jr.: *Modern Compressible Flow: With Historical Perspective*, 2d ed., McGraw-Hill, New York, 1990.
22. Kreyszig, E.: *Advanced Engineering Mathematics*, Wiley, New York, 1962.
23. Whitham, G. B.: *Linear and Nonlinear Waves*, Wiley, New York, 1974.
24. Ames, W. F.: *Nonlinear Partial Differential Equations in Engineering*, Academic, New York, 1965.
25. Courant, R., K. O. Friedrichs, and H. Lewy: "Uber die Differenzengleichungen der Mathematischen Physik," *Math. Ann.*, vol. 100, p. 32, 1928.
26. Thompson, Joe F. (ed.): *Numerical Grid Generation*, North-Holland, New York, 1982.

27. Thompson, Joe F., Z. V. A. Warsi, and C. Wayne Mastin: *Numerical Grid Generation: Foundations and Applications*, North-Holland, New York, 1985.
28. Viviand, H.: "Conservative Forms of Gas Dynamic Equations," *Rech. Aerosp.*, no. 1971-1, pp. 65–68, 1974.
29. Vinokur, M.: "Conservation Equations of Gas Dynamics in Curvilinear Coordinate Systems," *J. Comput. Phys.*, vol. 14, pp. 105–125, 1974.
30. Sullins, G. A., J. D. Anderson, Jr., and J. P. Drummond: "Numerical Investigation of Supersonic Base Flow with Parallel Injection," AIAA paper 82-1002, 1982.
31. Sullins, G. A.: "*Numerical Investigation of Supersonic Base Flow with Tangential Injection*," M.S. thesis, Department of Aerospace Engineering, University of Maryland, College Park, 1981.
32. Holst, T. L.: "Numerical Solution of Axisymmetric Boattail Fields with Plume Simulators," AIAA paper 77-224, 1977.
33. Roberts, B. O.: "Computational Meshes for Boundary Layer Problems," *Lecture Notes in Physics*, Springer-Verlag, New York, pp. 171–177, 1971.
34. Thompson, J. F., F. C. Thames, and C. W. Mastin: "Automatic Numerical Generation of Body-Fitted Curvilinear Coordinate Systems for Fields Containing Any Number of Arbitrary Two-Dimensional Bodies," *J. Comput. Phys.*, vol. 15, pp. 299–319, 1974.
35. Corda, Stephen: "*Numerical Investigation of the Laminar, Supersonic Flow over a Rearward-Facing Step Using an Adaptive Grid Scheme*," M.S. thesis, Department of Aerospace Engineering, University of Maryland, College Park, 1982.
36. Dwyer, H. A., R. J. Kee, and B. R. Sanders: "An Adaptive Grid Method for Problems in Fluid Mechanics and Heat Transfer," AIAA paper 79-1464, 1979.
37. Steinbrenner, John P., and Dale A. Anderson: "Grid-Generation Methodology in Applied Aerodynamics," in P. A. Henne (ed.), *Applied Computational Aerodynamics*, Progress in Astronautics and Aeronautics Series, vol. 125, AIAA, Washington, D.C., chap. 4, pp. 91–130, 1990.
38. Karman, S. L., Jr., J. P. Steinbrenner, and K. M. Kisielewski: "Analysis of the F-16 Flow Field by a Block Grid Euler Approach," *AGARD Conf. Proc. 412*, 1986.
39. Venkatakrishnan, V., and D. J. Mavriplis: "Implicit Solvers for Unstructured Meshes," AIAA paper 91-1537-CP, *Proc. AIAA 10th Comput. Fluid Dyn. Conf.*, pp. 115–124, June 24–27, 1991.
40. Hassan, O., K. Morgan, J. Peraire, E. J. Probert, and R. R. Thareja: "Adaptive Unstructured Mesh Methods for Steady Viscous Flow," AIAA paper 91-1538-CP, *Proc. AIAA 10th Comput. Fluid Dyn. Conf.*, pp. 125–133, June 24–27, 1991.
41. DeZeeuw, Darren, and Kenneth G. Powell: "An Adaptively-Refined Cartesian Mesh Solver for the Euler Equations," AIAA paper 91-1542-CP, *Proc. AIAA 10th Comput. Fluid Dyn. Conf.*, pp. 166–180, June 24–27, 1991.
42. Rubbert, Paul, and Dockan Kwak (eds): *AIAA 10th Computational Fluid Dynamics Conference*, June 24–27, 1991.
43. MacCormack, R. W.: "The Effect of Viscosity in Hypervelocity Impact Cratering," AIAA paper 69-354, 1969.
44. Kuruvila, G., and J. D. Anderson, Jr.: "A Study of the Effects of Numerical Dissipation on the Calculation of Supersonic Separated Flows," AIAA paper 85-0301, 1985.
45. Ames Research Staff: "Equations, Tables, and Charts for Compressible Flow," *NACA Rep. 1135*, 1953.
46. Abbett, M. J.: "Boundary Condition Calculation Procedures for Inviscid Supersonic Flow Fields," *Proc. 1st AIAA Comput. Fluid Dyn. Conf.*, pp. 153–172, 1973.
47. Shang, J. S., and S. J. Scherr: "Navier–Stokes Solutions for a Complete Re-Entry Configuration," *J. Aircr.*, vol. 23, no. 12, pp. 881–888, December 1986.
48. Schroder, W., and F. Mergler: "Comparative Study of Inviscid and Viscous Flows Over an STS," in C. Hirsch, J. Periaux, and W. Kordulla (eds.), *Computational Fluid Dynamics '92*, vol. 1, Elsevier, Amsterdam, 1992, pp. 323–330.
49. Turkel, E., R. C. Swanson, V. N. Vatsa, and J. A. White: "Multigrid for Hypersonic Viscous Two- and Three-Dimensional Flows," AIAA paper 91-1572-CP, *Proc. AIAA 10th Comput. Fluid Dyn. Conf.*, 1991.
50. Vilsmeier, R., and D. Hanel: "Adaptive Solutions for Compressible Flows on Unstructured,

Strongly Anisotropic Grids," in C. Hirsch, J. Periaux, and W. Kordulla (eds.), *Computational Fluid Dynamics '92*, vol. 2, Elsevier, Amsterdam, 1992, pp. 945–951.
51. Vassberg, J. C., and K. R. Dailey: "AIRPLANE: Experiences, Benchmarks and Improvements," AIAA paper 90-2998, 1990.
52. Petot, B., and A. Fourmaux: "Validation of Viscous and Inviscid Computational Methods Around Axial Flow Turbine Blades," in C. Hirsch, J. Periaux, and W. Kordulla (eds.), *Computational Fluid Dynamics '92*, vol. 2, Elsevier, Amsterdam, 1992, pp. 611–618.
53. Enomoto, S., and C. Arakawa: "2-D and 3-D Numerical Simulation of a Supersonic Inlet Flowfield," in C. Hirsch, J. Periaux, and W. Kordulla (eds.), *Computational Fluid Dynamics '92*, vol. 2, Elsevier, Amsterdam, 1992, pp. 781–788.
54. Borland, C. J.: "A Multidisciplinary Approach to Aeroelastic Analysis," in A. K. Noor and S. L. Venneri (eds.), *Computing Systems in Engineering*, vol. 1, Pergamon, New York, 1990, pp. 197–209.
55. Vandromme, D., and A. Saouab: "Implicit Solution of Reynolds-Averaged Navier–Stokes Equations for Supersonic Jets on Adaptive Mesh," in C. Hirsch, J. Periaux, and W. Kordulla (eds.), *Computational Fluid Dynamics '92*, vol. 2, Elsevier, New York, 1992, pp. 727–731.
56. Kandil, O. A., H. A. Kandil, and C. H. Liu: "Supersonic Quasi-Axisymmetric Vortex Breakdown," AIAA paper 91-3311-CP, *Proc. AIAA 9th Appl. Aerodyn. Conf.*, pp. 851–863, 1991.
57. Rai, M. M., and P. Moin: "Direct Numerical Simulation of Transition and Turbulence in a Spatially Evolving Boundary Layer," AIAA paper 91-1607-CP, *Proc. AIAA 10th Comput. Fluid Dyn. Conf.*, pp. 890–914, 1991.
58. Shaw, C. T.: "Predicting Vehicle Aerodynamics Using Computational Fluid Dynamics—A User's Perspective," *Research in Automotive Aerodynamics*, SAE Special Publication 747, pp. 119–132, February 1988.
59. Matsunaga, K., H. Mijata, K. Aoki, and M. Zhu: "Finite-Difference Simulation of 3D Vortical Flows Past Road Vehicles," *Vehicle Aerodynamics*, SAE Special Publication 908, pp. 65–84, February 1992.
60. Griffin, M. E., R. Diwaker, J. D. Anderson, and E. Jones: "Computational Fluid Dynamics Applied to Flows in an Internal Combustion Engine," AIAA paper 78-57, presented at AIAA 16th Aerospace Sciences Meeting, January 1978.
61. Mampaey, F., and Z. A. Xu: "An Experimental and Simulation Study of a Mould Filling Combined with Heat Transfer," in C. Hirsch, J. Periaux, and W. Kordulla (eds.), *Computational Fluid Dynamics '92*, vol. 1, Elsevier, Amsterdam, 1992, pp. 421–428.
62. Steijsiger, C., A. M. Lankhorst, and Y. R. Roman: "Influence of Gas Phase Reactions on the Deposition Rate of Silicon Carbide from the Precursors Methyltrichlorosilane and Hydrogen," in C. Hirsch, O. C. Zienkiewicz, and E. Onate (eds.), *Numerical Methods in Engineering '92*, Elsevier, Amsterdam, 1992, pp. 857–864.
63. Toorman, E. A., and J. E. Berlamont: "Free Surface Flow of a Dense, Natural Cohesive Sediment Suspension," in C. Hirsch, J. Periaux, and W. Kordulla (eds.), *Computational Fluid Dynamics '92*, vol. 2, Elsevier, Amsterdam, pp. 1005–1011, 1992.
64. Bai, X. S., and L. Fuchs: "Numerical Model for Turbulent Diffusion Flames with Applications," in C. Hirsch, J. Periaux and W. Kordulla (eds.), *Computational Fluid Dynamics '92*, vol. 1, Elsevier, Amsterdam, 1992, pp. 169–176.
65. McGuirk, J. J., and G. E. Whittle: "Calculation of Buoyant Air Movement in Buildings—Proposals for a Numerical Benchmark Test Case, *Computational Fluid Dynamics for the Environmental and Building Services Engineer—Tool or Toy?* The Institution of Mechanical Engineers, London, pp. 13–32, November 1991.
66. Alamdari, F., S. C. Edwards, and S. P. Hammond: "Microclimate Performance of an Open Atrium Office Building: A Case Study in Thermo-Fluid Modeling," *Computational Fluid Dynamics for the Environmental and Building Services Engineer—Tool or Toy?* The Institution of Mechanical Engineers, London, pp. 81–92, November 1991.
67. Patankar, S. V., and D. B. Spalding: "A Calculation Procedure for Heat, Mass and Momentum Transfer in Three-Dimensional Parabolic Flows," *Int. J. Heat Mass Transfer*, vol. 15, pp. 1787–1806, 1972.
68. Patankar, S. V.: *Numerical Heat Transfer and Fluid Flow*, Hemisphere, New York, 1980.

69. Oran, Elaine S., and Jay P. Boris: *Numerical Simulation of Reactive Flow*, Elsevier, New York, 1987.
70. Jacquotte, O. P., and G. Coussement: "Structural Grid Variation Adaption: Reaching the Limit?" in C. Hirsch, J. Periaux, and W. Kordulla (eds.), *Computational Fluid Dynamics '92*, vol. 2, Elsevier, Amsterdam, 1992, pp. 1077–1087.
71. Degani, D. and Y. Levy: "Asymmetric Turbulent Vortical Flows over Slender Bodies," *Proc. AIAA 9th Appl. Aerodyn. Conf.*, pp. 756–765, September 1991.
72. *TECPLOT Users Manual*, version 5, Amtec Engineering, Inc., Bellevue, Wash., 1992.
73. Selmin, V., E. Hettena, and L. Formaggia: "An Unstructured Node Centered Scheme for the Simulation of 3-D Inviscid Flows," in C. Hirsch, J. Periaux, and W. Kordulla (eds.), *Computational Fluid Dynamics '92*, vol. 2, Elsevier, Amsterdam, 1992, pp. 823–828.
74. MacCormack, R. W.: "Current Status of Numerical Solutions of the Navier–Stokes Equations," AIAA paper 88-0513, 1988.
75. Van Driest, E. R.: "Investigation of Laminar Boundary Layer in Compressible Fluids Using the Crocco Method," *NACA Tech. Note 2597*, January 1952.
76. Stollery, J. L.: "Viscous Interaction Effects and Re-entry Aerothermodynamics: Theory and Experimental Results," *Aerodynamic Problems of Hypersonic Vehicles*, vol. 1, AGARD Lecture Series 42, pp. 10-1–10-28, July 1972.
77. Beam, R. M., and R. F. Warming: "An Implicit Finite-Difference Algorithm for Hyperbolic Systems in Conservation Law Form," *J. Comput. Phys.*, vol. 22, pp. 87–110, 1976.
78. Briley, W. R., and H. McDonald: "Solution of the Three-Dimensional Navier–Stokes Equations by an Implicit Technique," *Proceedings of the Fourth International Conference on Numerical Methods in Fluid Dynamics, Lecture Notes in Physics*, vol. 35, Springer-Verlag, Berlin, 1975.
79. Peaceman, D. W., and H. H. Rackford: "The Numerical Solution of Parabolic and Elliptic Differential Equations," *J. Soc. Ind. Appl. Math.*, vol. 3, pp. 28–41, 1955.
80. Douglas, J., and H. H. Rackford: "On the Numerical Solution of Heat Conduction Problems in Two and Three Space Variables," *Trans. Am. Math. Soc.*, vol. 82, pp. 4231–4239, 1956.
81. Godunov, S. K.: "A Difference Scheme for Numerical Computation of Discontinuous Solution of Hydrodynamic Equations," *Math. Sb.*, vol. 47, pp. 271–306, 1959, in Russian; translated U.S. Joint Publications Research Service, JPRS 7226, 1969.

INDEX

Adaptive grid (*see* Grids)
ADI (*see* Alternating-direction-implicit technique)
Adiabatic wall condition, 81, 468
Adiabatic wall temperature, 81
Aeroelasticity, 526
Aircraft flowfields:
 generic fighter, 279
 hypersonic body, 274, 278
 Northrop F-20, 10–13, 209
 space transportation system, 519, 520
 X-24C, 517, 518
Alternating-direction-implicit (ADI) technique, 243–247, 494–495
Amplification factor, 160, 165
Analytical domain, 163, 164
Approximate factorization, 247, 492–496
Approximate Riemann solver, 499, 507, 528
Artificial viscosity, 236, 238–243
 for MacCormack's technique, 238, 363–364, 366–370
Automobile flowfields, 14–17

Backward difference (*see* Finite differences)
Base flow, 191
 (*See also* rearward-facing step)
Beam-Warming method (*see* Implicit methods)
Block tridiagonal matrices, 496–497
Blunt body, supersonic, 29, 30, 119, 120
Body forces, 61
Boundary conditions:
 Abbett's condition (inviscid flow over walls), 392–395
 for conservation form, 343
 no-slip, 80, 457
 physical, 90–92, 392
 for pressure correction method, 262, 263, 437
 reflection, 138
 for subsonic inflow, 303–306
 for subsonic outflow, 327–328
 for supersonic outflow, 305–307
Boundary-fitted grid (*see* Grids)
Boundary layer flows, 113, 114, 450, 472

Caloric equation of state, 79
Cartesian grid (*see* Grids)
Cell Reynolds number, 456–457
Central difference (*see* Finite differences)
CFD (*see* Computational fluid dynamics)
CFD-generated schlieren, 270, 271
CFL condition (*see* Courant-Friedrichs-Levy condition)
Characteristic lines, 97, 99, 162, 488
Civil engineering applications, 19, 20, 22
Compatibility equation, 101
Composite plots, 274, 278
Computational costs, 27
Computational fluid dynamics:
 definition of, 23, 25, 26
 as design tool, 9–13
 new approach in, 2–3
 as research tool, 6–9
Computational plane, definition of, 170, 171
Computer graphic techniques, 264–279, 516–517
Computer programming (*see* Programming procedures)
Conservation form:
 of continuity equations, 51, 55
 of energy equation, 74
 general discussion of, 42, 90, 225, 480–482
 generic form of, 83, 481
 of momentum equation, 65, 66
 for quasi-one-dimensional nozzle flow, 336–356
 strong form of, 88, 183, 185, 377, 452
 transformed form of, 185
 weak form of, 88
Consistent equation (*see* Finite differences)
Continuity equation, 49–60
 differential form, 55, 56
 integral form, 51, 53
 one-dimensional flow, 482
 quasi-one-dimensional flow, 286, 291
Contour plots:
 density, 272, 520, 525
 flooded, 266, 268
 general discussion of, 265–270, 516

543

Contour plots (*Cont.*):
    gray-scale color, 266
    Mach number, 523, 524, 527
    pressure, 11, 241, 272, 279, 521, 524
    velocity, 267, 268
    vorticity, 15, 530–533
Control surface, 41
Control volume, 41
Convective derivative, 45
Couette flow, 416–445
Courant-Friedrichs-Levy (CFL) condition, 162, 302, 395, 457
Courant number, 162, 324–325
Cramer's rule, 98, 178, 230
Crank-Nicolson method (*see* Implicit methods)
CRAY computers, 28, 533

Delta form, 496
Detonation wave, 266–268
Diagonalization, 500
Difference equation (*see* Finite differences)
Direct metrics (*see* Metrics)
Direct numerical simulation, 530–533
Dirichlet condition, 118
Discretization, 125, 165
Dispersion, 237
    (*See also* Numerical dissipation)
Dissipation (*see* Numerical dissipation)
Divergence form, 79
Divergence of velocity, 47, 48
Domain of dependence, 106, 107

Eigenvalues, 102, 482, 487–489
Eigenvector, 500
Elliptic nature:
    definition of, 100, 103
    equations, 30, 104, 105
    general discussion of, 117–119
    of pressure correction formula, 160
    regions in flow, 29, 277
Energy equation, 66–74
    differential form, 70, 71
    one-dimensional flow, 482
    quasi-one-dimensional flow, 286, 295, 296
Engine calculations, 14, 16, 17
Engineering Research Center for Computational Field Simulation, 276
Environmental engineering applications, 20–23, 25, 26
Errors:
    boundary condition, 321–322
    discretization, 154, 155
    general discussion of, 153–165

Errors (*Cont.*):
    high-frequency, 513
    low-frequency, 513
    round-off, 155
Euler equations, 77–79, 154
Euler explicit form, 162
Explicit methods, general discussion of, 145–153
    (*See also* Lax-Wendroff technique; MacCormack's technique)

Finite differences:
    based on Taylor's series, 128
    consistent difference equation, 144
    difference equations, 142–145
    first-order forward, 130
    first-order rearward, 131
    fourth-order central, 135
    general concept, 123, 127
    modules, 134–136, 147, 149, 494
    one-sided, 139
    second-order central, 132
    second-order second central, 132–134
    upwind, 499
Finite volumes:
    discretized equations, 167
    general concept, 123
Flat plate flow, 447–476
Flowcharts, 459–466
    (*See also* Programming procedures)
Fluid element model, 41, 42
Flux-corrected transport (FCT) method, 266
Flux limiters, 509, 510, 512
Flux terms, 84, 185, 339, 341, 380
Flux variables, 85
Flux-vector splitting, 500–502, 510, 511
Forward difference (*see* Finite differences)
Furnace applications, 21

Gauss-Seidel method, 231
Godunov schemes, 499, 502–508
Governing flow equations:
    generic form, 83
    introduction to, 38–40
    for quasi-one-dimensional flow, 296
    summary of, 75–80
    transformed generic flow, 185
Grid independence, 322–324, 355
Grid points, 126, 137, 299, 423
Grids:
    adaptive, 200–208
    boundary-fitted, 15, 18, 170, 192–200, 269
    C-type, 194
    compressed, 15, 186–192

Grids (*Cont.*):
  elliptically generated, 194-200
  finite volume, 20
  generation of, 124, 168, 171
  O-type, 194
  rectangular (cartesian), 16, 169, 212–214, 240
  staggered, 250–253, 436
  structured, 126, 210
  unstructured, 126, 210–212, 523

Heat conduction equation, 116, 121, 142, 145
High-resolution schemes, 508–510
Hyperbolic nature:
  definition of, 100, 103
  from eigenvalues, 488
  equations, 30, 104, 105
  general discussion of, 106–111, 416
  regions in flow, 29, 277

Implicit methods:
  Beam-Warming method, 490–492, 497
  Crank-Nicolson method, 148–151, 244, 420–425, 489, 491
  general discussion of, 145–153, 489
  lagging coefficients method, 490
  linearization, 490–492
Initial conditions, 307–308, 344, 362, 420
Initial data lines, 108, 226, 227, 379, 386, 387
Inverse metrics, 183
Inverse transformation (*see* Transformations)

Jacobi method, 231
Jacobian:
  determinant of the transformation, 179, 180, 206
  of the flux vector, 481, 493
  for one-dimensional flow, 486

Laminar flow, 7–9
Laplace's equation, 121, 176
Lax-Wendroff technique, 217–221
Local time stepping, 302–303

MacCormack's technique, 222–229, 238, 288, 330, 336, 375, 387, 448, 449, 453–455, 460, 461, 463–465, 474, 497, 517
Mach angle, 376
Mach disk, 528
Mach number profiles, flat plate flow, 474, 476
Mach wave, 376

Manufacturing applications, 17–19
Marching solutions:
  general, 146, 153
  space marching, 225–232, 375
  time marching, 30, 85, 119, 146, 221
Marching variables, 143
Mass source term in pressure correction method, 260, 443, 444
Mathematical behavior of equations, 95–121, 277
Mesh (*see* Grids)
Mesh plots, 273, 275–277
  (*See also* Computer graphic techniques)
Method of characteristics, 102
Metrics, 173, 178–183, 206, 207
Miley airfoil, 198, 199
Models of the flow, 40–42
Modified equation, 235
Momentum equation:
  differential form, 64
  general discussion of, 60–66
  one-dimensional flow, 482
  quasi-one-dimensional flow, 286, 294
Monotone variation, 499
Multigrid method, 513–514, 521

Naval architecture applications, 22, 23, 26
Navier-Stokes equations, 64, 66, 75–77, 79, 154, 225, 236, 239, 249, 250, 266, 417, 450, 451, 490
Neumann condition, 118
Nonconservation form:
  of continuity equation, 53, 56
  of energy equation, 70, 72
  general discussion of, 42
  of momentum equation, 64
Normal shock wave, 91, 357, 359
Nozzle flow (*see* Quasi-one-dimensional nozzle flows)
Numerical dispersion, 237
Numerical dissipation, 232–243
Numerical domain, 163, 164

One-dimensional flow, 482–489

Parabolic equations:
  boundary-layer equations, 113
  for Couette flow, 417, 421
  definition of, 100
  general discussion of, 111–117
  heat conduction equation, 116
  parabolized Navier-Stokes equations, 115
  regions in flow, 277

546 INDEX

Parabolized Navier-Stokes equations (*see* Parabolic equations)
Parallel processors (*see* Processors)
Particle paths, 15
Physical plane, definition of, 170, 171
Point-iterative method, 229
Poisson equation, 260
Prandtl-Meyer expansion wave, 374–415
Prandtl-Meyer function, 377
Pressure contours (*see* Contour plots)
Pressure correction formula, 260, 441
Pressure correction technique, 247–264, 435–445
Pressure profiles, 469, 475
Primitive variables, 85, 340, 380, 490
Processors:
　parallel, 28, 153
　vector, 28
Programming procedures, 459–467

Quasi-one-dimensional nozzle flows:
　general discussion of, 283–372
　with shock wave, 356–372
　subsonic isentropic flow, 325–336
　subsonic-supersonic isentropic flow, 285–325, 336–356

Rearward-facing step, 240–243
　(*See also* Base flow)
Rectangular grid (*see* Grids)
Region of influence, 106, 107, 110
Relaxation technique, 229–232
Residual, 316, 317
Riemann problem, 504
Runge-Kutta scheme, 521

Scatter plots, 273, 275
　(*See also* Computer graphic techniques)
Schlieren (*see* CFD-generated schlieren)
Shock-capturing method, 89–92, 356–372
Shock-fitting method, 89
Shock interaction, 268–270
Shock layer, 449
Shock tube problem, 502–504, 511, 512
SIMPLE algorithms, 248, 261–262
　(*See also* Pressure correction technique)
Solutions vector, 84, 87, 340
Source term, 84
Space marching (*see* Marching solutions)
Stability criterion, 151
Stability of solutions, 153–165
Staggered grid (*see* Grids)

Step size:
　spatial, 395–397
　time, 301–303, 455–457
Streamlines, 21, 271–274
Structured grid (*see* Grids)
Submarine flow field, 27
Substantial derivative, 43–46
Successive overrelaxation, 231
Supersonic nozzle flow, two-dimensional, 527

Taylor's series, 128
TECPLOT, 264
Temperature profiles, 471, 472
Thermal diffusivity, 116
Thermal equation of state, 79
Thomas' algorithm, 150, 243, 245, 246, 424, 426–429, 534–538
Time marching (*see* Marching solutions)
Time step calculation, 301–303, 455–457
　(*See also* Courant-Friedrichs-Levy condition)
Total-variation-diminishing schemes, 499, 509–510, 512
Transformations:
　of first derivatives, 173
　general discussion of, 124, 171–178
　inverse, 178
　of second derivative, 175, 176
Trapezoidal rule, 491, 493
Tridiagonal matrix, 150, 424, 496
Truncation error:
　of difference equation, 144
　of finite difference, 130
　for grid independence, 322
Turbulent flow, 7–9
TVD schemes (*see* Total-variation-diminishing schemes)

Unconditional instability, 162
Unconditional stability, 151, 425
Underrelaxation, 233
Unstructured grid (*see* Grids)
Upwind schemes, 497–508
Upwinding (*see* Upwind schemes)

Van Leer's flux splitting, 502
Vector plots, 270–274
　(*See also* Computer graphic techniques)
Vector processors (*see* Processors)
Velocity profiles:
　for flat plate flow, 473, 475
　for rearward-facing step, 242
　for unsteady Couette flow, 431

Viscous interaction, 470
von Neumann stability method, 161
Vortex-shock interaction, 528, 529

Wave diagram, 504, 505
Wave equation, 121, 161, 498

Wave number, 156–158
Well-posed problems, 120

xy plots, 264, 265
    (*See also* Computer graphic techniques)